MODERN ELECTRONIC DEVICES: CIRCUIT DESIGN AND APPLICATION

Milton Rosenstein
Paul Morris

Reston Publishing Company, Inc.
A Prentice-Hall Company
Reston, Virginia

Library of Congress Cataloging in Publication Data

Rosenstein, Milton.
Modern electronic devices.

Includes bibliographical references and index.
1. Electronic apparatus and appliances — Design and construction. 2. Semiconductors. I. Morris, Paul. II. Title.
TK7870.R74 1985 621.381 84-11513
ISBN 0-8359-4548-0

To our wives,
Rebecca and Alicia
for their love, patience, and encouragement

A Prentice-Hall Company
Reston, Virginia 22090

10 9 8 7 6 5 4 3 2 1

PRINTED IN THE UNITED STATES OF AMERICA

CONTENTS

CHAPTER 3: DIODE CIRCUITS

CHAPTER 4: BIPOLAR JUNCTION TRANSISTOR

CHAPTER 5: FIELD-EFFECT TRANSISTOR

CHAPTER 6: SPECIAL CIRCUITS

CHAPTER 7: OPERATIONAL AMPLIFIER

CHAPTER 8: SPECIAL DEVICES

CHAPTER 9: GATES AND DIGITAL CIRCUITS

CHAPTER 10: THEORY AND APPLICATIONS OF A/D AND D/A CONVERTERS AND AUXILIARY EQUIPMENT PERIPHERALS

CHAPTER 11: SEMICONDUCTOR MEMORIES

CHAPTER 12: MICROPROCESSORS AND MICROCOMPUTERS

PREFACE

This book is designed to meet the requirements of the electronic sequence for electrical technology students. It contains sufficient material for a three- or four-semester sequence, beginning with introductory material and ending with advanced analog and digital electronic circuits. For this work, the student requires a background in algebra and should be able to solve simple DC and AC circuits. Calculus is not required.

In our opinion this text fills a heretofore unsatisfied need in the education of technologists: the need to recognize the technologists' function in industry and to provide appropriate training. Put succinctly, technologists must be able to put their studies to practical use. This means having the ability to design circuits that work and to interface devices. Furthermore, this capability must include digital devices such as microprocessors and microcomputers. Toward this end, in almost every case, discussion of an electronic circuit or device in this text culminates in a practical design procedure. In addition, the real characteristics of devices as embodied in manufacturers' specification sheets are emphasized. Homework problems also reflect this design orientation. Finally, Chapters 9 through 12 are devoted to digital circuits and devices, including microprocessors and microcomputers.

We wish to thank all those who had a hand in this work: our typists, Mrs. Mollie Epstein and Ms. Betty Johnson; the several electronic firms who supplied us with updated information; our families for their patience and support; and the technology students at the New York Institute of Technology who served as guinea pigs.

Milton Rosenstein
Paul Morris

1

PROPERTIES OF SEMICONDUCTORS

1.1 INTRODUCTION

Many interesting and useful solid-state devices are available to the electronic technologist for industrial applications. The list, which is long, includes rectifying, zener and Schockley diodes; bipolar, unijunction, and field-effect transistors; silicon-controlled rectifiers, diacs, and triacs; and other types. A sampling of these devices is shown in Figure 1.1.

These semiconductor devices are made possible by the versatile properties of semiconductor materials in varied configurations. Therefore, a familiarity with these properties is essential to all engineering and technology students majoring in electronics or allied areas. This chapter is devoted to providing this familiarity. It presents accurate qualitative descriptions of these properties in terms of their underlying physics. The reader will emerge from a study of this chapter with the background needed for a subsequent study of solid-state devices and an appreciation of their potential for further development.

1.2 PURE STATE PROPERTIES

Diodes, transistors, and other solid-state devices, as we know them today, are made from semiconductor materials, which are so called because they lie between metals and insulators in their ability to conduct

FIGURE 1.1 Semiconductor devices. Courtesy of General Electric Semiconductor Products Department.

electricity. Figure 1.2 illustrates their location in the materials resistivity spectrum. Shaded areas, in this figure, are transition regions. Materials located in these areas may or may not be semiconductors depending on their chemical nature.

The two semiconductor materials presently used to make solid-state devices are germanium and silicon. Both are crystalline materials and lie between the conductors copper and silver and the insulators

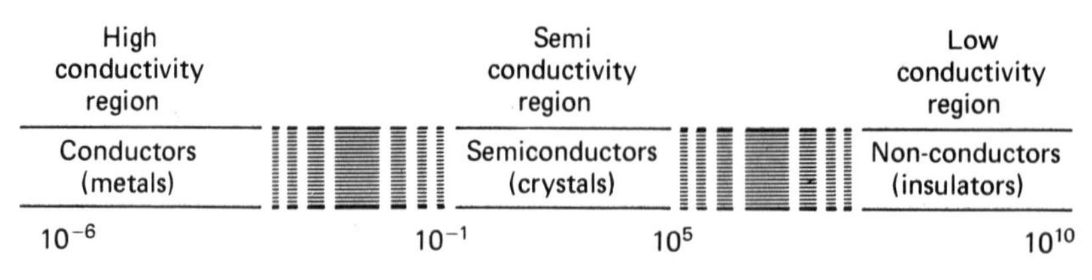

FIGURE 1.2 Semiconductors are located in region between conductors and insulators. Courtesy of General Electric Semiconductor Products Department.

TABLE 1.1 *Material and Resistance in Ohms per Centimeter Cubed for Conductor, Semiconductor, and Insulator*

Material	*Resistance in Ohms per Centimeter Cube (R/cm^3)*	*Category*
Silver	10^{-6}	Conductor
Aluminum	10^{-5}	
Pure germanium	50–60	Semiconductor
Pure silicon	50,000–60,000	
Mica	10^{12}–10^{13}	Insulator
Polyethylene	10^{15}–10^{16}	

Courtesy of the General Electric Semiconductor Products Department, Auburn, New York.

mica and polyethylene in terms of their electrical resistance. Table 1.1 compares the resistance of 1-cm cubes for the three categories.

1.2.1 Current Modes

A voltage applied across a rod or slab of relatively pure semiconductor material will cause an electric current to flow. This current will flow in two modes called *free electron* and *hole* flow. In both modes, electrons moving toward the more positive voltage are the sole carriers of energy. However, the electrons move through the semiconductor in two distinct ways. An understanding of free electron and hole flow is needed to understand solid-state devices. This, in turn, requires an understanding of the subatomic structure of crystals, which we now review.

Subatomic Structure of Crystals. All stable matter is composed of atoms. Atoms themselves, however, are dynamic structures containing a complex positively charged nucleus surrounded by electrons moving in circular and elliptical orbits (Figure 1.3).

The electron's orbit reflects its energy level; the greater its average orbital radius, the higher its energy level. At most two electrons can share the same orbit and possess exactly the same energy level. However, electrons with nearly the same energy level can be grouped together in *shells*. In the Bohr model of the atom (Figure 1.4), these shells are represented as circular orbits containing one or more electrons. They are denoted by the letters *K, L, M, N* . . . , the Kth shell being closest to the nucleus. The outermost shell of the atom is called the *valence shell*. When this shell is filled (contains a particular number of electrons), the atom is stable.

All crystals are composed of atoms that have completed their outermost shell by sharing electrons with neighboring atoms. This type

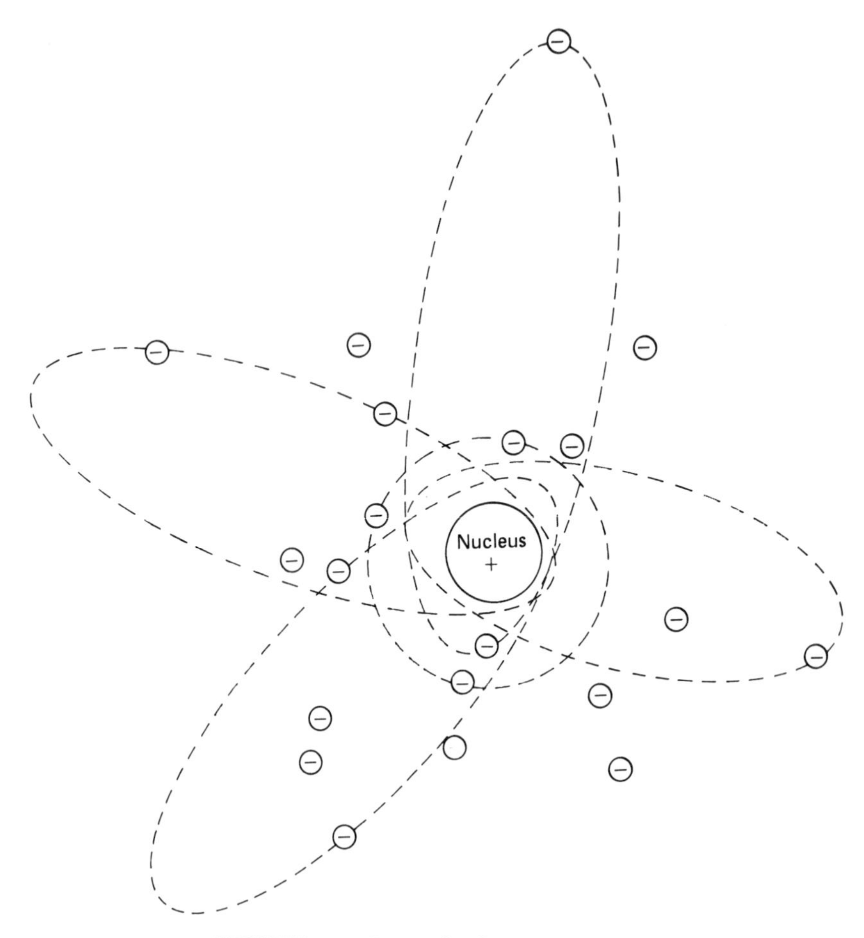

FIGURE 1.3 Typical orbits of electrons.

of atomic bonding is called *share* or *covalent* bonding, while the participating electrons are called *covalent electrons*. Covalent bonding produces the symmetric repeating lattice structure shown in Figures 1.5 and 1.6.

All atoms are in a constant state of agitation. Increases in energy within matter, whether thermal or radiant, will cause electrons to jump from one shell to another. Covalent electrons may escape their atoms completely, becoming free and leaving behind an atom that is shy one electron. This absence of an electron is called a *hole*. This process of electron–hole pair generation occurs continually and is the basis of the two modes of current flow in pure state semiconductors.

The movement of free electrons as a result of an external voltage

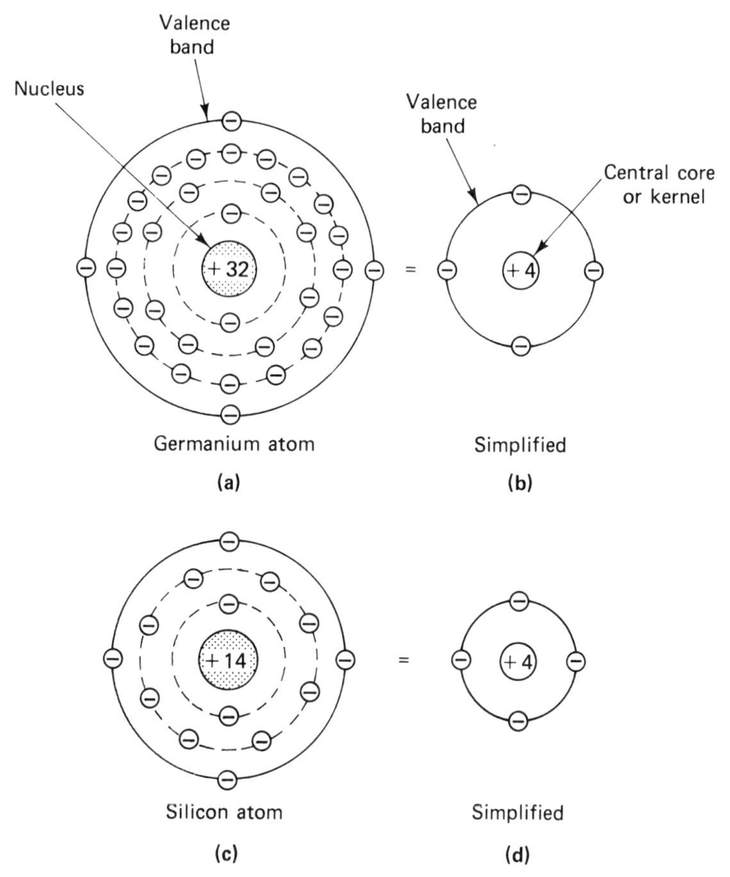

FIGURE 1.4 Bohr models of germanium and silicon atoms. Courtesy of General Electric Semiconductor Products Department.

across the semiconductor is called *electron flow.* These electrons move through the semiconductor in erratic paths interrupted by frequent collisions with atoms and other electrons. Nevertheless, in the aggregate they maintain a steady drift toward the more positive voltage. Conduction by electron flow in semiconductors is similar to current flow in conductors.

Conduction by holes is also a result of electron movement, but in this case it is the covalent electrons that move. These electrons leave their atom and jump into holes in neighboring atoms, leaving a hole in their previous location. Since, under the impetus of an external voltage, these bound electrons also move toward the more positive voltage, the holes appear to be moving in the opposite direction. Therefore, they may be thought of as positively charged carriers as opposed to the negatively

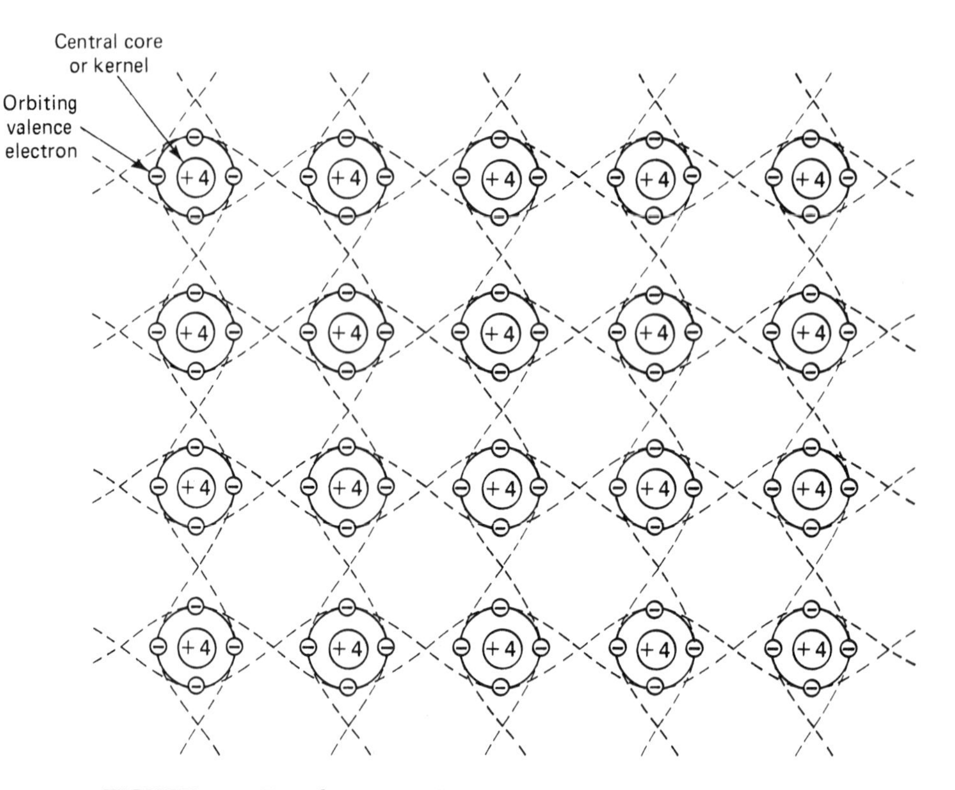

FIGURE 1.5 Two-dimensional crystal lattice structure. Courtesy of General Electric Semiconductor Products Department.

charged free electrons. Unlike free electron flow, which consists of highly mobile carriers, hole flow involves much less mobile, bound carriers and takes place entirely within the atom's valence band. Figure 1.7 illustrates free electron and hole flow in semiconductors.

1.2.2 Recombination

Free electrons are continually being recaptured by atoms with holes in their covalent shells. This process is called *recombination.* As the previously free electron fills the hole, both carriers effectively vanish. In the pure state semiconductor, free electron–hole pair generation and recombination occur continuously so that the average number of carriers remains unchanged. Therefore, the resistance of the semiconductor is unaffected. However, under special conditions, to be studied later,

FIGURE 1.6 Typical model of atomic crystal lattice structure. (Typifies spatial visualization.) Courtesy of General Electric Semiconductor Products Department.

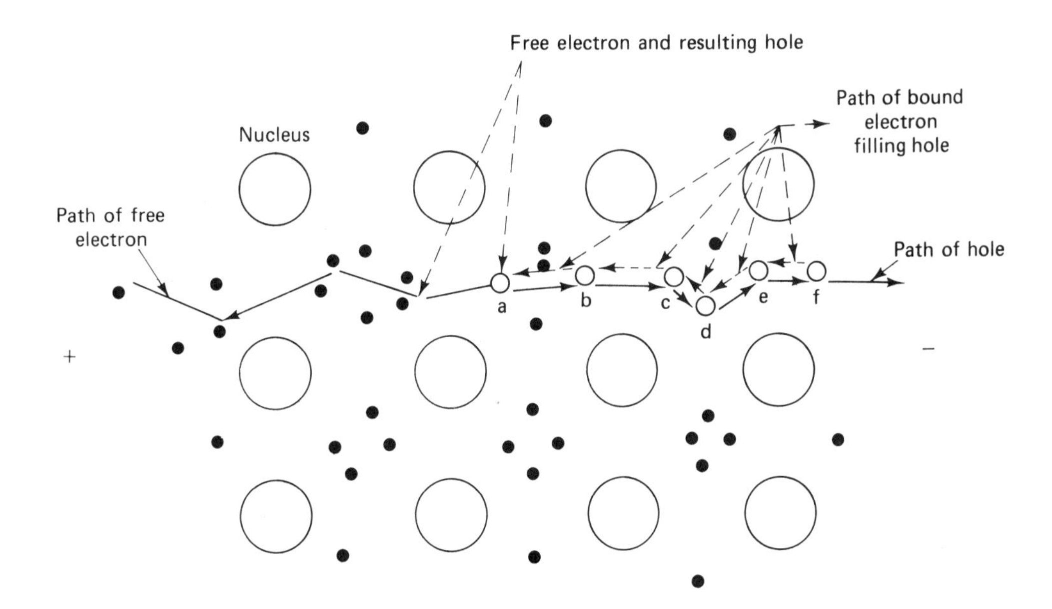

FIGURE 1.7 Electron and hole flow in semiconductors.

recombination plays an important role in the behavior of solid-state devices.

1.3 DOPED STATE PROPERTIES

The behavior of the doped semiconductor depends on the number and type of its carriers. In the pure state both carriers, free electrons and holes, are approximately evenly divided and relatively few. Impurities added to the semiconductor in its molten state can greatly increase the number of either type of carrier. If the added impurity is composed of atoms with five electrons in their valence shells (antimony, arsenic, phosphorus), then after a covalent bond is formed between each impurity atom and a semiconductor atom, one free electron will be left over. The resulting doped material will have excess free electrons and is therefore called *n*-type material. Impurity atoms with three covalent electrons (boron, gallium, indium), when added, will be unable to form filled valence shells with the semiconductor atoms. This will create an excess of holes. This doping procedure therefore creates *p*-type material. In all doped semiconductor material, both holes and free electrons coexist. In *n*-type material, free electrons are the majority carriers, and holes are the minority carriers. The reverse is true for *p*-type materials.

1.4 DRIFT AND DIFFUSION CURRENTS

There are two ways in which carriers (electrons or holes) may be made to move through a semiconductor. One way is to apply a voltage across the semiconductor, which causes the continuous migration of free electrons and holes described earlier. This kind of charge flow is called *drift current* (Figures 1.8 and 1.9). A second kind of charge movement results when a concentration of carriers occurs anywhere within the semiconductor (Figure 1.10). This could be the immediate result of a doping procedure, a momentary injection of carriers, or the bonding of *n*- and *p*-type materials. The concentrated carriers, being all of the same charge, repel each other and dispel themselves throughout the semiconductor. This charge flow is transient and ceases when the carriers are evenly distributed. It is called *diffusion current* and, unlike drift current, requires no external voltage. Both kinds of current, drift and diffusion, are basic to the operation of solid-state devices. Drift current in *n*- and *p*-type material is illustrated in Figures 1.8 and 1.9 and diffusion currents in Figure 1.10.

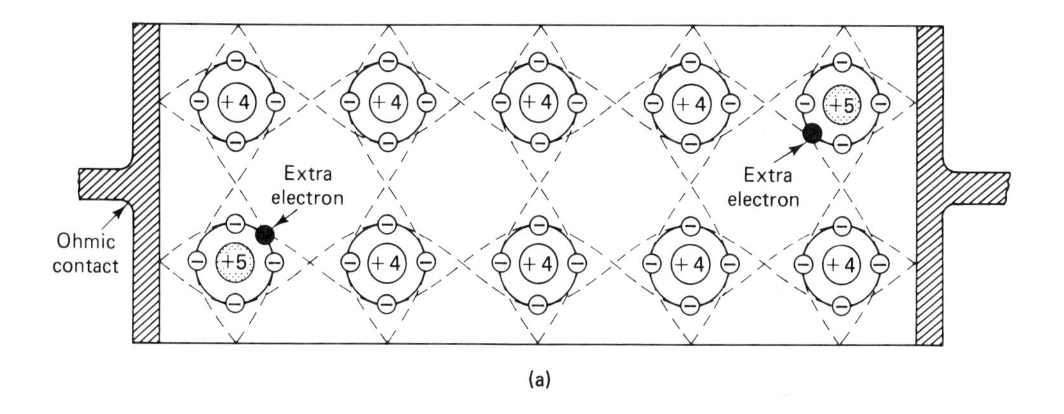

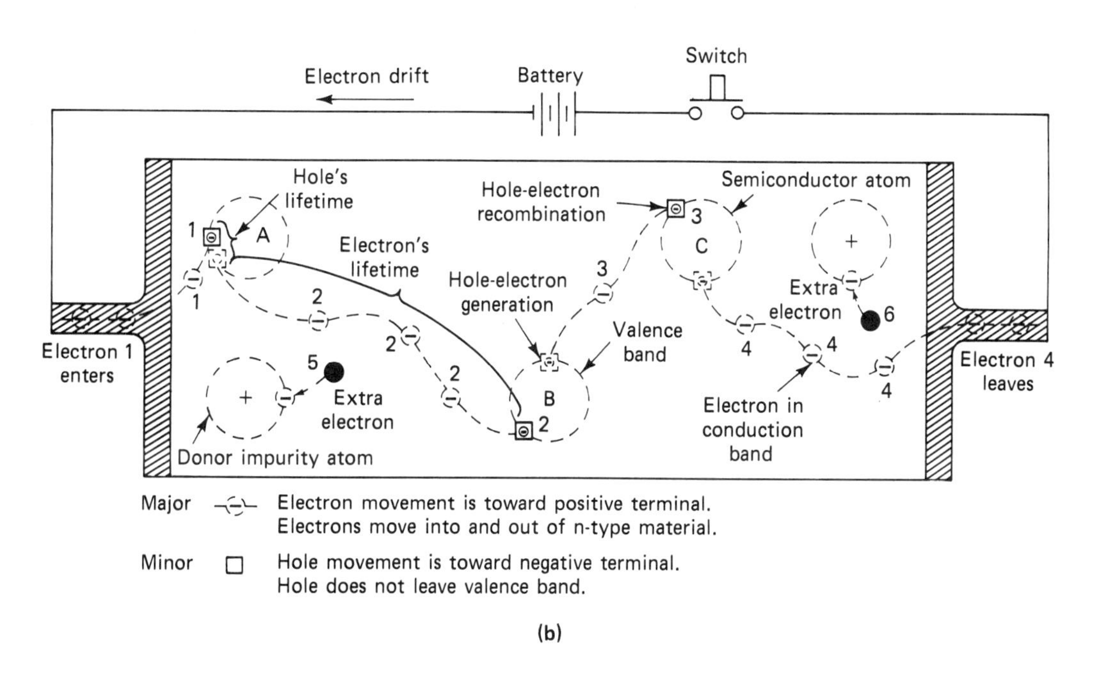

FIGURE 1.8 Drift in N-type semiconductor. Courtesy of General Electric.

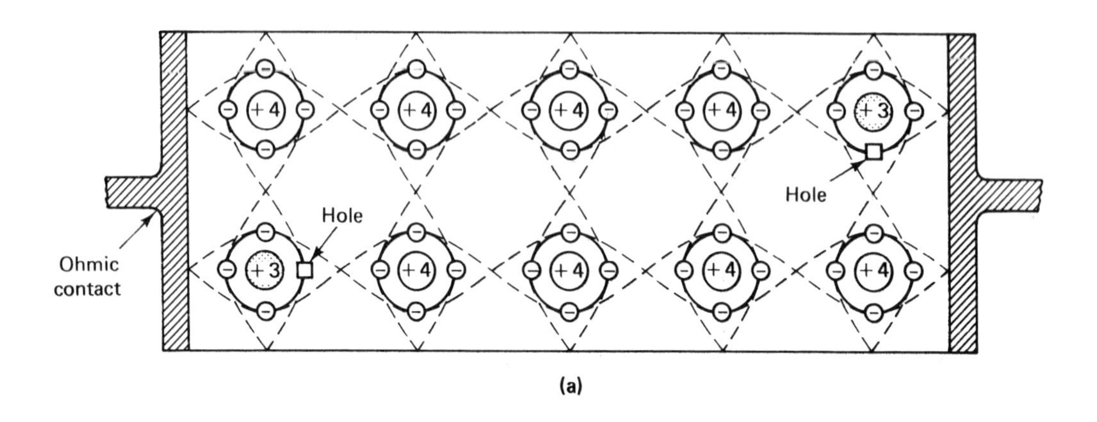

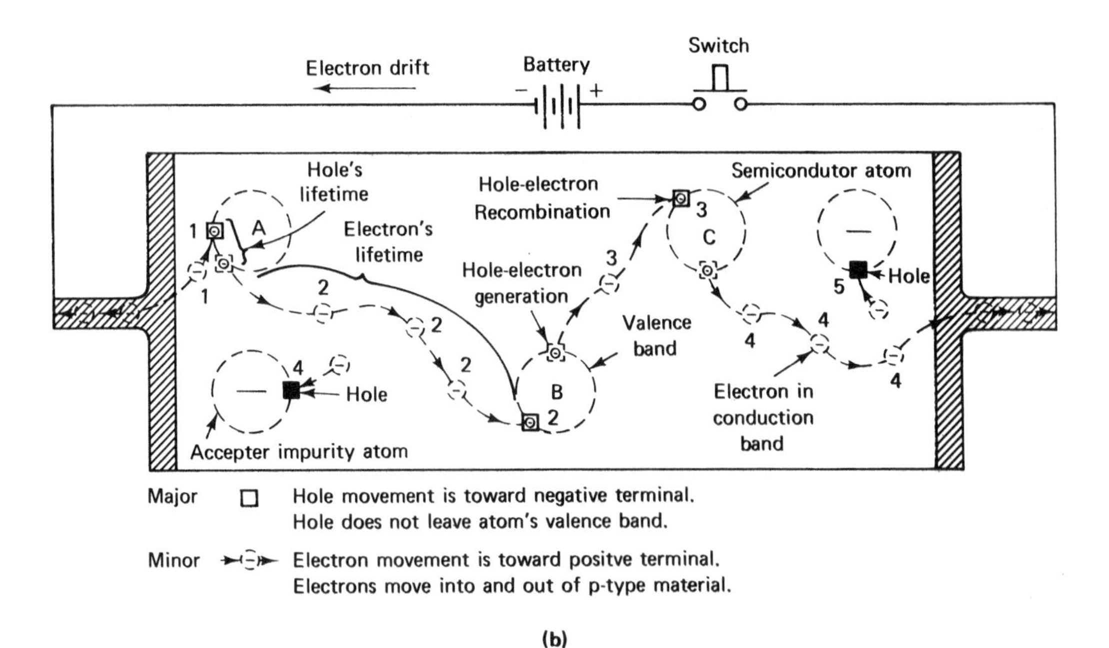

FIGURE 1.9 Drift in P-type semiconductor. Courtesy of General Electric.

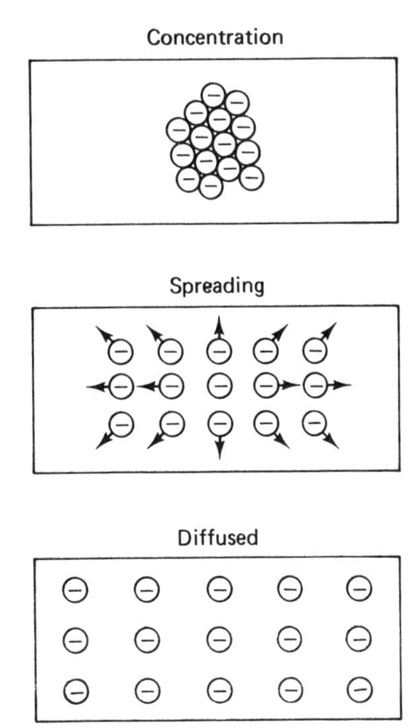

FIGURE 1.10 Electron diffusion in semiconductors. Courtesy of General Electric Semiconductor Products Department.

1.5 THE *p–n* JUNCTION

When *n*-type semiconductor material is brought into intimate contact with *p*-type material so that the two materials effectively form one crystal, a *p–n* junction is formed. The contact must be close enough to permit carriers to pass without hindrance from one material to the other. It may be achieved by adding appropriate impurities during the crystal growing process, by alloying the two materials together, or by a special solid or gaseous diffusion process. However created, this junction has interesting properties that can be exploited in the manufacture of solid-state devices. In the sections following, we study the properties of various configurations of these junctions. In later chapters this information will be used to explain the operation of pertinent solid-state devices.

1.5.1 Diode Action

The *p–n* junction offers much less resistance to current flowing in one direction as opposed to the other. This property is used to manufacture diodes. It is a result of diffusion and recombination occurring in the junction region.

The *p–n* junction begins to form immediately upon joining the two materials. Schematically, as a useful way of understanding *p–n* junction formation, we can posit diffusion currents flowing across the junction from both materials, holes from the *p* side and free electrons from the *n* side. Since both carriers exist, momentarily, most densely in a narrow region surrounding the junction, a high rate of recombination (carrier neutralization) ensues. This creates a region depleted of carriers called the *depletion region*. Also, the loss of electrons from the *n*-type material leaves it positive, and the loss of holes from the *p*-type material leaves it negative.

Although it is useful to think of holes and free electrons as positive and negative charges capable of independent motion, actually only electrons cross the junction. Excess (free) electrons leave their donor atoms in the *n*-type material, cross the junction, and are captured by acceptor atoms in the *p*-type material. The donor atom having lost electrons becomes positive; the acceptor atoms gaining electrons become negative. These atoms in the vicinity of the junction, having lost the ability to either supply or accept electrons, form the depletion region. These processes are illustrated by Figure 1.11.

Figure 1.12a and b are alternative ways to schematically represent the *p–n* junction after formation is complete. Figure 1.12a illustrates the carrier distribution within the joined semiconductors, and Figure 1.12b gives the resulting potential distribution.

To understand the effect of the potential distribution on carrier movement, think of the holes of helium-filled balloons and the electrons as lead spheres. Work must be done (energy added) to move the holes downward or to move the electrons upward. The potential energy (energy level) of each carrier therefore corresponds to its vertical position in Figure 1.12b. For holes to cross the depletion region, they must be lower than the potential hill; electrons must be higher. Majority current carriers cease to flow when the conditions of Figure 1.12b are established.

1.5.2 Forward Bias

Forward bias reduces the height of the potential hill and permits majority carriers to flow across the junction. Higher-energy electrons pass over the hill toward the plus voltage, and higher-energy holes pass below the hill in the opposite direction. Since the power supply continuously injects electrons into the *n*-type material and creates holes in the *p*-type material, continuous current passes through the semiconductor diode. Any increase in forward bias further reduces the potential hill, enabling more carriers to cross the junction and increasing the forward current (see Figure 1.13).

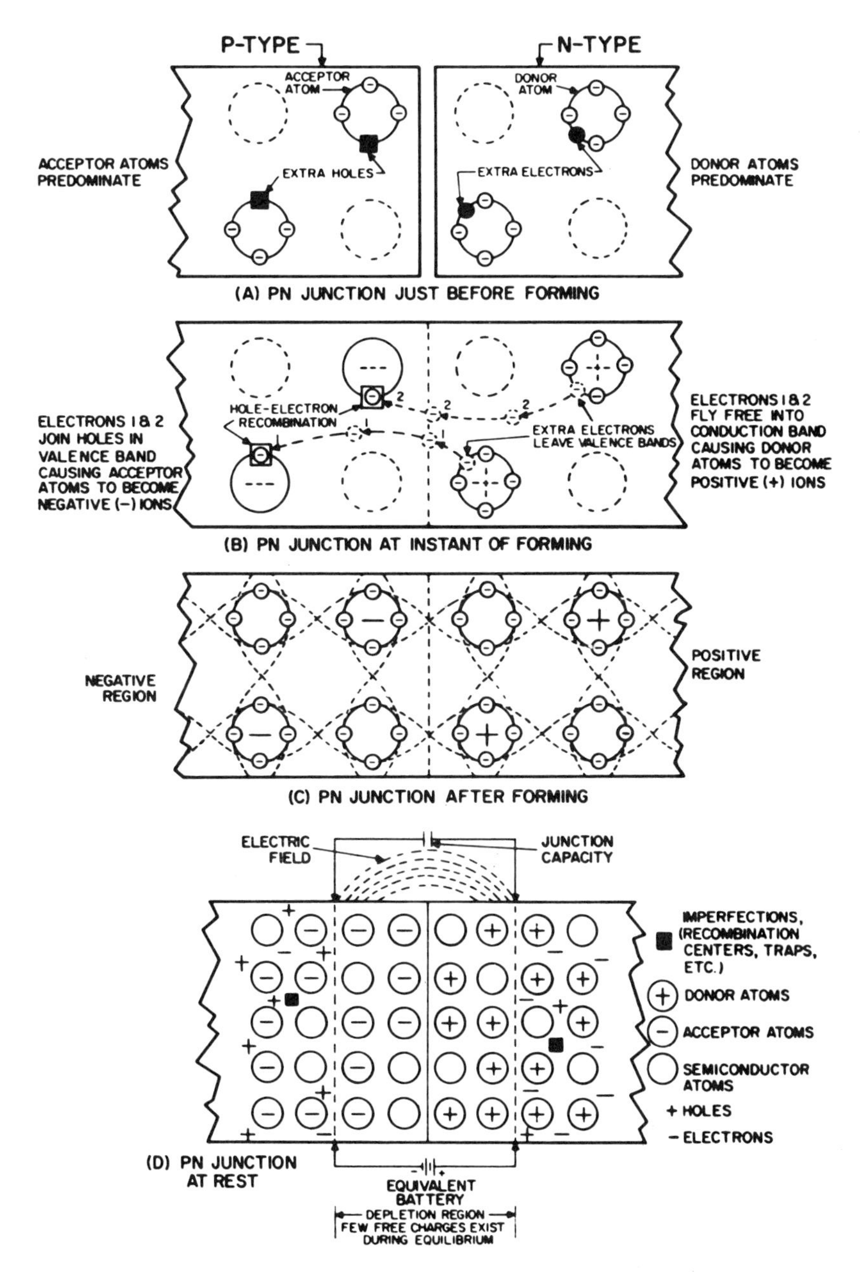

FIGURE 1.11 P–N junction forming. Courtesy of General Electric Semiconductor Products Department.

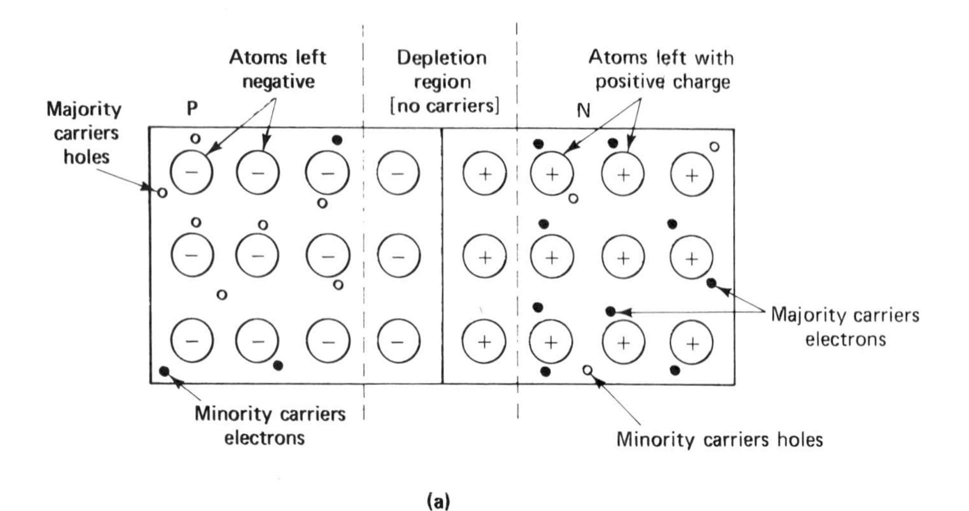

(a)

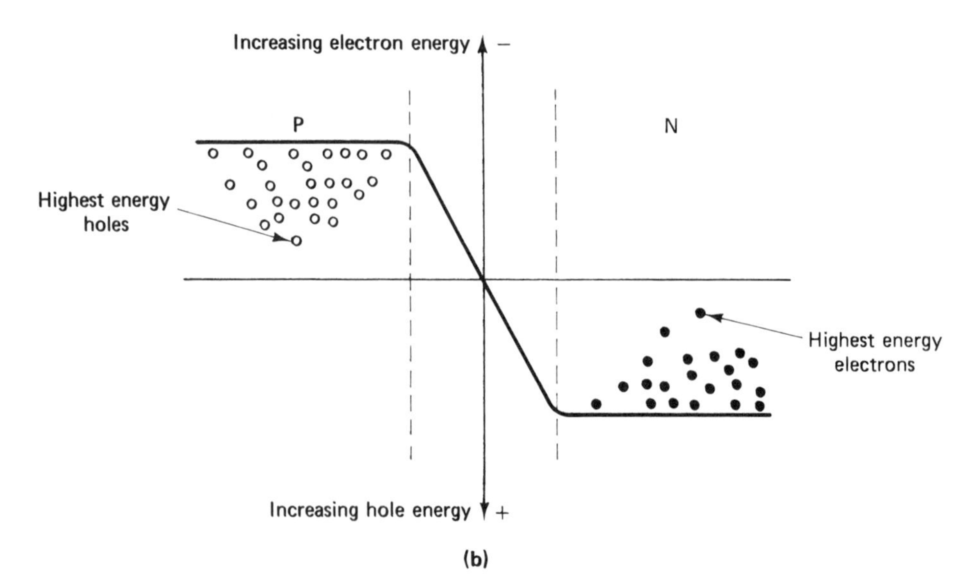

(b)

FIGURE 1.12 Depletion region and potential distribution for a passive P–N junction.

1.5.3 Reverse Bias

Increasing the potential difference between the *p*- and *n*-type materials by an external voltage (Figure 1.14) increases the height of the potential hill. The majority current carriers, virtually nil in the unbiased case, re-

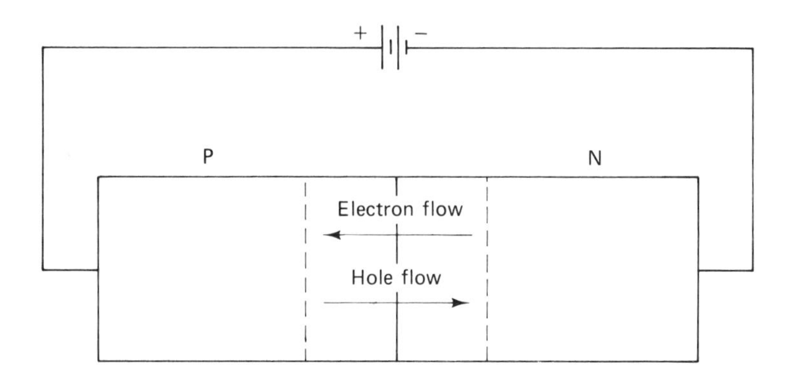

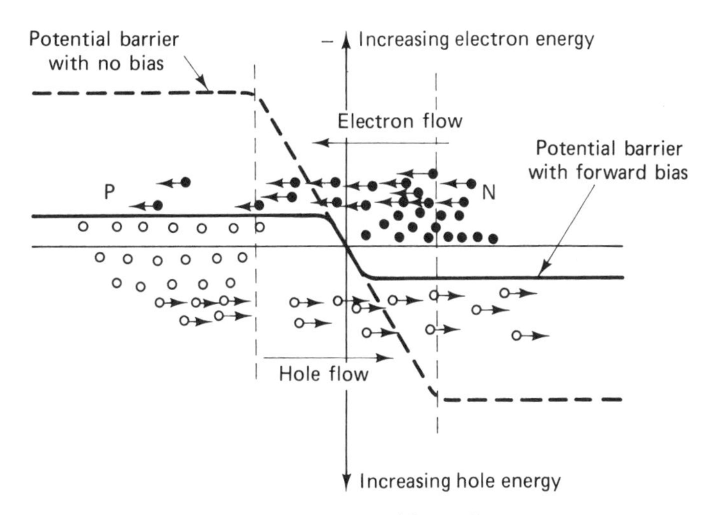

FIGURE 1.13 Forward-biased P–N junction.

main nil. However, effects due to the presence of minority carriers become prominent.

Minority Carrier Effects. *p*- and *n*-type semiconductors also contain minority carriers: holes in *n*-type material and free electrons in *p*-type material. These are created primarily by thermal generation of electron–hole pairs, but may also be created by collisions between carriers and atoms. Minority carriers move through the *p–n* junction in a direction opposite to that of the majority carriers. Reverse bias for majority carriers is forward bias for minority carriers. Therefore, instead of the junction current becoming zero when reverse biased, a small but potentially troublesome minority carrier current flows. This is called *leakage current.* At room temperature (20°C), this current is usually negligible

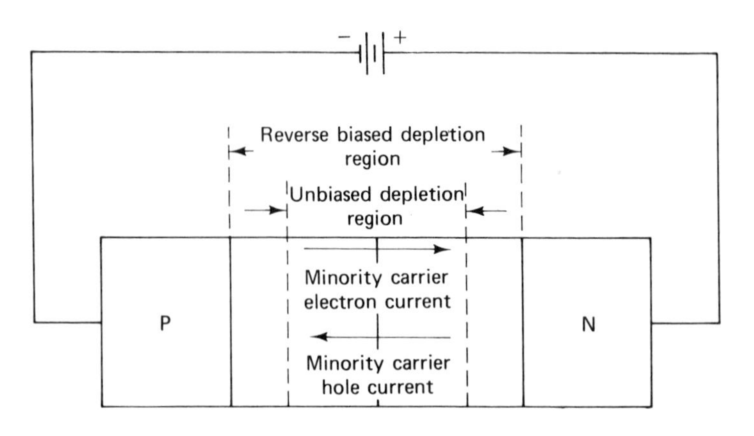

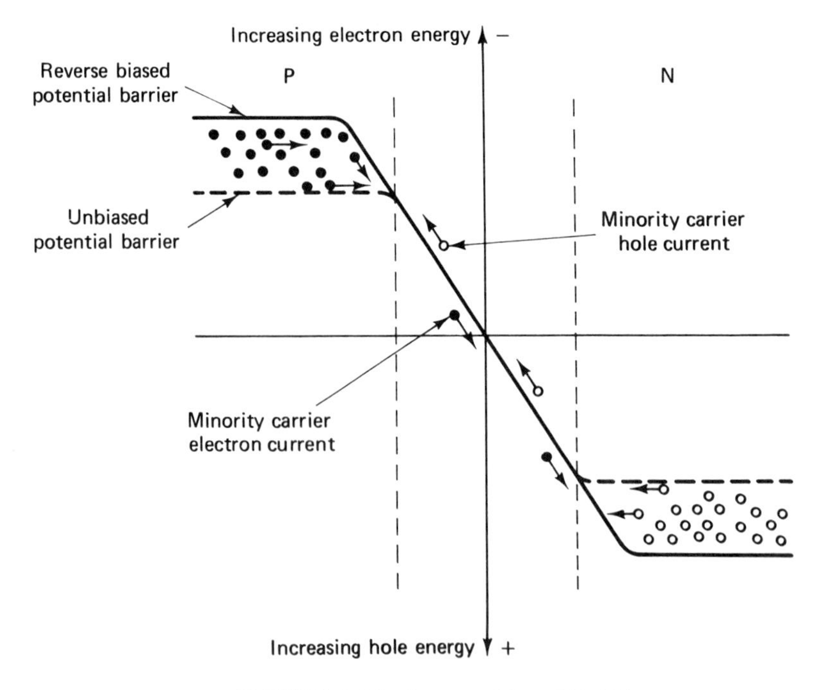

FIGURE 1.14 Reverse-biased P–N junction.

for silicon junctions (10^{-9} A), but it is much larger for germanium junctions (10^{-6} A). Figure 1.14 illustrates this leakage current.

Temperature Effect. Increased temperature increases the leakage current, doubling it for every 10°C increase in ambient temperature. At elevated temperatures the leakage current for the germanium junction diode, in particular, may become troublesome.

Reverse Bias Breakdown. At some level of reverse bias, the *p–n* junction will begin to break down. This is caused by several physical mechanisms acting in concert. To start, the depletion region, being an insulator, will have most of the bias voltage across it. Being relatively narrow, the electric field intensity (voltage/(width of depletion region)) can be extremely intense and exert strong forces on the bound electrons and the minority carriers. This can cause the junction region to break down in two ways: (1) ionization by electric field (zener breakdown) and (2) ionization by collision (avalanche breakdown). For rectifier diode junctions, both types of breakdown must be avoided. However, in specially designed diodes called *zener* diodes with junctions designed to withstand relatively large reverse currents, these breakdown mechanisms can be exploited.

Zener breakdown is a result of large numbers of bound electrons breaking away from their atoms under the influence of an intense electric field. The resulting profusion of electrons within the depletion region converts it into a conductor, and large reverse currents flow. Figure 1.15 illustrates this effect.

Avalanche breakdown is initiated by high-energy minority carriers colliding with the semiconductor atoms. At high reverse biases, these minority carriers will dislodge covalent electrons from these atoms and thus produce new minority carriers. These new minority carriers will continue the process. An avalanche of minority carriers results, and large potentially destructive current crosses the junction. Figure 1.16 illustrates this effect.

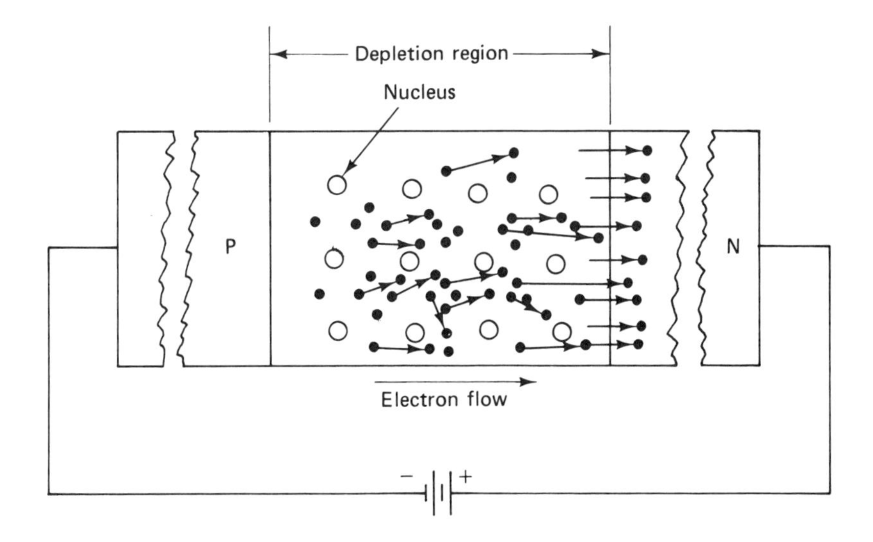

FIGURE 1.15 Ionization by electric field.

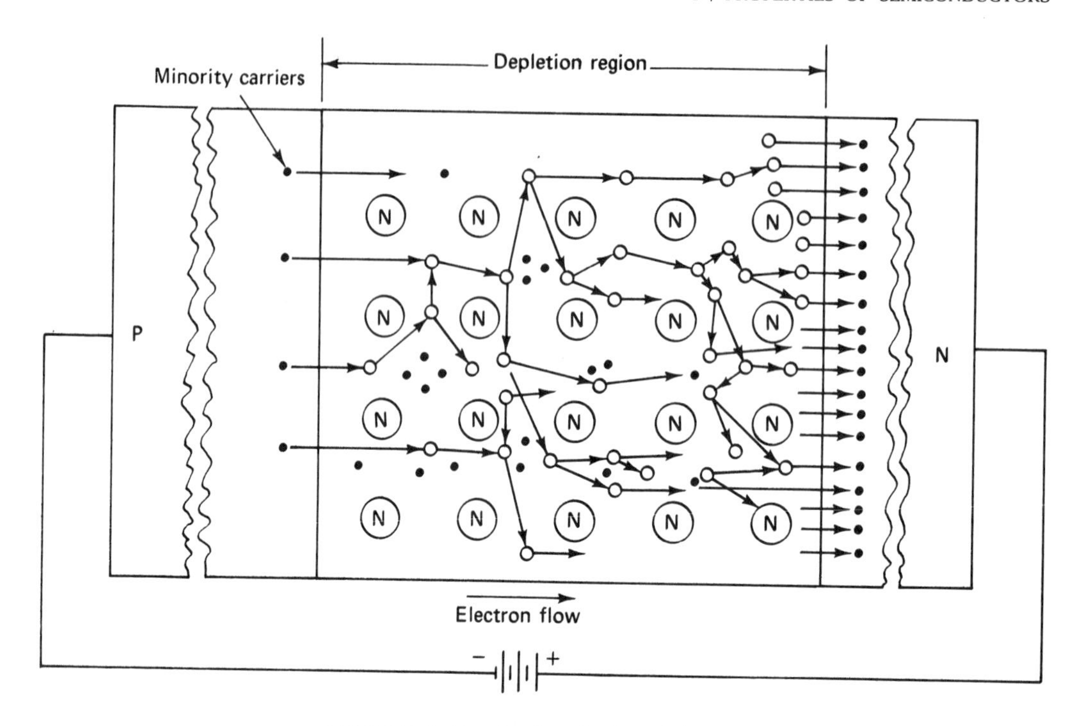

FIGURE 1.16 Avalanche breakdown.

1.6 THREE-LAYER JUNCTIONS

Npn or *pnp* sandwiches, when correctly proportioned, make possible the construction of solid-state amplifiers. For this amplifying action to occur, the central semiconductor must be thin enough to permit carriers to pass through it without substantial recombination. Widths of less than 10^{-3} cm are required. When this condition is satisfied, the sandwich acts as two *p–n* junctions back to back in which induced changes in carrier population in one junction affect the other. If the central semiconductor is too wide, the sandwich acts as two isolated junctions and no amplification occurs.

The mechanism by which a small current, injected into the central semiconductor, controls a much larger current flow between the outer semiconductors is illustrated in Figure 1.17. In anticipation of the use of the three-layer sandwich in transistors, the three parts of the configuration have been labeled *emitter, base,* and *collector.*

The potential distribution shown in Figure 1.17 exists when the base–emitter voltage (V_{BE}) is slightly negative, and the collector–base

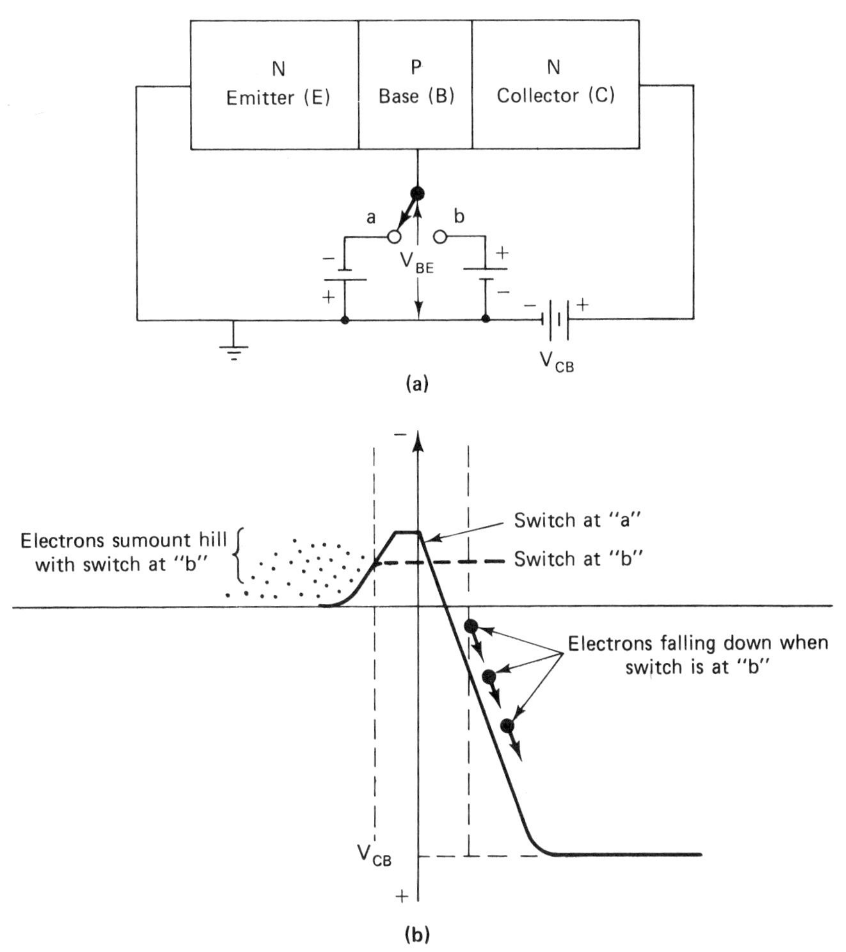

FIGURE 1.17 Potential distribution in *NPN*.

voltage (V_{CB}) is positive. Two depletion regions are formed, one at the weakly reverse biased base–emitter junction, and the other at the strongly reverse biased base–collector junction. Because the base–collector junction is more strongly biased than the base–emitter junction, its depletion region is much wider. Also, because the base is more lightly doped than the collector, the depletion region extends into it more deeply.

No emitter–collector current can flow when the conditions shown by the heavy line in Figure 1.17b exist (condition a). The majority emitter carriers (free electrons) are imprisoned behind the base–emitter hill. However, when V_{BE} is made positive (condition b), the height of the base–emitter hill is reduced and the more energetic emitter electrons will surmount it. (Note that making V_{BE} slightly positive still leaves a negative hill because of diffusion current effects.) Since the base is very thin and comparatively lightly doped, most emitter electrons escape recombination and are seized by the base–collector hill. They gain velocity and energy as they fall through the relatively high base–collector voltage difference and emerge from the collector as output current. Small variations of the base–emitter hill can produce large changes in collector current. Since only very small injections of base current are needed to significantly change the height of the base–emitter hill, high current amplification is realized.

1.7 FOUR-LAYER DEVICES

Devices composed of four alternating layers of *p*- and *n*-type materials have seen extensive application as high-power-handling latching switches. (A latching switch is a device that, when placed in a closed or open state by some activating signal, remains in that state upon removal of the signal.) The operation of the four-layer (three-junction) device is most easily understood by resolving it into two three-layer devices sharing two central layers. Figure 1.18 illustrates this.

The four-layer diode (Schockley diode) of Figure 1.18a exists either in an *off* or *on* state. When V_A is positive but below some critical voltage, called the *forward switching voltage* (V_S), the diode is off. In this state, junction J_2 is reverse biased so that most of V_A appears across it. Junctions J_1 and J_3 share the small fraction of V_A remaining and are so slightly forward biased that they are effectively open. As a result of these junction conditions, only a small leakage current flows.

In terms of the two three-layer devices (transistors) of Figure 1.18b, this means that the base current (I_{P_2}) of T_2 is too small to produce amplification. Therefore I_{N_1}, the collector current of T_2 and the base current of T_1, is also small, and the circuit is locked in the off state.

As V_A increases, I_{P_2} and, as a consequence, I_{N_1} increase. At V_A equal to V_S, I_{P_2} becomes large enough to produce amplification in T_2. I_{N_1} now begins to increase rapidly in turn, causing I_{P_2} to increase more rapidly. A flyaway regenerative process begins and continues until all three junctions are strongly forward biased. The voltage drop across the diode becomes small (0.5 to 2 V) and the diode is on. To "open" the diode, I_{P_2} must be dropped to a level insufficient to produce amplification in T_2. Since in the on state, diode resistance is very much less than in the off state, this occurs at a V_A much smaller than V_S given by V_H. The diode turned on at V_A equals V_S and therefore remains latched as long as V_A exceeds V_H.

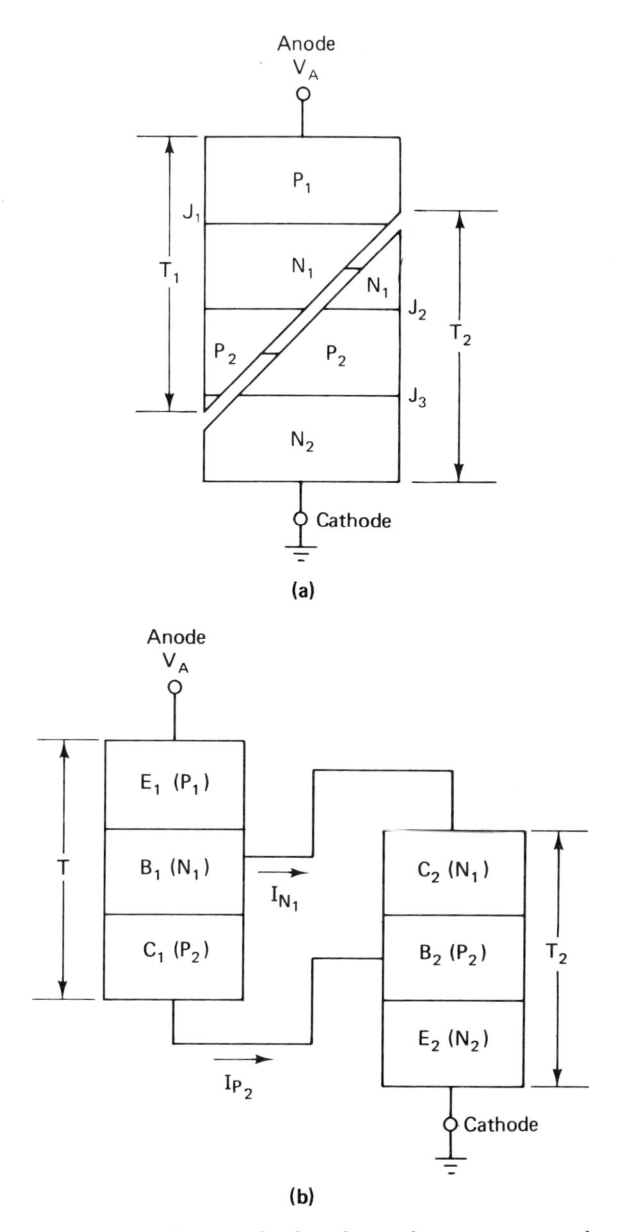

FIGURE 1.18 Resolution of a four-layer device into two three-layer ones.

The four-layer device may be switched from the off to the on state with V_A below V_S by an injection of current into P_2. This will drive T_2 into amplification and initiate regeneration. This action is called *gating* and is used in the silicon-controlled rectifier (SCR).

1.8 FIELD EFFECTS IN SEMICONDUCTORS

Field-effect action in semiconductors depends on the control of majority carrier flow by an electric field. Unlike the devices previously discussed, these carriers do not cross *p–n* junctions. Instead they pass between (or alongside) junctions and are affected by the way in which the junction depletion regions grow or diminish. Because depletion region widths are functions of voltage rather than current, field-effect devices are voltage controlled. The devices studied previously were, by contrast, current controlled. The advantages of voltage over current control will be studied later.

The field-effect structure is simpler and easier to fabricate than the *pnp* sandwich. In its basic form, it consists of a bar of *n*-type material (usually silicon) with two interconnected pieces of *p*-type material diffused or alloyed into its sides (Figure 1.19). In anticipation of the study of field-effect transistors (FETs), the terminals are labeled with FET nomenclature.

The operation of the FET can be explained by reference to Figure 1.19 as follows. With the gate tied to the source (switch in position a), and the drain–source voltage (V_{DS}) low, the *n*-channel acts as a simple resistor connected between drain and source. A free electron current (I_D) flows and develops a linear volt-drop within the *n*-channel. Because the gate is at ground and the section of the *n*-channel between the gates is at a more positive voltage, the *p–n* junctions formed by the gate and the *n*-channel are reverse biased. The depletion regions illustrated in Figure 1.19 are formed. The width of these regions, being a function of reverse bias, are greater at the drain side of the *p–n* junction than at the source side. Also, because the *n*-channel is less heavily doped than the gate, its depletion regions are wider.

Two antagonistic actions, both functions of V_{DS}, control the *n*-channel current. If the *p*-type material were absent, the *n*-channel would act as a fixed resistor and current would increase linearly with V_{DS}. However, with the *p*-type material present, an increase in V_{DS} also causes the reverse bias at the *p–n* junctions to increase and extends the depletion regions farther into the *n*-channel. This last action, acting alone, would reduce I_D by reducing the cross-sectional area of the *n*-channel, thereby increasing its resistance.

These two actions oppose each other to varying degrees as V_{DS} increases. At low V_{DS} the depletion regions have minor effect on I_D, and the *n*-channel acts as a fixed resistor. As V_{DS} is raised, the depletion regions extend into the *n*-channel and have greater effect on I_D. Above a critical voltage called the *pinch off voltage* (V_{PO}), any increase in current due to an increase in V_{DS} is almost entirely canceled by the throttling effect of

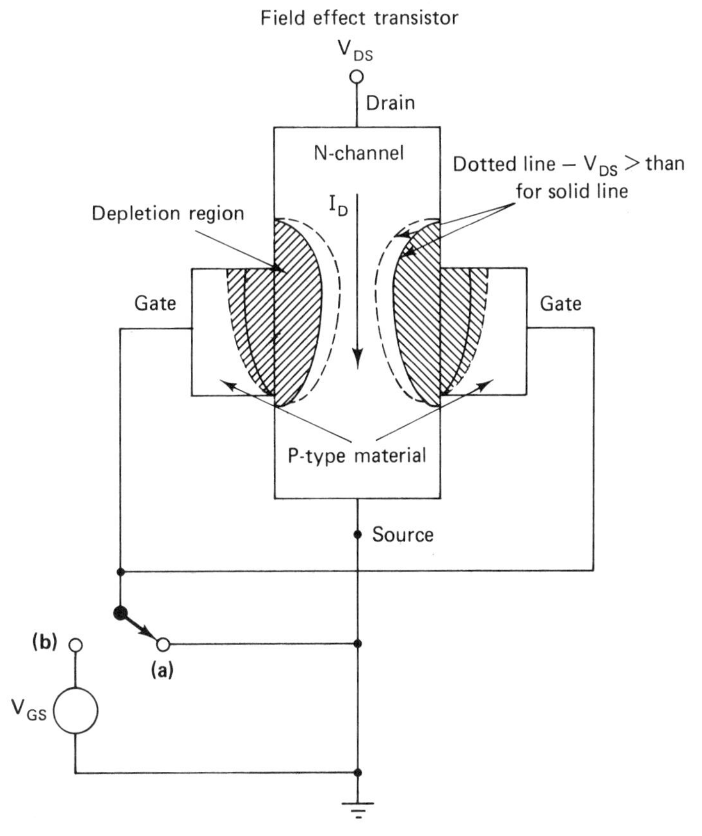

FIGURE 1.19 Field-effect action. In switch position a, $V_{GS} = 0$. In position b, V_{GS} is signal voltage.

the widening depletion regions. Therefore, further increases in V_{DS} above V_{PO} increase I_D only slightly. Above V_{PO}, I_D is a sensitive function of the depletion region width. Since this width varies with the gate-to-source voltage (V_{GS}), small changes in V_{GS} (switch at position b in Figure 1.19) will produce large changes in I_D. Voltage-controlled amplification results.

1.9 UNIJUNCTION ACTION

Unijunction action involves a property of semiconductors not exploited in the devices studied previously. This property derives from the fact that the resistance of a semiconductor depends on the density of its carriers. Therefore, this resistance may be controlled by an external source that can inject or remove carriers from the semiconductor. This property is exploited by the unijunction transistor (UJT).

The UJT, like the silicon-controlled rectifier, behaves as a latching switch. It consists of a bar of lightly doped *n*-type silicon with a small piece of heavily doped *p*-type material joined to one side. Figure 1.20 displays the structure of a bar and a cube UJT and Figure 1.21 gives a schematic representation of a UJT and its equivalent circuit. The upper and lower terminals of the bar (Figure 1.21a) are called base 2 (B_2) and base 1 (B_1), respectively, and the *p*-type terminal is called the emitter. Although superficially resembling the FET, the UJT operation depends on a forward-biased *p–n* junction instead of the reverse bias required by the FET.

UJT operation can be understood by considering it to be composed of two resistors and a diode (Figure 1.21b). The upper section of the *n*-bar (above *E*) is unaffected by emitter current and is therefore represented as a fixed resistor (R_{B_2}). The section of the *n*-bar below *E* is affected by emit-

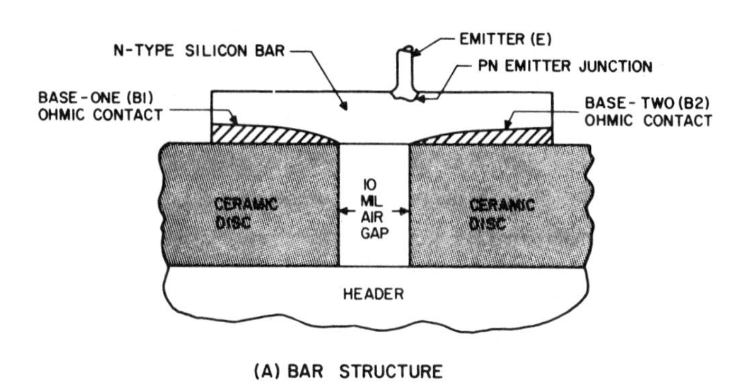

(A) BAR STRUCTURE

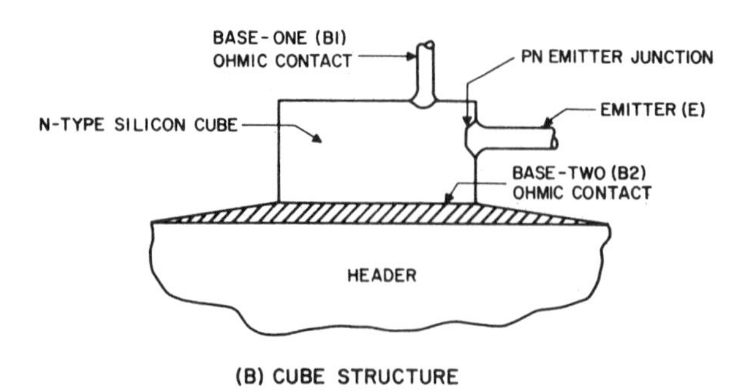

(B) CUBE STRUCTURE

FIGURE 1.20 Cross-sectional views of unijunction structures. Courtesy of General Electric Semiconductor Products Department.

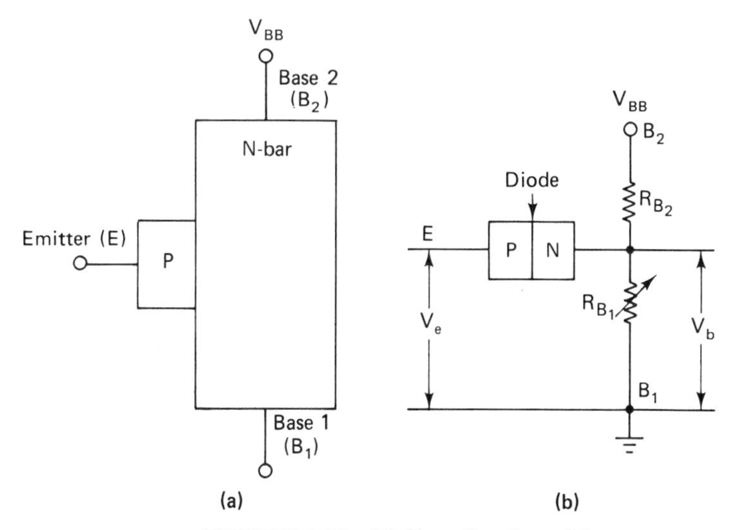

FIGURE 1.21 Unijunction transistor.

ter current and is therefore represented by the variable resistor R_{B_1}. The p–n junction diode is also included.

UJT action proceeds as follows. With E tied to B_1 (ground) and V_{BB} positive but low, the p–n junction is reverse biased (V_b positive, V_e at ground). Therefore, the n-bar acts as a single fixed resistor ($R_{B2} + R_{B1}$), typically between 4 and 10 kΩ. However, if V_e is made slightly higher than V_b, the p–n junction becomes forward biased. Two mutually reinforcing actions ensue. The emitter begins to inject hole current into the lower section of the n-bar, reducing R_{B_1}. Since V_{BB} and R_{B_2} are unchanged, V_b is reduced. This increases the forward bias of the emitter junction, which increases its hole current and thereby further reduces R_{B_1}. A regenerative flyaway action follows, ending only when the emitter current reaches a value limited by the circuit supplying V_e. The UJT is now in its on state. To switch to its off state, the emitter current must be turned off (or reduced to a low value) so that the p–n junction regains its reverse bias.

1.10 JUNCTION CAPACITANCES

1.10.1 Transition Capacitance

The reverse-biased p–n junction is surrounded by a region depleted of carriers bordered by regions containing excess oppositely charged carriers. The depletion region acts as an insulator, a barrier to current flow.

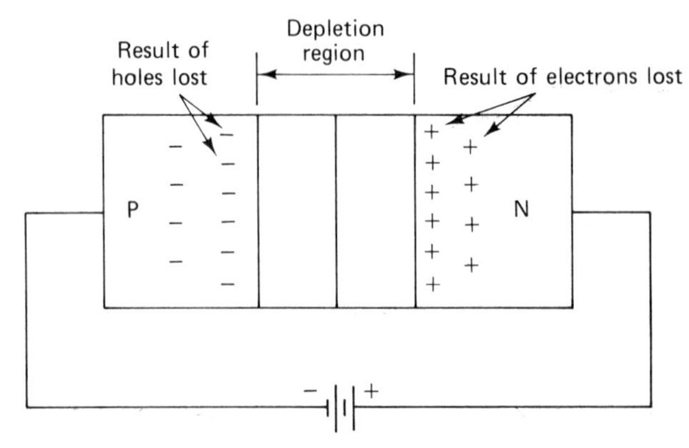

(a) Reverse biased junction

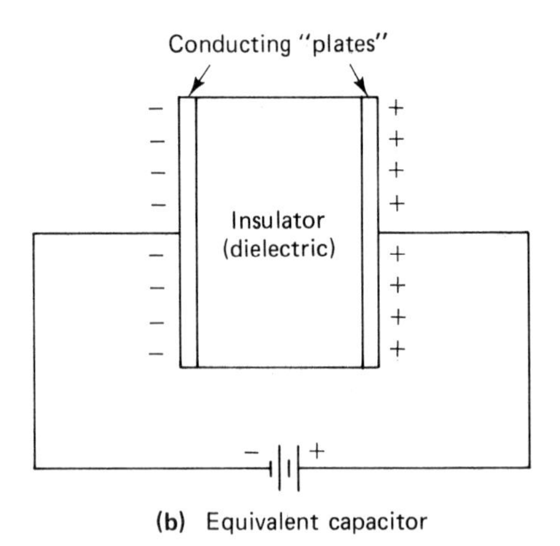

(b) Equivalent capacitor

FIGURE 1.22 Transition capacitance: (a) reverse-biased junction; (b) equivalent capacitance.

The oppositely charged border regions on each side of the barrier can be viewed as the plates of a capacitor separated by an insulator. Therefore, as shown in Figure 1.22, the reverse-biased *p–n* junction acts as a capacitor. This capacity is called the *transition capacitance* and its magnitude varies inversely with the width of the depletion region. Since the width of the depletion region is a direct function of its reverse bias, the *p–n* junction can be used as a voltage variable capacitor. Figure 1.23 shows how this capacity varies with reverse voltage for two different device junctions.

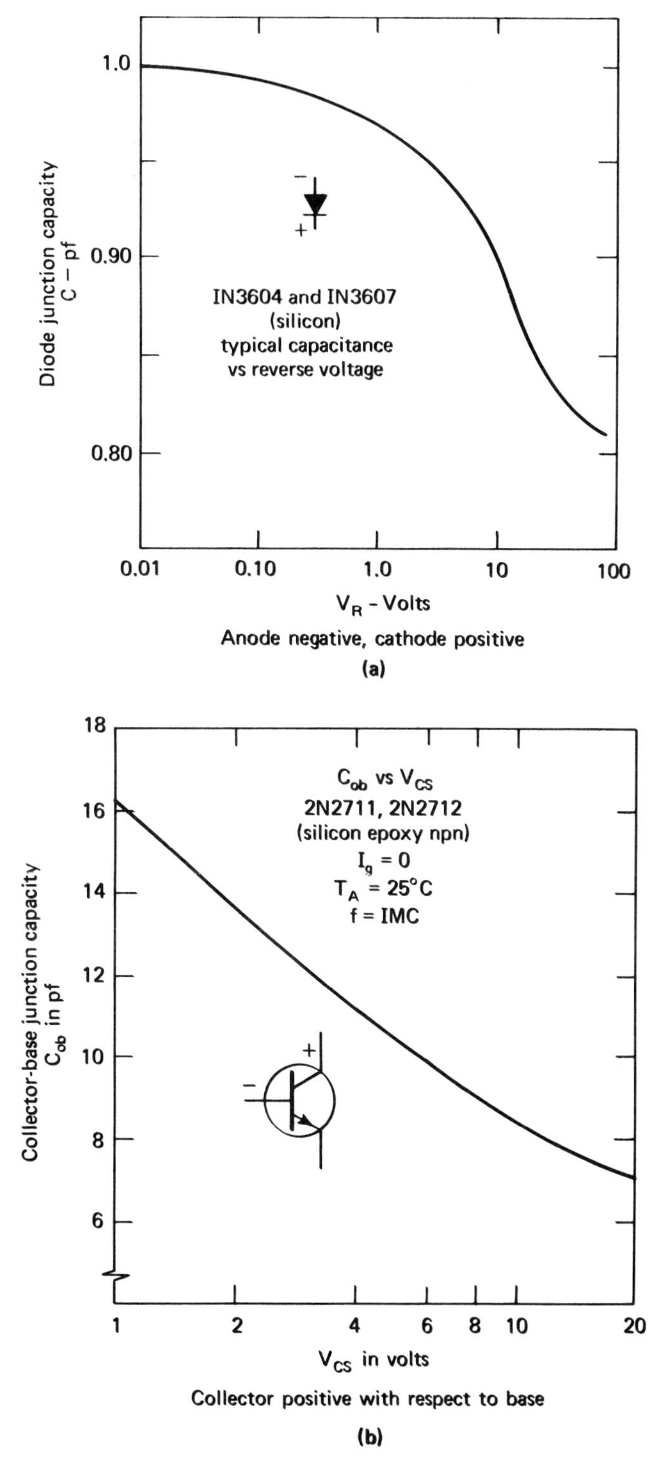

FIGURE 1.23 Transistor and diode transition capacity. Courtesy of General Electric Semiconductor Products Department.

1.10.2 Diffusion Capacitance

Forward-biased junctions cannot be turned off instantaneously by reversing the bias voltage. This is a result of the fact that the forward current fills the junction region with surplus majority carriers. Although the bias voltage may be abruptly reversed, forward current will continue until the junction region is emptied of carriers. Since this effect is similar to the discharging of a capacitor, it is represented by a symbolic capacitance called the *diffusion capacitance* (C_D). The effects of C_D are often given by specifying the time required to reverse a forward-biased junction. This time is called the *reverse recovery time* (t_{rr}). The presence of C_D is usually undesirable as it reduces the high-frequency response and switching speed of solid-state devices.

Problems

1. In your own words, distinguish between semiconductors, insulators, and conductors.
2. Discuss electron flow and hole flow in semiconductors.
3. Discuss the Bohr model of the atom.
4. Explain how doping affects the properties of semiconductors.
5. Differentiate between drift and diffusion currents.
6. Describe the diode action of a *p–n* junction.
7. Describe the effects of forward and reverse bias on a *p–n* junction.
8. Discuss minority carrier effects in semiconductors.
9. Describe the effects of increasing temperature on *p–n* junctions.
10. Describe zener and avalanche breakdown mechanisms.
11. Discuss the theory of operation of *npn* and *pnp* devices.
12. Draw the potential distribution diagram of a *pnp*.
13. Describe the operation of a Schockley diode.
14. Explain field effects in FETs.
15. Describe the theory of operation of the UJT.
16. Suggest two applications for a UJT.
17. Explain how reverse bias controls the transition capacitance.
18. Describe how C_D affects the reverse recovery time (t_{rr}) of diodes.

References

Bell, David. *Electronic Devices and Circuits,* 2nd ed., Reston Publishing Co., Reston, Virginia, 1980.

Cooper, William B. *Solid-State Devices: Analysis and Application,* Reston Publishing Co., Reston, Virginia, 1974.

G.E. Semiconductor Products Department, G.E. Transistor Manual, lightweight edition, Electronic Park, Syracuse, New York, 1969.

2

SEMICONDUCTOR DIODE

2.1 INTRODUCTION

The real diode is a complex device, as are all real electronic components. Fundamentally, the semiconductor diode passes current more readily in one direction than in the other, but it also possesses other characteristics; some are useful, some troublesome. For example, because semiconductor materials are heat sensitive, the diode's characteristics vary with temperature. At elevated temperatures, it suffers damage. Because *p–n* junctions are "grown," the manufacturer's control over diode parameters is incomplete. Therefore, the tolerances on parameters are much wider than desirable. Because reverse voltages stress semiconductor materials, excessively high voltages will damage the diode. The same is true for excessive forward currents. All these characteristics affect diode operation and must therefore be taken into account by the circuit designer.

2.2 DIODE AS A CIRCUIT ELEMENT

Obviously, all the characteristics of the real diode cannot be included in an initial study. We must first start with a simplified version. A good way to start is to consider the diode as an abstract circuit element that can only affect the rest of the circuit via the voltage–current relationship at its terminals. As such, its circuit effect can be completely defined by a graph that relates the current through the diode to the voltage across it. This graph is called the *diode terminal characteristic*. An example is given in Figure 2.1.

Because the diode characteristic is nonlinear, analytical (algebraic) solutions for circuits containing diodes are complicated. Therefore, simplified approximations of the characteristic are often used. These simplifications are arrived at by approximating the characteristic by two straight lines. Since characteristics composed of straight lines can be generated by linear circuits, these linear circuits, called *equivalent circuits*, can be substituted for the diode. This makes it possible to solve the circuit algebraically. Three useful simplified models of the diode are discussed next.

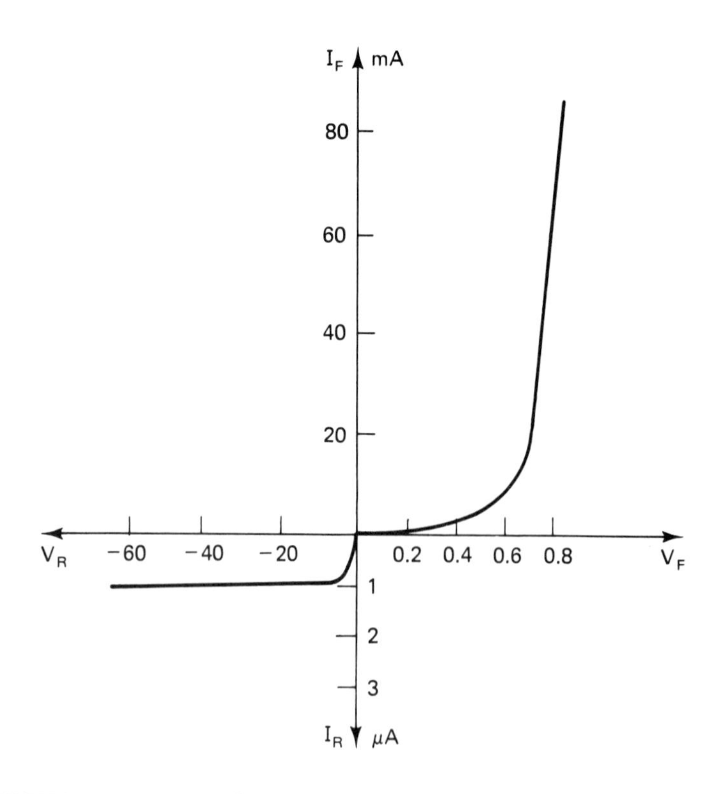

FIGURE 2.1 Forward and reverse characteristics for a typical low-current silicon diode. I_F, forward-bias current; V_F, forward-bias voltage; I_R, reverse-bias current; V_R, reverse-bias voltage.

2.2.1 Ideal Diode

In this, the simplest model, the diode is treated as a switch that is closed in the forward direction and open in the reverse direction. This model is called the *ideal diode*. Its characteristic and equivalent circuit are shown in Figure 2.2.

The ideal diode exists in one of two states. In the forward-biased state, it is a short circuit and the voltage across it (V_d) is zero for any current (I_d). I_d is controlled by the circuit connected to the diode. In the reverse-biased state, the diode is an open circuit. Its reverse current is always zero regardless of the voltage across it. In addition, the ideal

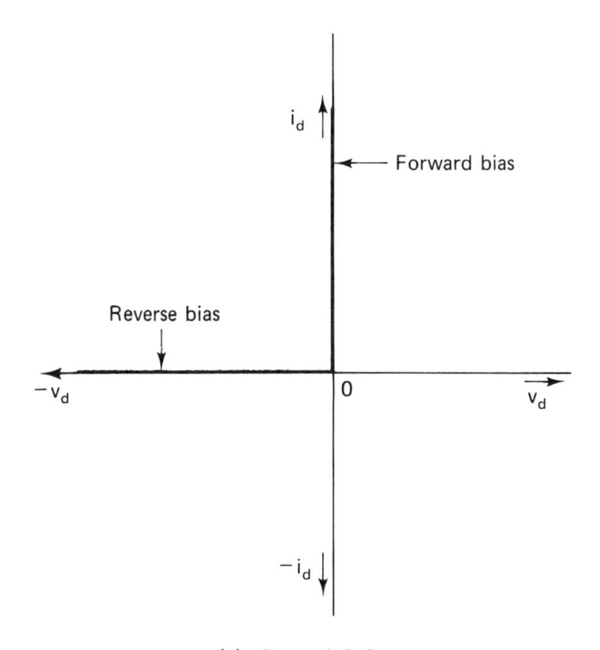

(a) Characteristic curve

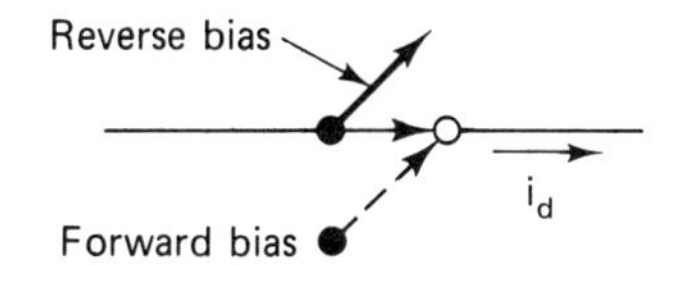

(b) Equivalent circuit

FIGURE 2.2 Ideal diode: (a) characteristic; (b) equivalent circuit.

diode is assumed to switch instantaneously from the reverse to the forward state whenever V_d reaches zero and to switch back instantaneously whenever V_d becomes negative.

2.2.2 Offset Ideal Diode

Another useful diode model is obtained by offsetting the ideal diode characteristic with a positive voltage (V_B) called the *break voltage*. Its characteristic and equivalent circuit are shown in Figure 2.3.

The equivalent circuit of Figure 2.3b can be shown to generate the offset ideal diode characteristic (Figure 2.3a) as follows: While V_d is less than V_B, the ideal diode is reverse biased since the voltage at its top end (anode) is lower than the voltage at its bottom end (cathode). Therefore, the ideal diode and the entire equivalent circuit are open and I_d is zero.

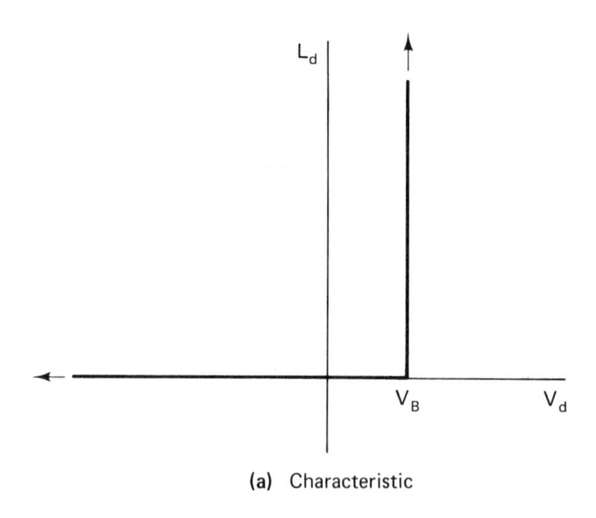

(a) Characteristic

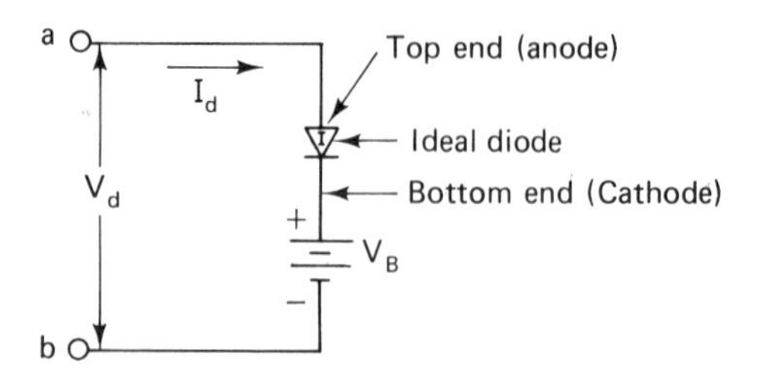

(b) Equivalent circuit

FIGURE 2.3 Offset ideal diode: (a) characteristic; (b) equivalent circuit.

The horizontal segment of the characteristic is generated. When V_d reaches V_B, the voltage across the ideal diode reaches zero. The ideal diode becomes a short circuit, and the voltage at the terminals of the equivalent circuit (V_d) is clamped at V_B. This action generates the vertical segment of the characteristic.

2.2.3 Piecewise Linear Model

The piecewise linear diode model provides the most accurate approximation of the actual characteristic so far. The nonideal model is derived from the true diode characteristic by first approximating it with two line segments, as shown in Figure 2.4a. Three values are then determined: R_F

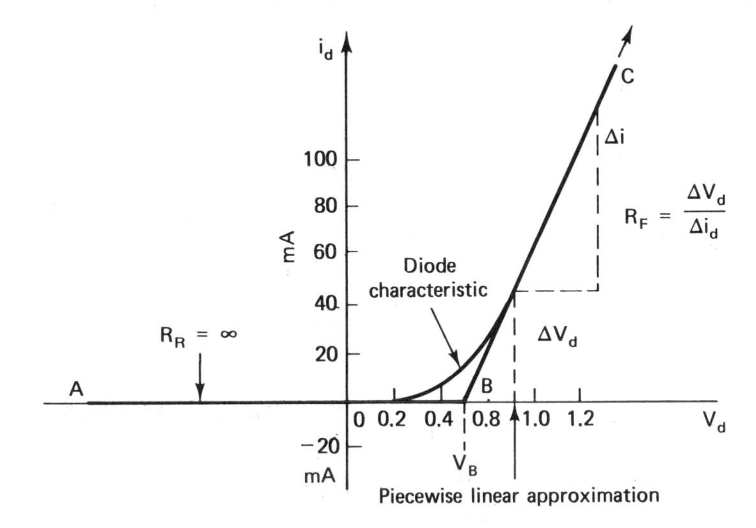

(a) Piecewise linear approximation of diode characteristic

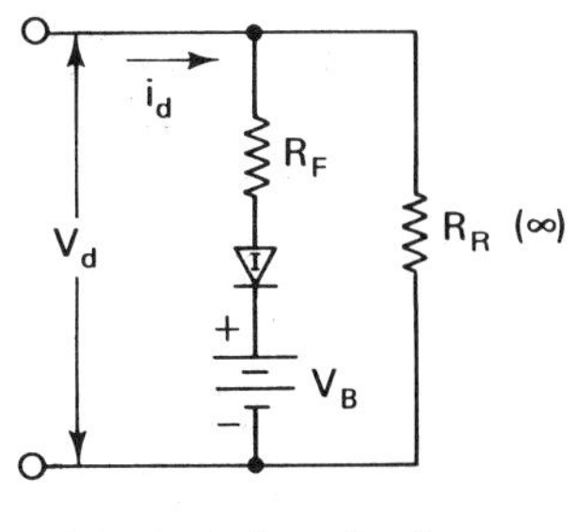

(b) Equivalent circuit

FIGURE 2.4 Piecewise linear approximation of a diode: (a) piecewise linear approximation of diode characteristic; (b) equivalent circuit.

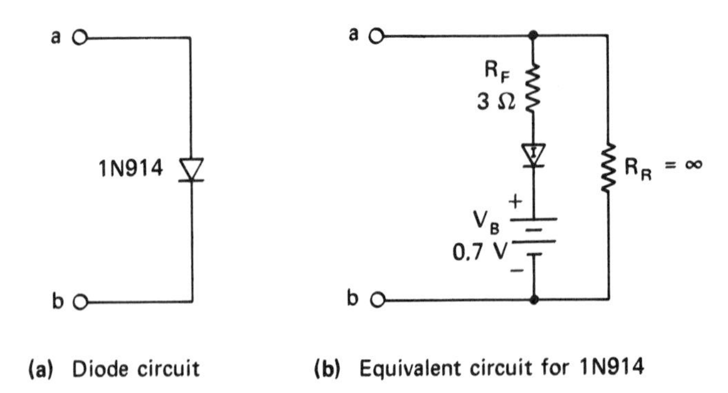

FIGURE 2.5 Illustration for Example 2.1: (a) diode circuit; (b) equivalent circuit for 1N914.

(the forward resistance), R_R (the reverse resistance), and V_B (the break voltage). R_F is the reciprocal of the slope of line segment *BC*, R_R is the reciprocal of the slope of *AB*, and V_B is the voltage at which *BC* crosses the V_d axis. These parameters are shown in Figure 2.4a. For silicon diodes R_R may be assumed to be infinite and V_B equal to 0.7 V. Figure 2.4b shows how these parameters are used to construct the equivalent circuit. This procedure is called *piecewise linear approximation.*

The equivalent circuit of Figure 2.4b generates the piecewise linear approximation of the diode characteristic as follows. While V_d is less than V_B, the ideal diode is open and the equivalent circuit consists of R_R alone. Since R_R is assumed to be infinite, the horizontal line segment *AB* is generated. When V_d equals or exceeds V_B, the ideal diode shorts and the equivalent circuit now consists of R_F in series with V_B. This circuit generates the line segment *BC*.

In practice, the parameters of the piecewise linear model are computed from numerical data supplied by the manufacturer. Since R_R is assumed to be infinite and V_B is known (0.7 V for silicon, 0.3 V for germanium), only R_F needs to be computed. To do this, we require only the coordinates of one point on the forward characteristic, a point above V_B. This information is provided by the manufacturer as a forward current (I_F) at a forward voltage (V_F). R_F is then given by

$$R_F = \frac{V_F - V_B}{I_F} \tag{2.1}$$

Example 2.1

For a diffused junction silicon switching diode (1N914), the manufacturer supplies the following data.

$$I_F \text{ (min. fwd. current, } V_F = 1\text{ V)} = 10\text{ mA}$$

Find R_F and draw the piecewise linear equivalent circuit.

$$R_F = \frac{V_F - V_B}{I_F} = \frac{1 - 0.7\text{ V}}{10\text{ mA}} = 30\ \Omega \quad (V_B = 0.7\text{ V for silicon})$$

The equivalent circuit for this diode is given in Figure 2.5.

2.3 ANALYZING DIODE CIRCUITS ALGEBRAICALLY

Circuits containing diodes are analyzed by first replacing the diode by the appropriate model. If the model is one of the equivalent circuits discussed previously, then the complete circuit can be analyzed algebraically. If, for reasons to be discussed later, the nonlinear graphical terminal characteristic is used, the complete circuit is solved graphically.

Each of the three diode models is a simplification of the terminal characteristic of the diode. The ideal model ignores both V_B and R_F. The offset ideal model includes V_B but ignores R_F. The piecewise linear equivalent circuit ignores neither but only approximates the true terminal characteristic. Even the terminal characteristic supplied by the manufacturer is only typical and may differ considerably from that of the diode in hand.

In the face then of all these diode representations, how does the technologist choose the one best suited for the analysis or design of a given circuit? The choice of a diode representation is governed by the error it introduces into the computation. The simplest model giving acceptable results should be used. Because of the normally large variability in diode parameters, errors up to $\pm 10\%$ are acceptable generally. The error introduced by a specific model depends on the magnitude of the signal voltage compared to the diode's break voltage and also on the ratio of circuit resistance in series with the diode to the forward resistance of the diode. Three general cases may be established based on the real diode circuit of Figure 2.6a. These cases, shown in Figures 2.6b, c, and d, represent this real diode as ideal, offset ideal, and piecewise linear equivalent circuits, respectively.

Case 1. V_{in} much larger than V_B, R_s (series resistance) much larger than R_F: Setting $V_{in} = 20$ V, $V_B = 0.7$ V, $R_s = 1\text{ k}\Omega$, and $R_F = 10\ \Omega$, we compute V_o (voltage across R_s) for models b, c, and d of Figure 2.6.

$$\text{Ideal diode model:} \quad V_o = V_{in} = 20\text{ V} \tag{2.2}$$

$$\text{Offset ideal model:} \quad V_o = V_{in} - V_B = 20 - 0.7 = 19.3\text{ V} \tag{2.3}$$

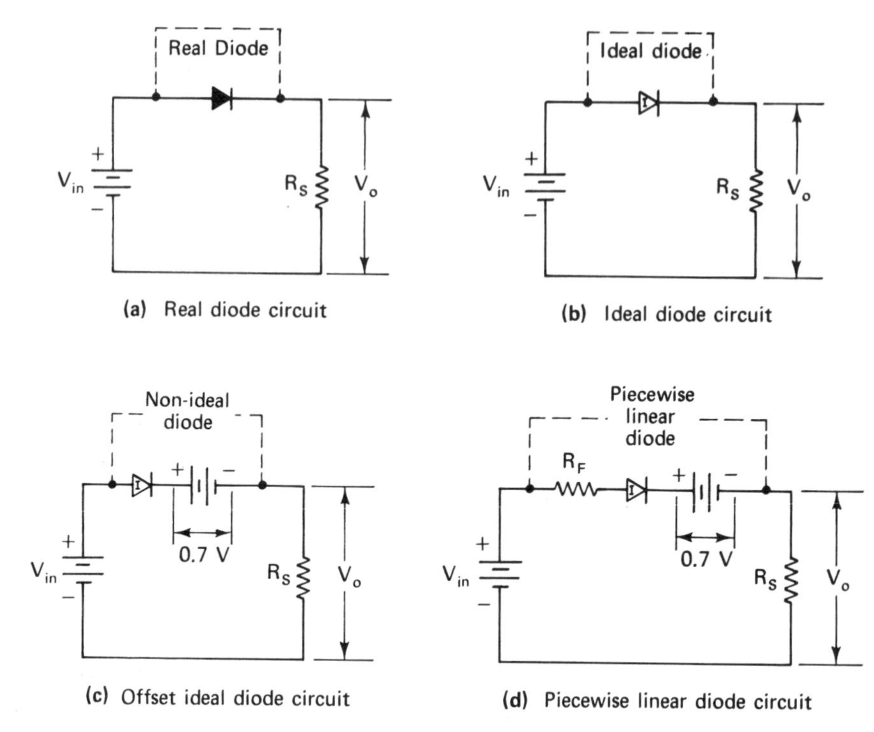

FIGURE 2.6 Forward-biased diode circuit using three diode models: (a) real diode circuit; (b) ideal diode circuit; (c) offset ideal diode circuit; (d) piecewise linear diode circuit.

Piecewise linear model: $$V_o = \frac{(V_{in} - V_B)R_s}{R_s + R_F} = \frac{(20 - 0.7) \times 1\ \text{k}\Omega}{1.01\ \text{k}\Omega}$$

$$= 19.1\ \text{V} \tag{2.4}$$

The simplest model, the ideal diode model, gives an answer within 5% of the most precise model, the piecewise linear model. The ideal diode model should be used.

Case 2. V_{in} not much larger than V_B, R_s much larger than R_F: Setting $V_{in} = 2$ V, $V_B = 0.7$ V, $R_s = 1$ kΩ, and $R_F = 10$ Ω, V_o for all models is given by

Ideal diode model: $$V_o = V_{in} = 2\ \text{V} \tag{2.5}$$

Offset ideal model: $$V_o = V_{in} - V_B = 2 - 0.7 = 1.3\ \text{V} \tag{2.6}$$

Piecewise linear model: $$V_o = \frac{(V_{in} - V_B)R_s}{R_s + R_F} = \frac{(2 - 0.7) \times 1\ \text{k}\Omega}{1.01\ \text{k}\Omega}$$

$$= 1.29\ \text{V} \tag{2.7}$$

The offset ideal model, which gives a result within 1% of the most accurate model, is the best choice. The ideal diode model is clearly unsatisfactory. V_B cannot be ignored as it subtracts significantly from V_{in}.

Case 3. V_{in} not very much larger than V_B, R_s not very much larger than R_F: Setting $V_{in} = 5$ V, $V_B = 0.7$ V, $R_s = 20\ \Omega$, and $R_F = 10\ \Omega$, V_o for all models is given by

$$\text{Ideal diode model:} \quad V_o = V_{in} = 5 \text{ V} \tag{2.8}$$

$$\text{Offset ideal model:} \quad V_o = V_{in} - 0.7 = 4.3 \text{ V} \tag{2.9}$$

$$\text{Piecewise linear model:} \quad V_o = \frac{(V_{in} - V_B)R_s}{R_s + R_F} = \frac{(5 - 0.7) \times 20\ \Omega}{20\ \Omega + 10\ \Omega}$$

$$= 2.87 \text{ V} \tag{2.10}$$

The piecewise linear model must be used. The other two models fail because they ignore the internal diode resistance (R_F), which significantly reduces V_o by voltage-divider action.

2.4 GRAPHICAL ANALYSIS OF DIODE CIRCUITS

When the diode operates in the region where its characteristic is extremely nonlinear (near V_B or in the reverse region), even the piecewise linear equivalent circuit will give inaccurate results. In this case, graphical analysis using the diode's terminal characteristic is necessary.

Figure 2.7 illustrates the concepts underlying the graphical analysis of diode circuits. The circuit for which I_d and V_d are to be found is given by Figure 2.7a. Figure 2.7b shows this circuit divided into two sections, the resistor and voltage on the left and the diode on the right. Each section has its own terminal characteristic. Both are shown on the same graph in Figure 2.7c. The diode characteristic is either provided by the manufacturer or determined experimentally. The voltage source–resistor characteristic can be computed.

When the two sections of Figure 2.7b are connected together (as in the original circuit), the circuit must operate at the intersection of the two characteristics for two reasons. First, V_d and I_d must be the same for both sections when connected. Second, the operating point must lie simultaneously on both characteristics. The operating voltage and current (V_d, I_d) are found by projecting the point of intersection onto both axes.

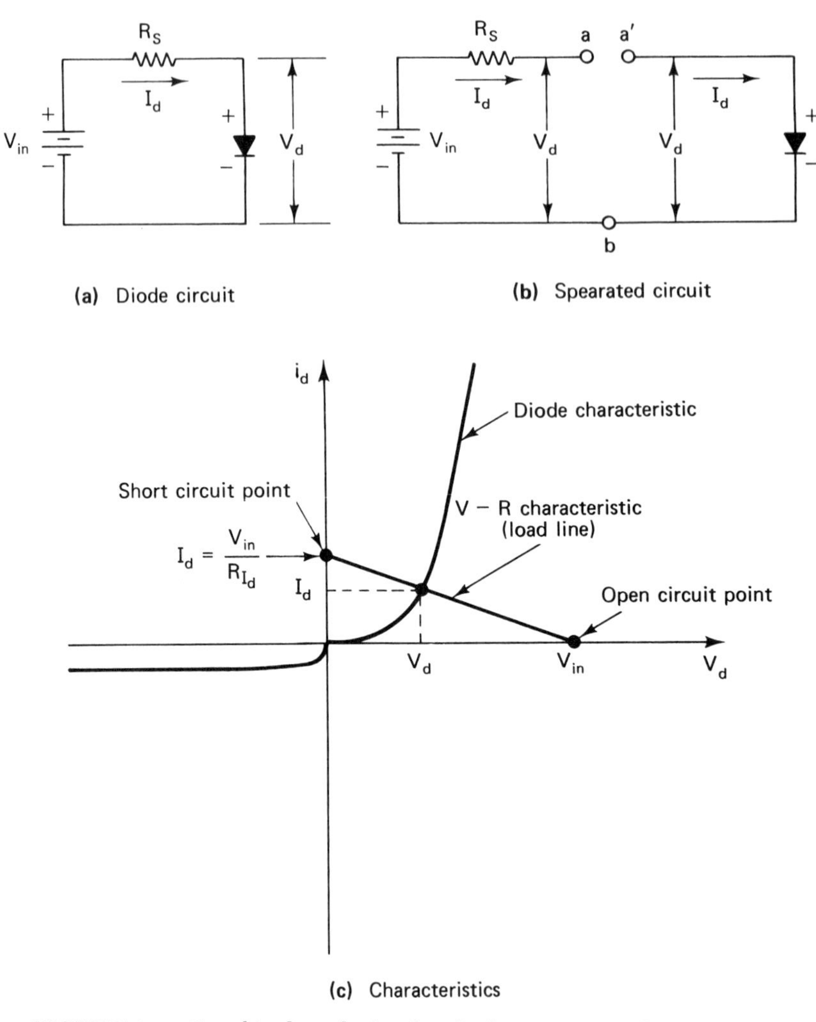

FIGURE 2.7 Graphical analysis of a diode circuit: (a) diode circuit; (b) separated circuit; (c) characteristics.

2.4.1 Finding the Load Line

Being linear, the terminal characteristic of the voltage source–resistor circuit (called the *load line*) can be constructed by locating any two points and drawing a straight line through them (Figure 2.8). Two convenient points are the open- and short-circuit points. For the open-circuit case, I_d is zero. Therefore, since no voltage drop occurs across R_s, V_d equals V_{in}. The coordinates then of the open-circuit points are

$$I_d = 0 \tag{2.11}$$

$$V_d = V_{in} \tag{2.12}$$

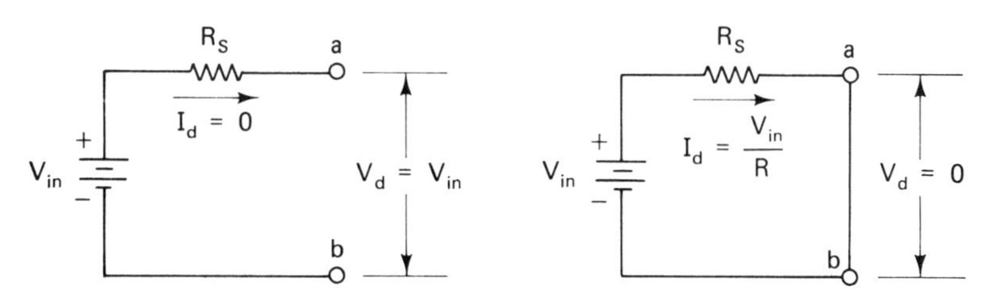

FIGURE 2.8 Finding the load line: (a) open-circuit coordinates; (b) short-circuit coordinates.

The coordinates of the short-circuit point are found by temporarily shorting points *a* and *b* (Figure 2.8). V_d becomes zero and I_d is given by

$$I_d = \frac{V_d}{R_s} \tag{2.13}$$

The short circuit coordinates are therefore

$$I_d = \frac{V_d}{R_s} \tag{2.14}$$

$$V_d = 0 \tag{2.15}$$

These results are summarized by Figure 2.8.

Example 2.2 *Graphical Solution of a Forward-Biased Diode Circuit*

Given the diode terminal characteristic of Figure 2.9b, find the operating point, I_d, V_d, for the circuit of Figure 2.9a.

(a) *Construct the load line on Figure 2.9b.*

Open circuit point: $V_{oc} = V_{in} = 2\text{ V}$, $\quad I_{oc} = 0$

Short circuit point: $V_{sc} = 0$, $\quad I_{sc} = \frac{V_{in}}{R_s} = \frac{2\text{ V}}{200\ \Omega} = 10\text{ mA}$

(b) *Find the intersection:* As shown in Figure 2.9b, the load line intersects with the diode characteristic at $I_d \approx 8$ mA and $V_d \approx 0.5$ V.

2.5 REAL DIODE

2.5.1 Physical Appearance of Typical Diodes

Figure 2.10 illustrates the physical aspects of typical discrete diodes (Figure 2.11 does the same for diode arrays). The forward current direction is usually indicated by one or more bands, a dot, or a flange at the

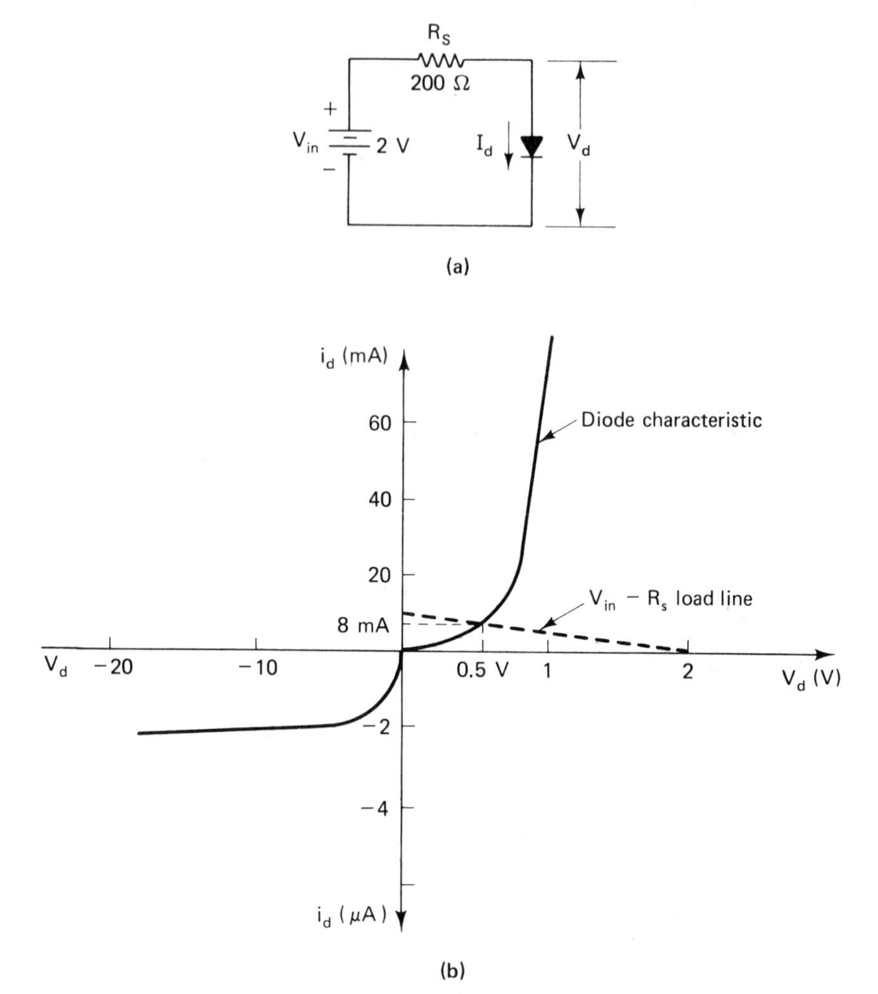

FIGURE 2.9 Graphical solution of a forward-biased diode circuit.

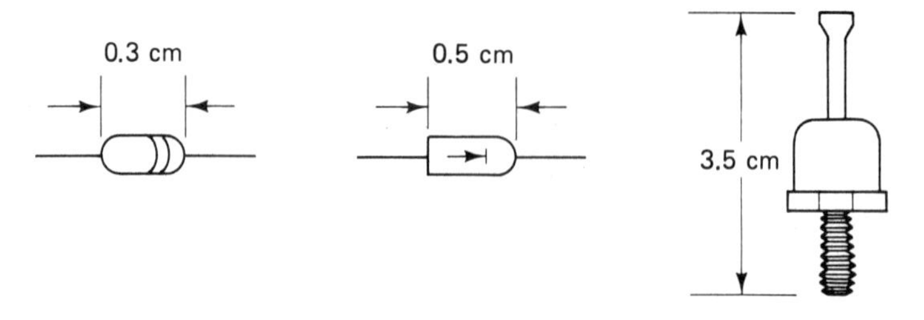

FIGURE 2.10 Physical appearance of diodes: (a) low-current diode; (b) medium-current diode; (c) high-current diode.

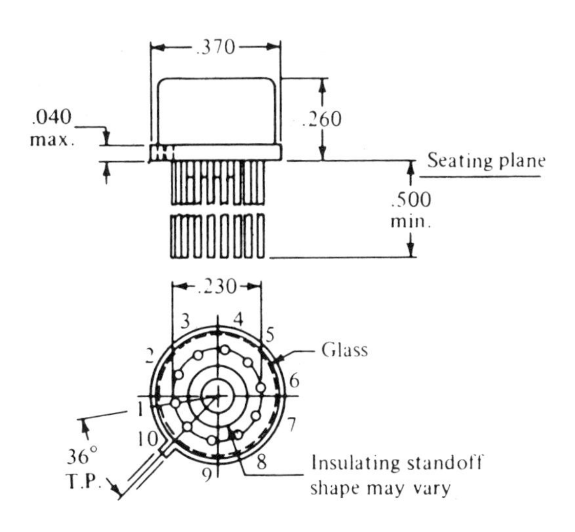

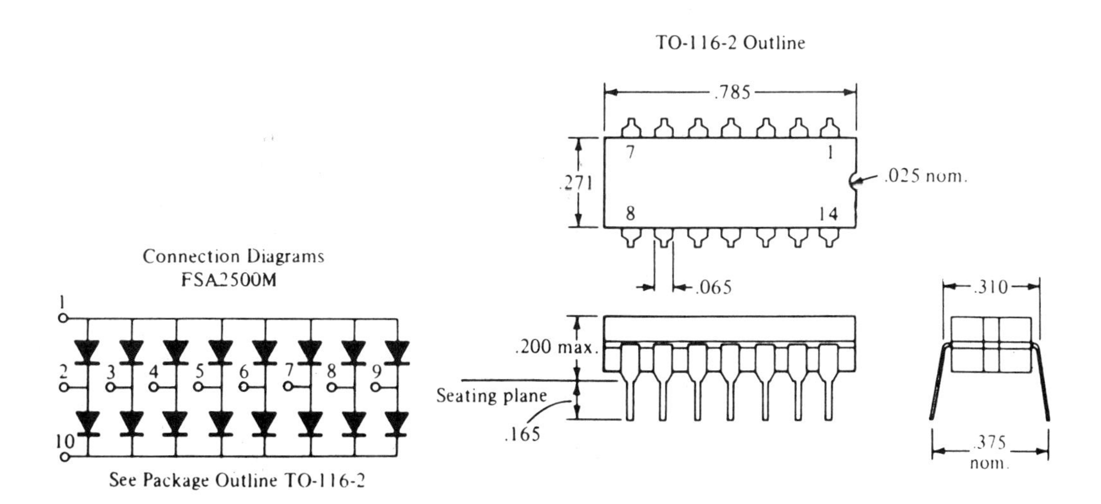

FIGURE 2.11 Multidiode array. Courtesy of Fairchild Camera and Instrument Corporation.

end from which the forward current emerges. This end is called the cathode, the other end, the anode. Sometimes (Figure 2.10b) the circuit symbol is displayed. High-current diodes are usually constructed so their case may be bolted to a heat-dissipating structure called a *heat sink*. Diodes vary widely in appearance, and reference to the manufacturer's data sheet is often necessary to determine proper connections.

Multidiode arrays in the form of integrated-circuit chips are also available. Figure 2.11 illustrates the physical appearance of a typical encased chip along with pin connections and a schematic of the diode array.

2.5.2 Manufacturer's Specifications: Using Data Sheets

Diode characteristics pertinent to design and applications are supplied by the manufacturer on *data sheets*, which range from short summaries of the salient characteristics of a whole family of diodes, through complete listings of all the controlled characteristics of a group of similar-purpose diodes, to an exhaustive description of a single diode, including curves.

A data sheet of the first type for a family of rectifier diodes is shown in Figure 2.12. This family handles rectified forward currents (I_o) from 1 to 50 A (topmost row) and withstands reverse voltages (V_{RRM}) from 100 to 1000 V (leftmost column). Maximum allowable forward surge currents lasting 1 second or less are also specified, as are maximum ambient operating temperature (T_A), maximum case temperature (T_C), and maximum junction temperature (T_{JRRM}). This summary provides sufficient data to make a choice of diode for low-frequency power supply applications.

Figure 2.13 provides an example of the second type of data sheet, a thorough listing of all controlled characteristics of a group of diffused-silicon high-speed switching diodes. Mechanical data, including body and lead dimensions for printed circuit insertion, are provided. A drawing or photo of the diode is also included. Following this is a section giving the absolute maximum ratings. Operation above these ratings may damage the diode and nullifies all guarantees. Below this is a section giving the electrical characteristics of the diode. The data required to represent the diode as a circuit element are provided here. Finally, a section on operating characteristics supplies the data pertinent to high-frequency or high-speed switching operation. We shall examine these data closely as they provide an excellent view of real diodes.

Absolute Maximum Ratings. Under this heading the limits on reverse voltages, forward currents, power dissipation, and operating and storage temperatures are given. Although specific diodes may not break

V_{RRM} (Volts)	I_O, AVERAGE RECTIFIED FORWARD CURRENT (Amperes)							
	1.0	1.5	3.0				6.0	12
	59-04 (DO-15) Plastic 		60 Metal 	70 Metal 	267 Plastic 		194-04 Plastic 	245 (DO-4) Metal 
50	†1N4001	**1N5391	1N4719	1N4997	**MR500	1N5400	**MR750	MR1120 1N1199,A,B
100	†1N4002	**1N5392	1N4720	1N4998	**MR501	1N5401	**MR751	MR1121 1N1200,A,B
200	†1N4003	1N5393 *MR5059	1N4721	1N4999	**MR502	1N5402	**MR752	MR1122 1N1202,A,B
400	†1N4004	1N5395 *MR5060	1N4722	1N5000	**MR504	1N5404	**MR754	MR1124 1N1204,A,B
600	†1N4005	1N5397 *MR5061	1N4723	1N5001	**MR506	1N5406	**MR756	MR1126 1N1206,A,B
800	†1N4006	1N5398	1N4724	1N5002	MR508		MR758	MR1128 1N3988
1000	†1N4007	1N5399	1N4725	1N5003	MR510		MR760	MR1130 1N3990
I_{FSM} (Amps)	30	50	300	300	100	200	400	300
T_A @ Rated I_O (°C)	75	$T_L = 70$	75	75	95	$T_L = 105$	60	
T_C @ Rated I_O (°C)								150
T_J (Max) (°C)	175	175	175	175	175	175	175	190

† Package Size: 0.120″ Max Diameter by 0.260″ Max Length.
* 1N5059 series equivalent Avalanche Rectifiers.
** Avalanche versions available, consult factory.

FIGURE 2.12 Rectifier diodes. Courtesy of Motorola, Inc.

I_O, AVERAGE RECTIFIED FORWARD CURRENT (Amperes)							
15	24	25		30	35	40	50
42A (DO-5) Metal	339 Plastic Note 1	193-03 Plastic Note 2	43-02 (DO-21) Metal		42A (DO-5) Metal		43-04 Metal
1N3208	MR2400	MR2500	1N3491	1N3659	1N1183	1N1183A	MR5005
1N3209	MR2401	MR2501	1N3492	1N3660	1N1184	1N1184A	MR5010
1N3210	MR2402	MR2502	1N3493	1N3661	1N1186	1N1186A	MR5020
1N3212	MR2404	MR2504	1N3495	1N3663	1N1188	1N1188A	MR5040
1N3214	MR2406	MR2506	MR328	Note 3	1N1190	1N1190A	Note 3
Note 3		MR2508	MR330	Note 3	1N3766	Note 3	Note 3
Note 3		MR2510	MR331	Note 3	1N3768	Note 3	Note 3
250	400	400	300	400	400	800	600
150	125	150	130	100	140	150	150
175	175	175	175	175	190	190	195

Note 1. Meets mounting configuration of TO-220 outline.
Note 2. Request Data Sheet for Mounting Information.
Note 3. Available on special order.

FIGURE 2.12 (continued) Rectifier diodes. Courtesy of Motorola, Inc.

TYPES 1N914, 1N914A, 1N914B, 1N915, 1N916, 1N916A, 1N916B and 1N917 DIFFUSED SILICON SWITCHING DIODES

- **Extremely Stable and Reliable High-Speed Diodes**

mechanical data

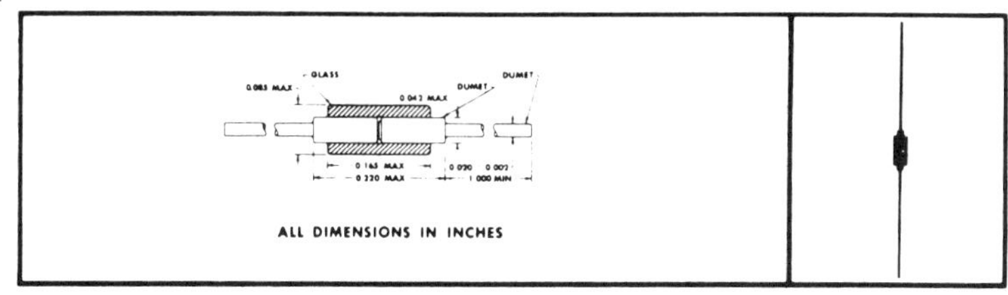

absolute maximum ratings at 25°C ambient temperature (unless otherwise noted)

		1N914	1N914A	1N914B	1N915	1N916	1N916A	1N916B	1N917	Unit
V_R	Reverse Voltage at — 65 to + 150°C	75	75	75	50	75	75	75	30	v
I_o	Average Rectified Fwd. Current	75	75	75	75	75	75	75	50	ma
I_o	Average Rectified Fwd. Current at + 150°C	10	10	10	10	10	10	10	10	ma
i_f	Recurrent Peak Fwd. Current	225	225	225	225	225	225	225	150	ma
$i_{f(surge)}$,	Surge Current, 1 sec	500	500	500	500	500	500	500	300	ma
P	Power Dissipation	250	250	250	250	250	250	250	250	mw
T_A	Operating Temperature Range	— 65 to + 175								°C
T_{stg}	Storage Temperature Range	. 200								°C

maximum electrical characteristics at 25°C ambient temperature (unless otherwise noted)

		1N914	1N914A	1N914B	1N915	1N916	1N916A	1N916B	1N917	Unit
BV_R	Min Breakdown Voltage at 100 μa	100	100	100	65	100	100	100	40	v
I_R	Reverse Current at V_R	5	5	5	5	5	5	5		μa
I_R	Reverse Current at — 20 v	0.025	0.025	0.025		0.025	0.025	0.025		μa
I_R	Reverse Current at — 20 v at 100°C	3	3	3	5	3	3	3	25	μa
I_R	Reverse Current at — 20 v at + 150°C	50	50	50		50	50	50		μa
I_R	Reverse Current at — 10 v				0.025				0.05	μa
I_R	Reverse Current at — 10 v at 125°C									μa
I_F	Min Fwd Current at V_F = 1 v	10	20	100	50	10	20	30	10	ma
V_F	at 250 μa								0.64	v
V_F	at 1.5 ma								0.74	v
V_F	at 3.5 ma								0.83	v
V_F	at 5 ma			0.72	0.73			0.73		v
V_F	Min at 5 ma				0.60					v
C	Capacitance at V_R = 0	4	4	4	4	2	2	2	2.5	pf

operating characteristics at 25°C ambient temperature (unless otherwise noted)

		1N914	1N914A	1N914B	1N915	1N916	1N916A	1N916B	1N917	Unit
t_{rr}	Max Reverse Recovery Time	••4 °8	••4 °8	••4 °8	°10	••4 °8	••4 °8	••4 °8	°3	nsec nsec
V_f	Fwd Recovery Voltage (50 ma Peak Sq. wave, 0.1 μsec pulse width, 10 nsec rise time, 5 kc to 100 kc rep. rate)	2.5	2.5	2.5	2.5	2.5	2.5	2.5	2.5	v

• Trademark of Texas Instruments

° Lumatron (10 ma I_F 10 ma I_R, recover to 1 ma)

•• EG&G (10 ma I_F, 6v V_R, recover to 1 ma)

FIGURE 2.13 Low-current diode data sheet. Courtesy of Texas Instruments, Inc.

down when these limits are slightly exceeded, many will, and the manufacturer's guarantees will not cover them. It is therefore poor practice to exceed these limits. Notice the following points. The reverse voltage (V_R) applies only within a $-65°$ to $+150°C$ temperature range. The average rectified forward current (I_o) is given at 25° and 150°C and is very much lower at the higher temperature. The recurrent peak forward current (i_f) can be much higher than I_o since its duration is briefer. The 1-s forward surge current, $i_{f(\text{surge})}$, is very large, but these values may be encountered when equipment is turned on. The power dissipation (P) applies only at 25°C. At higher temperatures, P must be decreased. The operating temperature range (T_A) is broad, but such ranges are encountered in practice. Finally, note that the storage temperature range (T_{stg}) must be complied with as storage above T_{stg} deteriorates semiconductor material.

Maximum Electrical Characteristics. This section of the data sheet gives the maximum (sometimes minimum) values of the diode parameters. Parameter values for specific diodes may be less than maximum (or exceed minimum) values. Notice the following points. The minimum breakdown voltage at 100 μA (BV_R) is higher than V_R. BV_R is provided to give another point on the characteristic at 100 μA, but operation at this point is not advisable. The reverse currents (I_R) given at several voltages and temperatures provide sufficient data to plot the reverse region characteristic. The minimum forward current (I_F) given at 1 V is used, together with the 0.7-V break voltage of silicon diodes to compute the forward resistance (R_F) of the diode (see Example 2.1). The forward voltages (V_F) given over a current range provide forward characteristic data. Note that many boxes are empty. This means that the manufacturer does not control or guarantee the characteristics in this region. Although these values can be measured for a specific diode, they may vary widely from diode to diode. Finally, the capacitance (C) taken at $V_R = 0$ is given. This is the maximum value of the transition capacity. The effect of V_R on C is discussed in Chapter 1 and an application is described in Chapter 3.

Operating Characteristics. This section gives data pertinent to high-speed operation of diodes. The maximum reverse recovery time (t_{rr}) is a measure of time required to turn off the diode. T_{rr} is here defined as the time required to switch from 10-mA forward current to 1-mA reverse current. When a large forward current is abruptly passed through a diode, V_F will rise above its steady-state value transiently. The forward recovery voltage specification gives the peak value of V_F for a specified input current waveform. Both specifications, the forward recovery time and the forward recovery voltage, are pertinent to diodes used as high-speed switches.

TYPICAL ELECTRICAL CHARACTERISTIC CURVES

AT 25°C AMBIENT TEMPERATURE UNLESS OTHERWISE NOTED

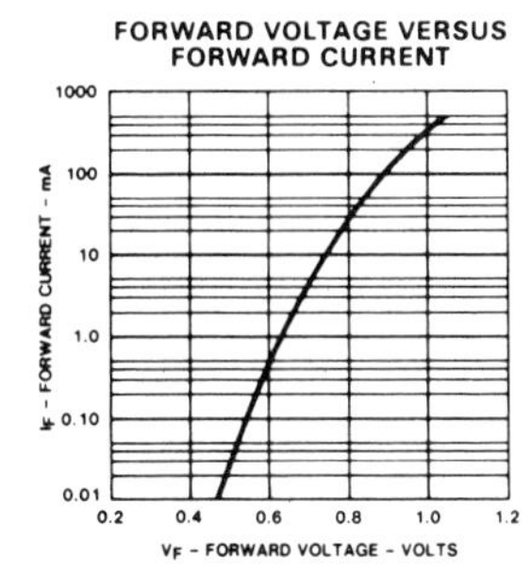

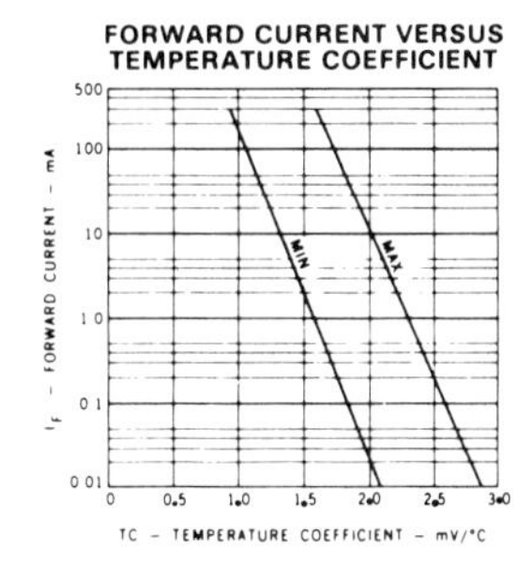

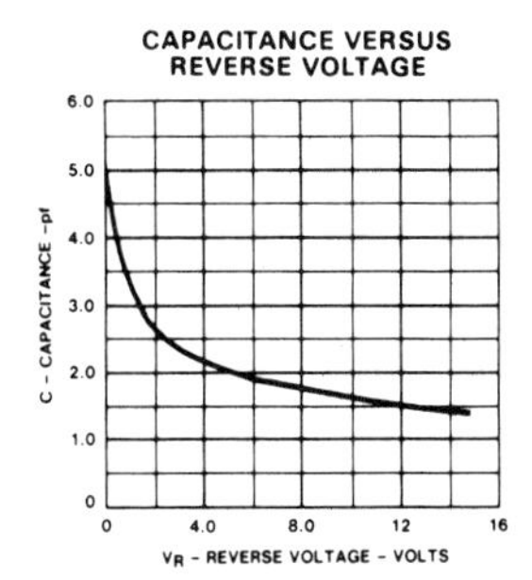

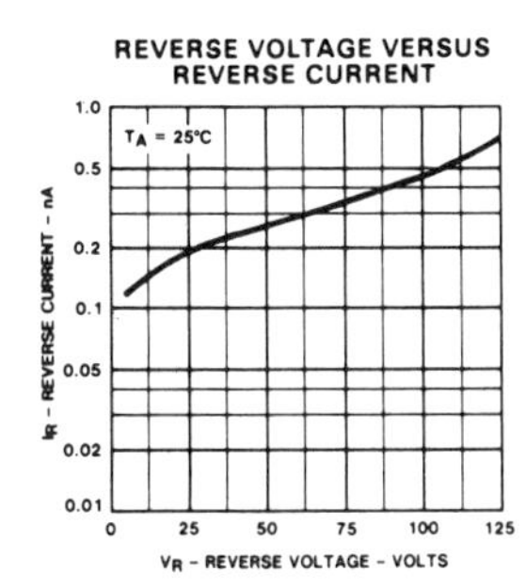

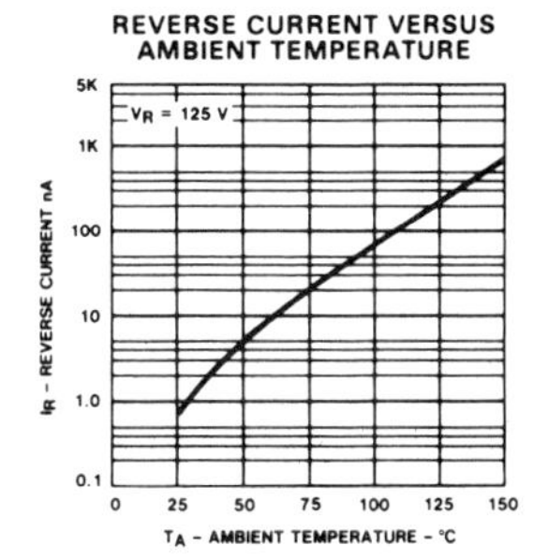

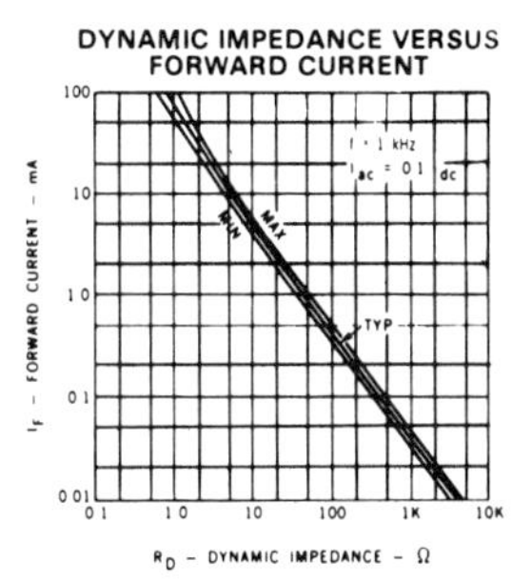

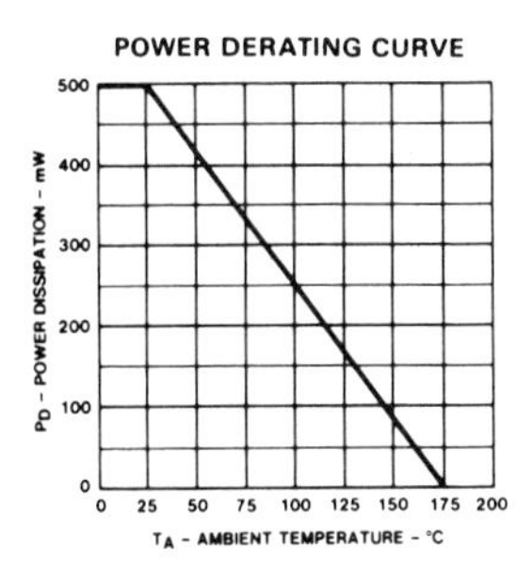

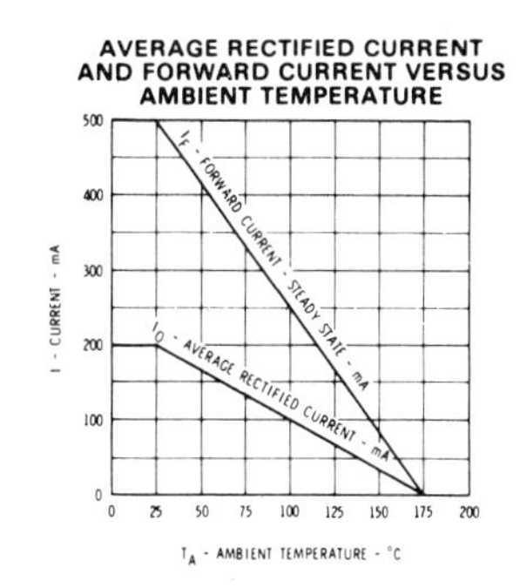

FIGURE 2.14 Terminal characteristics of the Fairchild Bay73–BA 129 high-voltage diodes. Courtesy of Fairchild Camera and Instrument Corporation.

Terminal Characteristics. In addition to supplying detailed tabular data, manufacturers also supply sets of terminal characteristics. These characteristics describe the variation of important diode parameters with temperature, voltage, and current. A typical set is shown in Figure 2.14.

2.6 TEMPERATURE EFFECTS

Temperature has a major effect on a diode's operation. It alters its terminal characteristics significantly and limits its allowable power dissipation. Figure 2.15 summarizes the temperature effects on the diode characteristic, and Figure 2.16 illustrates the reduction in allowable power dissipation with increasing temperature.

Figure 2.15 shows an improvement in the forward region characteristic of the diode and a deterioration in the reverse region with

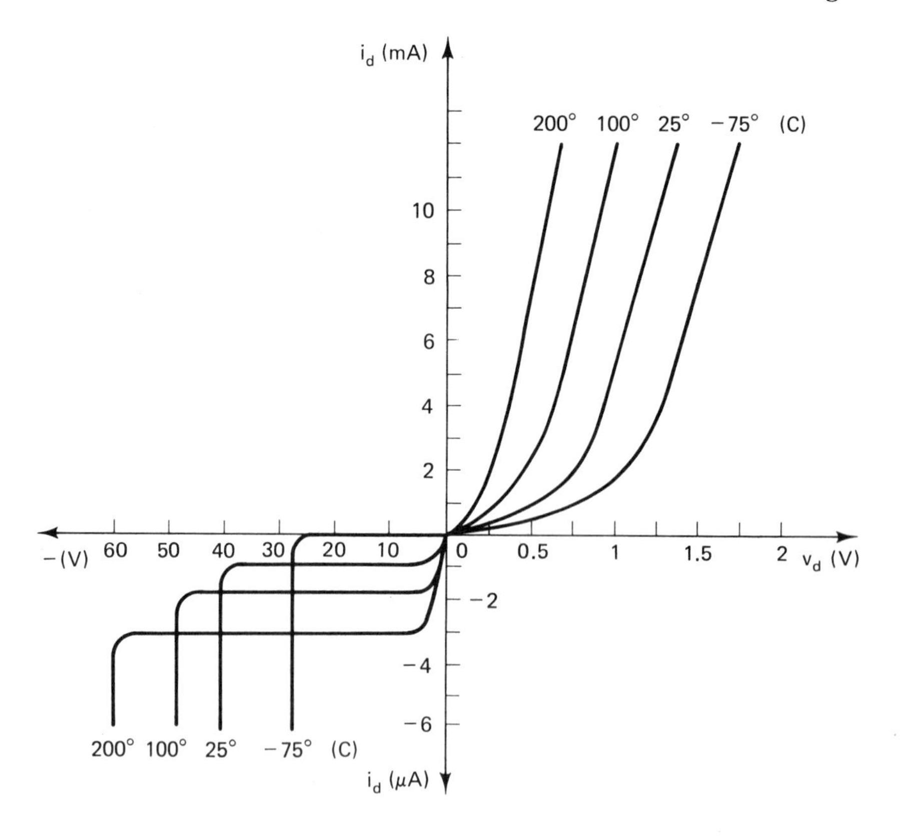

FIGURE 2.15 Temperature effects on diode characteristics.

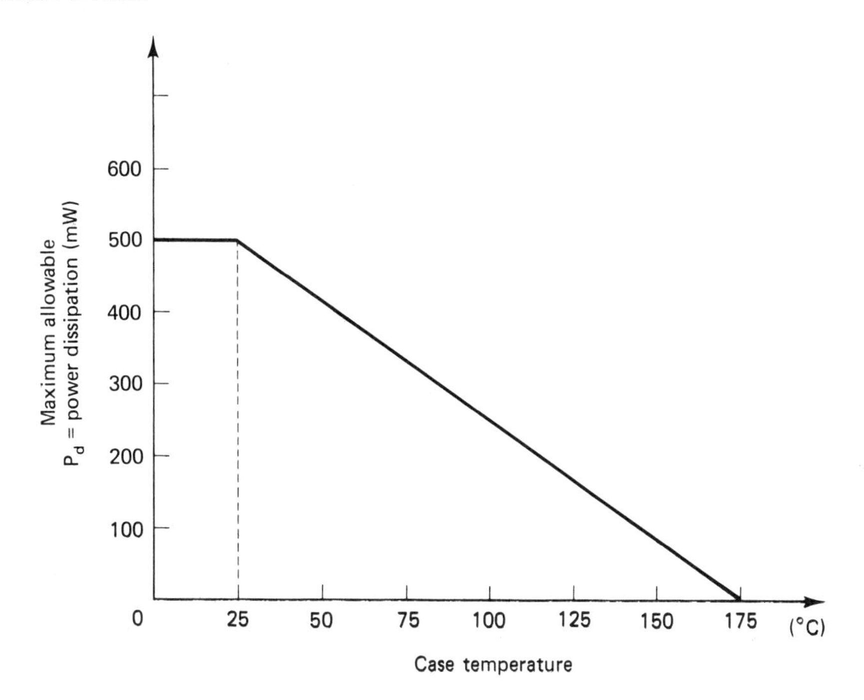

FIGURE 2.16 Power derating curve.

increasing temperature. Both effects result from an increase in majority and minority carriers. The improvement in the forward characteristic is seen in the decrease in V_B and R_F; the deterioration in the reverse characteristic is evidenced by the increase in reverse current. This increase can be large. The "reverse current versus ambient temperature" curve of Figure 2.14 shows reverse current increasing from 1 nA to 1000 nA from 25° to 150°C. If this increase is not taken into consideration in circuit design, it can have serious effects on circuit performance.

The effect of temperature on allowable power dissipation is particularly critical. Maximum allowable junction temperatures for silicon diodes range from 150° to 200°C. If this temperature is exceeded, the diode will be damaged. Junction temperature is a function of the power dissipated by the junction and the rate of heat removal from the junction to the environment. As the ambient temperature increases, the heat removal rate decreases and, as a consequence, the allowable power dissipation of the diode is reduced. Since junction temperature is not easily measured, manufacturers relate power dissipation to case temperatures for metal-encased diodes and to ambient (or lead) temperatures for glass-encased diodes. Figure 2.16 shows a power derating curve (maximum allowable power dissipation versus case temperature) for a metal-encased diode.

Power derating is a necessary consideration in all diode circuit design. The simplest way to keep diode dissipation within maximum ratings is to limit the maximum current (I_F) that flows through it. In the typical diode circuit, this current can be limited by external resistors. Therefore, the power derating task is primarily one of finding the maximum allowable I_F at a given temperature. This value can be read directly from curves of I_F versus temperature supplied by the manufacturer. Example 2.3 illustrates this procedure and also shows how V_F at maximum power dissipation can be found.

Example 2.3 *Diode Power Derating*

The manufacturer supplies the following data for the BAY 73:

$$P_D \text{ at } 25°\text{C} = 500 \text{ mW}$$

$$\text{Linear power derating factor (from } 25°\text{C}) = 3.33 \text{ mW/°C}$$

In addition, curves of maximum allowable power dissipation (P_D) and maximum allowable continuous forward current (I_F) versus ambient temperature (T_A) are given. Find I_F and V_F at $T_A = 100°\text{C}$.

(a) *Find* I_F: Using the forward current versus ambient temperature curve of Figure 2.14, at $T_A = 100°\text{C}$ we read $I_F = 250$ mA.

(b) *Find* V_F: We first find $P_{D(100)}$ (power dissipation at 100°C) by two methods.

(1) Using the linear power derating factor (LPD):

$$\begin{aligned} P_{D(100)} &= P_{D(25)} - \text{LPD}(100 - 25) \\ &= 500 \text{ mW} - 3.33 \text{ mW/°C} \times (100 - 25°\text{C}) \\ &= 250.25 \text{ mW} \end{aligned}$$

(2) Using the power derating curve (Figure 2.14), at 100°C we read $P_{D(100)} = 250$ mW.

$$P_D = V_F I_F$$

so

$$V_F = \frac{P_D}{I_F} = \frac{250 \text{ mW}}{250 \text{ mA}} = 1 \text{ V}$$

2.7 DIODE HEAT SINKS

Heat sinks remove heat from diode junctions more efficiently than the surrounding air. Therefore, they permit higher power dissipations at temperatures above 25°C. Small glass or ceramic encapsulated diodes are thermally connected to heat sinks through their leads (electrical isolation must of course be maintained). Larger metal-encased diodes are thermally connected to heat sinks through their cases. Electrical isolation is retained by using mounting materials that are good thermal conductors but electrical isolators. Figure 2.17 shows a metal-cased (top-hat) diode mounted on a finned heat sink. Heat flows readily from the diode case into the heat sink. It diffuses throughout the heat sink, whose extended surface area speeds dissipation into the surrounding air. Effective heat sinking can greatly increase the allowable power dissipation of a diode.

The junction temperature (T_j) of a metal-cased diode mounted on a heat sink is given by

$$T_j = P_j(\Theta_{jc} + \Theta_{cs} + \Theta_{sa}) + T_A \qquad (2.16)$$

where P_j is the power dissipated by the diode, Θ_{jc} is the junction-to-case thermal resistance, Θ_{cs} is the case-to-heat sink thermal resistance, Θ_{sa} is the heat sink-to-air thermal resistance, and T_A is the ambient temperature. Thermal resistance is a measure of the rate at which heat flows from a hotter region to a cooler one. The lower the thermal resistance is, the more rapid the heat transfer. Thermal resistance data are supplied by the diode and heat sink manufacturers. The selection of an appropriate heat sink is illustrated by Example 2.4.

Example 2.4 *Heat Sink Selection*

A top-hat diode with a maximum junction temperature ($T_{J\,max}$) of 150°C and a Θ_{Jc} of 3.0°C/W is to be mounted on a heat sink and operated at a P_D of 17 W at an ambient temperature (T_A) of 75°C ($\Theta_{cs} = 0.5$). Find Θ_{sa}.

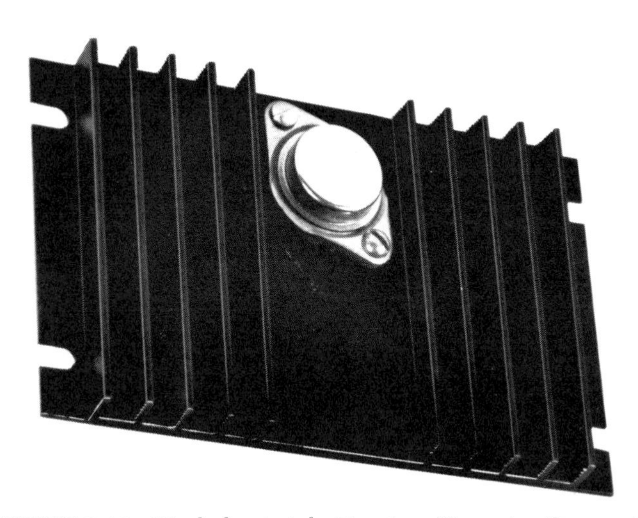

FIGURE 2.17 Diode heat sink. Courtesy Trans-tec Corporation.

Equation (2.16) can be manipulated to give

$$\Theta_{sa} = \frac{T_J - T_A - P_J(\Theta_{Jc} + \Theta_{cs})}{P_J} \tag{2.17}$$

Setting $T_J = T_{J\,max} = 150°C$, $T_A = 75°C$, and $P_J = 17$ W,

$$\Theta_{sa} = \frac{150 - 75 - 17(3.0 + 0.5)}{17} = 0.91°C/W$$

A Trans-Tec semiconductor cooler model 1128, $\Theta_{sa} = 0.8°C/W$, is satisfactory.

Problems

1. On one set of axes, draw the terminal characteristics (*i versus V*) of the following resistors: (a) 0, (b) ∞, (c) 1 kΩ, (d) 5 kΩ, (e) 25 kΩ, (f) 125 kΩ.
2. On one set of axes, draw the terminal characteristics for the circuits of Figure 2.18.
3. Using Figure 2.1, find the corresponding coordinate for the following: (a) $I_F = 60$ mA, (b) $I_F = 20$ mA, (c) $I_R = -0.5\ \mu A$, (d) $V_F = 0.6$ V, (e) $V_R = -60$ V.
4. Assuming all diodes to be ideal diodes, compute I and V for the circuits of Figure 2.19.
5. Assuming ideal diodes, sketch the wave shape of $v_{(t)}$ for the circuits of Figure 2.20.
6. Solve the circuits algebraically of Figure 2.21 for I and V using (a) the ideal diode model equivalent circuit; (b) the offset ideal model equivalent circuit; (c) the piecewise linear equivalent circuit. *Note:* $V_B = 0.7$ V, $R_F = 20\ \Omega$, and $R_R = \infty$.
7. Allowing an error of ±10%, which model is best suited for solving the circuits of Figure 2.21?
8. Using the diode characteristic of Figure 2.1, solve the circuits of Figure 2.22 graphically. Draw $i_{(t)}$ and $v_{(t)}$.
9. Derive the piecewise linear model equivalent circuit from Figure 2.1 and solve the circuits of Figure 2.22 algebraically. Draw $i_{(t)}$ and $v_{(t)}$.
10. Draw and label the piecewise linear equivalent circuit for the following diodes of Figure 2.13 (let $R_R = \infty$): (a) 1N914, (b) 1N914A, (c) 1N914B, (d) 1N915, (e) 1N916B.

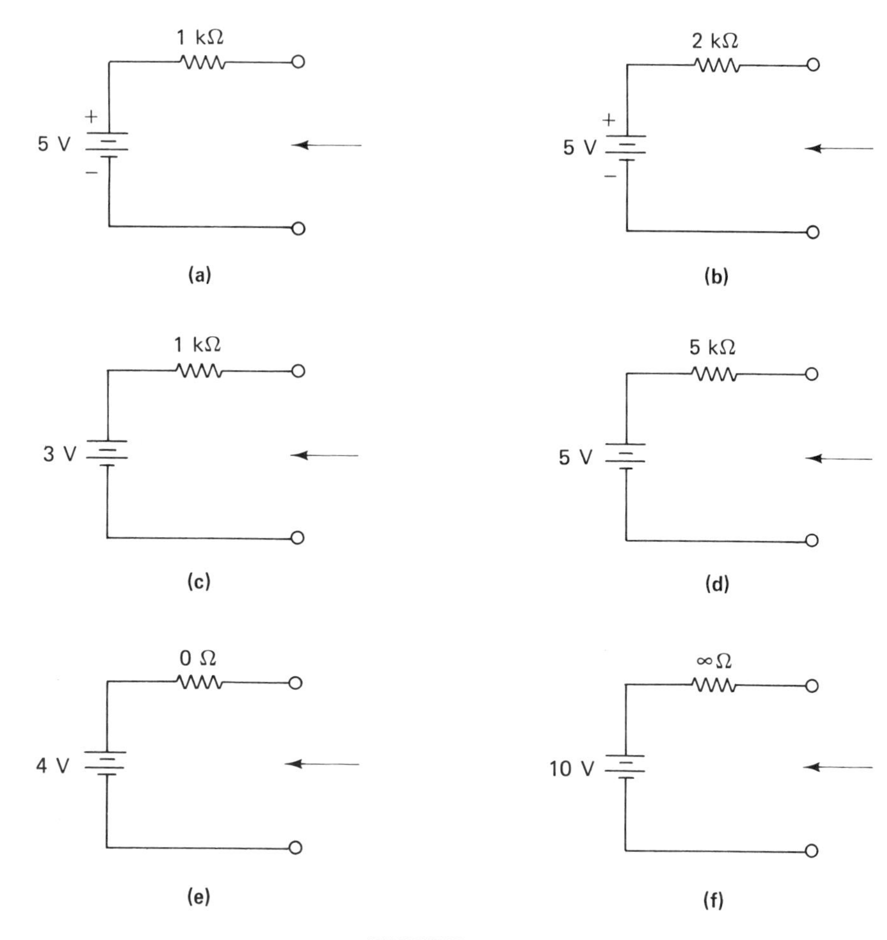

FIGURE 2.18

11. Choose the appropriate model for the diodes in each circuit of Figure 2.23. Explain and justify your choice.

12. Choose a general-purpose rectifier to meet the following specifications.
(a) $I_o = 3$ A, $V_{RRM} = 600$ V, $I_{FSM} = 300$ A, $T_A = 85°$C
(b) $I_o = 3$ A, $V_{RRM} = 600$ V, $I_{FSM} = 300$ A, $T_A = 70°$C
(c) $I_o = 3$ A, $V_{RRM} = 100$ V, $I_{FSM} = 300$ A, $T_A = 75°$C, $T_{case} = 60°$C
(d) $I_o = 3$ A, $V_{RRM} = 100$ V, $I_{FSM} = 700$ A, $T_A = 75°$C
(e) $I_o = 30$ A, $V_{RRM} = 1000$ V, $I_{FSM} = 400$ A, $T_A = 75°$C

13. Choose a diode to meet the following specs.
(a) $V_R = 65$ V, $I_o = 75$ mA, $P_D = 250$ mW, $R_F \leq 3\ \Omega$
(b) $V_R = 40$ V, $I_o = 50$ mA, $P_D = 200$ mW, $R_F \leq 30\ \Omega$, $t_{rr} \leq 3$ ns

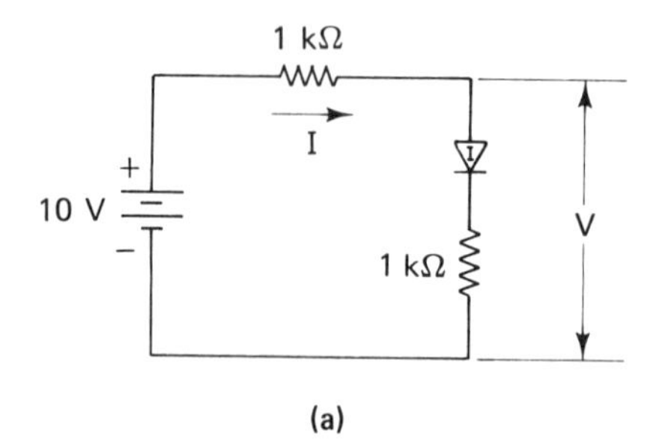

(a)

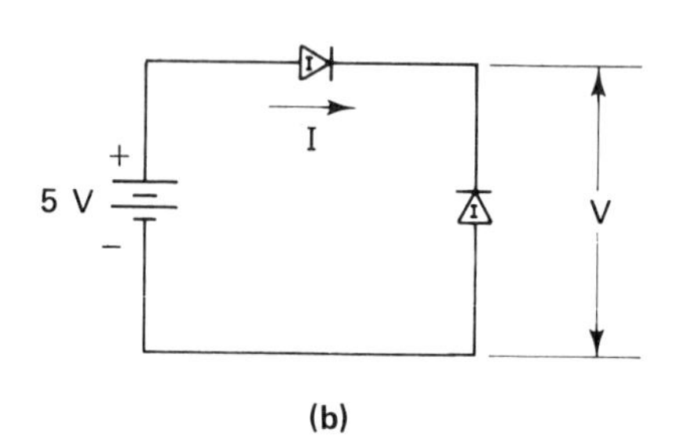

(b)

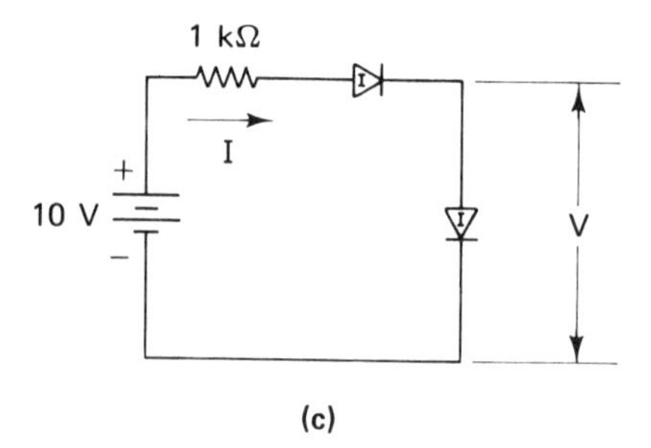

(c)

FIGURE 2.19

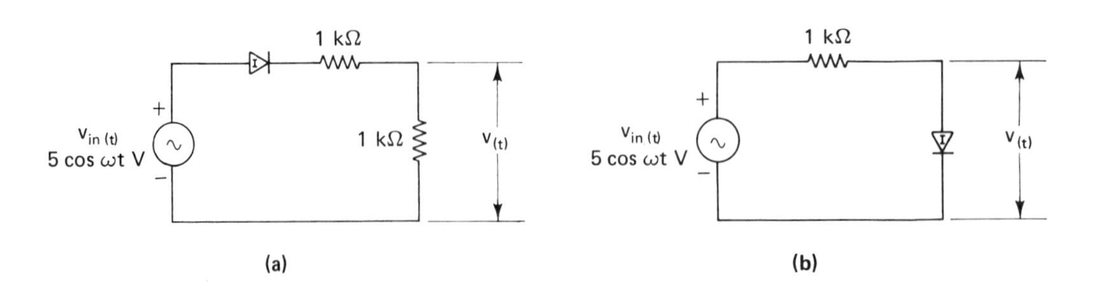

(a) (b)

FIGURE 2.20

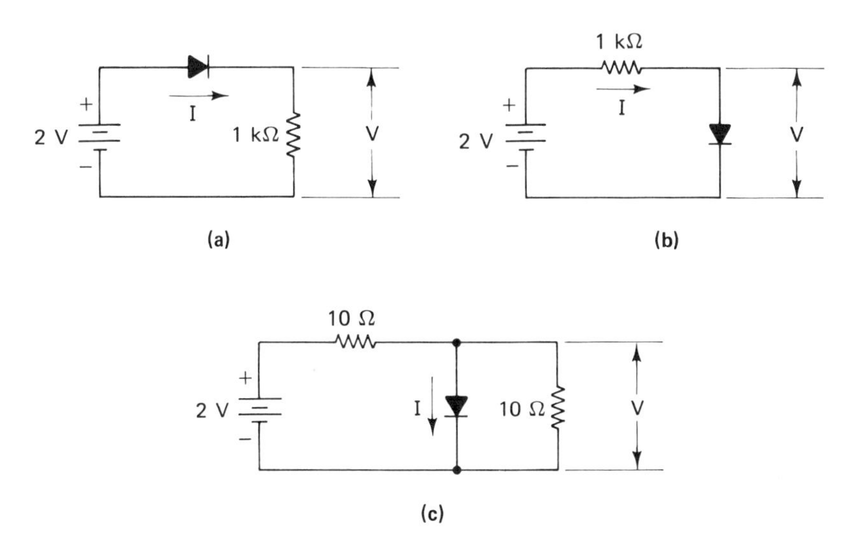

FIGURE 2.21

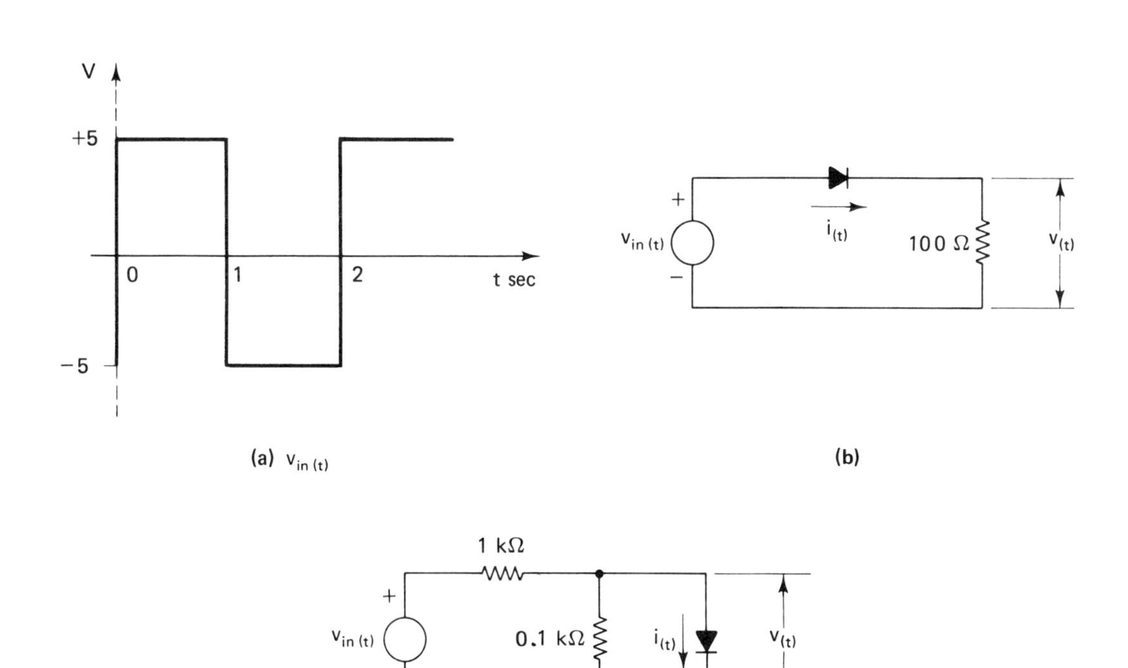

FIGURE 2.22

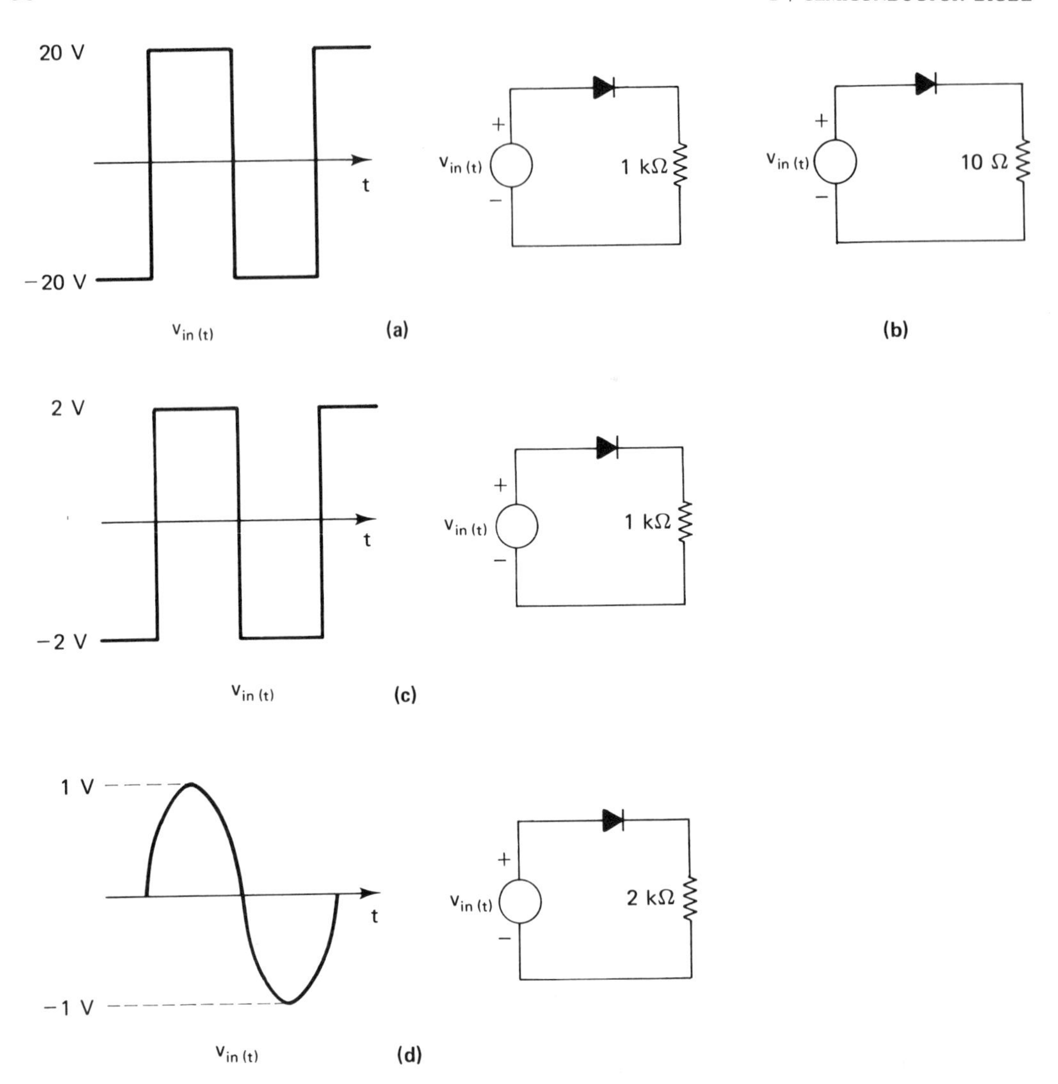

FIGURE 2.23

(c) $V_R = 25$ V, $I_o = 25$ mA, $P_D = 150$ mW, $R_F \leq 30\ \Omega$ (in addition, the forward region shape near 0.7 V should be specified)
(d) $V_R = 50$ V, $I_o = 50$ mA, $P_D = 100$ mW, V_F at 5 mA should lie between 0.73 and 0.60 V

14. Using Figure 2.14, find the following:
(a) I_R at $V_R = 25$ V, $T_A = 25°$C
(b) I_R at $V_R = 125$ V, $T_A = 25°$C
(c) I_R at $V_R = 125$ V, $T_A = 75°$C

(d) The change in I_F (ΔI_F) from $T_A = 25°$ to $100°C$
(e) P_D at 175°C
(f) C from $V_R = 4$ to 12 V

15. The manufacturer supplies the following specifications for a diode P_D at 25 °C = 250 mW

$$\text{Linear power derating factor (from 25°C)} = 2\ \text{mW/°C}$$

Letting $V_F \approx 0.7$ V for all I_F at all temperatures, find the maximum allowable I_F at 100°C.

16. Using the curves of Figure 2.14, find (a) P_D at 75°, (b) I_F at 75°, (c) I_o at 75°, (d) V_F for I_F at 25°C.

17. The temperature coefficient (TC) gives the decrease in V_F per degree centigrade above 25°. Using the two curves, forward voltage versus forward current and forward current versus temperature coefficient, given in Figure 2.14, compute the minimum and maximum V_F for the BAY 73 diode for $I_F = 100$ mA at 100°C.

18. A BAY129 diode is required to dissipate 400 mW. What is the highest T_A at which it can safely operate?

19. A metal-cased diode mounted on a heat sink ($\Theta_{JC} = 2.0$°C/W, $\Theta_{cs} = 0.5$°C/W) is to be operated with $P_D = 25$ W at $T_A = 75$°C ($T_{J\,max} = 175$°C). Find Θ_{sa}.

20. For glass-cased diodes, Θ_{ja} (junction-to-air thermal resistance) or the linear power derating factor is given. Junction temperature is given by

$$T_J = P_J\Theta_{ja} + T_A$$

Find Θ_{ja} for a glass-cased diode and prove that

$$\text{Linear power derating factor} = 1/\Theta_{ja}$$

Diode data: $P_{D\,max}$ (25°C): 500 mW; $T_{J\,max}$: 175°C; linear power derating factor (from 25°C) = 3.33 mW/°C.

References

Bell, David. *Electronic Devices and Circuits,* 2nd ed., Reston Publishing Co., Reston, Virginia, 1980.

Cooper, William D. *Solid State Devices: Analysis and Application,* Reston Publishing Co., Reston, Virginia, 1974.

Nashelsky, Louis, and Robert Boylestad. *Devices: Discrete and Integrated,* Prentice-Hall, Inc., Englewood Cliffs, New Jersey, 1981.

3
DIODE CIRCUITS

3.1 INTRODUCTION

Semiconductor diodes have wide applications in electronic circuits. They are used as rectifiers in dc power supplies; as switches in clippers, clampers, and logic gates; and as memory units in digital systems. All these applications exploit the unidirectional current flow property of the diode but for different purposes. DC power supplies convert alternating current and voltage to direct. Clippers eliminate unwanted segments of signals. Clampers shift the DC levels of signals. Gates are used to implement logic circuits, and diode arrays patterned into off–on configurations store data. In addition, diodes designed to operate in the zener-avalanche region are used to provide steady voltages from 1.8 to 200 V with power ratings up to 50 W. Forward-biased voltage reference diodes providing lower stable voltages from 0.5 to 2 V and constant current diodes providing currents from 0.2 to 4 mA are also available. Finally, there are many special-purpose diodes, including variable capacity diodes, tunnel diodes, and a family of optoelectronic diodes as well. Some of these diodes are discussed in this chapter; others will be left for appropriate sections of this text.

From a practical standpoint, the only valid reason for studying diode circuits is to learn to use them. Although an ability to analyze diode circuits is basic to an understanding of their operation, the working technologist is, after all, required to create circuit designs that will work when constructed. The remainder of this chapter and indeed the entire text are devoted to this endeavor.

3.2 DC POWER SUPPLIES

3.2.1 Half-wave Rectifier: Capacitor Smoothing

A half-wave rectifier circuit with capacitor smoothing is pictured in Figure 3.1a. Assuming the diode to be ideal, the circuit operates as follows. During the initial positive half-cycle of the input sinusoidal voltage (Figure 3.1b), the diode switches on and the capacitor is quickly charged to the peak voltage (V_p) of the input by a large current pulse (Figure 3.1c and d). Two actions follow. The voltage (v_a) at the anode side of the diode falls with the input voltage, while the voltage (v_b) at the cathode side falls relatively slowly as the capacitor discharges through the load resistor (R_L). Since v_b falls much more slowly than v_a, the diode, which turned off once the capacitor charged to V_p, remains off during the rest of the positive half of the input voltage and for the entire negative half as well. During this period v_b (the voltage across the capacitor and R_L) continues to decrease slowly as the load current (I_L) flows out of the capacitor into R_L. During the next positive excursion of the input voltage, the diode turns on again as soon as v_a equals v_b, and a pulse of current flows, which recharges the capacitor to V_p. This action, which repeats itself periodically, is illustrated in Figure 3.1b, c, and d.

3.2.2 Full-wave Rectification

Center-tapped Rectifier. A full-wave rectifier with capacitor smoothing is shown in Figure 3.2a. For this circuit, a center-tapped input transformer is required to provide the opposed input voltages needed to drive the diodes alternately. This transformer also performs the useful functions of isolating the output from the input (or supply) voltage and permitting voltages to be stepped up or down as needed.

The center-tapped rectifier may be looked upon as two half-wave rectifiers working alternately into one *RC* load. Its circuit operation is illustrated in Figure 3.2. During the positive half-cycle of the input, point *a* of Figure 3.2a is positive with respect to ground, while point *b* is negative. Therefore, during the positive half-cycle, the circuit acts as a half-wave rectifier, charging C through D_1 while D_2 remains open. During the negative half-cycle of the input, the reverse is true and C is charged through D_2.

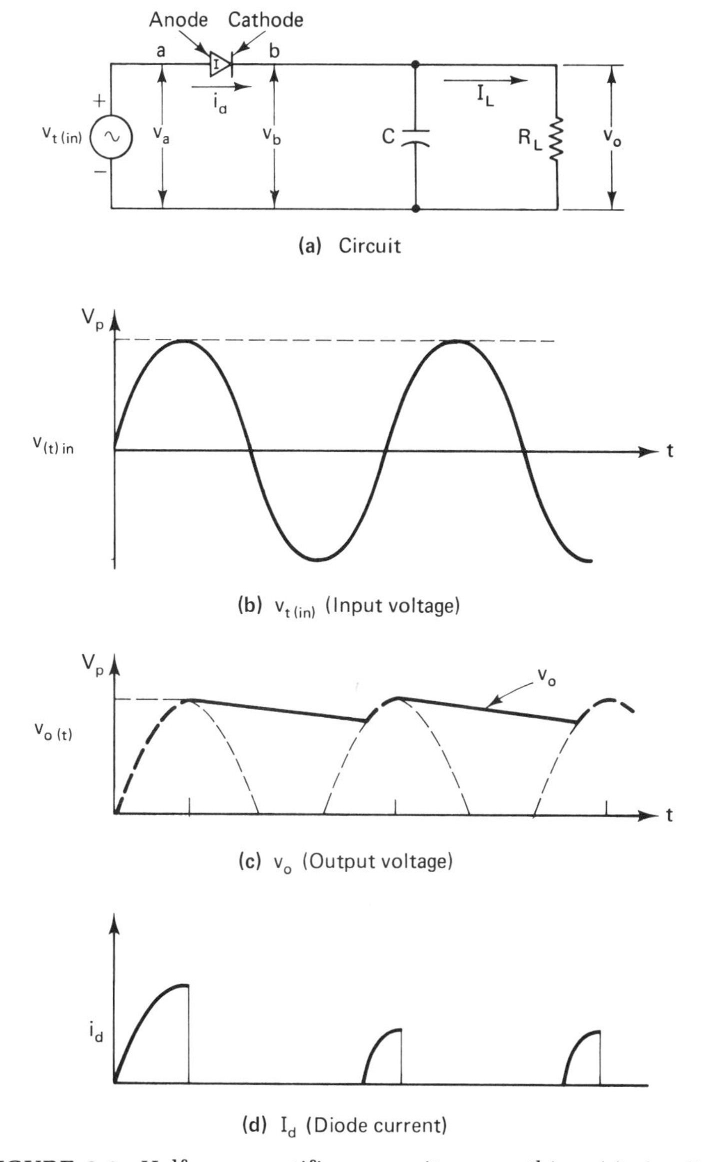

FIGURE 3.1 Half-wave rectifier: capacitor smoothing: (a) circuit; (b) $v_{t\ (in)}$, input voltage; (c) v_o, output voltage; (d) i_d, diode current.

Bridge Rectifier. Figure 3.2b shows another way to obtain full-wave rectification. This circuit, called a *bridge rectifier*, has advantages and disadvantages compared to the center-tapped rectifier. The advantages result from the fact that a center-tapped transformer is not needed.

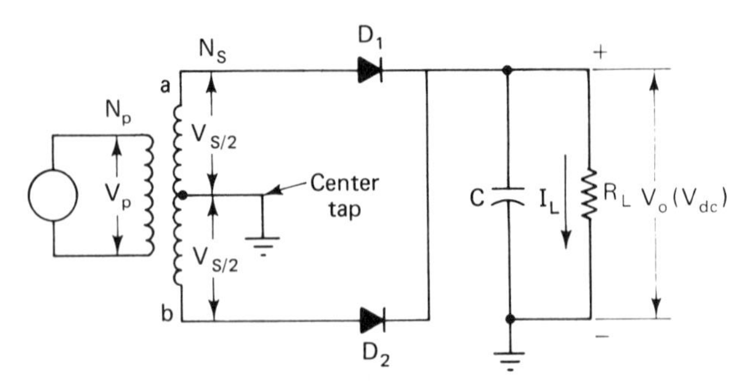

(a) Center tapped rectifier – capacitor smoothing

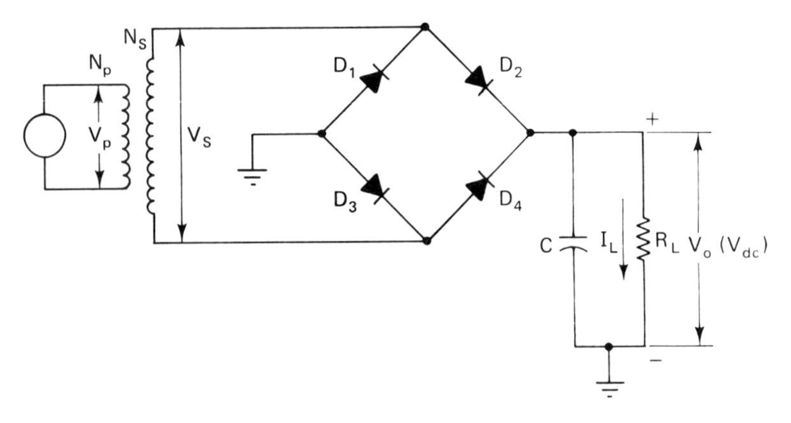

(b) Bridge rectifier – capacitor smoothing

FIGURE 3.2 Full-wave rectifier: capacitor smoothing; (a) center-tapped rectifier; (b) bridge rectifier.

Therefore, for the same number of primary/secondary turns, the bridge rectifier's DC voltage is almost twice that of the center-tapped circuit. Also, because the entire transformer secondary is used for both positive and negative input half-cycles, the transformer is utilized more efficiently. The primary disadvantages of the bridge circuit are the need for four diodes rather than two and the accompanying increasing voltage drops across the diodes.

The operation of the bridge rectifier is illustrated in Figure 3.3. Though R_L and C have been drawn differently in this figure than in Figure 3.2b, they are connected between the same two points and the circuit is unchanged. The circuit operates as follows. When the secondary voltage (v_s) is positive (Figure 3.3a), capacitor charging (and load) current flows through D_2, through the R_LC load, and returns to the negative end of the secondary through D_3. During this interval, the other two

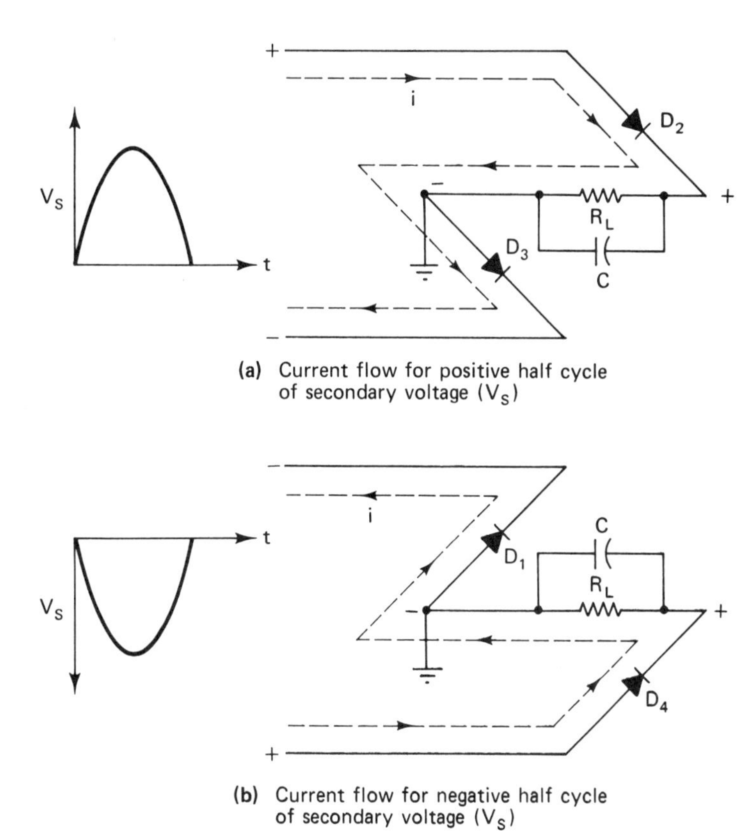

FIGURE 3.3 Bridge rectifier operation: (a) current flow for positive half-cycle of secondary voltage (V_s); (b) current flow for negative half-cycle of secondary voltage (V_s).

diodes are reverse biased. During the next half-cycle, when v_s is negative (Figure 3.3b), charging current flows through D_4, through R_L and C, and returns to the negative (top) end of the secondary through D_1. In both cases the current through and the voltage across the load are unidirectional. Hence full-wave rectification results.

Full-wave rectification of either type is superior to half-wave rectification because it charges the capacitor twice as often (every half-cycle as opposed to every cycle). Consequently, the full-wave output voltage decreases less between charging periods than does the half-wave for the same R_L and C. The full-wave output is therefore inherently smoother. In addition, because its output variation (ripple) is twice the frequency of the half-waves, it is easier to filter subsequently. Full-wave rectification is illustrated by Figure 3.4.

Half-wave rectifiers are used in inexpensive equipment drawing

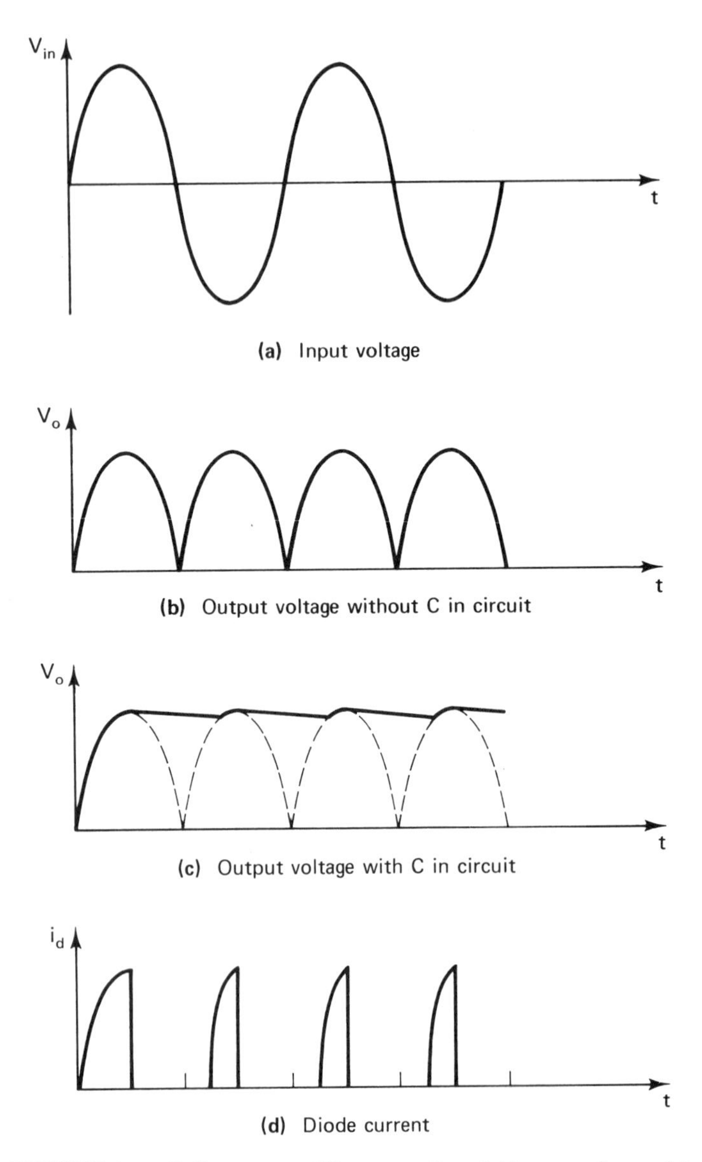

FIGURE 3.4 Full-wave rectifier operation: (a) input voltage; (b) output voltage without C in circuit; (c) output voltage with C in circuit; (d) diode current.

low current (small radios for example), where the savings gained by eliminating the transformer are important. Full-wave rectifiers are used in equipment calling for heavier load currents and smoother outputs. Since they are used much more often than half-wave rectifiers, our discussion will henceforth be limited to them.

3.2.3 Analysis of the Full-wave Rectifier Output Voltage: Capacitor Smoothing

The output voltage (V_o) of the full-wave rectifier fluctuates around its DC level as the capacitor is alternately charged and discharged, as shown in Figure 3.4c. Although this fluctuation is actually exponential, it can be approximated by a sawtooth wave for excursions of 10% or less around the DC level. Figure 3.5 illustrates this approximation. The parameters of the output voltage of interest to the technologist are the average level (V_{dc}), the peak value of the sawtooth wave ($V_{r\,peak}$), its rms value (V_r), and the maximum value of the entire output voltage (V_m). Also needed is the ratio (V_r/V_{dc}), called the *ripple factor* (r). These parameters are also illustrated in Figure 3.5.

For the case where the input is the AC power line (117 V, 60 Hz) and the ripple factor is less than 0.065, equations relating the parameters of Figure 3.5 are given by

$$r = \frac{0.0024 I_{dc}}{C V_{dc}} = \frac{V_r}{V_{dc}} \tag{3.1}$$

$$V_{r\,peak} = \sqrt{3}\, V_r \tag{3.2}$$

$$V_m = V_{DC} + V_{r\,peak} \tag{3.3}$$

3.2.4 Design of the Full-wave Rectifier with Capacitor Smoothing

The design of a full-wave rectifier with capacitor smoothing entails finding the transformer, capacitor, and diode specifications that will produce a circuit meeting the performance requirements. These performance requirements usually call for a DC voltage (V_{dc}) at a given load current (I_{dc}) and at a given ripple factor (r). The circuit designer must specify the following items:

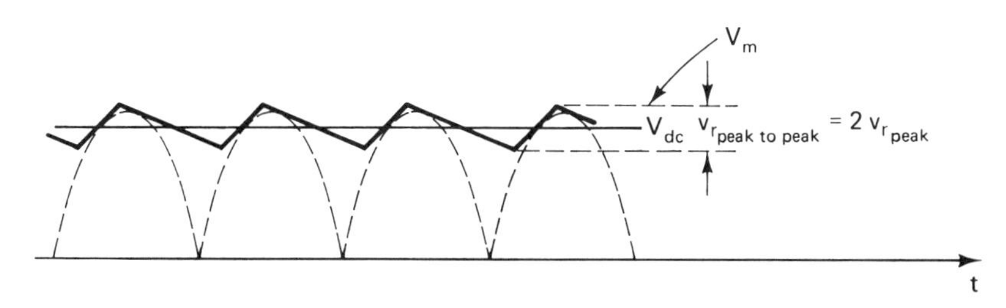

FIGURE 3.5 Sawtooth approximation of rectifier output.

1. Turns ratio (N_p/N_s) of the transformers
2. Capacitance of C
3. DC working voltage (WVDC) of C
4. Maximum rms current (I_c) of C
5. Average forward current (I_o) of the diodes
6. Maximum surge current (I_{FSM} or $I_{F\,\text{surge}}$) of the diodes
7. Maximum reverse working voltage (V_R or V_{RRM}) of the diodes

Equations (3.1) through (3.3) can be manipulated into the following convenient forms:

$$C = \frac{0.0024 I_{\text{DC}}}{r V_{\text{DC}}} \tag{3.4}$$

$$v_r = r V_{\text{DC}} \tag{3.5}$$

$$V_m = V_{\text{DC}} + \sqrt{3}\, v_r \tag{3.6}$$

The relationship between the primary (V_p) and the secondary (V_s) voltages of an ideal transformer and its turns ratio (N_p/N_s) is given by

$$\frac{N_p}{N_s} = \frac{V_p}{V_s} \tag{3.7}$$

where N_p is the number of turns in the primary winding of the transformer; N_s, the number of secondary turns; V_p, the peak primary voltage; and V_s, the peak secondary voltage. These parameters are illustrated in Figure 3.6.

Example 3.1 *Full-wave Center-tapped Rectifier Design*

Performance specifications:

$$V_{\text{DC}} = 12\text{ V}, I_{\text{DC}} = 100\text{ mA}, r = 0.05, V_{\text{in}} = 117\text{ V rms}, 60\text{ Hz}$$

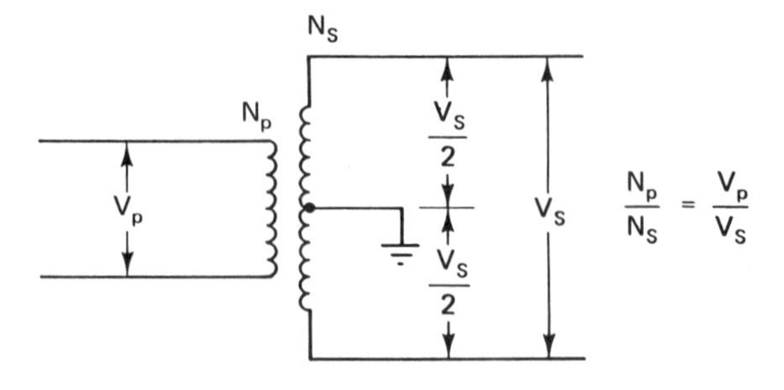

FIGURE 3.6 Ideal transformer.

circuit: center tapped, full wave

(a) *Finding the capacitor specifications:*

$$C = \frac{0.0024 I_{DC}}{r V_{DC}} = \frac{0.0024 \times 100\text{ mA}}{0.05 \times 12\text{ V}} = 400\,\mu\text{F}$$

During continuous operation the ripple voltage across C may be approximated as a sine wave of rms voltage V_r at 120 Hz. Therefore, the rms current through C is given by

$$V_r = r V_{DC} = 0.05 \times 12\text{ V} = 0.6\text{ V}$$

$$i_c = \frac{V_r}{1/2\pi f C} = V_r \times 2\pi f C = 0.6 \times 2\pi \times 120 \times 400\mu\text{F} = 180\text{mA}$$

Since C will be an electrolytic capacitor, it is necessary to specify its dc working voltage.

$$\text{WVDC} = V_{DC} = 12\text{ V}$$

(b) *Find the transformer turns ratio* (N_p/N_s):

$$V_r = 0.6\text{ V} \qquad \text{(previous calculation)}$$

$$V_m = V_{DC} + V_{r\text{ peak}} = V_{DC} + \sqrt{3}\, V_r = 12 + \sqrt{3} \times 0.6 = 13.04\text{ V}$$

$$\frac{V_s}{2} = V_m + V_{\text{diode}} = 13.04 + 0.7 = 13.74\text{ V}$$

$$V_s = 2 \times 13.74\text{ V} = 27.48\text{ V}$$

Since $V_{in} = 117$ V (rms), $V_{in(peak)} = V_p = 165.5$ V. Then

$$\frac{N_p}{N_s} = \frac{165.5\text{ V}}{27.48\text{ V}} = 6.02$$

(c) *Find the diode specifications:*

$$I_o = \frac{I_{DC}}{2} = \frac{100\text{ mA}}{2} = 50\text{ mA}$$

(each diode provides one half the DC load current)

The maximum reverse working voltage (V_{RRM}) can be determined by referring to Figure 3.2a. Note that V_s is across both D_1 and D_2. Since one diode is on while the other is off, the V_{RRM} across the off diode is given by

$$V_{RRM} = V_s - V_d = 27.48 - 0.7\text{ V} = 26.78\text{ V}$$

where the diode break voltage (V_B) may be used to approximate V_d.

The maximum diode forward surge current (I_{FSM}) occurs when the power supply is first turned on. Since the capacitor is uncharged, the entire $V_s/2$ exists briefly across one of the diodes. A large surge of current will flow through this diode, limited only by the forward resistance of the diode and the internal resistance of the transformer. Usually, the transformer resistance predominates. This resistance can be secured from potential manufacturers. Assume, for now, a total series resistance (R_{surge}) for diode plus transformer, of 10 Ω. Then

$$I_{FSM} = \frac{V_s/2 - V_d}{R_{surge}} = \frac{13.74 - 0.7\text{ V}}{10\ \Omega} = 1.304\text{ A}$$

Referring to Figure 2.12, we see that the smallest rectifier in the table, a 1N4002 with an I_o of 1 A, an I_{FSM} of 30 A, and a V_{RRM} of 100 V, is suitable. Further search might find a more closely matched general-purpose diode. However, it is sound practice to choose a diode designed for the specific application when possible.

Table 3.1 summarizes the design results.

Notes on Bridge Rectifier Design

Bridge rectifier design proceeds exactly like the center-tapped design of Example 3.1 except for the following parameters. Referring to Figure 3.3, we see that V_s includes two forward-biased diodes and the load in its loop rather than a single diode, as in the center-tapped case. Hence

$$V_s = V_m + 2V_d \tag{3.8}$$

Referring to Figure 3.2b, we see that V_s is also taken across one forward-biased and one reverse-biased diode. Therefore, V_{RRM} across the reverse-biased diode is given by

$$V_{RRM} = V_s - V_d \tag{3.9}$$

which turns out to be identical to the center-tapped case. The forward surge current flows through two diodes for the bridge rectifier case, but, unlike the center-tapped case, the entire V_s appears across the circuit. Hence

$$I_{FSM} = \frac{V_s - 2V_d}{2R_F + R_T} \tag{3.10}$$

TABLE 3.1 *Full-wave Center-tapped Rectifier Specifications for Example 3.1*

Capacitor	400 μF, 12 WVDC, 180 mA
Transformer	$N_p/N_s = 6.02$
Diodes	Two 1N4002s

where R_F is the diode forward resistance and R_T is the resistance presented by the transformer.

Also note that for the same supply voltage and transformer turns ratio, V_m and therefore V_{dc} are almost twice as great for the bridge rectifier compared to the center-tapped rectifier.

3.2.5 Ripple Factor Reduction with *LC* Filters

Often the required ripple factor (V_r/V_{dc}) cannot be secured with reasonable values of C using the circuits of Figure 3.2. It is then necessary to design the full-wave rectifier for a higher (achievable) ripple factor and to then interpose ripple-reducing *LC* sections between the input capacitor and the load. Such a circuit is shown in Figure 3.7.

The *LC* filter further reduces the ripple existing across the input capacitor, while having insignificant effect upon V_{dc}. If one section does not suffice, additional sections are used. For a full-wave rectifier (v_{in} = 117 V, 60 Hz), the reduction per *LC* section (L_1C_1) for the typical case, $X_L > 10X_c$ (or $\alpha_1 > 10$) is given by

$$\alpha_1 = \frac{r_{in}}{r_{out}} \approx \frac{X_L}{X_c} = \omega^2 LC = 5.68 \times 10^5 LC \tag{3.11}$$

where r_{in} is the ripple factor across the input capacitor, and r_{out} is the ripple factor across the load. For multiple sections composed of identical *LC*s, the overall ripple reduction (α_n) is given by

$$\alpha_n = \alpha_1^n \tag{3.12}$$

where n is the number of sections.

Example 3.2 *Using LC Filter Sections*

Suppose that in Example 3.1 the ripple factor had been 0.0005 instead of 0.05. The value of C in this case would have been very large (40000 μF).

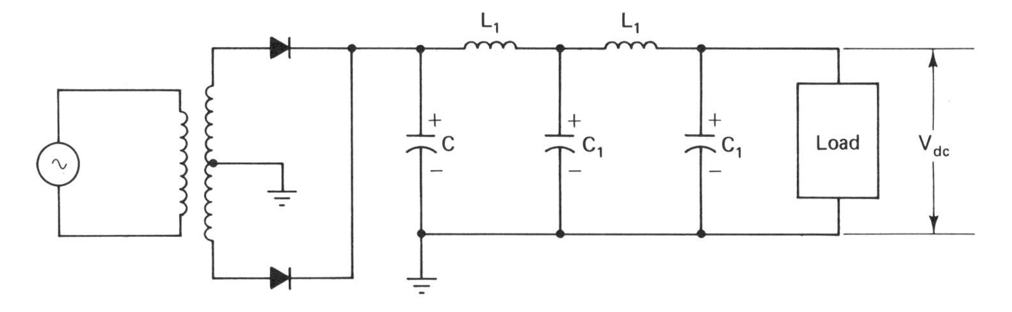

FIGURE 3.7 Center-tapped rectifier with *LC* filter.

A better procedure would be to design the circuit for a ripple factor of 0.05, as was done, and then to add an LC filter with

$$\alpha_1 = \frac{0.05}{0.0005} = 100$$

Then

$$\alpha_1 = 100 = 5.68 \times 10^5 LC$$

$$LC = \frac{100}{5.68 \times 10^5} = 1.76 \times 10^{-4}$$

Choosing a reasonable value for L ($L = 1$ H), we get $C = 176\,\mu$F. A single LC section is therefore satisfactory. If not, a double LC section would have been tried, and so on.

3.3 DIODE CLIPPERS

Clipper circuits are used to remove unwanted parts of a wave form. Two types of diode clipper circuits, series and shunt, are shown in Figure 3.8. (Note that the ideal diode model is used here.) These diode clippers operate as follows. For the negative-side series clipper (Figure 3.8b), $v_{\text{in}(t)}$ appears across R only when the ideal diode is in its *on* state. Since this occurs only during the positive half-cycle of the input, only this part of $v_{\text{in}(t)}$ appears in the output, $v_{\text{o}(t)}$. Because the ideal diode is a short circuit in its *on* state, $v_{\text{in}(t)}$ appears across R without attenuation. During the negative half-cycle of $v_{\text{in}(t)}$, the ideal diode is *open*. All of $v_{\text{in}(t)}$ appears across this open circuit, leaving zero voltage across R. This action is reversed for the positive-side series clipper (Figure 3.8c). For the negative- and positive-side shunt clippers (Figure 3.8d and e), $v_{\text{o}(t)}$ is zero when the diodes are *on*. When the diodes are *open*, no current flows in the circuit and no voltage drop occurs across R. Therefore, $v_{\text{o}(t)}$ equals $v_{\text{in}(t)}$. Therefore, only this half appears in the output. The positive-side shunt clipper reverses this effect.

Although the use of the ideal diode model simplifies the explanation of clipper action, it does not show the significant effect of the diode break voltage (V_B) on the output voltage wave shape. Because of this effect and because the forward resistance of the diode can usually be ignored relative to the circuit resistance R, the offset ideal model is more appropriate. Figure 3.9 shows an analysis of shunt clipper action using this model. Notice that the clipping level is at V_B instead of zero, as is the case for the ideal diode analysis.

The total break voltage (V_{BT}) can be increased or decreased by in-

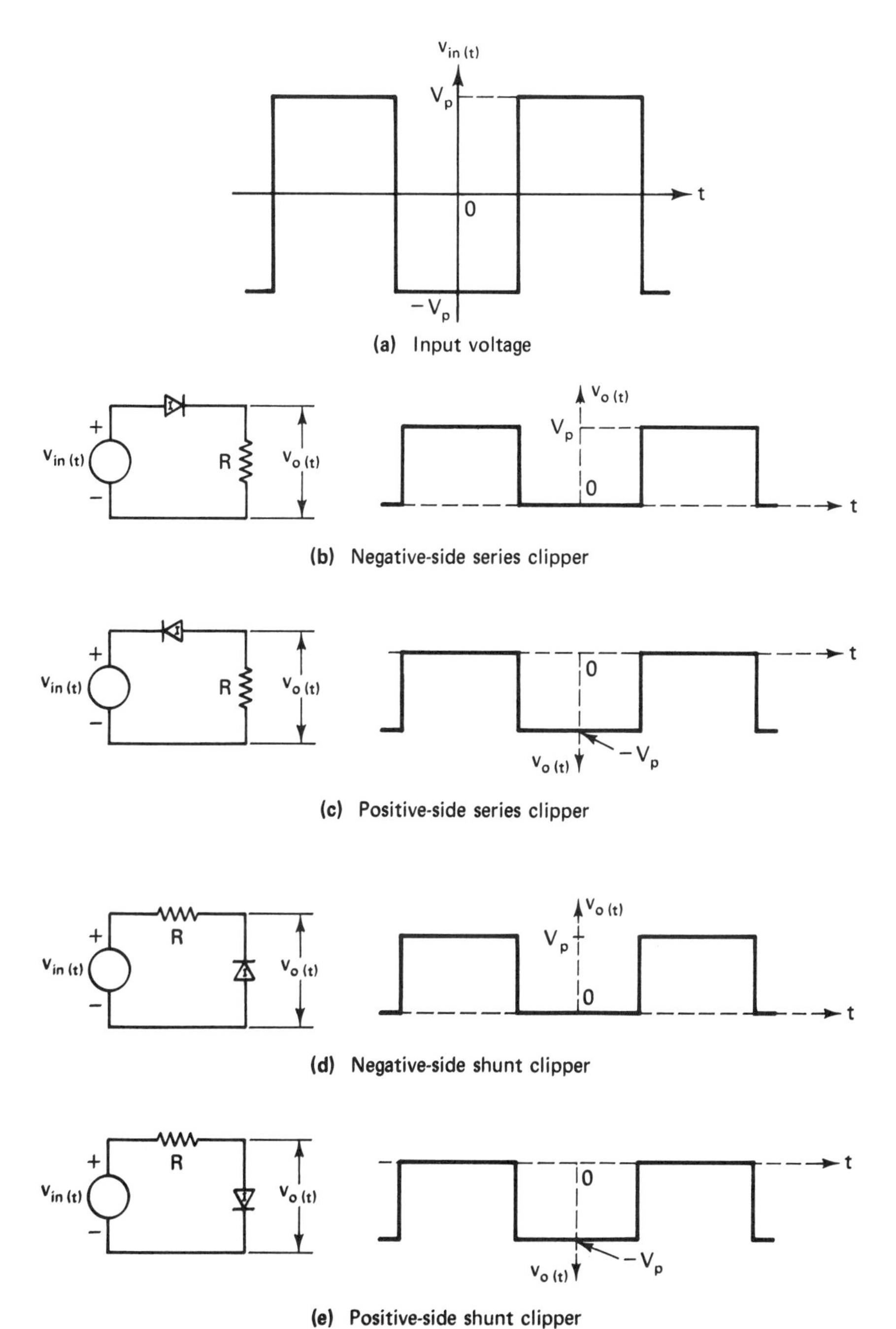

FIGURE 3.8 Diode clipper circuits: (a) input voltage; (b) negative-side series clipper; (c) positive-side series clipper; (d) negative-side shunt clipper; (e) positive-side shunt clipper.

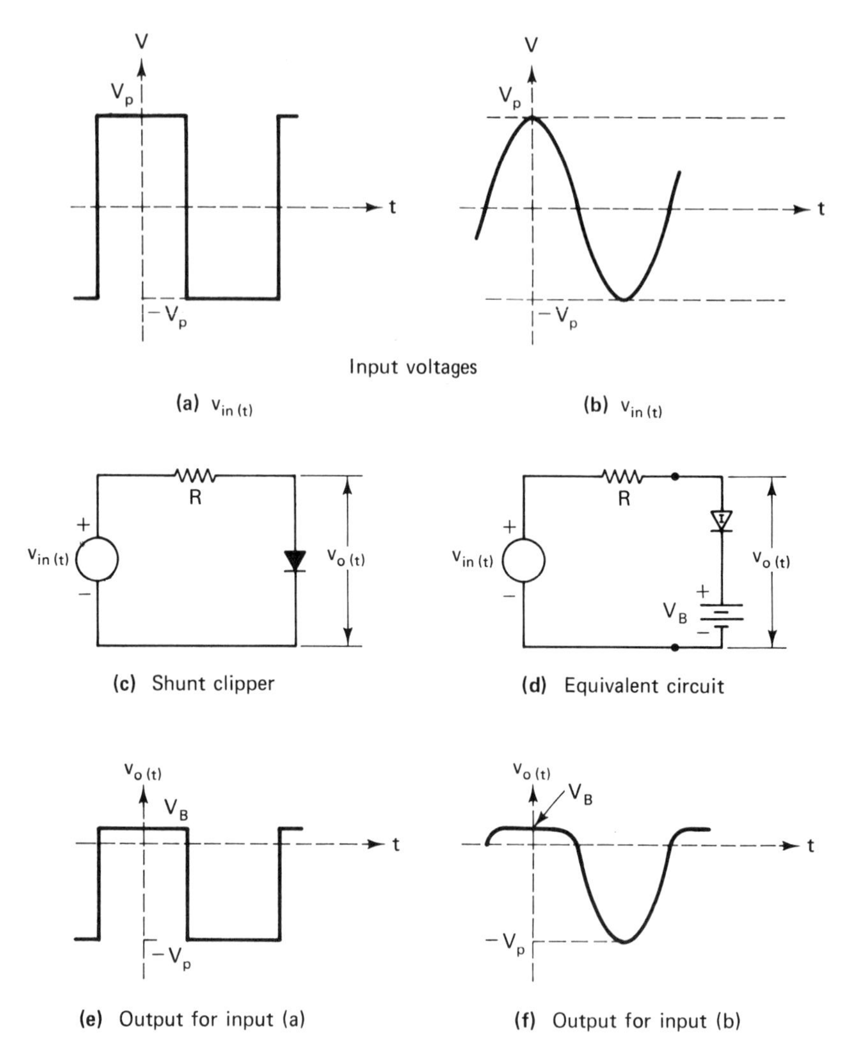

FIGURE 3.9 Clipper circuit action using the offset ideal diode model: (a) $v_{in(t)}$, input voltage; (b) $v_{in(t)}$, input voltage; (c) shunt clipper; (d) equivalent circuit; (e) output for input (a); (f) output for input (b).

troducing a DC voltage (V_{BO}) in series with the diode. The input signal will then be clipped at the V_{BT} level where V_{BT} is given by

$$V_{BT} = V_B + V_{BO} \tag{3.13}$$

Examples of this action are illustrated in Figure 3.10.

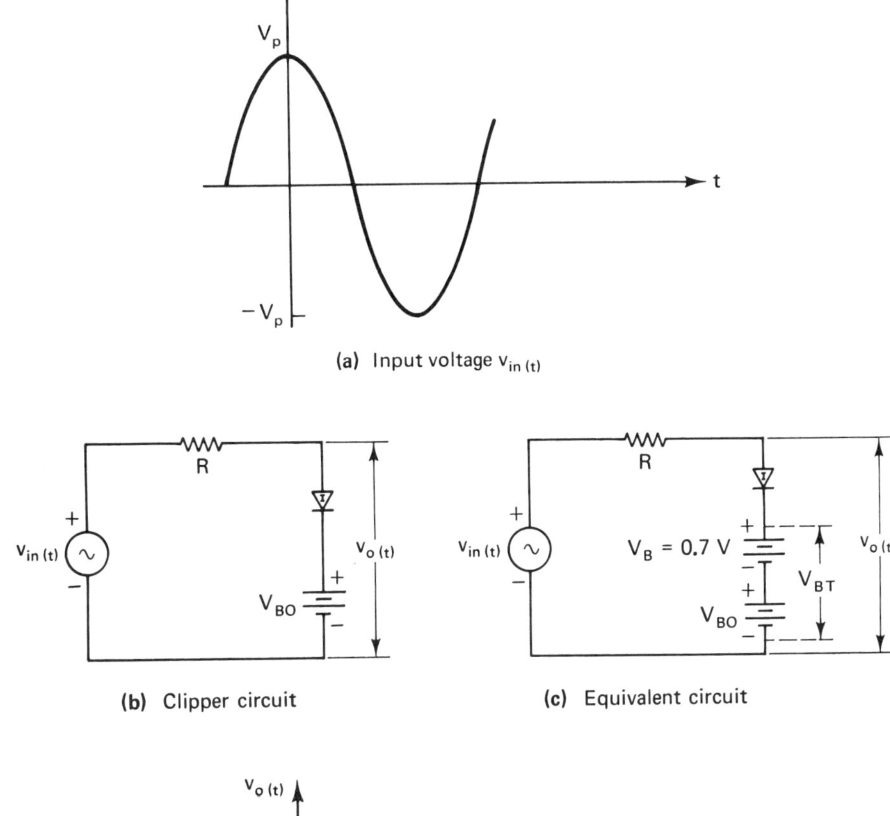

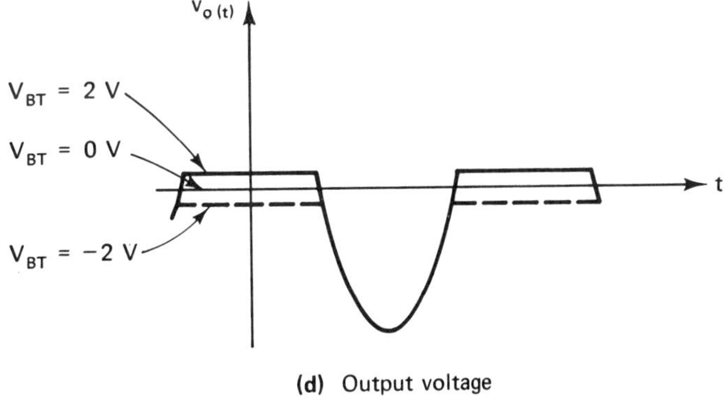

FIGURE 3.10 Controlling clipping levels: (a) input voltage $v_{in(t)}$; (b) clipper circuit; (c) equivalent circuit; (d) output voltage.

3.3.1 Clipper Circuit Design

Clipper circuit design using diodes alone is complicated by the effect of the diode break voltage and by the effects of source and load resistances. Later in this text we shall discuss a versatile device called an operational amplifier (op-amp). This device permits us to design active diode clippers that are superior to the diode clippers discussed so far. Also, because of the isolating property of the op-amp, source and load resistance effects can be ignored, thus simplifying clipper design. Therefore, we defer our discussion of clipper circuit design until then.

3.4 DIODE CLAMPERS

The diode clamper is used to shift the average (DC) level of a signal. This action is important in some amplifiers where it is necessary to maintain a bias proportional to signal magnitude. Clampers are also used in TV receivers to restore the brightness level of the video signal and in receivers generally to generate a bias for automatic volume control (AVC).

Clamper action is illustrated in Figure 3.11 for both a negative and positive clamper. Notice that the input signal is unchanged in the output except for its DC level. The top of the output voltage is clamped at V_{BT} (total break voltage) for the negative clamper; the bottom of the output is similarly clamped for the positive clamper. V_{BT} may be positive, negative, or zero for either clamper. The shifted DC level (V_{dc}) is given by

$$V_{DC} = V_{BT} - |V_p| \qquad \text{negative clamper} \tag{3.14}$$

$$V_{DC} = V_{BT} + |V_p| \qquad \text{positive clamper} \tag{3.15}$$

where $|V_p|$ is the absolute peak value of the input voltage.

The clamper circuits of Figure 3.11 operate as follows. The capacitor C, initially uncharged, is charged to the voltage

$$V_c = V_p - V_{BT} \tag{3.16}$$

during the first quarter-cycle, which forward biases the diode. (This is the first positive quarter-cycle for the negative clamper and the first negative quarter-cycle for the positive clamper.) During the remainder of the cycle, the diode is reverse biased. During this time, C can only discharge through R, but since the time constant (RC) is made very long relative to the signal period, this discharge is slight. (Any drop in voltage

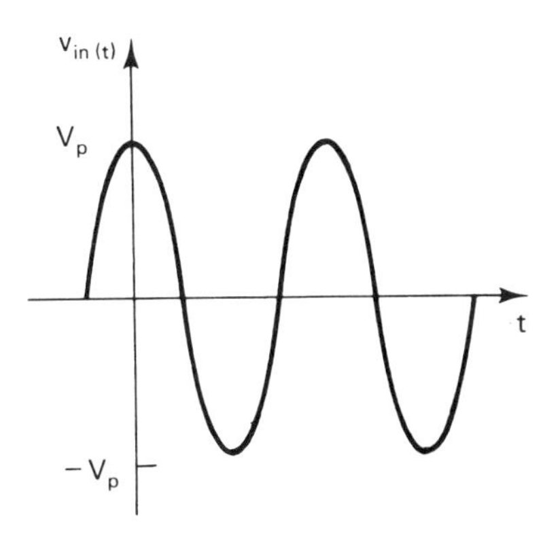

(a) Sine-wave input

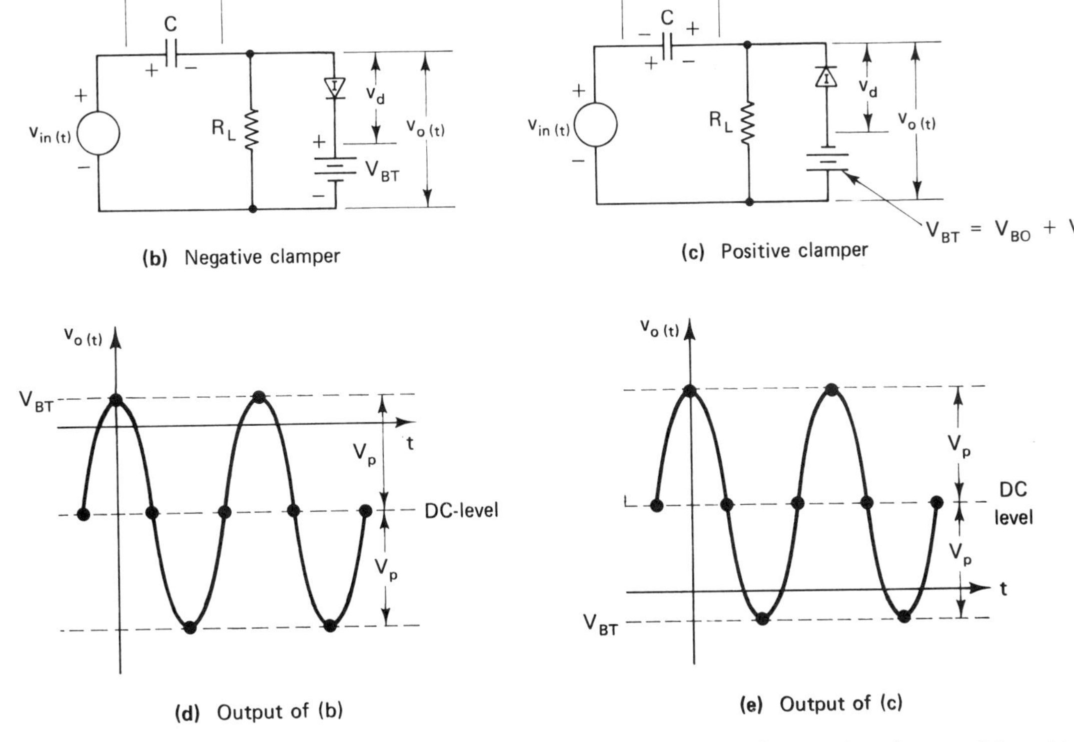

FIGURE 3.11 Clampers: (a) sine wave input; (b) negative clamper; (c) positive clamper; (d) output of (b); (e) output of (c).

is restored each cycle.) Assuming V_c to be constant and given by Eq. (3.16), then the output voltage $v_{o(t)}$ is given by

$$v_{o(t)} = v_{in(t)} - |V_c| \quad \text{negative clamper} \quad (3.17)$$

$$v_{o(t)} = v_{in(t)} + |V_c| \quad \text{positive clamper} \quad (3.18)$$

These outputs are shown in Figure 3.11d and e.

3.4.1 Diode Clamper Design

Performance requirements for a clamper specify the type (positive or negative), clamping level (V_{BT}), magnitude and frequency of signal, and source and load resistance. The circuit designer must specify the capacitance of C and its maximum working voltage, V_{BO}, the external part of V_{BT}, the device number of the diode, and the circuit configuration. Because clampers may be subject to relatively high input voltages, the reverse voltage across the diode and the forward surge current specifications must be established.

Example 3.3 *Diode Clamper Circuit Design*

Performance specifications:
Negative clamper
$v_{in(t)} = 20 \cos 200\,\pi\, t \times 10^3$
Clipping level: 6.9 V
Source resistance $(R_s) = 100\ \Omega$
Load resistance $(R_L) = 1\ k\Omega$

(a) *Find C:* The time constant $(\tau = CR_L)$ is set at 100 times the input period $(T = 1/f)$.

$$\tau = CR_L = 100 \times T = \frac{100}{f} = \frac{100}{100\ \text{kHz}} = 1\ \text{ms}$$

$$C = \frac{\tau}{R_L} = \frac{1\ \text{ms}}{1\ k\Omega} = 1\ \mu\text{F}$$

A paper or plastic film capacitor is suitable. Its maximum working voltage (MWV) must be greater than the peak input voltage; therefore,

$$\text{MWV} > V_p > 20\ \text{V}$$

(b) *Choose V_{BO}:*

$$V_{BO} = V_{BT} - V_B = 6.9 - 0.7\ \text{V} = 6.2\ \text{V}$$

(c) *Choose the diode:* The maximum reverse voltage (V_R) is given by

$$V_R = V_c + V_p + V_{BO} = 20 + 20 + 6.2\ \text{V} = 46.2\ \text{V}$$

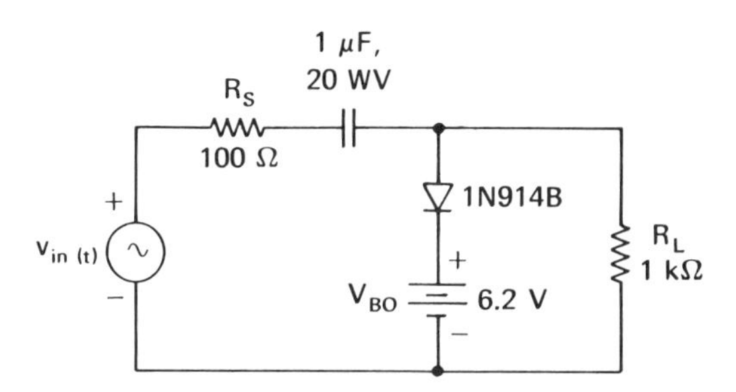

FIGURE 3.12 A clamper circuit design.

The maximum forward surge current ($I_{F\,\text{surge}}$) is

$$I_{F\,\text{surge}} = \frac{V_p - V_{BT}}{R_s} = \frac{20 - 6.9\ \text{V}}{100\Omega} = 131\ \text{mA}$$

A 1N914B diffused silicon diode (from Figure 2.13) with $V_R = 75$ V and $I_{F\,\text{surge}} = 500$ mA is suitable. Figure 3.12 summarizes the design results.

3.5 ZENER DIODE

Diodes designed to operate within the zener-avalanche region (see Chapter 1) are called zener diodes. These diodes maintain an almost constant terminal voltage within this region and consequently are widely used as constant voltage sources. The reverse voltage at which the diode enters the breakdown region is controlled by the doping level of the semiconductor materials. Diodes with nominal zener voltages (V_Z) from 1.8 to 200 V and with maximum power dissipations (P_Z) from 250 mW to 50 W are available.

As a circuit element, the zener diode is adequately represented by the offset ideal diode model as a first approximation. For more precise calculations taking into account the small change in V_Z as a function of zener current (I_Z), the piecewise linear equivalent circuit is needed. Zener characteristics and equivalent circuits are displayed in Figure 3.13.

3.5.1 Real Zener

The electrical characteristics of a 1N961 zener diode are given in Table 3.2.

These characteristics are defined as follows: V_Z is the nominal ter-

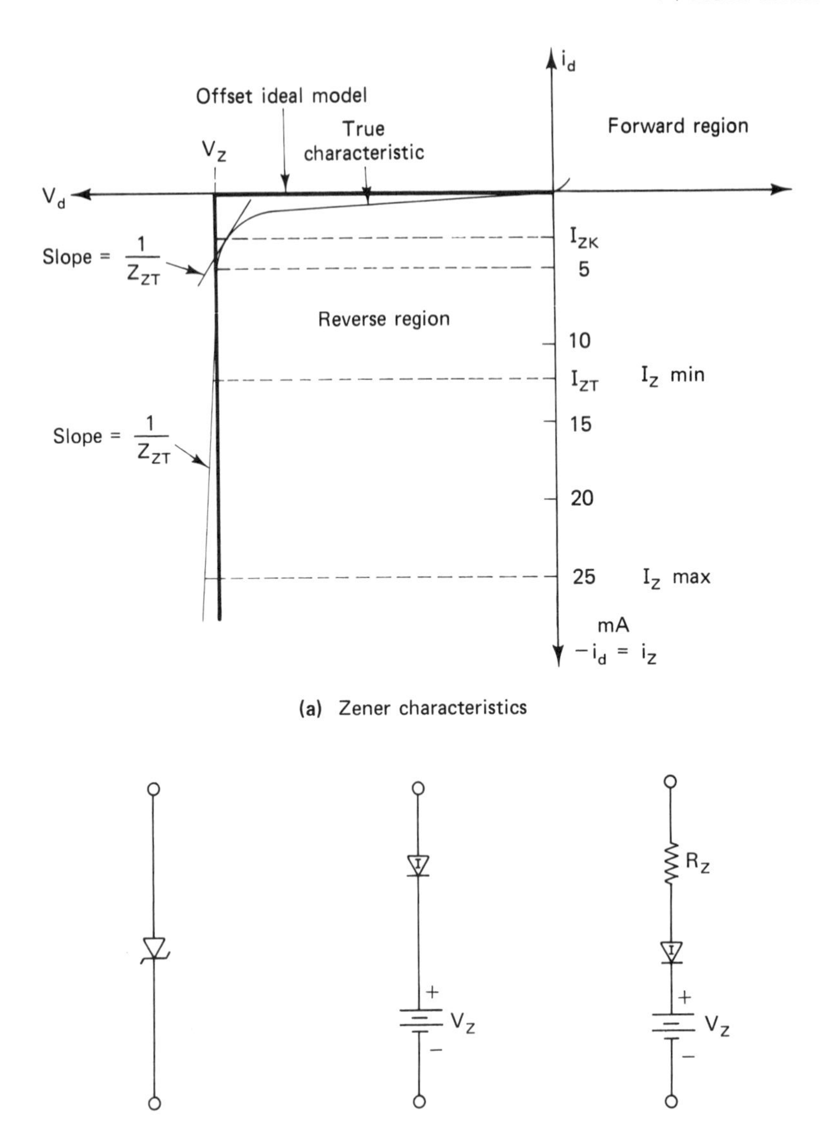

(a) Zener characteristics

(b) Zener symbol

(c) Equivalent offset ideal diode circuit

(d) Piecewise linear equivalent circuit

FIGURE 3.13 Zener diode: (a) zener characteristics; (b) zener symbol; (c) equivalent offset ideal diode circuit; (d) piecewise linear equivalent circuit.

minal voltage of the zener measured at a test current I_{ZT}. V_Z is given with tolerances ranging typically from ±1% to ±20%. Z_{ZT} is the maximum dynamic zener impedance measured at I_{ZT}. It is the inverse of the slope of the zener characteristic. Z_{Zk} is the maximum knee impedance measured

TABLE 3.2 *Electrical Characteristics of the 1N961 Zener Diode*

Jedec Type	Zener Voltage Nominal, V_Z (V)	Test Current, I_{ZT} (mA)	Max Dynamic Impedance, Z_{ZT} at I_{ZT} (Ω) (mA)	Max Knee Impedance, Z_{Zk} at I_{Zk} (Ω)	Max Reverse Current, I_R at V_R (μA)	Test Voltage, V_R (V)	Max Regulator Current I_Z (mA) M	Typical Temperature Coefficient (%/°C)
1N961	10	12.5	8.5	700 at 0.25	10	7.2	32	+0.072

25°C ambient temperature unless otherwise noted.

at the current I_{Zk}, which places the zener operating point at the knee of the characteristic. Z_{Zk} is too large for useful zener action, so zener current must be kept above I_{Zk}. I_R is the maximum reverse zener current between 0 and the knee. I_{ZM} is the maximum current the zener can handle without damage. Instead of I_{ZM}, some manufacturers will supply a maximum power dissipation P_Z equal to $I_{ZM}V_Z$. The typical temperature coefficient gives the increase in V_Z per degree centigrade increase in ambient temperature. Zeners are available with temperature coefficients as low as 5×10^{-3} %/°C.

3.5.2 Zener Circuits

Zener Voltage Reference Circuits. Circuits based on zener diodes are used whenever a stable low-source-resistance DC voltage is needed. For example, recall the clamper circuit of Figure 3.12. In this circuit a low-source-resistance, 6.2-V supply was needed to establish the clamping level. This voltage can be provided by the zener circuit designed in Example 3.4 and illustrated in Figure 3.14.

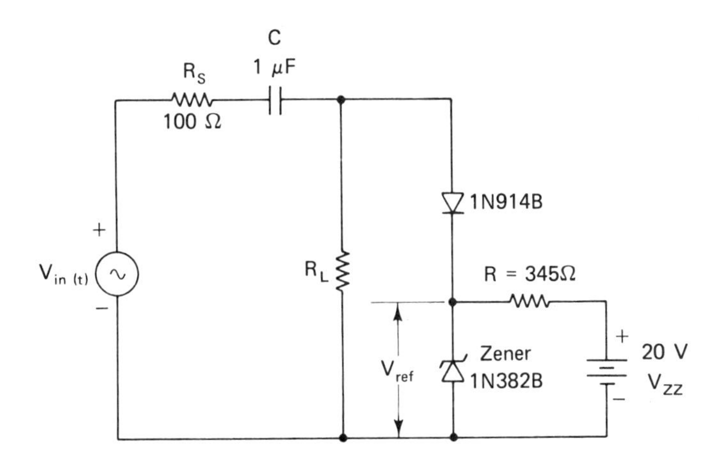

FIGURE 3.14 Clamper circuit with zener reference.

Example 3.4 *Design of a Zener Voltage Reference Circuit*

Performance specifications:

$$V_{ref} = 6.2\text{ V} \pm 5\%$$

(a) *Choose the zener:* The nominal zener voltage (V_Z) is given at a specific I_{ZT}. Since this I_{ZT} must be maintained as constant as possible to prevent changes in V_Z, we pick a zener giving a relatively high I_{ZT} (40 mA). Although small capacitor recharging currents will flow through the zener as a result of clamping action, they will be swamped by I_{ZT} and have insignificant effect upon V_Z. From a table of zeners (not shown), we choose the 1N382B zener that has a V_{ZT} of 6.2 V ± 5% at an I_{ZT} of 40 mA.

(b) *Choose R:* Assuming a supply voltage (V_{ZZ}) of 20 V, we have

$$R = \frac{V_{ZZ} - V_{ZT}}{I_{ZT}} = \frac{20 - 6.2\text{ V}}{40\text{ mA}} = 345\,\Omega$$

The completed circuit is shown in Figure 3.14.

Zener Voltage Regulator. Zener diodes have wide application in DC voltage regulators. Because of the almost constant voltage characteristic of the diode in its zener region, these regulators maintain excellent regulation over wide variations in supply voltages and load currents. A zener regulator is shown in Figure 3.15.

To understand the operation of the zener regulator, consider the case where the supply voltage V_S is fixed, but I_L varies (Figure 3.15). Because both V_S and V_Z are fixed, the following set of relations is established.

1. The volt drop across R_S (V_{RS}) is fixed because

$$V_Z = V_S - V_{RS} \tag{3.19}$$

and

$$V_{RS} = V_Z - V_S \tag{3.20}$$

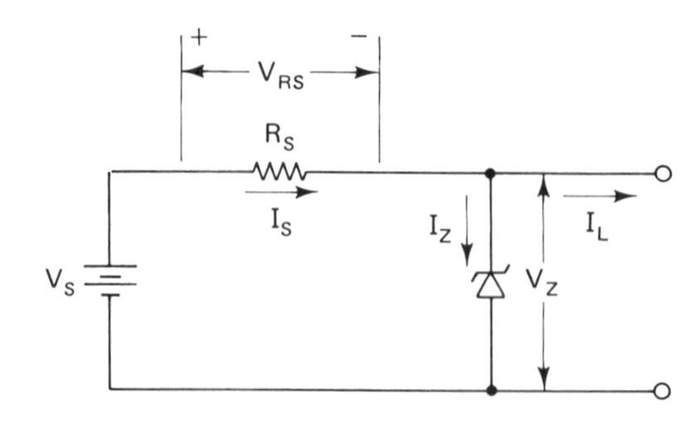

FIGURE 3.15 Zener voltage regulator.

Since both V_Z and V_S are fixed, V_{RS} is fixed.

2. I_S must remain constant because

$$V_{RS} = I_S R_S \tag{3.21}$$

3. $I_Z + I_L$ must remain constant since

$$I_S = I_Z + I_L \tag{3.22}$$

Therefore, I_Z must compensate for changes in I_L by decreasing by the amount I_L increases, and vice versa. Understand that the zener diode has no load current sensing ability. It simply maintains a fixed V_Z and all the above consequences follow.

In the case where I_L is held constant but V_S changes, I_Z increases or decreases so as to keep the relation

$$V_S - I_S R_S = V_Z \tag{3.23}$$

where V_Z is constant. Similar compensating changes in I_Z occur when both V_S and I_L change.

The design procedure for the zener regulator consists of choosing an R_S such that the zener current range is sufficient to carry out the required compensation. At the same time, the P_Z or I_{ZM} of the diode and the power dissipation of R_S (P_{RS}) must be computed. At no time during operation must the zener current be permitted to come too close to I_{Zk} since at (or near) this valve zener regulator action ceases. For safe regulation, the minimum I_Z should be three to five times I_{Zk}. I_{Zk} is a function of zener power dissipation and, when not provided by the manufacturer, can be estimated from Table 3.3.

Example 3.5 *Design of a Zener Voltage Regulator*

Performance specifications:
I_L: 2 to 20 mA
V_S: 14 V
$V_{out} = V_Z$: 10 V ± 20% at 25°C

TABLE 3.3 *Minimum Zener Currents*

P_Z (W)	I_{Zk} (mA)
50	5
10	1
1	0.25
0.25	0.1

Design procedure:

(a) *Choose R_S:* To choose R_S, we must first know $I_{Z\,min}$. Since $I_{Z\,min}$ depends on P_Z (or on the zener diode we end up with), which we do not yet know, we make an arbitrary choice of $I_{Z\,min} = 10\%$ of $I_{L\,max}$. $I_{Z\,min}$ will be verified later, but this engineering rule-of-thumb choice is usually satisfactory. Therefore,

$$I_{Z\,min} = 0.10 \times I_{L\,max} = 0.10 \times 20\,\text{mA} = 2\,\text{mA}$$

Then

$$I_S = I_L + I_Z = 20 + 2\,\text{mA} = 22\,\text{mA}$$

and

$$R_S = \frac{V_S - V_Z}{I_s} = \frac{14 - 10\,\text{V}}{22\,\text{mA}} = 181.8\,\Omega$$

Maximum power dissipation in R_s is given by

$$P_{RS} = \frac{(V_S - V_Z)^2}{R_s} = \frac{(14 - 10)^2}{181.8} = 88\,\text{mW}$$

(b) *Choose the zener:*

$$V_Z = V_{out} = 10\,\text{V} \pm 20\%\text{ at }25°\text{C}$$

$I_{Z\,max} = I_Z$ is computed at the worst condition. This would occur if the load were accidentally disconnected and I_Z would equal I_s. Then

$$I_Z = 22\,\text{mA}$$

Consulting Table 3.2, we see that the 1N961 zener with $V_Z = 10$ V $\pm$ 20%, $I_Z = 32$ mA, and $I_{Zk} = 0.25$ mA is suitable. Note that $I_{Z\,min}$ is greater than $5I_{Zk}$, as required. The final circuit is shown in Figure 3.16.

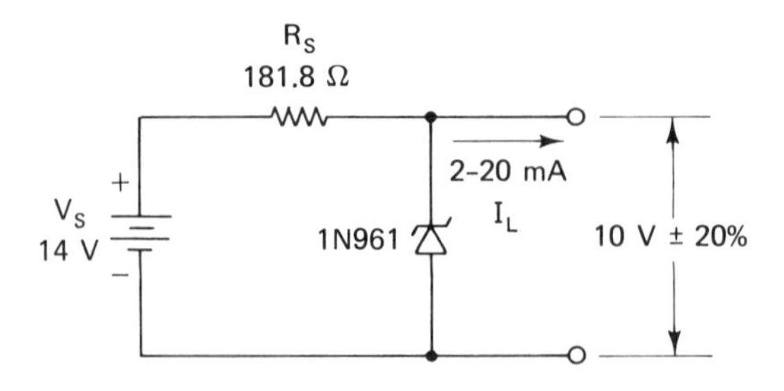

FIGURE 3.16 Zener voltage regulator.

3.6 SCHOTTKY BARRIER DIODE

The Schottky barrier diode (hot carrier diode) is characterized by an extremely low break voltage (0.2 V or less), a forward resistance similar to the *p–n* junction diode, and a higher reverse current. Its chief advantage over the junction diode is its shorter reverse recovery time (typically 0.1 ns, as compared to 4 ns for a high-speed junction diode). This short t_{rr} makes the Schottky diode suitable for switching at frequencies approaching 25 GHz (25×10^9 Hz). Figure 3.17 shows a specification sheet for some Motorola hot carrier diodes.

The short reverse recovery time of the Schottky diode is a result of its unique structure. Instead of a *p–n* junction, the diode contains a metal to *n*-type material junction. On the formation of this junction, high-energy electrons from the *n*-type material move into the metal. Because there are virtually no holes in the metal, recombination rarely occurs. Instead, the additional electrons create a negative voltage barrier between the two materials, and further current flow is prevented. Placing a forward bias across the diode (metal positive) pulls electrons out of the metal, reducing the barrier and allowing current to flow. Reversing the bias increases the barrier. The low reverse recovery time of the Schottky

		ELECTRICAL CHARACTERISTICS					
Device Type	Case	$V_{(BR)R}$ Reverse Breakdown Voltage I_R = 10 μA Volts Min	C_T Diode Capacitance V_R = 0 V, f = 1.0 MHz (1) V_R = 20 V, f = 1.0 MHz (2) pF Max	V_F Forward Voltage I_F = 10 mA Volts Max	I_R Reverse Leakage V_R = 3.0 V (3) V_R = 25 V (4) V_R = 35 V (5) μA Max	NF Noise Figure dB Max	t_{rr} (Note 1) Reverse Recovery ps Max
MBD101	182-01	4.0	1.0 (1)	0.6	0.25 (3)	7.0	–
MBD501	↓	50	1.0 (2)	1.2	0.20 (4)	–	100
MBD701		70	1.0 (2)	1.2	0.20 (5)	–	100
MBD102	226	4.0	1.0 (1)	0.6	0.25 (3)	7.0	–
MBD502	↓	50	1.0 (2)	1.2	0.20 (4)	–	100
MBD702		70	1.0 (2)	1.2	0.20 (5)	–	100
MICRO-I HOT-CARRIER DIODE							
MBI-101	166-02	4.0	1.0 (1)	0.6	0.25 (3)	7.0	–
CERAMIC HOT-CARRIER DIODE							
MBD103	45	4.0	1.0(1)	0.6	0.25	7.0	–

Note 1: Krakauer Method

FIGURE 3.17 Hot carrier diodes. Courtesy Motorola Semiconductor Products Inc.

diode is due to the fact that carriers do not have to be drained from a depletion region to cut off current. Instead, it is only necessary to reestablish the barrier. This requires much less time.

The break voltage of the diode is lower than for the *p–n* junction diode, because the barrier is at a lower potential than the *p–n* junction hill. For low-current Schottky diodes, V_B is almost zero. For high- or medium-current Schottky diodes, it is near 0.2 V. However, the weaker barrier also offers less impedance to reverse current than the *p–n* junction hill. As a result, reverse currents are higher and reverse breakdown voltages lower.

3.6.1 Effects of Reverse Recovery Time

Diodes used as switches are limited in switching speed by their reverse recovery time (t_{rr}). Although most diodes will switch rapidly from reverse to forward bias, the opposite action usually takes considerably longer. As noted previously, t_{rr} is a measure of this switch-off time.

The effect of the finite reverse recovery period is shown in Figure 3.18. Changing abruptly from forward to reverse bias (Figure 3.18a) does

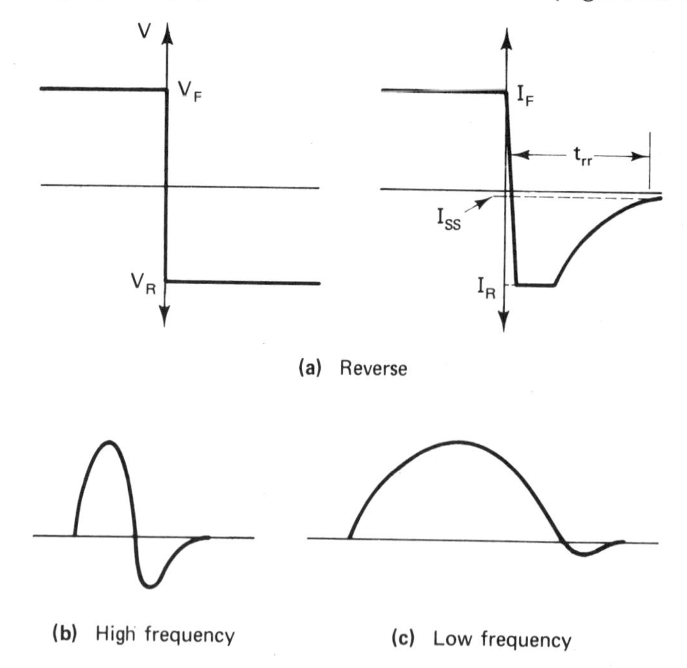

FIGURE 3.18 Reverse recovery time and its effects on high-and low-frequency inputs: (a) reverse; (b) high frequency; (c) low frequency.

not switch the diode to its off state. Instead, the diode conducts almost as well in the reverse direction for a portion of the reverse recovery time and then falls off exponentially during the remainder of the reverse recovery period. Normal diode action resumes only after this period. Figure 3.18b shows the effect of t_{rr} on rectification of a sinusoidal voltage with a period equal to twice the t_{rr}. Instead of passing current in the forward direction only, the diode passes negative current almost as well with some trailing edge distortion. Only when the period of the sinusoid is increased to 10 or more times the diode's t_{rr} is normal diode action obtained (Figure 3.18c). In practice, then, the upper frequency (f_u) for diode operation is given by

$$f_u = \frac{1}{10t_{rr}} \tag{3.24}$$

Example 3.6 *Comparison of the Upper Frequencies of Two Diodes*

In this example, we compute f_u for the 1N914, a high-speed diffused silicon switching diode, and for the MBD–702 hot carrier diode.

(a) f_u for the 1N914 ($t_{rr} = 4$ ns):

$$f_u = \frac{1}{10 \times 4 \text{ ns}} = 25 \text{ MHz}$$

(b) f_u for the MBD702 ($t_{rr} = 0.1$ ns):

$$f_u = \frac{1}{10 \times 0.1 \text{ ns}} = 1000 \text{ MHz}$$

3.7 VOLTAGE VARIABLE CAPACITOR DIODE (VVC)

The VVC diode, also called the varactor, varicap, epicap, and tuning diode, exploits the fact that the depletion region of a diode widens as its reverse bias increases. As a result, the capacity across the diode (transition capacity) decreases with increasing reverse voltage (see Chapter 1), and the diode can be used as a voltage-controlled capacitor.

By controlling the level and spatial distribution of the doping (doping profile) in the vicinity of the junction, various ranges of capacity and rates of capacity versus voltage change are obtainable. Abrupt profiles are produced by maintaining a uniform doping level throughout. Capacity ranges of 2 to 1 or 3 to 1, with maximums of about 400 pF, are obtained. Hyper-abrupt profiles are produced by a doping profile that increases toward the junction. As a result, narrow, highly voltage sensitive depletion regions are formed. Capacity ranges of 10 to 1, with maximums of 550 pF, are obtainable. Typical capacitance–voltage characteristics for abrupt and hyper-abrupt VVCs are shown in Figure 3.19.

Manufacturer's specifications for a typical VVC are given in Figure 3.20. Electrical characteristics not previously explained include the following:

1. C (capacitance): This is given at two V_R's of 3 and 25 V. A curve relating C to V_R is also given in Figure 3.21c.
2. C_3/C_{25} (capacitance ratio): This is the ratio of C as measured at $V_R = 3$ V and at $V_R = 25$ V.
3. Q (figure of merit): This is given by X_c/R_s, where R_s is the effective resistance in series with C. The higher the Q, the better. A curve relating Q to frequency is given in Figure 3.21d.
4. L_S (series inductance): All physical bodies possess some inductance. This gives the effective inductance in series with C and R_s.
5. f_o (series resonant frequency): At this frequency, C_{25} resonates with L_S and the diode is no longer capacitive. Instead, it looks like a resistance R_S. The VVC must be operated below f_o.
6. TC_c (capacitance temperature coefficient): This is used to compute the change in capacity with temperature. Figure 3.21b relates TC_c to V_R.

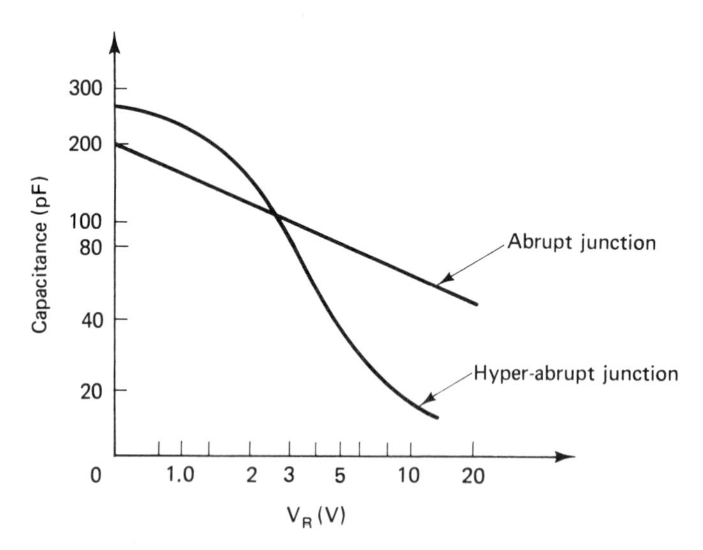

FIGURE 3.19 Capacitance–voltage characteristics for abrupt and hyper-abrupt junction devices.

BB139

VHF / FM VARACTOR DIODE

DIFFUSED SILICON PLANAR

- C_3/C_{25} . . . 5.0–6.5
- **MATCHED SETS** (Note 2)

ABSOLUTE MAXIMUM RATINGS (Note 1)

Temperatures

Storage Temperature Range	−55°C to +150°C
Maximum Junction Operating Temperature	+150°C
Lead Temperature	+260°C

Maximum Voltage

WIV	Working Inverse Voltage	30 V

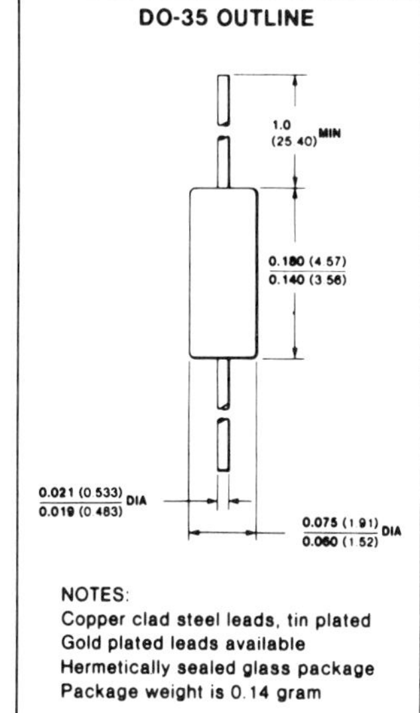

ELECTRICAL CHARACTERISTICS (25°C Ambient Temperature unless otherwise noted)

SYMBOL	CHARACTERISTIC	MIN	TYP	MAX	UNITS	TEST CONDITIONS
BV	Breakdown Voltage	30			V	I_R = 100 μA
I_R	Reverse Current		10 0.1	50 0.5	nA μA	V_R = 28 V V_R = 28 V, T_A = 60°C
C	Capacitance	 4.3	29 5.1	 6.0	pF pF	V_R = 3.0 V, f = 1 MHz V_R = 25 V, f = 1 MHz
C_3/C_{25}	Capacitance Ratio	5.0	5.7	6.5		V_R = 3 V/25 V, f = 1 MHz
Q	Figure of Merit		150			V_R = 3.0 V, f = 100 MHz
R_S	Series Resistance		0.35		Ω	C = 10 pF, f = 600 MHz
L_S	Series Inductance		2.5		nH	1.5 mm from case
f_o	Series Resonant Frequency		1.4		GHz	V_R = 25 V

NOTES:
1. These ratings are limiting values above which the serviceability of the diode may be impaired.
2. The capacitance difference between any two diodes in one set is less than 3% over the reverse voltage range of 0.5 V to 28 V.

FIGURE 3.20 Varactor specifications. Courtesy Fairchild Camera and Instrument Corporation.

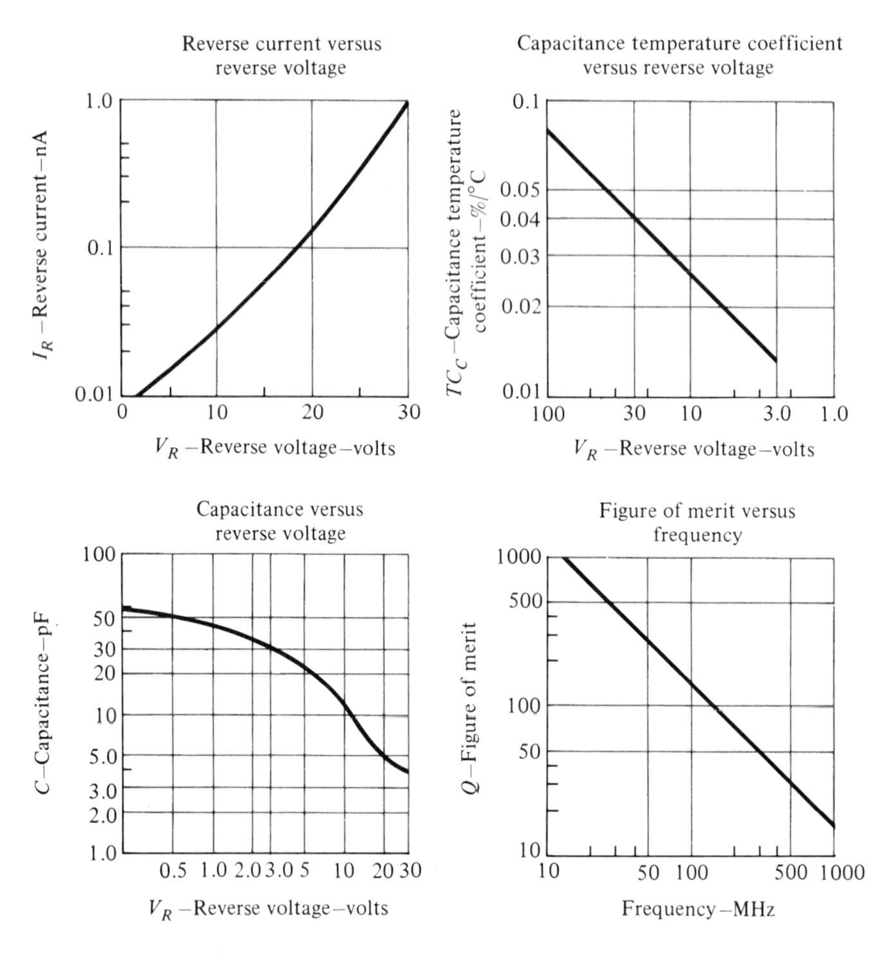

FIGURE 3.21 Characteristic curves for a VHF/FM Fairchild Varactor Diode. Courtesy Fairchild Camera and Instrument Corporation.

3.7.1 Design of a Voltage Variable Tuned Circuit

Figure 3.22 shows a transistor amplifier with a tuned circuit controlled by a VVC diode. The center frequency of the tuned circuit is a function of the voltage across the diode. This voltage depends on V_{CC}, R_1, R_2, and the position of the slider arm of R_2. R_3 is an isolating resistor that prevents R_1 and R_2 from loading the tuned circuit. C_C is a large capacitor that does not affect tuning but prevents the inductance from creating a dc short across the diode. R_4 represents the total resistance, whatever its source, in parallel with L and C. It is selected on the basis of bandwidth considerations. Design of the circuit consists of choosing a VVC, computing

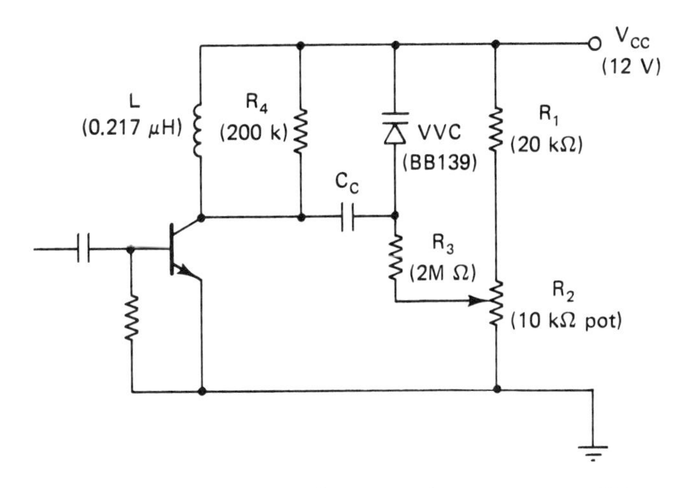

FIGURE 3.22 Voltage-tuned transistor amplifier.

the required voltage swing, computing the resistive network to provide this voltage swing, and computing L.

Example 3.7 *Design of a Voltage Variable Tuned Circuit*

Performance specifications:

Frequency range: 88 to 108 MHz (FM band)

$V_{CC} = 12, \qquad R_4 = 200\ \text{k}\Omega$

(a) *Choose the VVC diode:* We choose the BB139 VHF/FM varactor diode. (Devices designed for the required application should be used when possible.)

(b) *Find the required voltage swing:* First we find the ratio of capacitance for the lower and upper frequencies, f_L and f_H. Since

$$f = \frac{1}{2\pi\sqrt{LC}}$$

then

$$\frac{f_H}{f_L} = \frac{1/(2\pi\sqrt{LC_H}}{1/(2\pi\sqrt{LC_L}} = \frac{\sqrt{LC_L}}{\sqrt{LC_H}} = \frac{\sqrt{C_L}}{\sqrt{C_H}}$$

and the capacitor ratio C_R is given by

$$C_R = \frac{C_L}{C_H} = \frac{f_H^2}{f_L^2} = \frac{108^2}{88^2} = 1.51$$

Next we compute C_L and C_H. From Figure 3.21, we see that at $V_R = V_{CC} = 12$ V, $C_H = 10$ (the smallest C occurs at the highest V_R). Then

$$C_L = C_R \times C_H = 1.5 \times 10\,\text{pF} = 15.1\,\text{pF}$$

From Figure 3.21c we read

$$C_L = 15.1\,\text{pF} \qquad \text{occurs at } V_R = 8\,\text{V}\,(V_{R\,\text{min}})$$

$$C_H = 10\,\text{pF} \qquad \text{occurs at } V_R = 12\,\text{V}\,(V_{R\,\text{max}})$$

Therefore, the resistive network must be designed to provide the $V_{R\,\text{min}}$ to $V_{R\,\text{max}}$ voltage swing.

(c) *Choosing resistors:* To prevent tuned circuit loading, let $R_3 = 10R_4$.

$$R_3 = 10 \times 200\,\text{k}\Omega = 2\,\text{M}\Omega$$

From Figure 3.21a, we see that I_R is always less than 0.1 nA, so the voltage drop across R_3 is negligible. Let $R_1 = 20$ kΩ (arbitrary choice to prevent excessive current from V_{CC}). R_2 is now computed to give $V_{R\,\text{min}}$.

$$V_{R\,\text{min}} = \frac{V_{CC}R_1}{R_1 + R_2} \qquad \text{(voltage-divider equation)}$$

$$R_2 = \frac{R_1(V_{CC} - V_{R\,\text{min}})}{V_{R\,\text{min}}} = \frac{20\,\text{k}\Omega(12 - 8)\,\text{V}}{8\,\text{V}} = 10\,\text{k}\Omega \quad (10\,\text{k}\Omega\text{ pot.})$$

(d) *Computing L:* Either C_H or C_L may be used.

$$L = \frac{1}{(2\pi f_H)^2 C_H} = \frac{1}{(2\pi \times 108\,\text{MHz})^2 \times 10\,\text{pF}} = 0.217\,\mu\text{H}$$

$$\left(\text{From } f_o = \frac{1}{2\pi\sqrt{LC}}.\right)$$

The results of this design are shown in Figure 3.22.

3.8 CONCLUSION

This chapter has presented some basic diode circuits. Upon completion, you should have more than just an analytical understanding of the various diode applications. In addition, you should be capable of producing practical circuit designs as well. Diligent performance of the following problems will help you to achieve this important goal.

Problems

1. The output voltage of a full-wave rectifier (V_{in} = 117 V, 60 Hz) has the following characteristics: r = 0.05, V_{dc} = 12 V (r = ripple factor). Compute V_r, $V_{r\,peak}$, and V_m.

2. A full-wave rectifier (V_{in} = 117 V, 60 Hz) outputs 100 mA at 12 V DC with a ripple factor of 0.06; compute V_r, $V_{r\,peak}$, V_m, and C.

3. A full-wave rectifier (V_{in} = 117 V, 60 Hz) outputs 12 V DC into a load resistor of 100. V_m = 12.2 V. Compute V_r, $V_{r\,peak}$, r, I_{dc}, and C.

4. A transformer has a sinusoidal primary voltage of 117 V rms and a secondary voltage of 12 V peak. Find its turns ratio, N_p/N_s.

5. A transformer has a turns ratio of 4. If $V_{primary}$ is a sinusoidal voltage of 117 V rms, what is $V_{s\,peak}$?

6. Choose a satisfactory general-purpose rectifier for a circuit design that requires a diode with the following characteristics: I_o = 25 A, V_{RRM} = 500 V, and I_{FSM} = 200 A.

7. Choose a satisfactory general-purpose rectifier for a circuit design that requires a diode with the following characteristics: I_o = 500 mA, V_R = 150 V, and I_{FSM} = 2 A.

8. Design a full-wave rectifier (input 117 V rms, 60 Hz) to meet the following specifications: V_{dc} = 20 V, I_{dc} = 100 mA, and r = 0.05. Specify C, WVDC, and I_c of C, N_p/N_s, I_o, V_{RRM}, I_{FSM}, and the device number of the diode.

9. Repeat problem 8 for V_{dc} = 40 V, I_{dc} = 100 mA, and r = 0.02.

10. (a) Repeat problem 8 for V_{dc} = 30 V, I_{dc} = 100 mA, and r = 0.05.
    (b) Repeat problem 10(a) with r = 0.10 and compare the results.

11. Using the required number of LC filters (at least one), design a full-wave rectifier (V_{in} = 117 V rms, 60 Hz) to meet the following specifications: V_{dc} = 30 V, I_{dc} = 100 mA, and $r = 5 \times 10^{-4}$.

12. Repeat problem 11 with $r = 5 \times 10^{-6}$.

13. Using the negative-side shunt clipper of Figure 3.8, draw $v_{o(t)}$ versus t for the input voltages of Figure 3.23.

14. Using the positive-side series clipper of Figure 3.8, draw $v_{o(t)}$ versus t for the input voltages of Figure 3.23.

15. Using the shunt clipper of Figure 3.9, draw $v_{o(t)}$ versus t for the input voltages of Figure 3.24. Let V_B = 0.7 V.

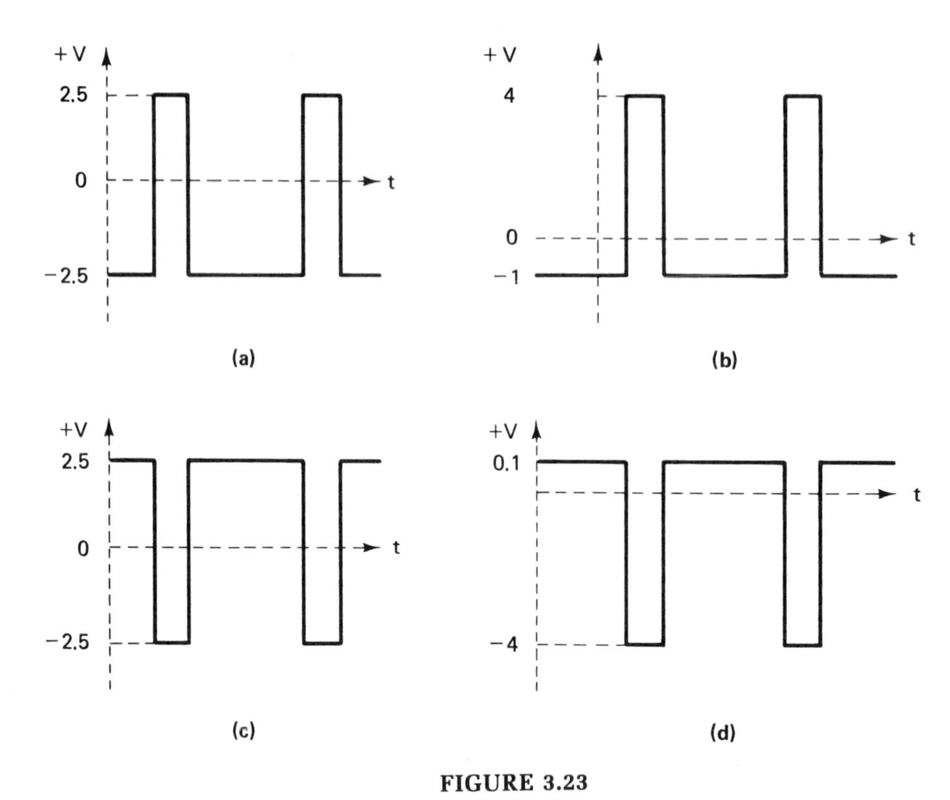

FIGURE 3.23

16. Using the shunt clipper of Figure 3.10, draw $v_{o(t)}$ for the input voltage of Figure 3.25 for (a) $V_{BO} = -0.7$ V, (b) $V_{BO} = 4.7$ V, and (c) $V_{BO} = 2.3$ V.

17. Using the equivalent negative clamper circuit, $V_B = 0.7$, of Figure 3.11, a $V_{BT} = 1$ V, and a sinusoidal input $v_{(t)} = 10 \cos 200\pi t \times 10^3$, (a) compute V_{BO}, V_{dc}, V_{max}, and V_{min} of $v_{o(t)}$; (b) draw $v_{o(t)}$.

18. Repeat problem 17 using the equivalent positive clamper circuit of Figure 3.11.

19. For $v_{(t)} = 10 \cos 200\pi t \times 10^3$ and $R_L = 10$ kΩ, what must be the value of C for proper clamper action?

20. In a negative clamper containing a silicon diode ($V_B = 0.7$), what must be the value of V_{BO} to clamp at 0 V?

21. Design a negative diode clamper to meet the following specifications: $v_{in(t)} = 30 \sin 20\pi t \times 10^3$, clamping level $= +4$ V, $R_L = 10$ kΩ, and R_S (source resistance) $= 100$ Ω. Specify C, its maximum working voltage, V_R, $I_{F\,surge}$, and the device number of the diode.

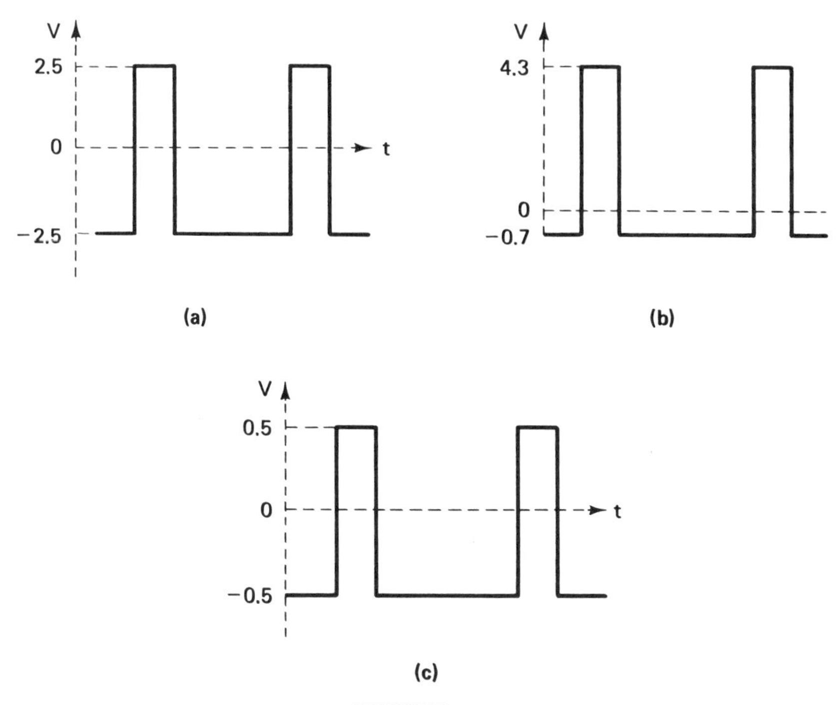

FIGURE 3.24

22. Design a positive diode clamper to meet the following specifications: $v_{in(t)} = 2 \sin \pi\ 10t \times 10^3$, clamping level = 0 V, R_L = 1 kΩ, and R_S = 10 Ω. Specify as in problem 21.

23. Repeat problem 21 for $v_{in(t)} = 10 \sin 20t \times 10^6$, $V_{dc} = -4$ V, R_L = 100 kΩ, and R_S = 100 Ω.

24. Compute the maximum power dissipation for the 1N961 of Table 3.2.

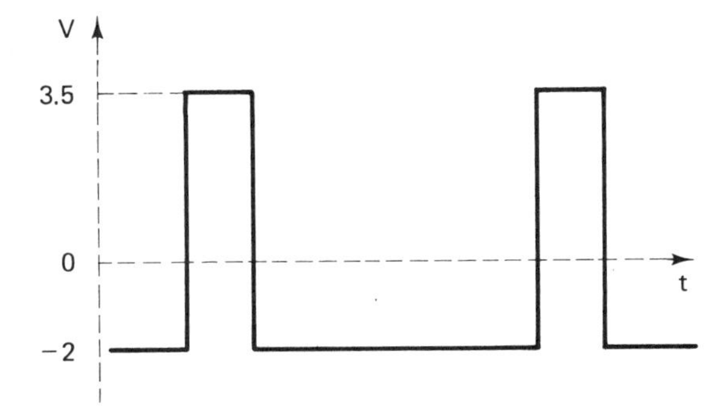

FIGURE 3.25

25. A zener diode with a V_Z of 20 V operates with an I_Z of 100 mA. How much power is it dissipating?

26. A 1N961 has a terminal voltage of 10.5 V with an I_Z of 12.5 mA. What is its terminal voltage if I_Z increases to 15 mA? (Use a piecewise equivalent circuit.)

27. A Motorola zener reference diode 1N4929 has a V_Z of 19.2 V at 25°C at I_{ZT} = 7.5 mA, a Z_{ZT} = 15, and a TC_c = 0.01% per °C. What is its terminal voltage if I_Z increases to 10 mA and the ambient temperature increases to 75°C?

28. Design a voltage reference circuit operating from a supply voltage of 20 V to provide 9.1 V ± 0.05 V from 0° to 75°C. Choose a diode from Table 3.4.

29. Given a 1N5347 zener rated at 5 W with V_Z = 10 V measured at $I_{ZT} = \frac{1}{4} I_{Z\,max}$, design a voltage reference circuit to give 10 V operating from a supply voltage of 20 V. (Note that some manufacturers specify zeners in this way.)

30. Given the zener voltage regulator circuit of Figure 3.26 in which I_L = 10 to 100 mA and I_Z = 100 to 10 mA, compute R_S, P_{RS}, and $P_{Z\max}$ for (a) V_Z = 10 V, V_S = 20 V; (b) V_Z = 6 V, V_S = 20 V.

31. Design a zener voltage regulator for a V_o = 39 V and I_L from 50 to 250 mA. Specify R_S, P_{RS}, V_Z, P, $I_{Z\,min}$, and $I_{Z\,max}$. Let V_S = 70 V.

32. Repeat problem 31 for V_S = 60 V.

33. Repeat problem 31 for V_S varying from 60 to 70 V.

34. From Figure 3.17, select a hot carrier diode to satisfy the following specifications: $V_{(BR)R}$ = 25 V, t_{rr} = 250 ps.

35. Compute the upper operating frequency of the diode selected in problem 34.

TABLE 3.4

Reference Voltage (V)	*Temperature Range (°C)*	*Test Current (mA)*	ΔV_z *(V)*	*Device No.*
9.0	0, 25, 75	7.5	0.067	1N936
9.0	−55, 0, 25, 75, 100	7.5	0.139	1N936A
9.0	−55, 0, 25, 75, 100	7.5	0.184	1N936B
9.1	0, 25, 75	0.5	0.068	1N4766
9.1	−55, 0, 25, 75, 100	0.5	0.141	1N4766A
9.1	0, 25, 75	1.0	0.068	1N4771
9.1	−55, 0, 25, 75, 100	1.0	0.141	1N4771A

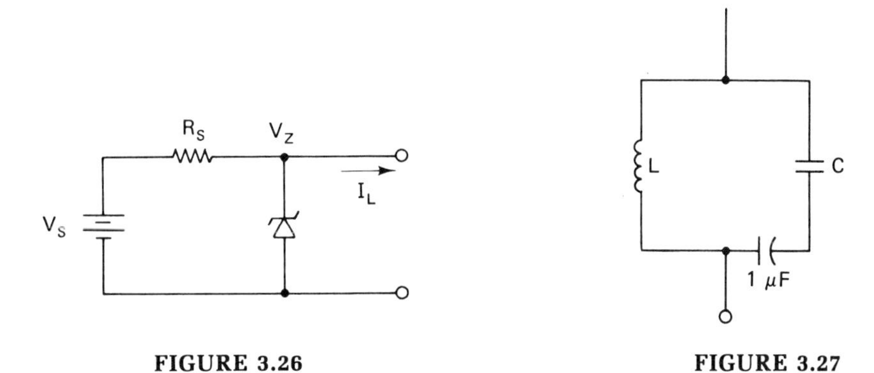

FIGURE 3.26

FIGURE 3.27

36. Compare the MBD101 (Figure 3.17) with the 1N914 (Figure 2.14) with respect to maximum reverse voltage, reverse current, V_F at $I_F = 10$ mA, and t_{rr}.

37. For the circuit of Figure 3.27, calculate the following:
(a) f_o if $L = 1\ \mu$H, $C = 5.1\ \mu$F.
(b) L if $f_o = 60$ MHz, $C = 29\ \mu$F.
(c) C if $f_o = 108$ MHz, $L = 0.2\ \mu$F.

38. Using the circuit of Figure 3.22 (ignore values) and the capacitance–voltage characteristic for the hyper-abrupt junction (Figure 3.19), design a voltage tuned transistor amplifier to tune from 10 to 45 MHz; let $V_{CC} = 12$ V and $R_4 = 100$ kΩ.

39. Using the BB139, design a voltage tuned amplifier to tune from 55 to 200 MHz.

References

Bell, David A. *Electronic Devices and Circuits,* 2nd ed., Reston Publishing Co., Reston, Virginia, 1980.

Cirovic, Michael. *Basic Electronics: Devices, Circuits and Systems,* 2nd ed., Reston Publishing Co., Reston, Virginia, 1978.

Cooper, William D. *Solid State Devices: Analysis and Application,* Reston Publishing Co., Reston, Virginia, 1982.

International Rectifier Corporation. *Engineering Handbook,* 1959.

4

BIPOLAR JUNCTION TRANSISTOR

4.1 INTRODUCTION

The bipolar junction transistor (BJT) has wide application as an amplifier and as a switch. As an amplifier it has virtually replaced its predecessor, the vacuum tube, because of its smaller size, lower supply voltage, and absence of a filament. As a switch the transistor is superior to the diode, because it permits higher switching rates and lower closed-switch voltage drops. In this era of integrated circuits a major advantage of the transistor lies in its simple structure, which makes it possible to etch hundreds of transistors into a single chip. The physics of BJT operation has already been covered in Chapter 1. In this chapter we concern ourselves strictly with circuit applications.

4.2 BJT PARAMETERS

Fundamentally, the transistor is a device that permits electronic control of current flow. Figure 4.1 illustrates this flow for an *npn* and *pnp* transistor. (Note that the middle character, *p* in *npn* and *n* in *pnp*, specifies the base material, and the arrowhead on the emitter indicates the direction of conventional current flow.)

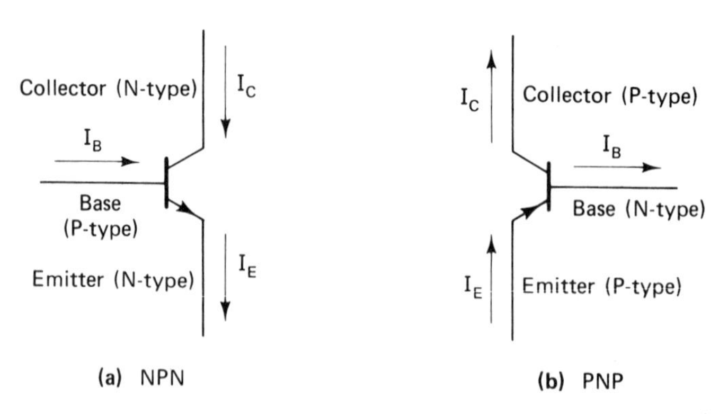

FIGURE 4.1 Current flow in the BJT: (a) *NPN*; (b) *PNP*.

For both transistors

$$I_E = I_C + I_B \tag{4.1}$$

I_C is composed of two components, a majority carrier emitter current (αI_E) diffusing through the base into the collector and a collector–base current I_{CB}. Therefore,

$$I_C = \alpha I_E + I_{CB} \tag{4.2}$$

When the transistor is used as an amplifier, its base–collector junction is reverse biased, and I_{CB} is a small minority carrier temperature-sensitive current called I_{CBO}. However, for switching operation, the base–collector junction may be forward biased and I_{CB} may be large.

Two basic parameters describing currents through the transistor are alpha (α) and beta (β, also called h_{FE}). These are defined next.

Alpha is derived from Eq. (4.2):

$$\alpha = \frac{I_C - I_{CB}}{I_E} \tag{4.3}$$

When I_{CB} is small, as in amplifier operation,

$$\alpha = \frac{I_C}{I_E} \tag{4.4}$$

Beta is defined by

$$\beta = \frac{\alpha}{1 - \alpha} \tag{4.5}$$

Substituting Eq. (4.1) into Eq. (4.2) and assuming a small I_{CB}, we get

$$\beta = \frac{I_C}{I_B} \tag{4.6}$$

Typically, α's range from 0.90 to 0.99 and β's from 10 to 500.

4.3 TRANSISTOR CHARACTERISTICS

All electronic devices may be represented by a set of one or more terminal characteristics. In Chapter 2 the diode, a one-port (two terminal) device, was represented by a single terminal characteristic. The transistor, being a two-port (three terminal) device, requires two terminal characteristics, one for the input port and the other for the output port. Since there are three ways in which transistors may be incorporated into amplifier and switching circuits, three pairs of terminal characteristics are required.

The most widely used configuration is called the *common emitter* (CE) because the emitter is common to both input and output. As an amplifier, the CE configuration is the only one that can give both current and voltage gain. It provides the highest sensitivity as a switch. Figure 4.2 gives the CE terminal characteristics and shows a typical circuit.

4.3.1 Input Port (CE)

The input characteristic (Figure 4.2b), aside from a small effect due to V_{CE}, does not differ in shape from that of a diode. This result is to be expected, since the base–emitter junction functions as a forward-biased diode in normal transistor operation. The V_{CE} effect occurs because increased levels of V_{CE} decrease I_B by permitting the collector to capture more of the carriers diffusing from the emitter, thus leaving fewer carriers to flow out of the base. Typically, the series resistance in the base–emitter circuit is much larger than the forward resistance of the base–emitter diode. Therefore, the offset ideal diode equivalent circuit is appropriate here (this is true for DC only; AC circuits will be discussed later).

Figure 4.3 illustrates the development of the equivalent DC input circuit for the CE configuration of Figure 4.2a. Figure 4.3a shows the circuit as it appears to DC (all capacitors are open circuits to DC). The circuit to the left of points *a* and *b* in Figure 4.3a is redrawn in Figure 4.3b as a Thevenin's equivalent. In Figure 4.3c, the base–emitter circuit of the transistor is replaced by the offset ideal diode model. Notice that

$$I_E = (\beta + 1)I_B \tag{4.7}$$

flows through R_E. If we wish to consider the circuit current to be I_B throughout, we associate $\beta + 1$ with R_E instead of I_B. This gives the circuit of Figure 4.3d (here we have replaced the forward-biased ideal diode by a short). Figures 4.3c and d are perfectly equivalent and give the following loop equation:

$$V_{BB} = I_B R_B + V_B + I_B(\beta + 1)R_E \tag{4.8}$$

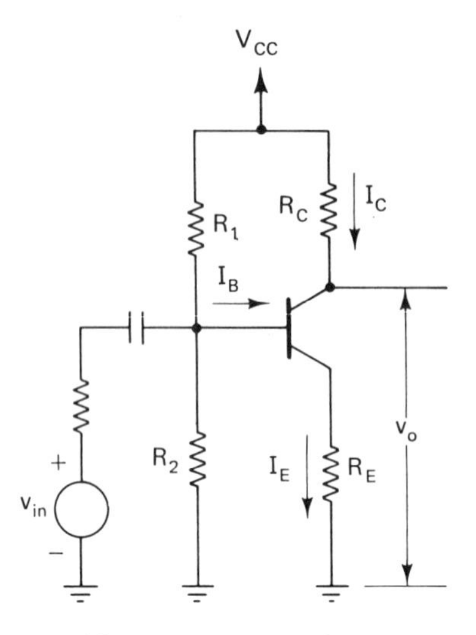

(a) Typical CE amplifier

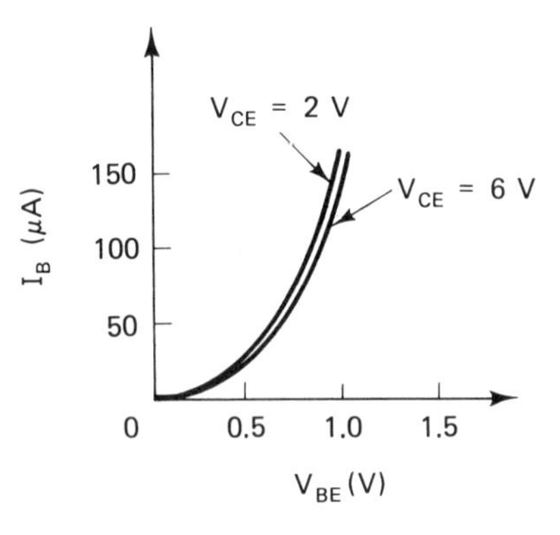

(b) Input characteristic

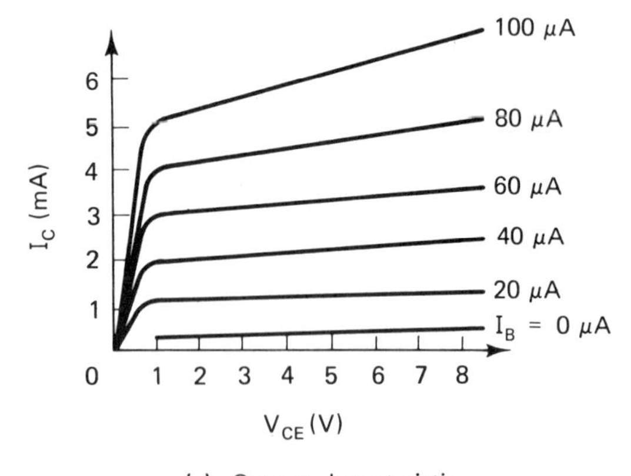

(c) Output characteristic

FIGURE 4.2 Terminal characteristics and typical circuit for the CE configuration: (a) typical CE amplifier; (b) input characteristic; (c) output characteristic.

Example 4.1 *Computing I_B for the CE Circuit*

For the circuit of Figure 4.2a, set $V_B = 0.7$ V, $V_{CC} = 12$ V, $R_1 = 40\ \Omega$, $R_2 = 20$ kΩ, and $\beta = 100$. Find I_B.

(a) *Find the Thevenin's base emitter circuit:*

$$V_{BB} = \frac{V_{CC} \times R_2}{R_1 + R_2} = \frac{12\text{ V} \times 20\text{ k}\Omega}{20\text{ k}\Omega + 40\text{ k}\Omega} = 4\text{ V}$$

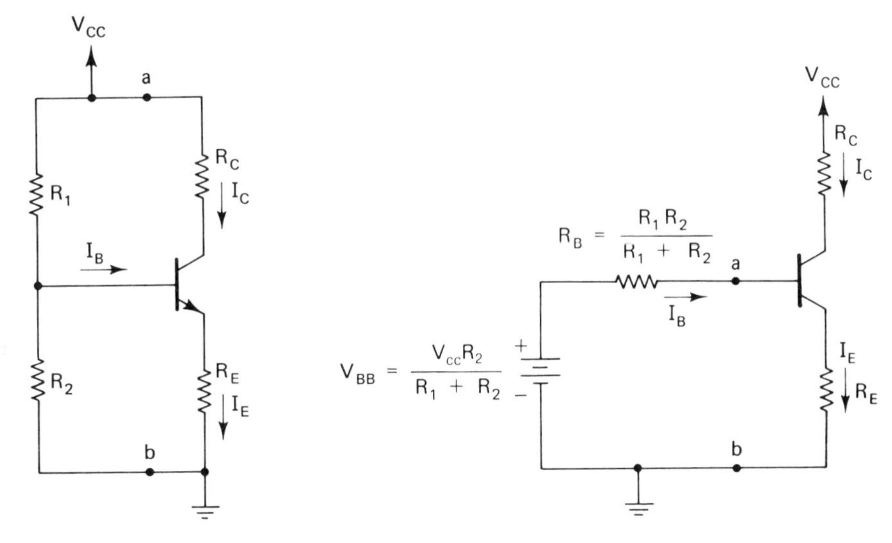

(a) DC circuit

(b) Circuit redrawn with Thevenin's equivalent left of a–b.

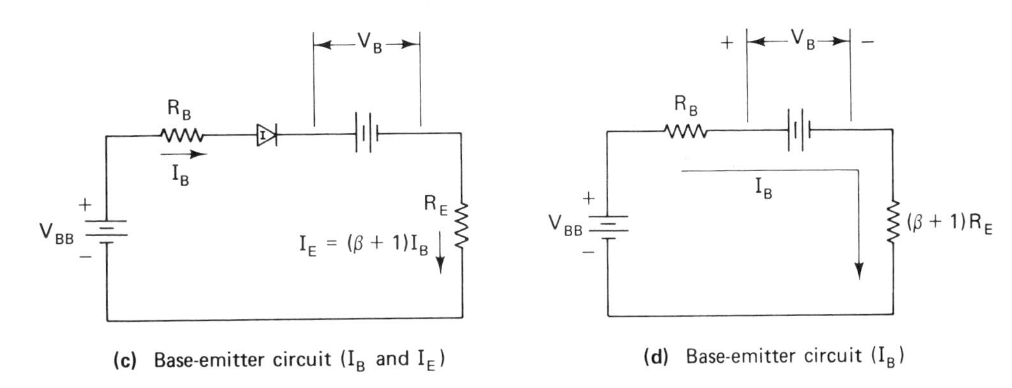

(c) Base-emitter circuit (I_B and I_E)

(d) Base-emitter circuit (I_B)

FIGURE 4.3 DC input circuit for the CE circuit of Figure. 4.2(a): (a) DC circuit; (b) circuit redrawn with Thevenin's equivalent left of a − b; (c) base-emitter circuit (I_B and I_E); (d) base-emitter circuit (I_B).

$$R_B = \frac{R_1 R_2}{R_1 + R_2} = \frac{20\text{ k}\Omega \times 40\text{ k}\Omega}{20\text{ k}\Omega + 40\text{ k}\Omega} = 13.3\text{ k}\Omega$$

(b) *Set up Kirchhoff's loop equation and solve for I_B:*

$$V_{BB} = I_B R_B + V_B + I_B(\beta + 1)R_E$$

$$I_B = \frac{V_{BB} - V_B}{R_B + (\beta + 1)R_E} = \frac{4\text{V} - 0.7\text{ V}}{13.3\text{ k}\Omega + 101\text{ k}\Omega} = 28.9\,\mu\text{A}$$

4.3.2 Output Port (CE)

The DC model for the collector–emitter circuit of the CE configuration is obtained by approximating the output terminal characteristic with a set of equally spaced horizontal lines. Figure 4.4a illustrates this. This approximation is not too inexact since the CE curves are straight and almost horizontal down to about 0.3 V. The equivalent circuit generating this set of lines consists of a current source in which

$$I_C = \beta I_B$$

This equivalent circuit is shown in Figure 4.4b. Notice that R_E has been replaced by R_E/α in this circuit. This has been done to account for the fact that I_E, rather than I_C, flows through R_E.

4.3.3 Input-Output DC CE Model

Figure 4.5b shows the complete input–output DC equivalent circuit for the CE configuration. Having established this model, there is no reason to resort to graphical analysis using the transistor characteristics. Transistor parameters may vary by as much as ±300%. The actual characteristics for a specific transistor may therefore be very different from the typical or average curve supplied by the manufacturer. In the face of such divergence, very little is to be gained from a graphical analysis. Analytical analysis based on the linear model is completely adequate for computing dc conditions in a transistor circuit. Example 4.2 illustrates its application.

Example 4.2 *Computing DC Conditions in a CE Circuit*

For the circuit of Figure 4.5a, let $R_1 = 60$ kΩ, $R_2 = 30$ kΩ, $V_{CC} = 12$ V, $R_C = 2$ kΩ, and $R_E = 2$ kΩ. Find I_B, I_E, I_C, and V_{CE} for (a) $\beta = 50$ and (b) $\beta = 200$.

(a) *Set up the model:*

$$R_B = \frac{R_1 R_2}{R_1 + R_2} = \frac{60\text{ k}\Omega \times 30\text{ k}\Omega}{60\text{ k}\Omega + 30\text{ k}\Omega} = 20\text{ k}\Omega$$

$$V_{BB} = \frac{V_{CC} R_2}{R_1 + R_2} = \frac{12\text{ V} \times 30\text{ k}\Omega}{90\text{ k}\Omega} = 4\text{ V}$$

$$R_E(\beta + 1) = 102\text{ k}\Omega \qquad (\beta = 50)$$

$$R_E(\beta + 1) = 402\text{ k}\Omega \qquad (\beta = 200)$$

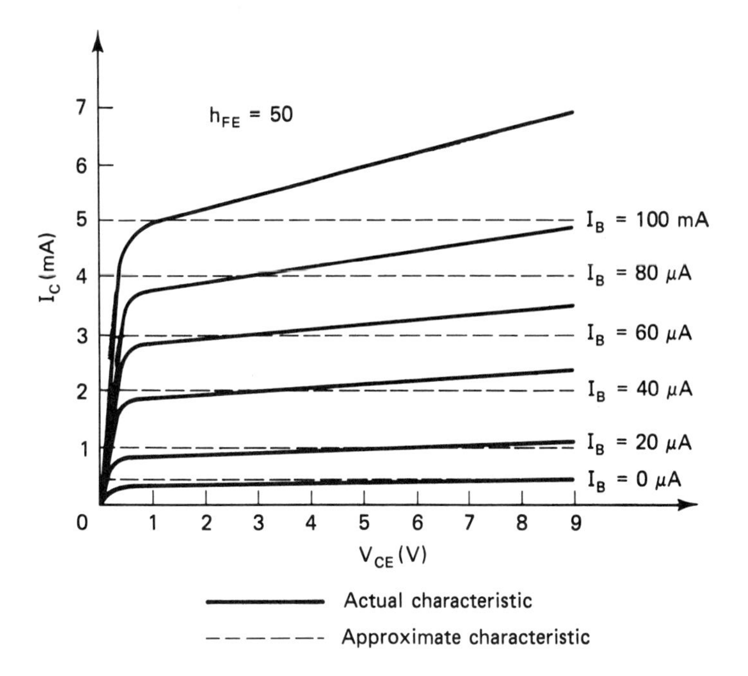

(a) Approximating a CE characteristic

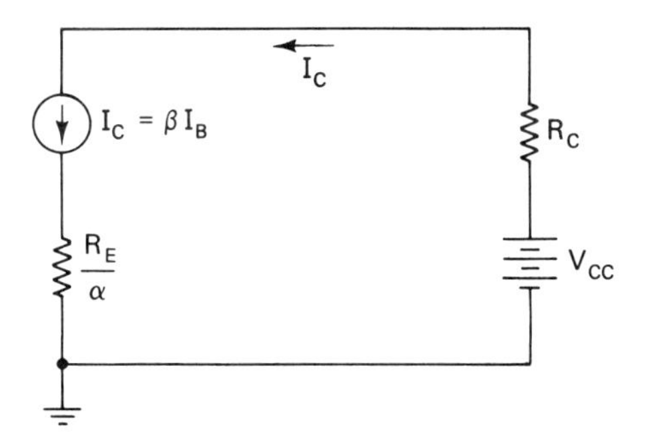

(b) DC equivalent circuit for CE circuit

FIGURE 4.4 Development of the DC equivalent circuit for a CE circuit: (a) approximating a CE characteristic; (b) DC equivalent circuit for CE circuit.

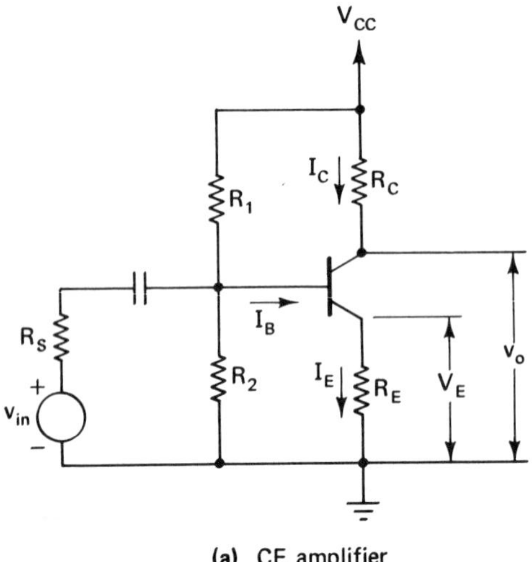

(a) CE amplifier

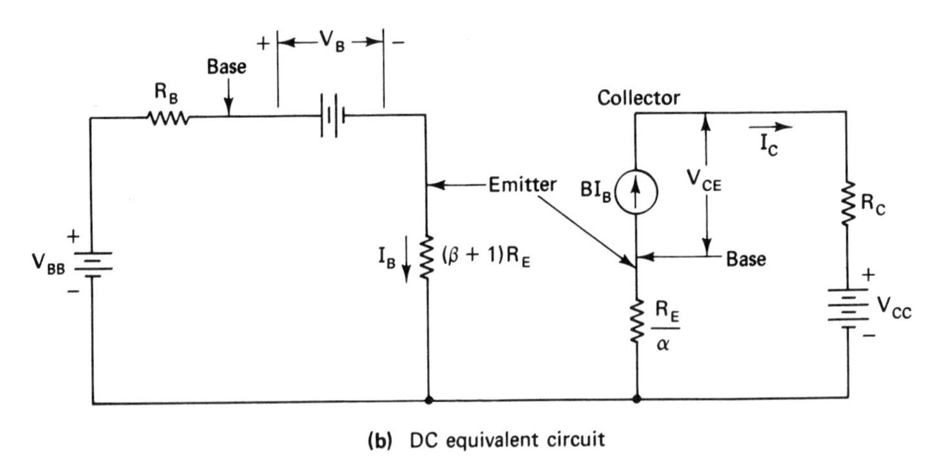

(b) DC equivalent circuit

FIGURE 4.5 Input-output DC CE model: (a) CE amplifier; (b) DC equivalent circuit.

(b) *Find* I_B:

$$I_B = \frac{V_{BB} - V_B}{R_B + (\beta + 1)R_E} = \frac{4\text{ V} - 0.7\text{ V}}{20\text{ k}\Omega + 102\text{ k}\Omega} = 0.027\text{ mA} \qquad (\beta = 50)$$

$$I_B = \frac{4\text{ V} - 0.7\text{ V}}{20\text{ k}\Omega + 402\text{ k}\Omega} = 0.0078\text{ mA} \qquad (\beta = 200)$$

(c) *Find I_C:*

$$I_C = \beta I_B = 50 \times 0.027\,\text{mA} = 1.35\,\text{mA} \qquad (\beta = 50)$$

$$I_C = 200 \times 0.0078\,\text{mA} = 1.56\,\text{mA} \qquad (\beta = 200)$$

(d) *Find V_{CE}:*

$$V_{CE} = V_{CC} - I_C\left(R_C + \frac{R_E}{\alpha}\right)$$

Using $\alpha = \dfrac{\beta}{(\beta + 1)}$,

$$\alpha = \frac{50}{51} = 0.98 \qquad (\beta = 50)$$

$$\alpha = \frac{200}{201} = 0.995 \qquad (\beta = 200)$$

$$V_{CE} = 12\,\text{V} - 1.35\,\text{mA}\left(1\,\text{k}\Omega + \frac{2\,\text{k}\Omega}{0.98}\right) = 7.89\,\text{V} \qquad (\beta = 50)$$

$$V_{CE} = 12\,\text{V} - 1.56\,\text{mA}\left(1\,\text{k}\Omega + \frac{2\,\text{k}\Omega}{0.995}\right) = 7.3\,\text{V} \quad (\beta = 200)$$

Notice that despite the wide variation in β and the resulting wide variation in I_B, I_C and V_{CE} show a relatively small variation. This, as we shall see, is the goal of good design.

4.4 BIASING THE CE AMPLIFIER

The first step in the design of a CE amplifier is establishing a stable DC bias. When amplifying, an input signal into the base will cause the collector operating point (V_{CE}, I_C) to deviate about this bias, providing voltage and/or current gain. This action is illustrated in Figure 4.6, where the idealized collector characteristics are shown as horizontal lines and the terminal characteristic of the R_C–V_{CC} circuit is given by the superimposed load line (see Section 2.4).

The position of the DC bias point (quiescent point) affects the drain on the power supply, the maximum undistorted voltage and current output, and the stability of the amplifier. For example, if it is necessary to get the maximum output current or voltage swing, then the bias point should be placed at the center of the load line (point *a*). If, however, the output signal is to be small, placing the bias at point *c* will minimize the current drain. Since, as will be seen, the DC bias affects the dynamic in-

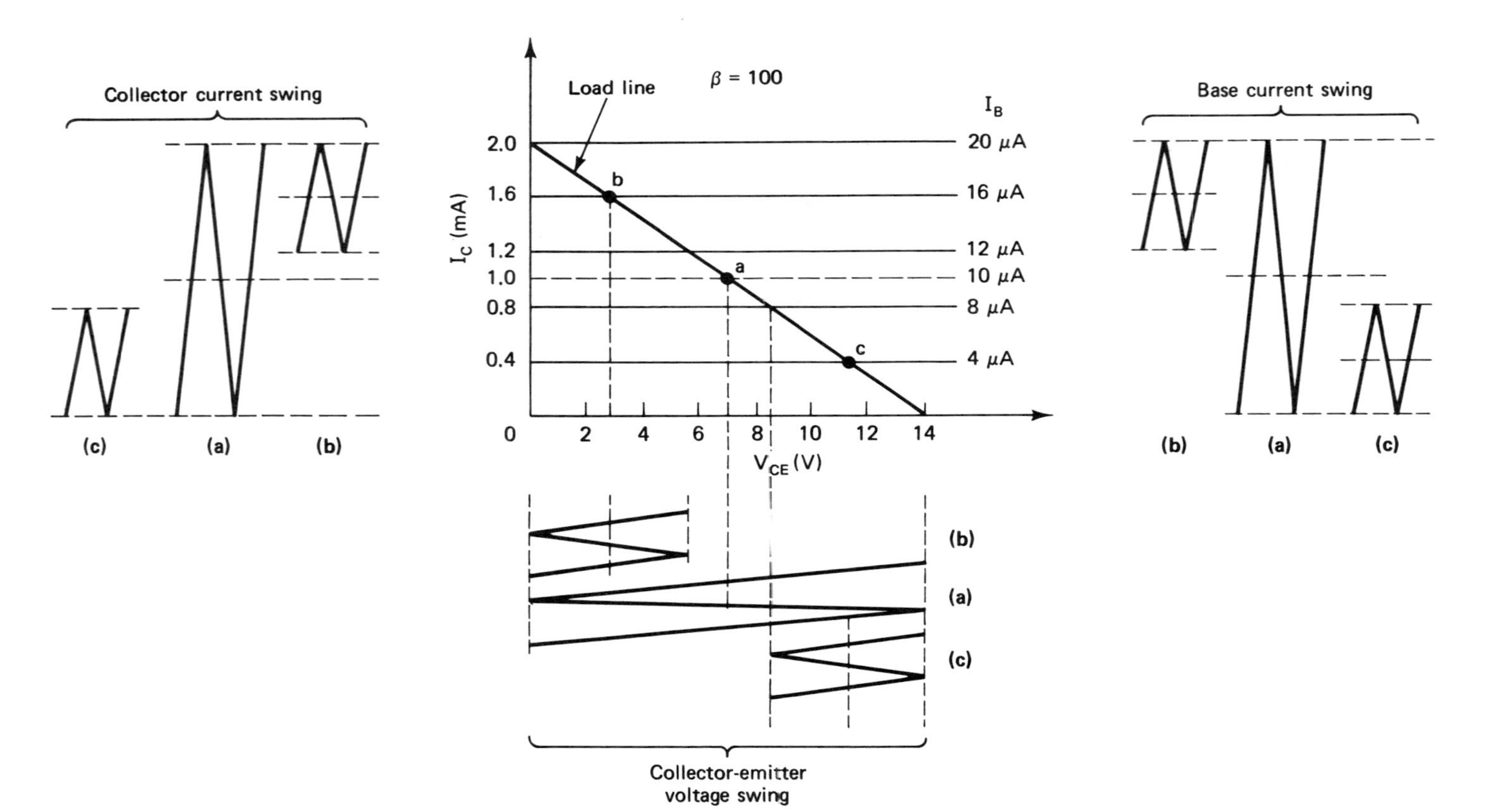

FIGURE 4.6 Bias point effects on the CE amplifier.

put impedance of the transistor, a high current bias (point *b*), giving a reduced input impedance, may be needed. Finally, when stable operation despite transistor interchange and variable ambient temperature is the main concern, a different bias may be required. At present we postpone this decision and concentrate on the most widely used CE amplifier design.

4.4.1 Biasing the Emitter-Current-Biased CE Amplifier

Design of the emitter-current-biased CE amplifier of Figure 4.5a consists of finding the values of R_1 and R_2 that give a desired I_{CQ}. The design procedure reverses the analysis of Example 4.2. First, using the input circuit of Figure 4.5b, an R_B is chosen and V_{BB} is computed to give I_{CQ}. Then an inverse Thevenin's is performed to transform this circuit to its original form.

The goal of good bias design is a stable I_{CQ}. If I_{CQ} is permitted to wander, an amplifier working under one set of conditions will not work under others. For example, the base drive (a) in Figure 4.6 produces an undistorted output only when the bias point is at position *a*. If the same drive is input when the bias point wanders to position *c*, the output will be distorted (clipped). This result is illustrated in Figure 4.7.

Bias points change for two reasons: temperature effects and variations in β. Rising ambient temperature increases the minority carrier current and decreases V_B. Both effects increase I_{CQ} slightly. More significant is the wide variation of β encompassed within any transistor type. For example, the β of a 2N2017 transistor is permitted to range from 50 to 200 at 25°C. Variations up to 600% in other types are not uncommon.

To minimize the effects of β changes, the voltage across the emitter resistor must be kept as constant as possible. A constant V_E means a constant I_E and a nearly constant I_{CQ}. Referring to Figure 4.5b, we see that V_E is most constant when $R_B = 0$, giving

$$V_E = V_{BB} - V_B \tag{4.9}$$

If V_{BB} is several times larger than V_B, this circuit also offers protection against the decrease in V_B with rising temperature. However, since R_B transforms into R_1 and R_2 in the emitter-biased configuration, and since R_1 in series with R_2 determines the power supply drain, R_B cannot be very small. A good engineering compromise is to make R_E as large as other considerations permit (discussed later) and to let

$$R_B = 0.1(\beta + 1)R_E \tag{4.10}$$

This restricts bias point changes to roughly ±10% with wide variations in β. V_{BB} is then given by

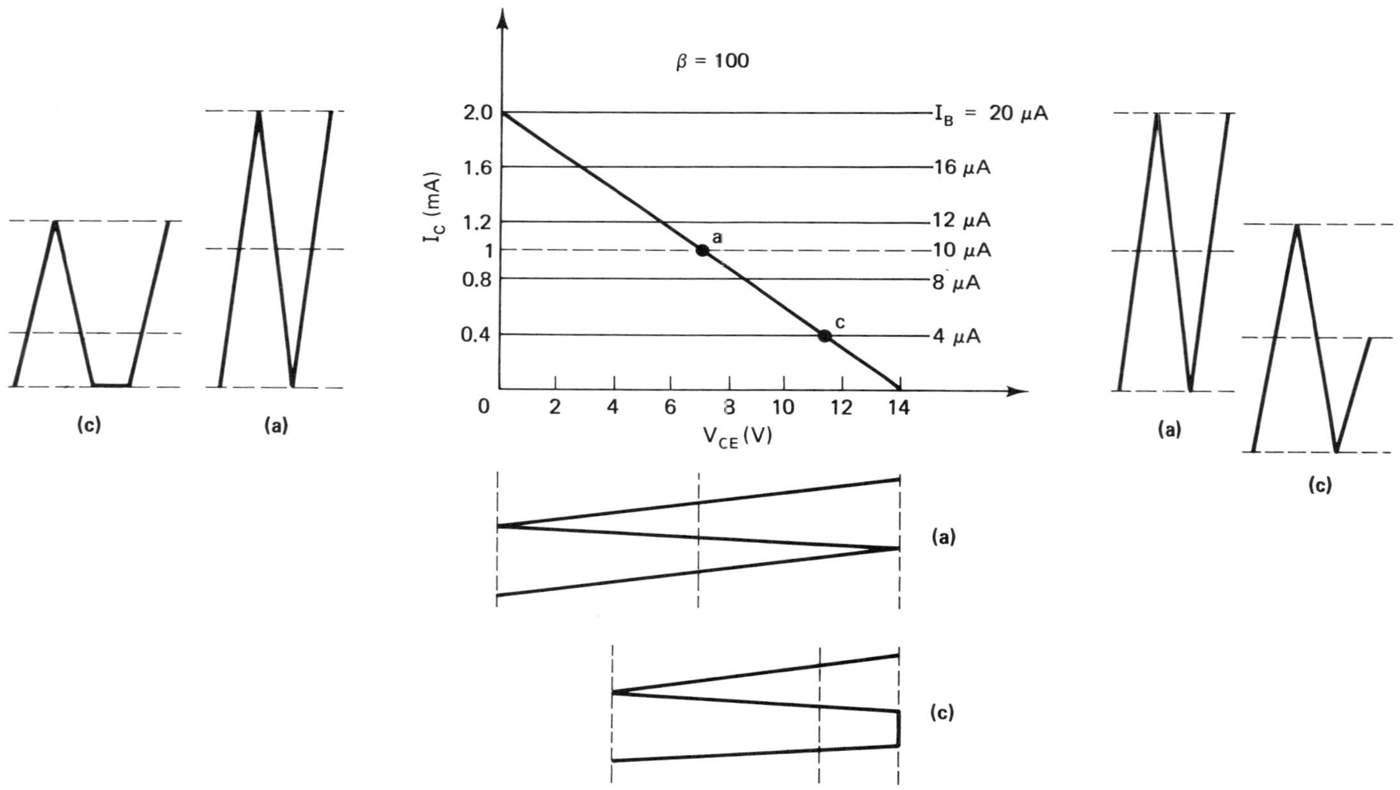

FIGURE 4.7 Effect of bias point shift.

$$V_{BB} = I_B[R_B + (\beta + 1)R_E] + V_B \quad (4.11a)$$

$$= I_B[0.1(\beta + 1)R_E + (\beta + 1)R_E] + V_B \quad (4.11b)$$

$$= 1.1I_b(\beta + 1)R_E + V_B \quad (4.11c)$$

$$= 1.1I_E R_E + V_B \quad (4.11d)$$

Letting $I_{EQ} \approx I_{CQ}$, we get the final equation:

$$V_{BB} = 1.1I_{CQ}R_E + V_B \quad (4.11e)$$

The derivation of the inverse Thevenin's transformation is left to the reader (see problem 11). The results are

$$R_1 = \frac{V_{CC}R_B}{V_{BB}} \quad (4.12)$$

$$R_2 = \frac{V_{CC}R_B}{V_{CC} - V_{BB}} \quad (4.13)$$

Example 4.3 illustrates the design of a CE amplifier using the preceding procedure.

Example 4.3 *Design of an Emitter-Current-Biased CE Amplifier*

Given the circuit of Figure 4.5a with $V_{CC} = 12$ V, $\beta = 100$, and $R_E = 1$ kΩ, find R_1 and R_2 to give $I_{CQ} = 2$ mA. Also check I_{CQ} at $\beta = 50$ and 200.

(a) $V_{BB} = 1.1I_C R_E + V_B = 1.1 \times 2\text{ mA} \times 1\text{ k}\Omega + 0.7\text{ V} = 2.9\text{ V}$

$$R_B = 0.1(\beta + 1)R_E = 0.1 \times 101 \times 1\text{ k}\Omega = 10.1\text{ k}\Omega \approx 10\text{ k}\Omega$$

$$R_1 = \frac{V_{CC}R_B}{V_{BB}} = \frac{12\text{ V} \times 10\text{ k}\Omega}{2.9\text{ V}} = 41.4\text{ k}\Omega$$

$$R_2 = \frac{V_{CC}R_B}{V_{CC} - V_{BB}} = \frac{12\text{ V} \times 10\text{ k}\Omega}{12\text{ V} - 2.9\text{ V}} = 13.2\text{ k}\Omega$$

(b) *Check I_{CQ} at $\beta = 50$ and 200:*

$$I_{CQ} = \beta I_B = \beta \frac{(V_{BB} - V_B)}{R_B + (\beta + 1)R_E}$$

$$= 50 \frac{(2.9\text{ V} - 0.7\text{ V})}{10\text{ k}\Omega + 51\text{ k}\Omega} = 1.803\text{ mA} \qquad (\beta = 50)$$

$$= 200 \frac{(2.9\text{ V} - 0.7\text{ V})}{10\text{ k}\Omega + 201\text{ k}\Omega} = 2.085\text{ mA} \qquad (\beta = 200)$$

4.4.2 Choosing I_{CQ} for the Emitter-Current-Biased CE Amplifier

We consider three different criteria for choosing I_{CQ} for the CE amplifier. In the first case, the output circuit contains no capacitors. This is the DC output case shown in Figure 4.8a. I_{CQ} in this case is to be chosen for maximum undistorted output. In the second case, maximum undistorted output is still the goal, but the output circuit contains bypass and coupling capacitors. This is the AC output case shown in Figure 4.8b. Finally, we consider the case where the input signal is very small and the design goal is minimum I_{CQ}.

Case 1: DC Output, Maximum Undistorted Output. For maximum undistorted output, the quiescent voltage (V_{CEQ}) and current (I_{CQ}) must be equal to one-half their maximums. This bias will permit the output to swing from maximum to zero (assuming ideal characteristics). Therefore,

$$I_{CQ} = \frac{I_{c\,max}}{2} \tag{4.14}$$

$$V_{CEQ} = \frac{V_{CC}}{2} \tag{4.15}$$

Since $I_{c\,max}$ occurs when $V_{CE} \approx$ zero, and the entire supply voltage is across the output circuit resistors, $I_{c\,max}$ for the circuit of Figure 4.8a is given by

$$I_{c\,max} = \frac{V_{CC}}{R_C + R_E} \tag{4.16}$$

and

$$I_{CQ} = \frac{V_{CC}}{2(R_C + R_E)} \tag{4.17}$$

Example 4.4 *Finding I_{CQ} for the DC Output Case*

For the circuit of Figure 4.8a, $R_C = 2$ kΩ, $R_E = 1$ kΩ, and $V_{CC} = 12$ V. Find I_{CQ} for maximum undistorted output.

$$I_{CQ} = \frac{V_{CC}}{2(R_C + R_E)} = \frac{12\text{ V}}{2(2\text{ k}\Omega + 1\text{ k}\Omega)} = 2\text{ mA}$$

Case 2: AC Output, Maximum Undistorted Output. Because all capacitors may be considered open circuits for DC and short circuits for higher frequencies, the output circuit of the CE amplifier sees two load lines each with its own Thevenin's resistance. One is the familiar DC load line with a resistance R_{dc} found by open circuiting all output circuit capacitors and tracing out the total series resistance connected across

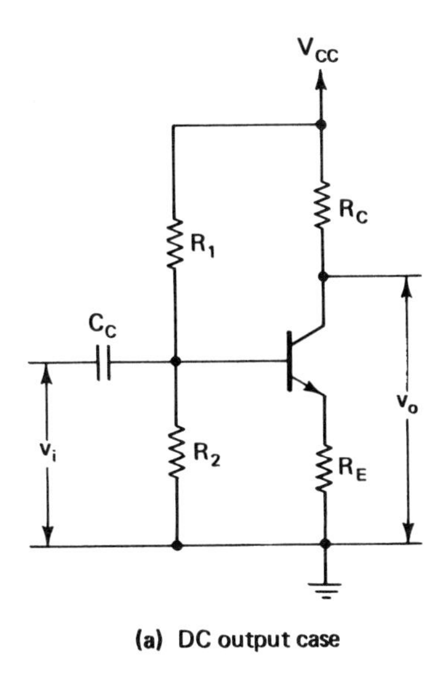

(a) DC output case

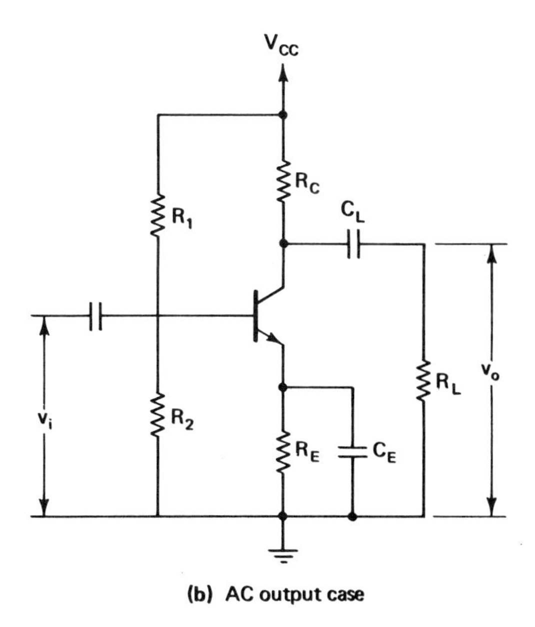

(b) AC output case

FIGURE 4.8 DC and AC output CE amplifiers: (a) DC output case; (b) AC output case.

the C–E terminals of the transistor. The other is the AC load line with a Thevenin's resistance R_{ac} found by short circuiting all output circuit capacitors, assuming V_{CC} to be a short circuit, and tracing out the total series resistance connected across the transistor output. Once R_{dc} and R_{ac} are found, a straightforward analysis shows that to obtain maximum undistorted output I_{CQ} is given by

$$I_{CQ} = \frac{V_{CC}}{R_{dc} + R_{ac}} \tag{4.18}$$

Example 4.5 illustrates this procedure.

Example 4.5 *Finding I_{CQ} for the AC Output Case*

Find I_{CQ} for maximum undistorted output for the circuit of Figure 4.9a.

(a) *Find R_{dc}:* Tracing all resistive paths from C through the power supply back to E of the transistor, with all capacitors open circuited, we get Figure 4.9b. Therefore,

$$R_{dc} = R_E + R_C = 1\ \text{k}\Omega + 2\ \text{k}\Omega = 3\ \text{k}\Omega$$

(b) *Find R_{ac}:* Consider all capacitors and V_{CC} as short circuits. Then tracing all external paths from C to E gives the circuit of Figure 4.9c. Therefore,

$$R_{ac} = \frac{R_C R_L}{R_C + R_L} = \frac{2\ \text{k}\Omega \times 2\ \text{k}\Omega}{2\ \text{k}\Omega + 2\ \text{k}\Omega} = 1\ \text{k}\Omega$$

(c) *Find I_{CQ}:*

$$I_{CQ} = \frac{V_{CC}}{R_{dc} + R_{ac}} = \frac{12\ \text{V}}{3\ \text{k}\Omega + 1\ \text{k}\Omega} = 3\ \text{mA}$$

Case 3: Low Output Swing, Minimum I_{CQ}. To keep the battery drain low in battery-operated equipment or to reduce heat generation in large-scale equipment, it may be desirable to minimize I_{CQ}. Although there is a cost associated with low I_{CQ}'s, this procedure is especially feasible in amplifiers where the maximum output swing is very small. The major danger to be guarded against is the possibility that I_{CQ} will fall because of a decrease in β or an increase in V_B causing distortion or cut off. Because the two preceding designs gave a centered I_{CQ}, they were tolerant of these changes. This design, however, must specifically take into account the expected extremes of h_{FE} and V_B. Although especially applicable to low I_{CQ}'s, it is suitable for restricting the range of any I_{CQ}.

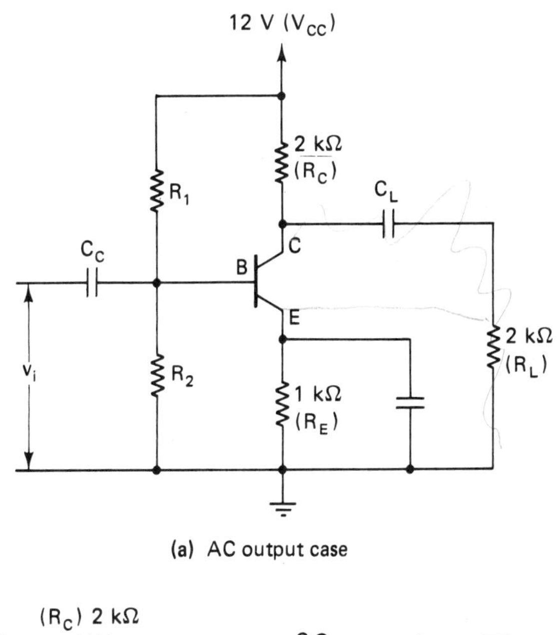

(a) AC output case

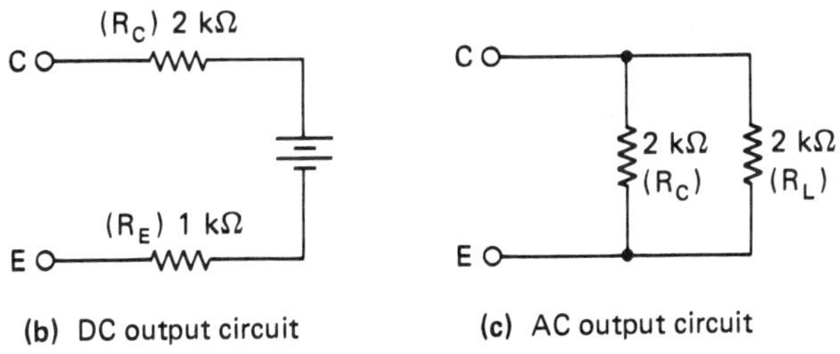

(b) DC output circuit

(c) AC output circuit

FIGURE 4.9 Finding R_{DC} and R_{AC} for the AC output CE amplifier: (a) AC output case; (b) DC output; (c) AC output circuit.

Beta (h_{FE}) varies with temperature, I_{CQ}, and among transistors of the same type number. Table 4.1 illustrates the effects of I_{CQ} on h_{FE} and gives the range of intrafamily variation. Figure 4.10 illustrates the temperature effects.

TABLE 4.1 *Effects of I_{CQ} on h_{FE}*

Forward Current Transfer Ratio (h_{FE})	*Minimum (2N2193)*	*Maximum*
I_C = 150 mA, V_{CE} = 10 V	40	120
I_C = 10 mA, V_{CE} = 10 V	30	
I_C = 1000 mA, V_{CE} = 10 V	15	
I_C = 0.1 mA, V_{CE} = 10 V	15	
I_C = 500 mA, V_{CE} = 10 V	20	

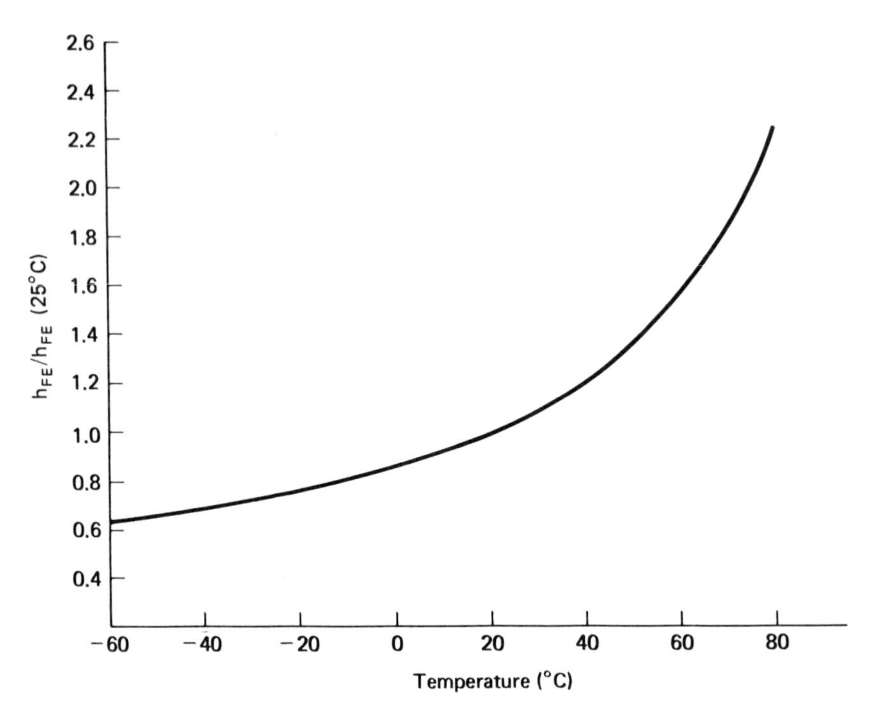

FIGURE 4.10 Typical variation of h_{FE} with temperature.

The break voltage (V_B) has a negative temperature coefficient (TC) given by

$$TC = -2.5\,\text{mV/°C} \tag{4.19}$$

Therefore, V_B falls with increasing temperature and since

$$V_{BB} = [R_B + (h_{FE} + 1)R_E]I_{BQ} + V_B \tag{4.20}$$

which upon substituting

$$\frac{I_{EQ}}{h_{FE} + 1} = I_{BQ} \tag{4.21}$$

becomes

$$V_{BB} = \left(\frac{R_B}{h_{FE} + 1} + R_E\right)I_E + V_B \tag{4.22}$$

Then for constant V_{BB}, I_{EQ} (and I_C) increase as V_B decreases.

Increasing temperature will also increase I_{CQ} by raising the collector leakage current I_{CO}. For those designs where I_{CO} becomes significant, its minimum value is determined and its maximum found by assuming

I_{CO} to double every 10°C increase. However, for low-level operation, especially with silicon transistors, I_{CO} may be neglected except when I_{CQ} is in the microampere region.

In the design procedure for case 3, we cannot set R_B at $0.1(h_{FE} + 1)R_E$ since there are 2 h_{FE}'s, $h_{FE\,min}$ and $h_{FE\,max}$. Also, we require a larger R_B to compensate for the reduction in V_B with temperature. (As I_B tends to increase with lower V_B, a large R_B minimizes this change.) Instead we note that I_E will have its minimum value at the lowest operating temperature, for which case h_{FE} is minimum and V_B is maximum. Therefore,

$$V_{BB} = \left[\frac{R_B}{h_{FE\,min} + 1} + R_E\right] I_{E\,min} + V_{B\,max} \tag{4.23}$$

I_E will be maximum at the highest temperature, for which case h_{FE} is maximum and V_B is minimum. Therefore,

$$V_{BB} = \left[\frac{R_B}{h_{FE\,max} + 1} + R_E\right] I_{E\,max} + V_{B\,min} \tag{4.24}$$

Since V_{BB} is always fixed, we can equate Eqs. (4.23) and (4.24) and solve for R_B. This gives

$$R_B = \frac{(I_{E\,max} - I_{E\,min})R_E + V_{B\,min} - V_{B\,max}}{\dfrac{I_{E\,min}}{h_{FE\,min} + 1} - \dfrac{I_{E\,max}}{h_{FE\,max} + 1}} \tag{4.25}$$

Entering Eq. (4.25) with $I_{E\,min}$, $I_{E\,max}$, $V_{B\,min}$, $V_{B\,max}$, $h_{FE\,min}$, $h_{FE\,max}$, and a selected R_E, we find R_B. Then using either Eq. (4.23) or (4.24), we find V_{BB}. (Knowing R_B, V_{BB}, and V_{CC}, we can find R_1 and R_2 of the inverse Thevenin's circuit using Eqs. 4.12 and 4.13.) An example follows.

Example 4.6 *Design of a Low I_{CQ} Emitter-Current-Biased CE Amplifer*

The circuit of Figure 4.9a uses a 2N2193 transistor and must operate from 0° to 50°C. Since the output current swing is restricted to a few microamperes, it is desired to restrict I_{CQ} between 0.1 and 0.2 mA. Find R_B and V_{BB}. ($V_{CC} = 12$ V.)

(a) *Find $h_{FE\,min}$ and $h_{FE\,max}$:* According to Table 4.1, $h_{FE\,min}$ at 25°C at 0.1 mA = 15. Although no maximum value is given, we can safely assume a range of 3 to 1, giving $h_{FE\,max}$ at 25°C equal to 45. We assume these values hold for 0.2 mA as well (the effect of V_{CE} is ignored in this example). From Figure 4.10, at 0°,

$$h_{FE\,min} = 0.8 h_{FE\,min} \text{ at } 25°\text{C} = 0.8 \times 15 = 12$$

At 50°,

$$h_{FE\max} = 1.3h_{FE\max} \text{ at } 25°\text{C} = 1.3 \times 45 = 58.5 \approx 60$$

(b) *Find $I_{E\,max}$ and $I_{E\,min}$:*

$$I_E = \frac{I_C}{\alpha} = \frac{(h_{FE} + 1)I_C}{h_{FE}} \qquad \left(\text{since } \alpha = \frac{h_{FE}}{h_{FE} + 1}\right)$$

$$I_{E\,\min} = \frac{(h_{FE\,\min} + 1)I_{C\,\min}}{h_{FE\,\min}} = \frac{(12 + 1)}{12}\,0.1\,\text{mA} = 0.108\,\text{mA}$$

$$I_{E\,\max} = \frac{(h_{FE\,\max} + 1)I_{C\,\max}}{h_{FE\,\max}} = \frac{(60 + 1)0.2\,\text{mA}}{60} = 0.203\,\text{mA}$$

(c) *Find $V_{B\,max}$ and $V_{B\,min}$:*

$$V_{B\max} = V_{B(25°\text{C})} - (25°\text{C} - T_{\min}) \times \text{TC}$$
$$= 0.7 - (25°\text{C})(-2.5\,\text{mV/°C}) = 0.763\,\text{V}$$

$$V_{B\,\min} = V_{B(25°\text{C})} - (25°\text{C} - T_{\max}) \times \text{TC}$$
$$= 0.7 - (25°\text{C} - 50°\text{C})(-2.5\,\text{mV/°C}) = 0.638$$

(d) *Choose R_E:* R_E should be chosen so that V_E at minimum is about one-sixth to one-quarter of V_{CC}.

$$R_E = \frac{12/6}{0.1\,\text{mA}} = 20\,\text{k}\Omega$$

(e) *Find R_B:*

$$R_B = \frac{(I_{E\,\max} - I_{E\,\min})R_E + V_{B\,\min} - V_{B\,\max}}{\dfrac{I_{E\,\min}}{h_{FE\,\min} + 1} - \dfrac{I_{E\,\max}}{h_{FE\,\max} + 1}}$$

$$= \frac{(0.203 - 0.108)\,\text{mA} \times 20\,\text{k}\Omega + (0.638 - 0.763)\,\text{V}}{(0.108\,\text{mA}/13) - (0.203\,\text{mA}/61)}$$

$$= 356.44\,\text{k}\Omega$$

(f) *Find V_{BB}:*

$$V_{BB} = \left(\frac{R_B}{h_{FE\,\max} + 1} + R_E\right) I_{E\,\max} + V_{B\,\min}$$

$$= \left(\frac{356.44\,\text{k}\Omega}{61} + 20\,\text{k}\Omega\right) 0.203\,\text{mA} + 0.638\,\text{V} = 5.83\,\text{V}$$

4.5 SMALL-SIGNAL CIRCUIT

Although the transistor is a nonlinear device whose parameters vary with its operating point, for small output swings it can be considered to be a linear circuit with fixed parameters. Under these conditions it may be replaced by an equivalent circuit and computations of voltage, current, and power gains made analytically. This equivalent circuit is called the *small-signal circuit*.

There are several small-signal circuits in the literature, but the hybrid (h) parameter circuit is the most widely used. Figure 4.11 gives these equivalent circuits for the three principal transistor configurations. There are four h-parameters for each configuration, each defined by its subscript. The first character of the subscript describes the function of the h-parameter and the second, the configuration to which it applies. For example, h_{ie} is the input parameter for the CE while h_{ob} is the output parameter for the common base (CB). When it is necessary to refer to all configurations at once, the first character is used alone, h_i for example.

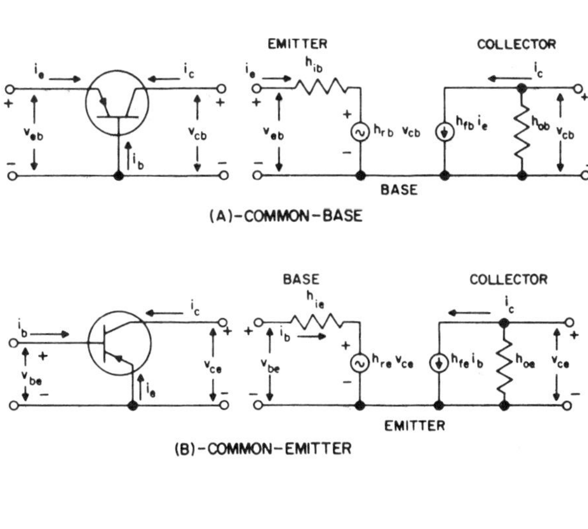

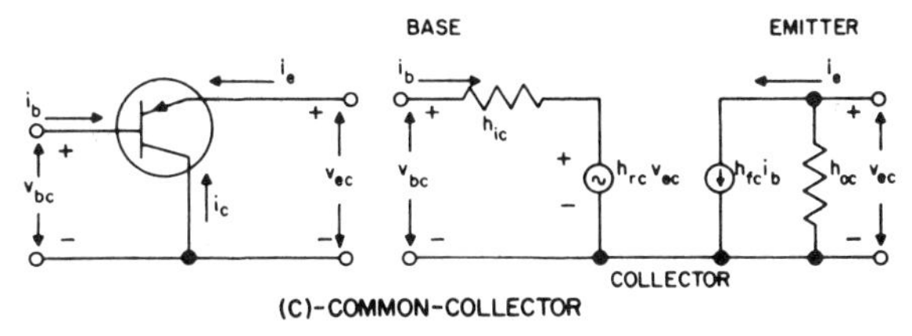

FIGURE 4.11 Hybrid equivalent circuits. Courtesy of General Electric.

Definitions of each parameter are most easily realized in conjunction with Figure 4.11. Shorting all outputs on the equivalent circuits zeroes v_o ($v_o = v_{cb}$, v_{ce}, or v_{ec}). This zeroes $h_r v_o$ so that h_i becomes the input resistance. This is given analytically by

$$h_i \quad \text{(short-circuit input resistance)} = \frac{v_{in}}{i_{in}} \bigg|\; v_o = 0 \tag{4.26}$$

where $v_{in} = v_{eb}$, v_{be}, or v_{bc}, and $i_{in} = i_e$ or i_b. Also, with v_o shorted all of $h_f i_{in}$ flows through the short. Therefore,

$$h_f \quad \text{(short-circuit forward transfer ratio)} = \frac{i_o}{i_{in}} \bigg|\; v_o = 0 \tag{4.27}$$

where $i_o = i_b$ or i_e. Open circuiting the input of the equivalent circuits by removing v_{in} zeroes i_{in} and sets v_{in} equal to $h_r \times v_o$. Therefore,

$$h_r \quad \text{(open-circuit reverse voltage ratio)} = \frac{v_{in}}{v_o} \bigg|\; i_{in} = 0 \tag{4.28}$$

In modern transistors h_r is so small (10^{-4}) that it can almost always be ignored. Also, since $i_{in} = 0$, $h_f i_{in} = 0$, and h_o, which is given as an admittance, is defined by

$$h_o \quad \text{(short-circuit output admittance)} = \frac{i_o}{v_o} \bigg|\; i_{in} = 0 \tag{4.29}$$

For the CE and CC, $1/h_{oe}$ is typically 100 to 400 kΩ, and for the CB, 1 to 2 MΩ, so it may often be ignored. All h-parameters vary with I_{EQ} and V_{CEQ}. Figure 4.12 shows these variations for a small transistor, the 2N525.

4.5.1 Finding Gains and Impedances (CE Amplifier)

The small-signal equivalent circuit is used to find the current, voltage, and power gains of the transistor amplifier and its input and output impedances. Before this circuit can be derived, the actual transistor amplifier circuit, including source and load impedances, must be known. The h-parameters can then be evaluated. The manufacturer provides a minimum and maximum h_{fe} for some specific V_{CB}, I_E, and frequency. Knowing or computing I_{EQ} and V_{CEQ}, Figure 4.12, or its equivalent, can be used to find h_{fe} for the actual operating point, $h_{fe\,op}$. Then h_{ie} may be computed by

$$h_{ie} = \frac{(h_{fe\,op} + 1)25 \times 10^{-3}}{|I_{EQ}|} \tag{4.30}$$

H_{re} and h_{oe} are usually ignored.

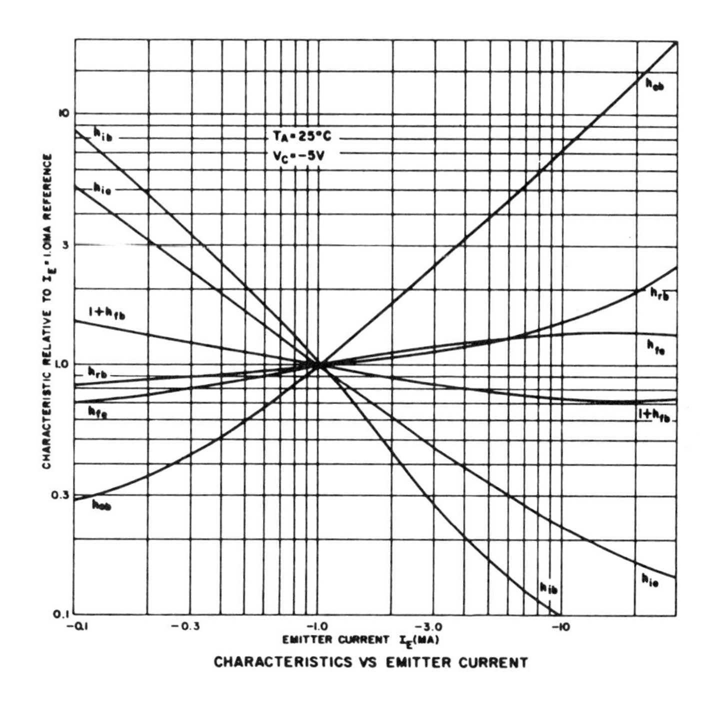

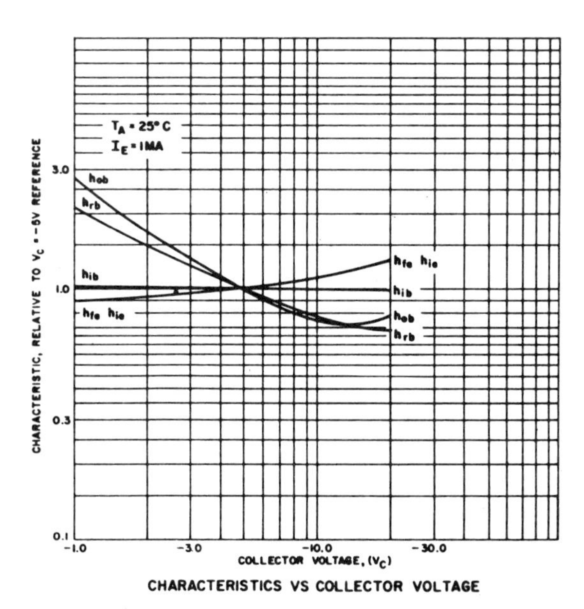

FIGURE 4.12 Hybrid parameters versus I_E and V_C V_{CE}: (a) h-parameters versus I_E; (b) h-parameters versus V_{CE}.

Figure 4.13 shows a complete CE amplifier circuit and its small-signal equivalent. (The values for this equivalent circuit are calculated in Example 4.7.) To derive the small-signal circuit, it is necessary to consider V_{CC} as ground (short out all DC voltage supplies) and to short all capacitors. Consequently, R_1 and R_2 are in parallel from base to ground and equal to R_B, and R_L is in parallel with R_C. R_E is shorted and out of the circuit. The signal source, whatever its actual configuration, is most conveniently represented by its Norton's equivalent, as shown.

The current, voltage, and power gains are computed as follows:

Current Gain, A_I:

$$A_I = \frac{i_L}{i_{in}} \tag{4.31}$$

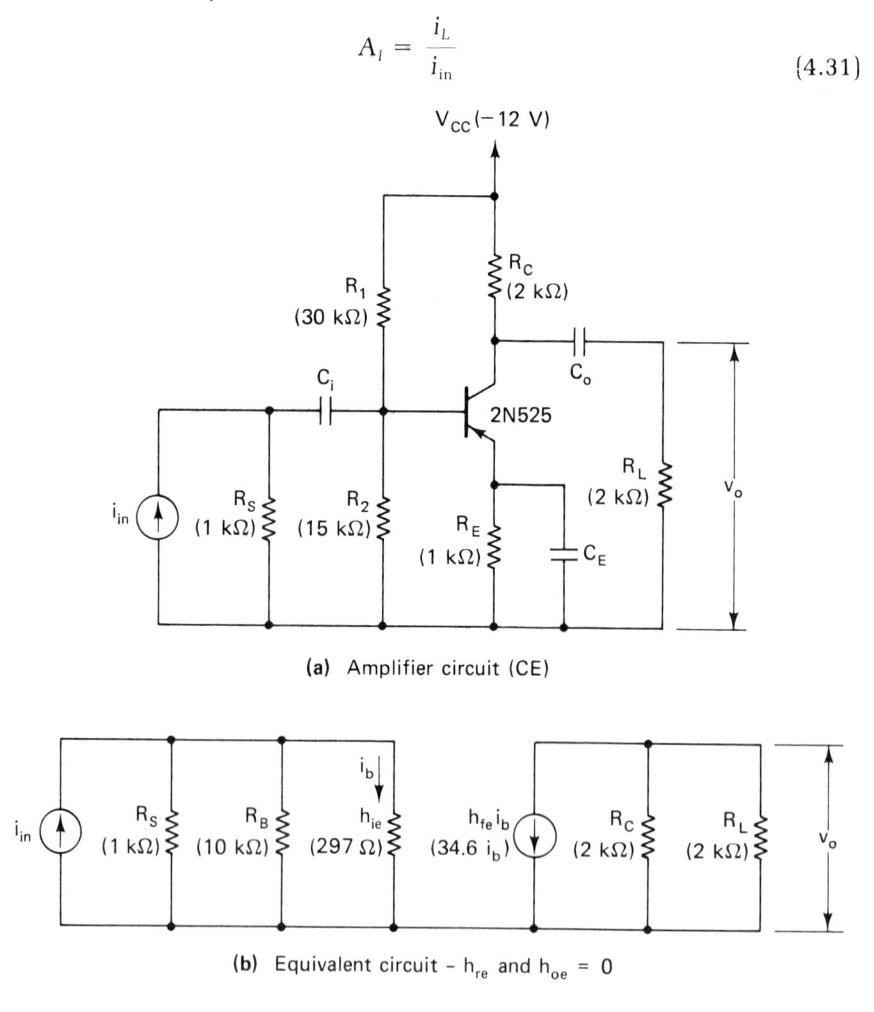

FIGURE 4.13 CE amplifier and its small-signal equivalent: (a) CE amplifier circuit; (b) equivalent circuit, h_{re} and $h_{oe} = 0$.

Multiplying by i_b/i_b and regrouping to divide the expression into an input and output expression gives

$$A_I = \frac{i_L}{i_{in}} \times \frac{i_b}{i_b} = \frac{i_b}{i_{in}} \times \frac{i_L}{i_b} \tag{4.32}$$

where i_b/i_{in} is the source to base expression and i_L/i_b, the collector to load expression. By the current-division equation, the ratio of the current (i_x) in a resistor (R_x) in a parallel circuit to the input current i_{in} is given by

$$\frac{i_x}{i_{in}} = \frac{1/R_x}{1/R_T} \tag{4.33}$$

Therefore,

$$A_I = \frac{1/h_{ie}}{1/R_S + 1/R_B + 1/h_{ie}} \times \frac{h_{fe\ op} \times 1/R_L}{1/R_C + 1/R_L} \tag{4.34}$$

Voltage Gain A_V:

$$A_V = \frac{v_L}{v_{in}} \tag{4.35}$$

where v_{in} is the open-circuit voltage of the source and not its loaded value. (This definition facilitates the comparison of voltage amplifiers since the user wishes to know the actual v_o obtainable for a given input voltage.) As $v_L = i_L R_L$ and $v_{in} = i_{in} R_5$,

$$A_V = \frac{i_L R_L}{I_{in} R_S} = A_I \frac{R_L}{R_S} \tag{4.36}$$

Power Gain A_P: The power gain is the product of the voltage and current gains, given by

$$A_P = A_V A_I \tag{4.37}$$

The input impedance (z_{in}) to the right of R_s is given by

$$z_{in} = h_{ie} || R_B = \frac{h_{ie} R_B}{h_{ie} + R_B} \tag{4.38}$$

where $R_B = R_1 R_2/(R_1 + R_2)$. The output impedance ($z_o$) to the left of R_L is given by

$$z_o = R_c || h_{oe} \approx R_C \tag{4.39}$$

Example 4.7 *Finding Gains and Input–Output Impedances for a CE Amplifier*

For the amplifier of Figure 4.13a operating at 25°C, find A_I, A_V, A_P, z_{in}, and z_o.

(a) *Find* $h_{fe\,op}$: Manufacturer's data for the 2N525, a 225-mW *pnp* transistor, give a min h_{fe} of 30 at $V_{CB} = -5$ V (since $V_{BE} = 0.7$ V, we may safely assume that $V_{CE} = V_{CB}$). Analysis of the circuit gives I_{EQ} ($I_{EQ} \approx I_{CQ}$) = 3 mA and $V_{CEQ} = -3$ V. Figure 4.12 gives

$$h_{fe\,op} = h_{fe} \times \text{(emitter current factor)} \times \text{(collector voltage factor)}$$

$$= 30 \times 1.2 \times 0.96 = 34.6$$

(b) *Find* h_{ie}:

$$h_{ie} = \frac{(h_{fe} + 1) \times 25 \times 10^{-3}}{I_{CQ}} = \frac{35.6 \times 25 \times 10^{-3}}{3\text{ mA}} = 297\,\Omega$$

(c) *Find gains:* The current gain is

$$A_I = \frac{1/h_{ie}}{1/R_S + 1/R_B + 1/h_{ie}} \times \frac{h_{fe\,op} \times 1/R_L}{1/R_C + 1/R_L}$$

$$= \frac{1/297\,\Omega}{1/1\text{ k}\Omega + 1/10\text{ k}\Omega + 1/297\,\Omega} \times \frac{34.6 \times 1/2\text{ k}\Omega}{1/2\text{ k}\Omega + 1/2\text{ k}\Omega} = 13.04$$

The voltage gain is

$$A_V = A_I\,\frac{R_L}{R_S} = 13.04 \times \frac{2\text{ k}\Omega}{1\text{ k}\Omega} = 26.08$$

The power gain is

$$A_p = A_I \times A_V = 13.04 \times 26.28 = 342.7$$

(d) *Find input–output impedances:*

$$z_{in} = h_{ie}||R_B = 297||10\text{ k}\Omega = 297$$

$$z_o = R_c = 2\text{ k}\Omega$$

4.6 SMALL-SIGNAL FREQUENCY ANALYSIS

The calculations of the preceding section do not apply at relatively high and low frequencies where the capacitors can no longer be treated as short or open circuits. As the signal frequency decreases, the input and output coupling capacitors and the emitter resistor bypass capacitor offer increasing impedance and, as a consequence, the low-end gain decreases. At the high end, the base–emitter diffusion capacity ($C_{b'e}$), the collector–base transition capacity ($C_{b'c}$), and the various stray capacities also decrease the gain (see Section 1.10). The lower cutoff frequency (f_L) is defined as the lowest frequency at which the mid-frequency voltage and current gains decrease by a factor of $1/\sqrt{2}$. The upper cutoff frequency

(f_H) is the highest frequency at which this same reduction in mid-band gain occurs. The bandwidth (B_f) is the difference and is given by

$$B_f = f_H - f_L \tag{4.40}$$

Figure 4.14 illustrates this behavior for the RC coupled amplifier.

4.6.1 Low-Frequency Response

The f_L of the simple high-pass filter shown in Figure 4.15 for the voltage transfer ratio v_o/v_{in} is given by

$$f_L = \frac{1}{2\pi R_T C} \tag{4.41}$$

where R_T is the total series resistance, $R_1 + R_2$. Therefore, to compute the f_L of the input and output circuits of the transistor due to the coupling capacitors, it is necessary to determine the total series R and to apply Eq. (4.41). Figure 4.16 shows how this is done using the Thevenin's to Norton's transformations.

In Figure 4.16, the input–output circuits of the amplifier are transformed into the circuit of Figure 4.15. The ratios v_o/v_{in} are proportionate to the ratios i_b/i_{in} and i_L/i_b, the gain factors. Therefore,

$$f_{L\,in} = \frac{1}{2\pi C_i\,(R_S + R_B||R_{in})} \tag{4.42}$$

$$f_{L\,out} = \frac{1}{2\pi C_o(R_C + R_L)} \tag{4.43}$$

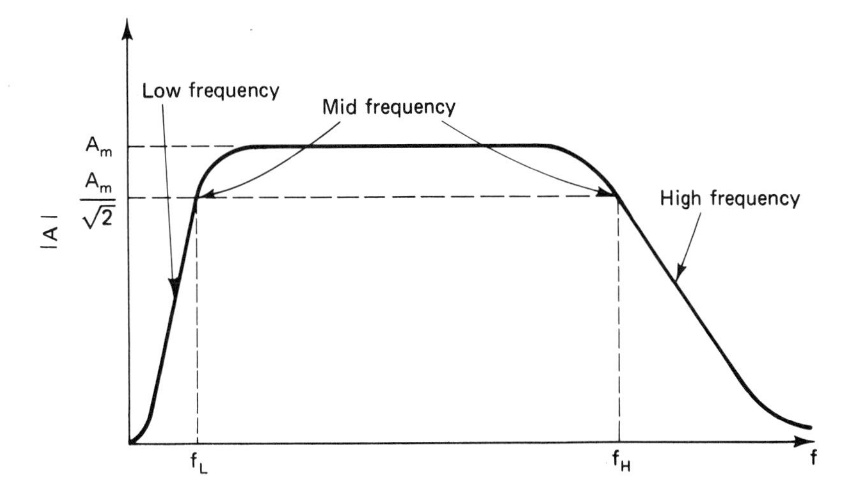

FIGURE 4.14 Gain versus frequency for the RC-coupled amplifier.

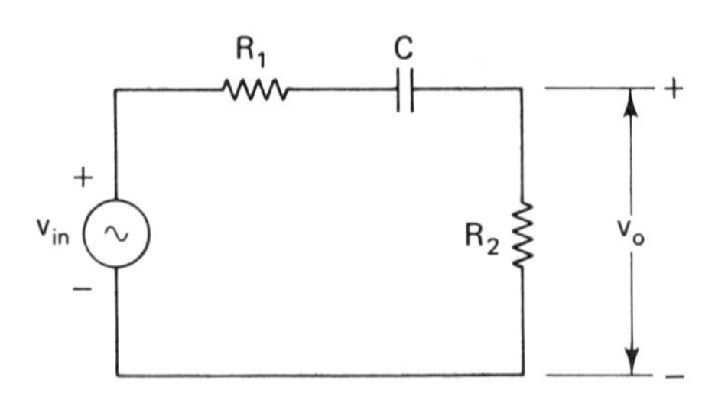

FIGURE 4.15 Simple series *RC* circuit.

When f_L is low enough so that R_E is effectively unbypassed, $R_{in} = (h_{fe} + 1)R_E$. However, in the worst case f_L is high enough (or C_E is big enough) so that R_E remains bypassed and $R_{in} = h_{ie}$. In this case, $R_S + R_B || h_{ie}$ (and $R_C + R_L$) are usually about 1 kΩ or more. In a design for a low f_L of 20 Hz, C_i and C_o would then be given by

$$C_o = C_i = \frac{1}{2\pi \times 20 \times 1\text{ k}\Omega} = 7.9\mu\text{F}$$

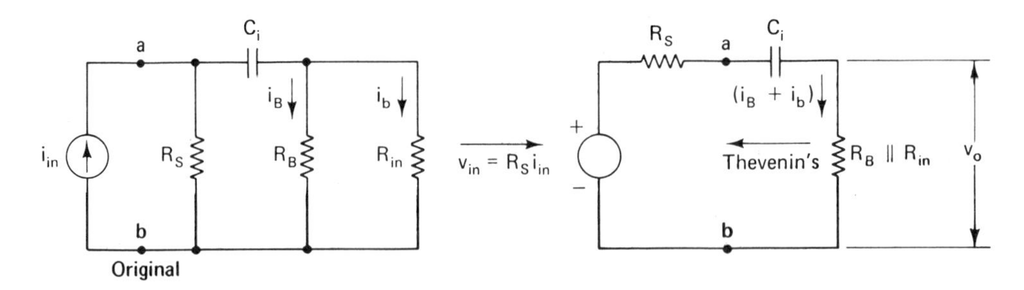

(a) Input circuit

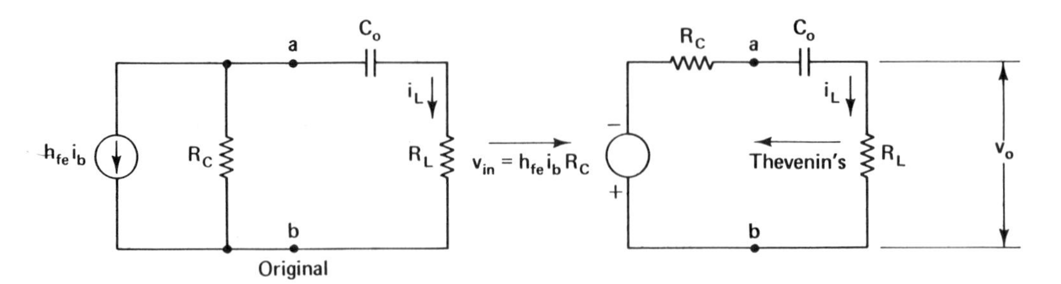

(b) Output circuit

FIGURE 4.16 Input–output transformations for the CE amplifier: (a) input circuit; (b) output circuit.

Since low-voltage electrolytic capacitors of this value (and larger) are readily available, C_i and C_o do not typically limit f_L.

Instead the low-frequency cutoff of the transistor amplifier is primarily determined by the emitter bypass capacitor (C_E), which must usually be much larger than C_i or C_o for the same f_L. Figure 4.17 shows the low-frequency small-signal input circuit for the CE amplifier. R_E and C_E have been included since both offer significant impedance at low frequencies near f_L, but C_i has been assumed large enough to act as a short. Since the effect of $(h_{fe} + 1)i_b$ flowing through R_E and C_E is to increase their apparent impedance (Section 4.3.1), R_E becomes $(h_{fe} + 1)R_E$ and C_E becomes $C_E/(h_{fe} + 1)$.

The parameter of interest in Figure 4.17 is i_b, since the output current and voltage are proportional to it. As the signal frequency decreases, the impedance of C_E increases, and the current flowing into branch b decreases. The cutoff frequency f_L is given by

$$f_L = \frac{1}{\dfrac{2\pi C_E}{h_{fe} + 1} \times \dfrac{(h_{fe} + 1)R_E R'}{(h_{fe} + 1)R_E + R'}} = \frac{1}{2\pi C_E \times \dfrac{R_E R'/(h_{fe} + 1}{R_E + R'/(h_{fe} + 1)}} \qquad (4.44)$$

where R' equals h_{ie} in series with R_B in parallel with R_S given by

$$R' = h_{ie} + \frac{R_B R_S}{R_B + R_S} \qquad (4.45)$$

The combination

$$\frac{R_E R'/(h_{fe} + 1)}{R_E + R'/(h_{fe} + 1)}$$

is the total resistance in parallel with C_E. Example 4.8 shows that large C_E's are required for low f_L's in the usual case because of the low value of this resistance.

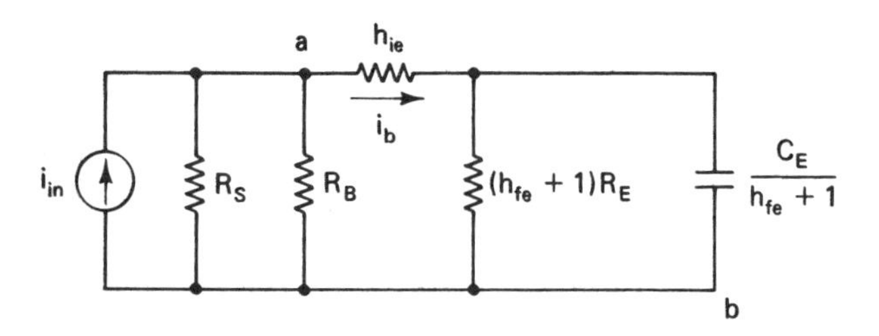

FIGURE 4.17 Circuit for finding f_L due to C_E.

Example 4.8 *Finding C_E for a Given f_L*

For a CE amplifier with R_E, R_S, and h_{ie} = 1 kΩ, R_B = 10 kΩ, and h_{fe} = 99, find C_E for f_L = 20 Hz (ignore C_i and C_o).

(a) *Find the total parallel resistance (R_T):*

$$R' = h_{ie} + \frac{R_S R_B}{R_S + R_B} = 1\text{k}\Omega + \frac{1\text{ k}\Omega \times 10\text{ k}\Omega}{1\text{ k}\Omega + 10\text{ k}\Omega} = 1.9\text{ k}\Omega$$

$$R_T = \frac{R_E R'/(h_{fe} + 1)}{R_E + R'/(h_{fe} + 1)} = \frac{1\text{ k}\Omega \times 1.9\text{ k}\Omega/(100)\,\Omega}{1\text{ k}\Omega + 1.9\text{ k}\Omega/(100)\,\Omega} = 18.6\,\Omega$$

(b) *Find C_E:*

$$C_E = \frac{1}{2\pi f_L R_T} = \frac{1}{2\pi \times 20 \times 18.6\,\Omega} = 427\,\mu\text{F}$$

4.6.2 High-Frequency Response of the CE Amplifier

The high-frequency response of the CE amplifier is limited mainly by the intrinsic transistor capacitances, $C_{b'e}$ and $C_{b'c}$, by the transistor transconductance, gm, and by the total parallel resistance in the input circuit. (Stray capacities may also become significant, but this is unusual and will be ignored.) Figure 4.18 illustrates the derivation of an equivalent circuit useful for finding f_H. Figure 4.18a shows a typical CE amplifier with the significant intrinsic transistor capacitances added. Figure 4.18b is its small-signal equivalent. This last circuit is similar to the mid-frequency equivalent circuit (Figure 4.13) except that $C_{b'e}$ and $C_{b'c}$ have been added, and $h_{fe}i_b$ has been replaced by a new term, $g_m v_{b'e}$, which is a current source, as is $h_{fe}i_b$, but it is dependent on v_{be}. Since $i_b = v_{be'}/h_{ie}$ and $g_m = h_{fe}/h_{ie}$, the two current sources are entirely equivalent. The circuits of Figure 4.18b and c are also equivalent since $C_{b'c}$ is connected between $v_{b'e}$ and v_{ce} in both. The input circuit of Figure 4.18c, which now contains all the circuit capacity, determines the f_H of the amplifier. It is redrawn in Figure 4.18d with the $C_{b'c}m - v_{ce}$ branch replaced by its equivalent Miller effect capacitance C_M.

Although $C_{b'c}$ (called C_{cbo}, C_{bc}) is typically small, C_M is often significant because of the Miller effect multiplication. This occurs because the voltage across $C_{b'c}$ is $v_{b'c} + v_{ce}$ instead of just $v_{b'e}$. For a sinusoidal input the current (I_c) flowing through the apparent capacitor, C_M, is given by

$$i_c = v_{b'e} \times J\omega C_M \tag{4.46}$$

In terms of $C_{b'c}$, I_c is given by

$$i_c = (v_{b'e} + v_{ce})J\omega C_{b'c} = (v_{b'e} + g_m R_o)J\omega C_{b'c} \tag{4.47}$$

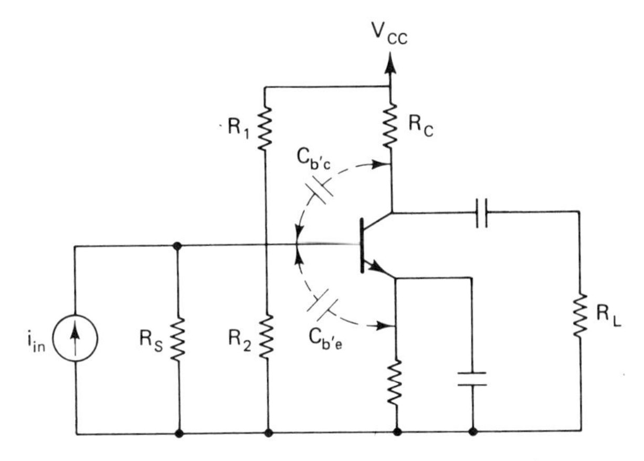

(a) Original circuit with intrinsic transistor capacitance

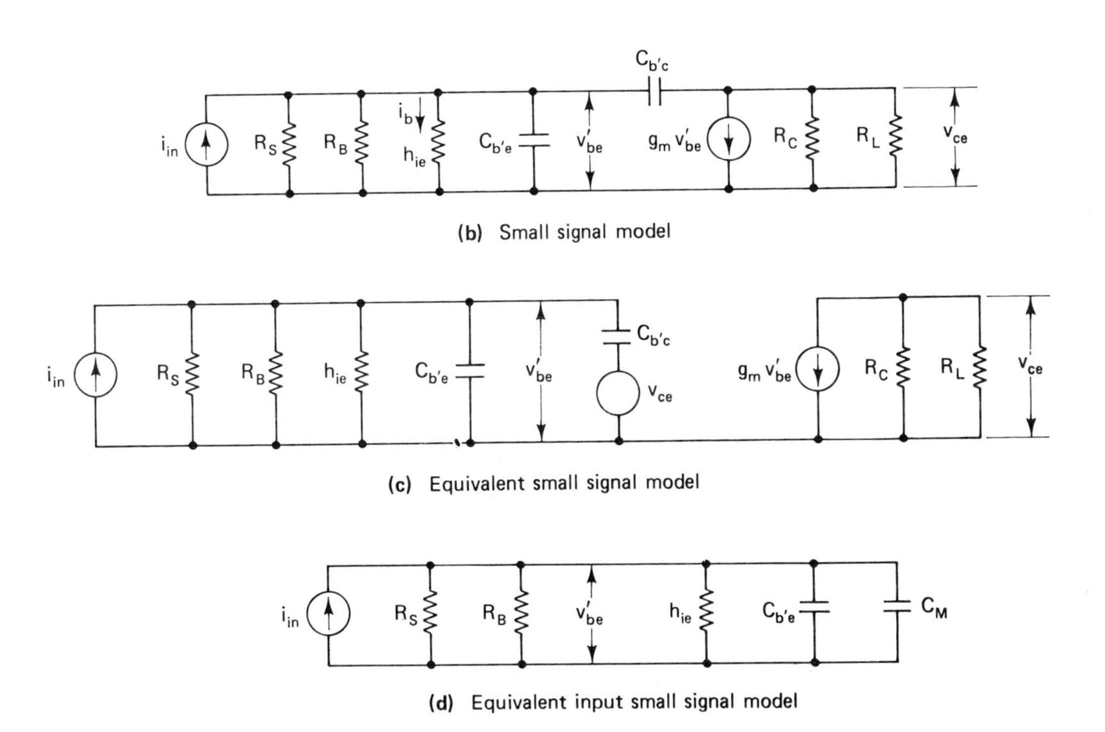

(b) Small signal model

(c) Equivalent small signal model

(d) Equivalent input small signal model

FIGURE 4.18 Small-signal, high-frequency model for the CE amplifier: (a) original circuit with intrinsic transistor capacitance; (b) small-signal model; (c) equivalent small-signal model; (d) equivalent input small-signal model.

where $R_o = R_c||R_L$. Equating Eqs. (4.46) and (4.47) and canceling the $J\omega$'s and $V_{b'e}$'s gives

$$C_M = (1 + g_m R_o)C_{b'c} \tag{4.48}$$

The output voltage and current for the amplifier of Figure 4.18 are proportional to $v_{b'e}$. Therefore, f_H for the transfer ratio $v_{b'e}/i_{in}$ is the f_H for the amplifier. This is given by

$$f_H = \frac{1}{2\pi R_T(C_{b'e} + C_M)} \tag{4.49}$$

where $R_T = R_s||R_B||h_{ie}$.

To compute f_H, R_T, $C_{b'e}$, $C_{b'c}$, and g_m are needed. R_T is computable and manufacturers supply $C_{b'e}$ (which they call C_{bc}, C_{cb}, C_{cbo}, etc.), but not $C_{b'e}$ or g_m. However, they supply a value for the frequency at which the unloaded CE current gain is 1. This is called f_T which is given by

$$f_T = \frac{g_m}{2\pi C_{b'e}} \tag{4.50}$$

Also equating $h_{fe}i_b$ to $g_m v_{b'e}$ since they represent the same current source (Figure 4.18c), we can show (problem 29) that

$$g_m = 40 I_{EQ} \tag{4.51}$$

Example 4.9 illustrates how the preceding equations are used to compute f_H.

Example 4.9 *Finding f_H*

In Figure 4.18b, let R_S and $R_E = 1$ kΩ, $R_B = 10$ kΩ, R_C and $R_L = 2$ kΩ, $I_{CQ} = 2.5$ mA, and $V_{CC} = 14$ V. Find f_H for a 2N3858A transistor operating at 25°C.

From a General Electric transistor list we find the following:

TABLE 4.2

Type No.	h_{FE} (V_{CE} = 4.5 V) (I_c = 2 mA)	C_{cbo} (pF) V_{CB} = 10 V	f_T MHz
2N3858A	60–120	2.5	135

(a) *Find h_{ie}, R_T, R_o, g_m, $C_{b'e}$, and C_M:* To find h_{ie} we need h_{fe}. The manufacturer supplies only h_{FE}. This may be used for h_{fe} since the two are close. H_{FE} is given at $V_{CE} = 4.5$ V and 2 mA. The 2.5 mA is close enough to 2 mA to ignore its effect on h_{fe}. And

$$V_{CE} = V_{CC} - I_{CQ}(R_C + R_E) = 14 - 2.5\,\text{mA} \times 4\,\text{k}\Omega = 4\,\text{V}$$

which is very close to 4.5 V. Therefore, the manufacturer's values for h_{fe} may be used. We choose the maximum value of h_{fe} since this gives the highest h_{ie} and the lowest f_H.

$$h_{ie} = \frac{121 \times 25 \times 10^{-3}}{2.5 \times 10^{-3}} = 1210\,\Omega \approx 1.2\,\text{k}\Omega$$

$$R_T = \frac{1}{1/R_S + 1/R_B + 1/h_{ie}} = \frac{1}{1/1\,\text{k}\Omega + 1/10\,\text{k}\Omega + 1/1.2\,\text{k}\Omega} = 517\,\Omega$$

$$R_o = \frac{R_C R_L}{R_C + R_L} = \frac{2\,\text{k}\Omega \times 2\,\text{k}\Omega}{4\,\text{k}\Omega} = 1\,\text{k}\Omega$$

$$g_m = 40 I_{EQ} = 40 I_{CQ} = 40 \times 2.5\,\text{mA} = 100\,\text{mS}$$

$$C_{b'e} = \frac{g_m}{2\pi \times f_T} = \frac{100 \times 10^{-3}}{2\pi \times 135\,\text{MHz}} = 117.9\,\text{pF}$$

To find C_M, we need to know $C_{b'c}$ at $V_{CE} = 4.5$ V. The manufacturer supplies $C_{b'c}$ (C_{cbo}) at $V_{CB} = 10$ V. We use the fact that $C_{b'c}$ is proportional to $1/V_{cb}^p$, where p is one-third for the grown junction (one-half for the alloyed junction). The 2N3858A is a grown junction transistor, and

$$V_{CB} = V_{CE} - 0.7 = 4.5 - 0.7 = 3.8\text{ V}$$

$$C_{b'c(\text{op}-3.8\text{V})} = C_{b'c(\text{op}-10\text{V})} \times \frac{V^p_{cb(\text{op}-10\text{V})}}{V_{cb\,(\text{op}-3.8\text{V})}} = 2.5\,\text{pF} \times \frac{10^{1/3}}{3.8^{1/3}} = 3.45\,\text{pF}$$

and

$$C_M = C_{b'c(\text{op})}(1 + g_m R_o) = 3.45\,\text{pF}(1 + 100\,\text{mS} \times 2\,\text{k}\Omega) = 693\,\text{pF}$$

(b) *Find f_h:*

$$f_h = \frac{1}{2\pi R_T(C_{b'e} + C_M)} = \frac{1}{2\pi \times 517\,\Omega(370\,\text{pF} + 693\,\text{pF})} = 289.6\,\text{kHz}$$

4.7 OTHER CONFIGURATIONS

The two other transistor amplifier configurations, the common collector (CC) and the common base (CB), are widely used for impedance transformation. The CC amplifier has a much higher input impedance and a much lower output impedance than the CE. It is therefore used to connect a device requiring a high-impedance load to one with a low-impedance input, an FET to a CE amplifier, for example. The CB has a much lower input impedance than the CE and a higher output impedance. Also, there is no phase reversal between input and output voltages for the CB. Figure 4.19 shows both configurations.

4.7.1 Biasing the CC and CB

The dc circuit of the CC amplifier (all capacitors open) is identical to a CE with $R_C = 0\,\Omega$. Therefore, the equations and the discussion of Section

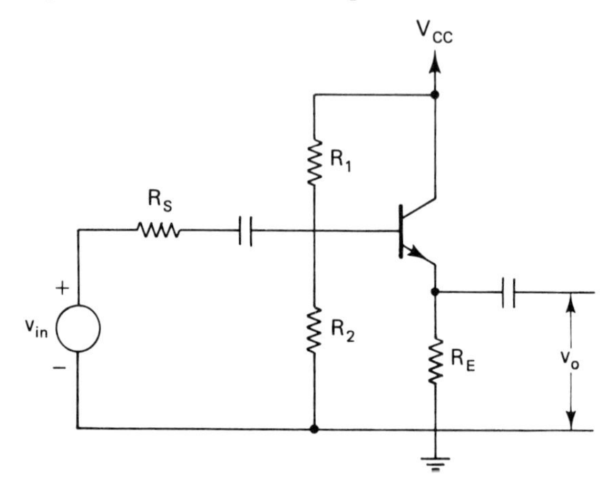

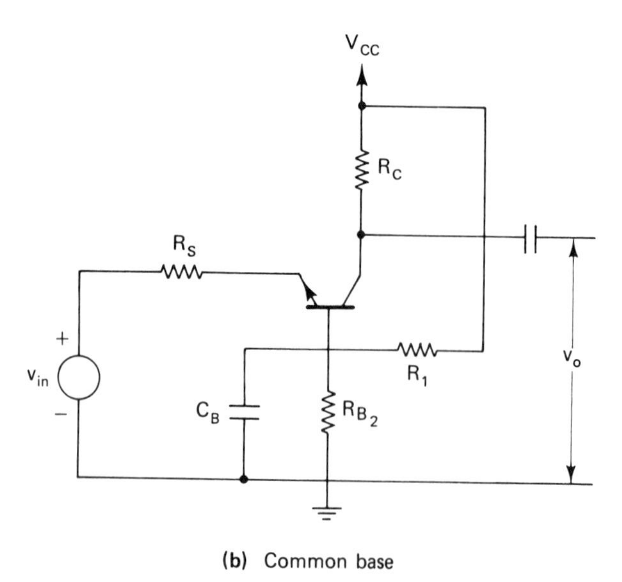

FIGURE 4.19 CC and CB amplifiers: (a) common collector; (b) common base.

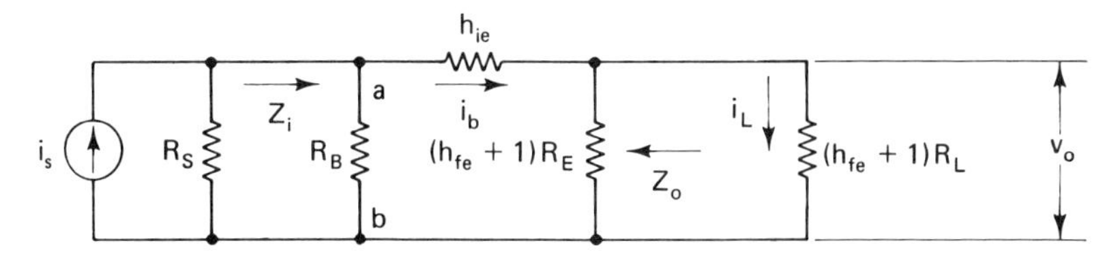

(a) CC circuit – current source

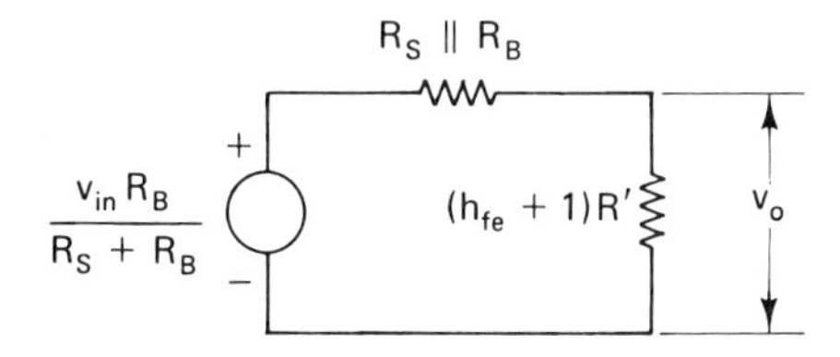

(b) Simplified CC circuit ($v_{in} = i_S R_S$, $R' = R_E \| R_L$, $h_{ie} = 0$)

FIGURE 4.20 CC small-signal circuit: (a) CC circuit-current source; (b) simplified CC circuit ($V_{in} = isRs$, $R' = R_E||R_L$, $h_{ie} = 0$).

4.4 apply without change to biasing the CC. The DC circuit of Figure 4.19b, the CB amplifier (all capacitors open and V_{in} set to 0), also reduces to the CE circuit with R_E replaced by R_S (problem 31). Therefore, Section 4.4 applies to the CB as well.

4.7.2 CC Small-Signal Circuit

Figure 4.20 shows two useful small-signal circuits for the CC amplifier. Figure 4.20a is identical to the input circuit for the CE amplifier with unbypassed R_E except that the load (R_L) is in parallel with R_E. Since R_L is part of the input circuit, the collector circuit is not required. Figure 4.20b is derived from 4.20a by ignoring h_{ie} as it is in series with the much larger $(h_{fe} + 1)R'$ ($R' = R_E||R_L$) and by carrying out successive Thevenin's simplifications of the section to the left of a–b.

Using Figure 4.20b, we see that the voltage gain referred to the open-circuit source voltage V_{in} is given by

$$A_V = \frac{v_o}{v_{in}} = \frac{R_B}{R_S + R_B} \times \frac{(h_{fe} + 1)R'}{\dfrac{R_S R_B}{R_S + R_B} + (h_{fe} + 1)R'} \tag{4.52}$$

When, as is often the case, $R_S||R_B$ is considerably smaller than $(h_{fe} + 1)R'$, the equation reduces to

$$A_V = \frac{R_B}{R_S + R_B} \tag{4.53}$$

which is always less than 1.

The current gain is given by

$$A_I = \frac{i_L}{i_{in}} = \frac{v_{in}A_V/R_L}{v_{in}/R_S} = A_V \frac{R_S}{R_L} \tag{4.54}$$

where i_{in} is the short-circuit source current. A_I, unlike A_V, can exceed 1.

The input impedance (z_i) to the right of R_s is given by

$$z_i = R_B||(h_{fe} + 1)R' \approx R_B \tag{4.55}$$

when $R_B << (h_{fe} + 1)R'$.

To find z_o for the CC, h_{ie} must be reinserted into Figure 4.20b, and it must be transformed into a circuit containing R_E rather than $(h_{fe} + 1)R_E$. This can be accomplished as follows: Write the Kirchhoff voltage loop equation for Figure 4.20b, reinserting h_{ie} and letting $R' = R_E$.

$$\frac{v_{in}R_B}{R_S + R_B} = i_b(R_S||R_B + h_{ie}) + i_b(h_{fe} + 1)R_E \tag{4.56}$$

Replace i_b by $i_e/(h_{fe} + 1)$:

$$\frac{v_{in}R_B}{R_S + R_B} = i_e \frac{(R_S||R_B + h_{ie})}{h_{fe} + 1} + i_eR_E \tag{4.57}$$

This transformed circuit, shown in Figure 4.21, is the input circuit as seen from the emitter (or referred to the emitter). The output impedance z_o looking into R_E is seen to be

$$z_o = R_E|| \frac{(h_{ie} + R_S||R_B)}{h_{fe} + 1} \tag{4.58}$$

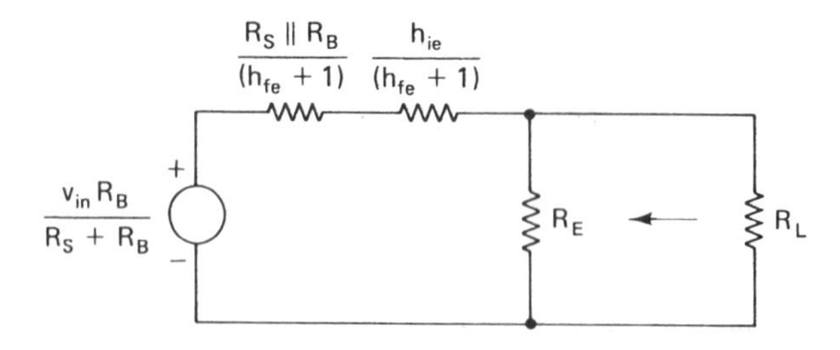

FIGURE 4.21 CC small-signal input circuit referred to the emitter.

Example 4.10 *Design of a CC Impedance Transformer*

Design a CC impedance transformer (buffer) for audio frequencies. z_i must be $\geq$ 10 kΩ and $z_o \leq 100$. Specify maximum R_s.

We choose a general-purpose small-signal *NPN* silicon transistor, 2N2711, from a table of transistors with each entry similar to Table 4.2. For the 2N2711, $h_{FE} = 30$ to 90 at $V_{CE} = 5$ V, $I_c = 2$ mA in an ambient temperature of $T_A = 25°$C. To keep the design procedure simple, we design for the given conditions. Also, we use $h_{fe} = 30$ as this gives the highest z_o and has insignificant effect on z_i.

(a) *Establishing bias:* Since $R_B = 0.1h_{FE}R_E$ (for good β stability) and $z_i = R_B = 10$ kΩ,

$$R_E = \frac{R_B}{0.1h_{fe}} = \frac{10 \text{ k}\Omega}{0.1 \times 30} = 3.3 \text{ k}\Omega$$

For $I_{CQ} = 2$ mA, $V_{CE} = 5$ V with $R_E = 3.3$ kΩ:

$$V_{CC} = I_{CQ}R_E + V_{CE} = 2 \text{ mA} \times 3.3 \text{ k}\Omega + 5 \text{ V} \approx 11 \text{ V}$$

R_1 and R_2, the bias resistances, are calculated as in Section 4.4 for maximum undistorted output.

(b) *Check z_i:*

$$z_i = R_B||[(h_{fe} + 1)R_E] = 10 \text{ k}\Omega \; || \; [(30 + 1)3.3 \text{ k}\Omega] \approx 10 \text{ k}\Omega$$

(Note that if the R_L across R_E is not very much larger than R_E then $R_E' = R_E||R_L$ must replace R_E.)

(c) *Find the maximum R_S for $z_o = 100$:*

$$z_o = R_E|| \frac{(h_{ie} + R_S||R_B)}{h_{fe} + 1} \approx \frac{h_{ie}}{h_{fe} + 1} + \frac{R_S||R_B}{h_{fe} + 1}$$

since, as will be seen, $R_E >> (h_{ie} + R_S||R_B)/(h_{fe} + 1)$.

$$\frac{h_{ie}}{h_{fe} + 1} = \frac{25 \times 10^{-3}}{I_{CQ}} = \frac{25 \times 10^{-3}}{2 \text{ mA}} = 12.5$$

Then for $z_o \geq 100$,

$$z_o = 100 = 12.5 + \frac{R_S||10 \text{ k}\Omega}{31}$$

Solving for R_S we get

$$R_{S\max} = 3.72 \text{ k}\Omega$$

4.7.3 CB Small-Signal Circuit

Figure 4.22 shows a CB small-signal circuit simplified by the omission of h_{rb} (usually ≈ 0) and h_{ob} ($1/h_{ob}$ usually > 1 MΩ). The circuit is similar to the CE's. It differs only in the designation of some parameters and the absence of R_B. The parameters h_{ib} and i_e replace h_{ie} and i_b in the input circuit, and h_{fb} replaces h_{fe} in the output. The two parameters h_{ib} and h_{fb} are defined by

$$h_{ib} = \frac{h_{ie}}{h_{fe} + 1} = \frac{25 \times 10^{-3}}{I_{CQ}} \tag{4.59}$$

$$h_{fb} = \frac{h_{fe}}{h_{fe} + 1} = \alpha \tag{4.60}$$

The gain and impedance equations for the CB can be obtained directly from those of the CC by making the appropriate substitutions.

$$A_I = \frac{i_L}{i_{in}} = \frac{i_e}{i_{in}} \times \frac{i_L}{i_e} = \frac{1/h_{ib}}{1/R_S + 1/h_{ib}} \times \frac{h_{fb} \times 1/R_L}{1/R_C + 1/R_L} \tag{4.61}$$

(since $h_{fb} < 1$, A_I is always less than 1).

$$A_V = \frac{v_L}{v_{in}} = \frac{i_L R_L}{i_{in} R_S} = A_I \frac{R_L}{R_S} \tag{4.62}$$

(A_V can exceed 1).

$$z_i \text{ (Figure 4.22)} = h_{ib} \tag{4.63}$$

$$z_o \text{ (Figure 4.22)} = R_C \tag{4.64}$$

The design of a CB low-input, high-output impedance transformer is similar to the CC design. The I_{CQ} for a specified z_{in} (h_{ib}) is computed, R_c is set equal to z_o, and V_{CC} and R_E are computed to give I_{CQ}. (If one is fixed, the other is computed.) This problem is left for the reader (problem 34).

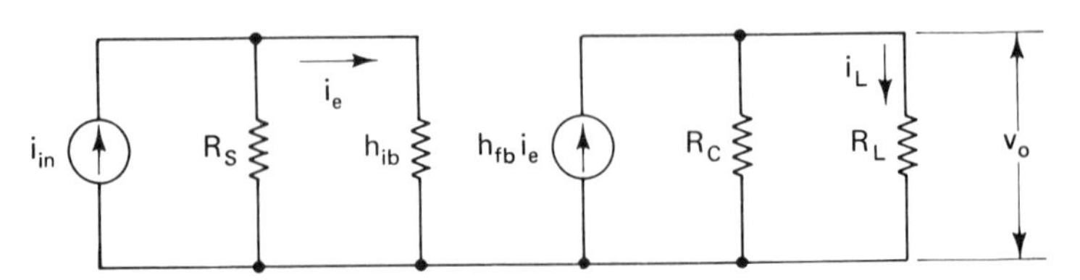

FIGURE 4.22 Small-signal circuit for the CB.

4.8 TRANSISTOR SPECIFICATIONS

Transistor specifications are available as (1) comprehensive listings covering many (sometimes all) transistors, providing only the salient characteristics, and (2) a detailed extensive description of a single closely related family.

Figure 4.23 is an example of a comprehensive listing. The h_{FE} range (also assumed to be h_{fe} in the absence of other specification) is given. V_{CEO}, the collector–emitter breakdown voltage under the most stringent condition, base floating, follows. Data pertinent to switching applications V_{CE} (S_{AT}) at a specified I_c and I_B are next. The maximum power dissipation (P_T) at 25°C, given typically by $V_{CEQ} \times I_{CQ}$, is given. As in the diode case, P_T must be reduced as the ambient temperature (T_A) increases. Data required for computing f_H, that is, C_{cbo} ($C_{b'c}$) and f_T, follow. Comments describing the transistor's category and typical applications complete the summary. The concise comprehensive listing is useful for transistor selection and often sufficient for design purposes as well.

An example of the detailed data provided for a family of transistors is given by Figure 4.24. The case designation, application note, and maximum rating sections are similar to those for diodes, except that maximum ratings for collector–base, collector–emitter, and emitter–base are included. Also, data needed for power derating are given. Maximum ratings should never be exceeded or even approached too closely.

The following section, "Electrical Characteristics," gives the value of the transistor parameters under stated conditions. The section labeled "Cutoff Characteristics" gives the various breakdown voltages. The O in BV_{CBO}, BV_{CEO}, and BV_{EBO} means the unnamed element is floating. Thus BV_{CBO} applies for open emitter, BV_{CEO} for open base, and BV_{EBO} for open collector. The "On Characteristics" comprise h_{FE} at various I_C's and V_{CE}'s and two saturation voltages, $V_{CE(SAT)}$ and $V_{BE(SAT')}$ pertinent to switching applications.

The two small-signal characteristics not yet discussed are the high-frequency current gain, $|h_{fe}|$ and the noise figure, NF. The high-frequency current gain is related to f_T. In fact, f_T, which is called the gain-bandwidth product, is the frequency at which $|h_{fe}|$ equals 1. Sometimes f_{hfe} or f_{hfb} are given in place of $|h_{fe}|$ or f_T. Known as the beta cutoff frequency, f_{hfe} is the common-emitter operating frequency at which the current gain falls to 70.7% of its mid-frequency value and f_{hfb} (the alpha cutoff frequency) is defined in the same way in terms of the CB circuit. The relationship between f_T and h_{fe} is given by

$$f_T = h_{feo} \times h_{hfe} \tag{4.65}$$

General Purpose NPN Amplifiers and Switches — Epitaxial (D32C, D32D, D32B, D16B, D16N, D16E, D16X, D16D, D16C, D33E Product Lines)

Type	h_{FE} $V_{CE}=4.5V$ $I_C=2$ mA	V_{CEO} $I_C=1$ mA Volts	$V_{CE(SAT)}$ $I_C=10$ mA $I_B=1$ mA Volts max.	P_T @ $T_A=25°C$ mW	C_{obo} $V_{CB}=10V$ Typical pF	f_t Typical mHz	Comments
2N3858A (D16E)	60-120						
2N3859A (D16E)	100-200	60	0.125	360	2.5	135	For higher voltage applications, 2 to 1 h_{FE} spreads.
2N3860A (D16E)	150-300						
2N3877 (D16X)	20 Min.	70	0.125	360	2.5	135	Especially suited for driving high voltage indicating devices.
2N3877A (D16X)	20 Min.	85					
2N5174 (D16D)	40-600(7)	75					
2N5175 (D16D)	55-160(7)	100	0.95	360	2.5	135	Especially suited for driving high voltage indicating devices.
2N5176 (D16D)	• 140-300(7)						
2N4256 (D16E)	100-500	$V_{CES}=30$	0.125	360	2.5	135	High h_{FE}, low saturation voltage with guaranteed Q_{SR}.
2N3973 (D16C)	35-100(8)						Low $V_{CE(SAT)}$, good h_{FE} linearity with current. Turn-on time is 60nsec max.; Turn-off is 110nsec max.(10)
2N3974 (D16C)	55-200(8)						Like2N3973except turn-off time is 200nsec max.(10)
2N3975 (D16C)	35-100(8)	30	0.3(9)	360	5.2	350	Like2N3974except turn-off time is 250nsec max(10)
2N3976 (D16C)	55-200(8)						
2N4951 (D16C)	60-200(4)						$t_{on}=40$nsec., $t_{off}=350$nsec(11)
2N4952 (D16C)	100-300(4)						$t_{on}=40$nsec., $t_{off}=350$nsec(11)
2N4953 (D16C)	200-600(4)	30 @ $I_C=10$mA	0.3(9)	360	8.0 max.	350	$t_{on}=40$nsec., $t_{off}=400$nsec(11)
2N4954 (D16C)	60-600(4)						$t_{on}=40$nsec., $t_{off}=400$nsec(11)
2N3605 (D33E)	30 Min.(8)	14					
2N3605A (D33E)	30-120(8)	15					
2N3606 (D33E)	30 Min.(8)	14	0.25	360	6.0	300	Low storage times:(12) $t_s=20$nsec. (2N3605, A) $t_s=35$nsec. (2N3606, A) $t_s=45$nsec (2N3607)
2N3606A (D33E)	30-120(8)	15					
2N3607 (D33E)	30 Min.(8)	14					

NOTES:
(1) h_{FE} @ $V_{CE}=5V$, $I_C=10\mu A$.
(2) h_{FE} @ $V_{CE}=5V$, $I_C=150mA$.
(3) $V_{CE(SAT)}$ @ $I_C=150mA$, $I_B=15mA$.
(4) h_{FE} @ $V_{CE}=10V$, $I_C=150mA$.
(5) Max. switching times @ $I_C=150mA$, $I_{B1}=I_{B2}=15mA$.
(6) $V_{CE(SAT)}$ @ $I_C=50mA$, $I_B=3mA$.
(7) h_{FE} @ $V_{CE}=5V$, $I_C=10mA$.
(8) h_{FE} @ $V_{CE}=1V$, $I_C=10mA$.
(9) $V_{CE(SAT)}$ @ $I_C=150mA$, $I_B=15mA$.
(10) Switching times @ $I_C=100mA$, $I_{B1}=I_{B2}=10mA$.
(11) Max. switching times @ $I_C=150mA$, $I_{B1}=I_{B2}=15mA$.
(12) Storage time @ $I_C=10mA$, $I_{B1}=3mA$, $I_{B2}=-1mA$.

FIGURE 4.23 Example of a comprehensive transistor listing. Courtesy of General Electric Semiconductor Products Department.

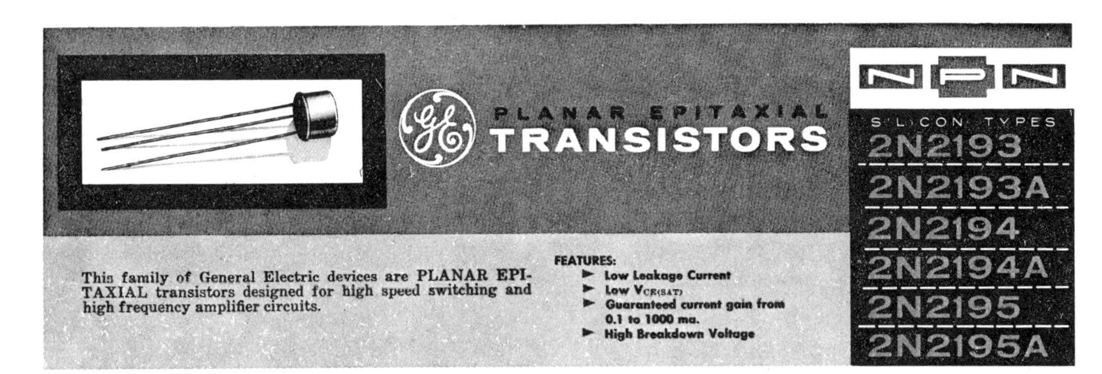

absolute maximum ratings (25°C unless otherwise specified)

		2N2193 2N2193A	2N2194 2N2194A	2N2195 2N2195A	
Voltage					
Collector to Base	V_{CBO}	80	60	45	volts
Collector to Emitter	V_{CEO}	50	40	25	volts
Emitter to Base	V_{EBO}	8	5	5	volts
Current					
Collector	I_C	1.0	1.0	1.0	amp
Transistor Dissipation					
(Free Air 25°C) *	P_T	0.8	0.8		watts
(Free Air 25°C) **	P_T			0.6	watts
(Case Temperature 25°C) ***	P_T	2.8	2.8	2.8	watts
(Case Temperature 100°C) ***	P_T	1.6	1.6	1.6	watts
Temperature					
Storage	T_{STG}	←	−65 to +300	→	°C
Operating Junction	T_J	←	−65 to +200	→	°C

*Derate 4.6 mw/°C increase in ambient temperature above 25°C
**Derate 3.4 mw/°C increase in ambient temperature above 25°C
***Derate 16.0 mw/°C increase in case temperature above 25°C

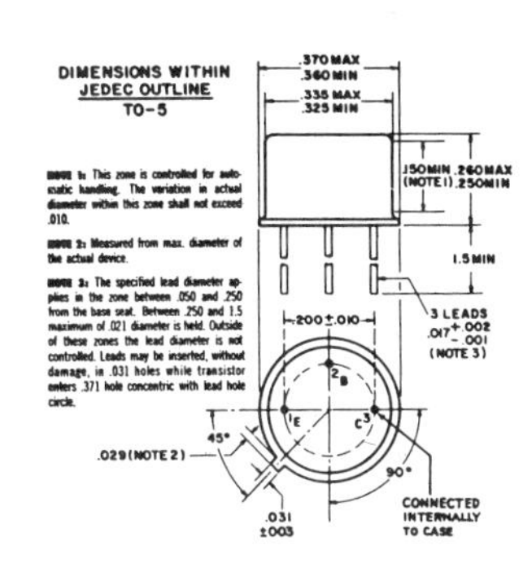

electrical characteristics: (25C° unless otherwise specified)

		2N2193 2N2193A Min.	Max.	2N2194 2N2194A Min.	Max.	2N2195 2N2195A Min.	Max.	
DC CHARACTERISTICS								
▶ **Collector to Base Voltage** ($I_C = 100\ \mu a$)	V_{CBO}	80		60		45		volts
▶ **Collector to Emitter Voltage** ($I_C = 25$ ma)†	V_{CEO}	50		40		25		volts
Emitter to Base Voltage ($I_E = 100\ \mu a$)	V_{EBO}	8		5		5		volts
Forward Current Transfer Ratio								
($I_C = 150$ ma, $V_{CE} = 10$ V)†	h_{FE}	40	120	20	60	20		
($I_C = 10$ ma, $V_{CE} = 10$ V)	h_{FE}	30		15				
($I_C = 1000$ ma, $V_{CE} = 10$ V)†	h_{FE}	15						
($I_C = 0.1$ ma, $V_{CE} = 10$ V)	h_{FE}	15						
($I_C = 500$ ma, $V_{CE} = 10$ V)†	h_{FE}	20		12				
($I_C = 10$ ma, $V_{CE} = 10$ V, $T_A = -55$°C)	h_{FE}	20						
Base Saturation Voltage ($I_C = 150$ ma, $I_B = 15$ ma)	$V_{BE(SAT)}$		1.3		1.3		1.3	volts
▶ **Collector Saturation Voltage** ($I_C = 150$ ma, $I_B = 15$ ma)	$V_{CE(SAT)}$	(0.35 volts max., 2N2193, 94, 95 only)						
	$V_{CE(SAT)}$	(0.25 volts max., 2N2193A, 94A, 95A only)						
	$V_{CE(SAT)}$	(0.16 volts typ., 2N2193A, 94A, 95A only)						
CUTOFF CHARACTERISTICS								
▶ **Collector Leakage Current** ($V_{CB} = 30$ V)	I_{CBO}				10		100	mμa
($V_{CB} = 30$ V, $T_A = 150$°C)	I_{CBO}				25		50	μa
($V_{CB} = 60$ V)	I_{CBO}		10					mμa
($V_{CB} = 60$ V, $T_A = 150$°C)	I_{CBO}		25					μa
Emitter Base Cutoff Current ($V_{EB} = 5$ V)	I_{EBO}		50					mμa
▶ **Emitter Base Leakage Current** ($V_{EB} = 3$ V)	I_{EBO}				50		100	mμa
HIGH FREQUENCY CHARACTERISTICS								
Current Transfer Ratio ($I_C = 50$ ma, $V_{CE} = 10$ V, f = 20 mc)	h_{fe}	2.5		2.5		2.5		
Collector Capacitance ($I_E = 0$, $V_{CB} = 10$ V, f = 1 mc)	C_{ob}		20		20		20	pf
SWITCHING CHARACTERISTICS (See Figure 1) ($V_{IN} = 15$V, $V_L = 15$V)								
Rise Time	t_r		70		70			nsec
Storage Time	t_s		150		150			nsec
Fall Time	t_f		50		50			nsec

†Pulse width ≤300 μsec, duty cycle ≤2%

GENERAL ELECTRIC

FIGURE 4.24 Complete transistor data sheet. Courtesy of General Electric.

where h_{feo} is the mid-frequency h_{fe}. The alpha cutoff frequency (f_{hfb}) is close to and usually somewhat above f_T. The noise figure NF is the ratio of the total noise output by the device to the total noise input. It is expressed in decibels (dB) and defines the amount of noise added by the transistor.

Under "Switching Characteristics," data pertinent to transistor switches are given. Rise time (t_r) is a measure of the maximum rate at which the output can rise. Storage time (t_s) is a measure of the time required to turn the transistor off, and fall time (t_f) defines the maximum rate at which the output can fall. These characteristics will be studied later.

Problems

1. A transistor has an $\alpha = 0.98$. If $I_B = 0.02$ mA, compute I_E and I_C (assume I_{CB} is negligible).
2. Letting $I_{CB} = 0$ and using Eqs. (4.1), (4.2), and (4.6), prove that $\beta = \alpha/(1 - \alpha)$.
3. Show that $I_E = (\beta + 1)I_B$.
4. For Figure 4.2a, let $V_{CC} = 12$ V, $R_E = 1$ kΩ, $R_1 = 20$ kΩ, $R_2 = 10$ kΩ, and $\beta = 49$. Find I_B.
5. Prove that the circuit of Figure 4.3c is equivalent to the input circuit of Figure 4.3b.
6. Using the approximate characteristic of Figure 4.4a, determine β.
7. In Figure 4.4b, let $\beta = 100$, $R_C = 2$ kΩ, $V_{CC} = 12$ V, $R_E = 1$ kΩ, and $I_B = 0.01$ mA. Find I_C, V_R, and V_{CE}.
8. If for Figure 4.4b, $V_{CC} = 9$ V, $R_C = R_E = 1$ kΩ, $V_{RC} = 4$ V, and $\beta = 50$, find V_{CE} and I_B.
9. In Figure 4.5a, let $R_1 = 40$ kΩ, $R_2 = 10$ kΩ, $V_{CC} = 12$ V, $R_C = 2$ kΩ, $R_E = 1$ kΩ, and $\beta = 100$. Find I_B, I_E, I_C, and V_{CE}.
10. In Figure 4.5b, let $V_{BB} = 4$ V, $R_B = 10$ kΩ, $V_B = 0.7$ V, $\beta = 100$, $R_E = 1$ kΩ, and $R_C = 3$ kΩ. Find I_B, I_E, I_C, V_{CE}, and V_R for $V_{CC} = 12$ V.
11. Derive Eqs. (4.12) and (4.13).
12. In the circuit of Figure 4.5a, $V_{CC} = 9$ V, $R_E = 1$ kΩ, and $I_{CQ} = 1$ mA. Find R_1 and R_2 for $\beta = 100$. Check I_{CQ} in this circuit for $\beta = 50$, 150, 200, and 400.

13. Repeat problem 12 for $I_{CQ} = 4$ mA. Find the power dissipated in R_1 and R_2.

14. Repeat problem 12, but instead of using $R_B = 0.1(\beta + 1)R_E$, let $R_B = (\beta + 1)R_E$. Comment on changes in I_{CQ} versus β.

15. In the circuit of Figure 4.8a, let $R_C = 4$ kΩ, $R_E = 2$ kΩ, and $V_{CC} = 16$ V. Find I_{CQ} for maximum undistorted output. Also find V_{CEQ} and $I_{C\,max}$.

16. On an appropriate set of axes (ordinate I_C, abscissa V_{CE}) draw load lines for the following conditions: (a) $V_{CC} = 12$ V, $R_C = 1$ kΩ, 2 kΩ, 4 kΩ; (b) $V_{CC} = 9$ V, $R_C = 1$ kΩ, 2 kΩ, 4 kΩ.

17. In the circuit of Figure 4.8a, let $R_C = 10$ kΩ, $R_E = 4$ kΩ, and $V_{CC} = 12$ V. Find I_{CQ}, V_{CEQ}, and $I_{C\,max}$ for maximum undistorted output.

18. Find R_{ac} and R_{dc} for the circuits of Figure 4.25.

19. Find I_{CQ} for maximum undistorted output for the circuits of Figure 4.25.

20. Verify Eq. (4.25).

21. The circuit of Figure 4.9a uses a 2N2193 transistor and operates from $-20°$ to $+65°$C. For $V_{CC} = 9$ V, find R_1 and R_2 such that I_{CQ} is between 0.2 and 0.4 mA. Analyze the circuit to verify the design.

22. Repeat problem 21 for an I_{CQ} between 3 and 3.5 mA.

23. In problem 19, I_{CQ} was found for the circuits of Figure 4.25. Using this I_{CQ}, find R_1 and R_2 using the method of Section 4.4.2. Let $T_A = 0°$ to $50°$C, $V_{CC} = 12$ V, and use a 2N2193 transistor.

24. For the circuit of Figure 4.13a, let $R_C = 4$ kΩ, $R_L = 1$ kΩ, $R_S = 1$ kΩ, $V_{CC} = -12$ V, and $R_E = 1$ kΩ. Find I_{CQ}, V_{CEQ}, R_1, and R_2 for maximum undistorted output, and find A_V, A_I, A_p, z_i, and z_o. Use a 2N2195 transistor.

25. Repeat problem 24 using a 2N3858A (assume that Figure 4.12 applies with negative abscissas changed to positive values).

26. In the circuit of Figure 4.13a, assume that R_E is shorted by C_E to 0 Hz and calculate C_i and C_o for $f_L = 20$ Hz.

27. Repeat problem 26 with C_E removed and $\beta = 99$.

28. For the circuit of Figure 4.13a, assuming that C_i and C_o are so large that they do not affect f_L, find C_E for an f_L of 30 Hz.

29. Show that $g_m = 40\, I_{EQ}$.

30. In Figure 4.18a, R_S and $R_E = 1.5$ kΩ, $R_C = 4$ kΩ, $R_L = 2$ kΩ, and $V_{CC} = 12$ V. The circuit uses a 2N3973 transistor, and I_{CQ} is chosen for maximum undistorted output at 25°C. Find f_H.

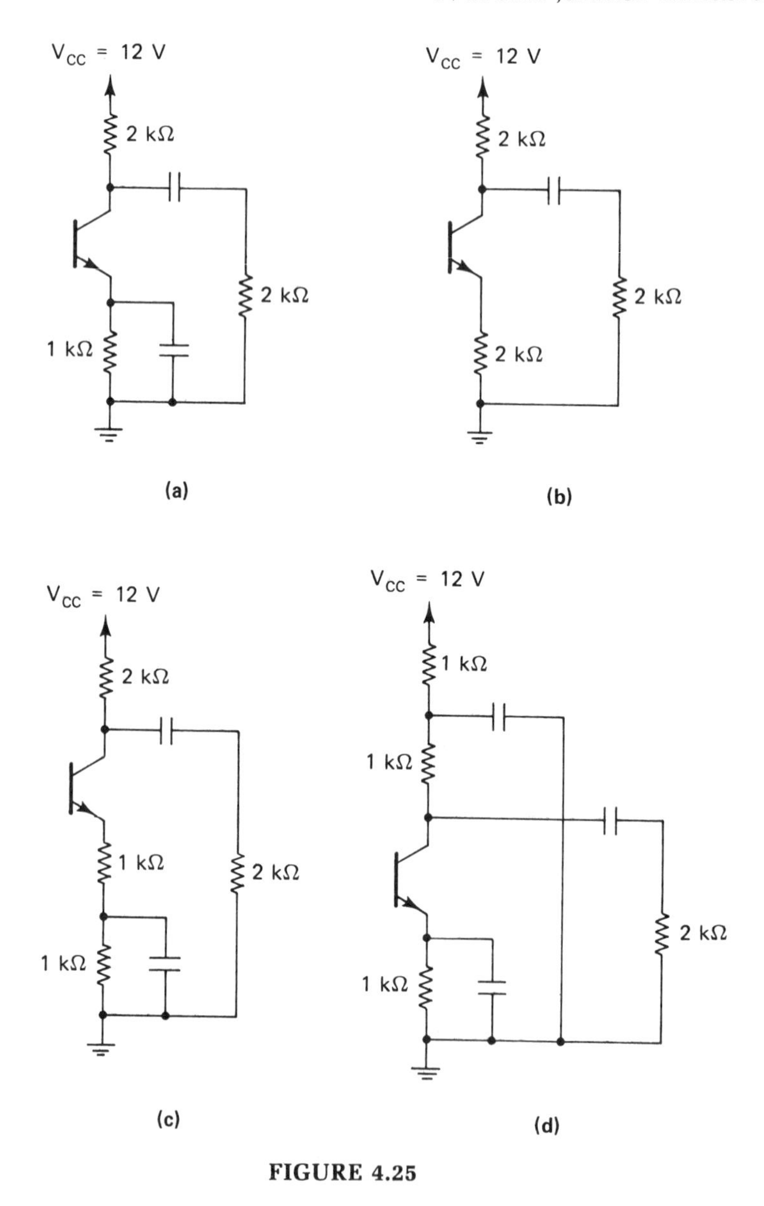

FIGURE 4.25

31. Show that the DC circuit of the CB in Figure 4.19b is identical to the CE circuit (Figure 4.19a) with R_E replaced by R_S.

32. Design a CC buffer for audio frequencies using a 2N2711. $z_i \geq 20$ kΩ and $z_o \leq 150$ Ω. Also find A_V and A_I.

33. Repeat problem 32, choosing the most appropriate transistor from Figure 4.23. Let $z_i \geq 30$ kΩ and $z_o \leq 50$ Ω. Explain your selection.

34. Design a CB circuit using a 2N2711 with $z_i \leq 30$ Ω, $z_o \geq 5$ kΩ, $R_S = 500$Ω, and $R_L = 500$. Also find A_V and A_I.

References

Bell, David A. *Electronic Devices and Circuits,* 2nd ed., Reston Publishing Co., Reston, Virginia, 1980.

G. E. Semiconductor Products Department, *G. E. Transistor Manual,* lightweight edition, Electronic Park, Syracuse, New York, 1969.

Schilling, D. L., and C. Belove. *Electronic Circuits: Discrete and Integrated,* 2nd ed., McGraw-Hill Book Co., New York, 1979.

5

FIELD-EFFECT TRANSISTOR

5.1 INTRODUCTION

The field-effect transistor (FET) is a three-terminal solid-state device that, like the junction transistor (BJT), can function as an amplifier or switch. Unlike the BJT, which is current controlled, the FET is a voltage-controlled device. This gives it three important advantages over the BJT. First, requiring only a voltage as its control element and virtually zero current, it possesses an input impedance as high as $10^{14}\ \Omega$. Second, since its input power (vi) is virtually zero, it produces much higher power gains than the BJT. This is important in power amplification and switching applications. Third, the FET (particularly the MOSFET) is simpler and smaller than the BJT. An etched MOSFET occupies 20% to 30% of the chip area needed for the BJT. On the negative side, however, the FET is not as linear as the BJT, nor does it permit as wide an output swing for a given supply voltage.

There are at present two basic types of FETs, the junction FET (JFET) and the metal-oxide-semiconductor FET (MOSFET). Although the two differ in structure, their basic theory of operation is the same, that is, an output circuit channel whose cross section and, consequently, conductance are controlled by a signal voltage (see Section 1.8).

5.2 THE JFET

Figure 5.1 shows some JFETs and their circuit symbols. Figure 5.2 shows a typical JFET and a typical BJT amplifier. Notice the correspondence between the gate of the JFET and the base of the BJT, the drain of the JFET and the collector of the BJT, and the source of the JFET and the emitter of the BJT. Unlike the BJT, however, FETs are usually made so that the drain and source are interchangeable. This is often advertised as "symmetrical operation." *N*-channel and *p*-channel JFETs are distinguished in circuit diagrams by the direction of the arrow on the gate connection (Figures 5.1b and d).

A typical data sheet for a family of JFETs is shown in Figure 5.3. As for the BJT, the maximum interterminal voltages are given, as is the maximum device power dissipation and temperature derating coefficient. A note beneath the case designation at the upper left sanctions symmetrical operation. An important JFET parameter, the gate–source cutoff voltage, $V_{GS(off)}$, is included in the electrical "Off Characteristics." This is the gate-to-source voltage, which effectively cuts off the drain-to-source current (I_D). Notice, however, that $V_{GS(off)}$ is a function of the drain–source voltage (V_{DS}) and that even at cutoff a small I_{DS} flows. The "On Characteristics" specifies the important parameter I_{DSS}, the drain current I_D for $V_{GS} = 0$ V. The "Dynamic Characteristics" gives the forward transfer admittance $|y_{fs}|$, also called the $V_{GS} = 0$ V transconductance (g_{mo}), plus the device capacities. The transconductance is used to compute the gain of the JFET, the device capacities, and the frequency response.

5.2.1 JFET Parameters

The information given on the data sheet defines the output characteristics of the JFET. (The input at mid-frequencies is treated as an open circuit and needs no further consideration.) The I_D versus V_{GS} characteristic is given approximately by

$$I_D = I_{DSS}\left(1 - \frac{|V_{GS}|}{V_{po}}\right)^2 \tag{5.1}$$

where V_{po} is equal to $|V_{GS(off)}|$ and is called the pinch-off voltage. The transconductance (g_m) of the JFET is given by the slope of the I_D–V_{GS} curve. Therefore, it is also a function of V_{GS} given by

$$g_m = \frac{\Delta I_D}{\Delta V_{GS}}\bigg|_{V_{DS\ \text{constant}}} = g_{mo}\left(1 - \frac{|V_{GS}|}{V_{po}}\right) \tag{5.2}$$

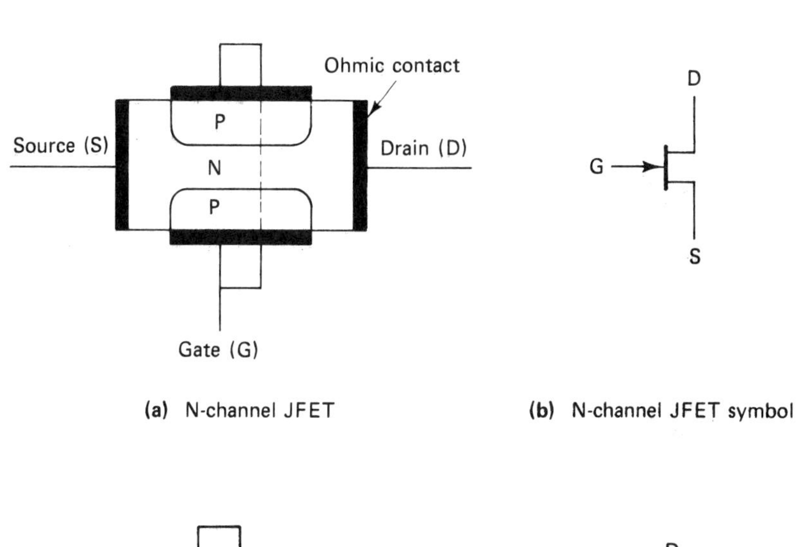

(a) N-channel JFET (b) N-channel JFET symbol

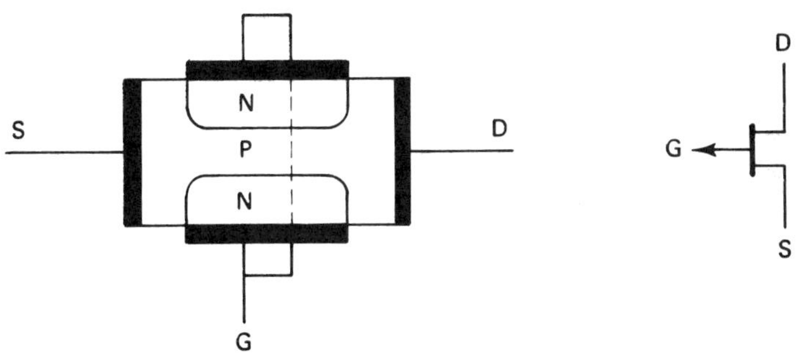

(c) P-channel JFET (d) P-channel JFET symbol

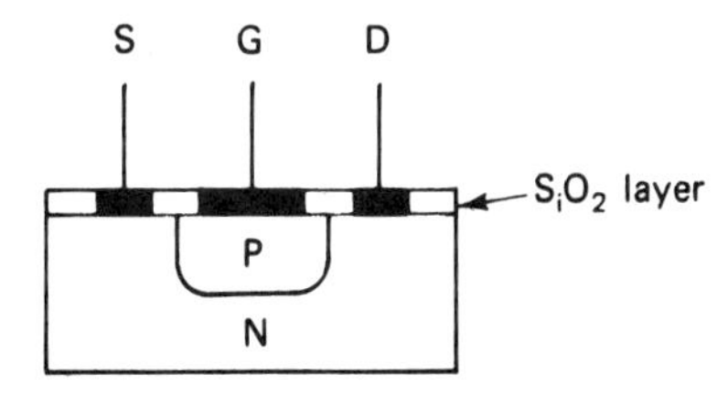

(e) Flat-construction N-channel JFET (circuit symbol same as (b))

FIGURE 5.1 JFET constructions and symbols: (a) *N*-channel JFET; (b) *N*-channel JFET symbol; (c) *P*-channel JFET; (d) *P*-channel JFET symbol; (e) flat-construction *N*-channel JFET (circuit symbol same as (b)).

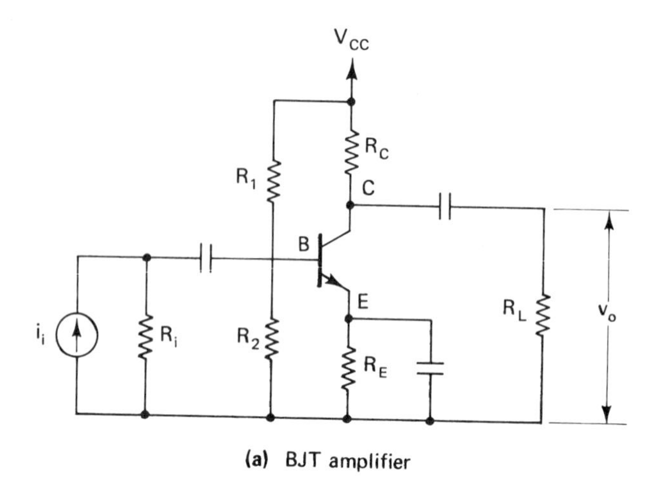

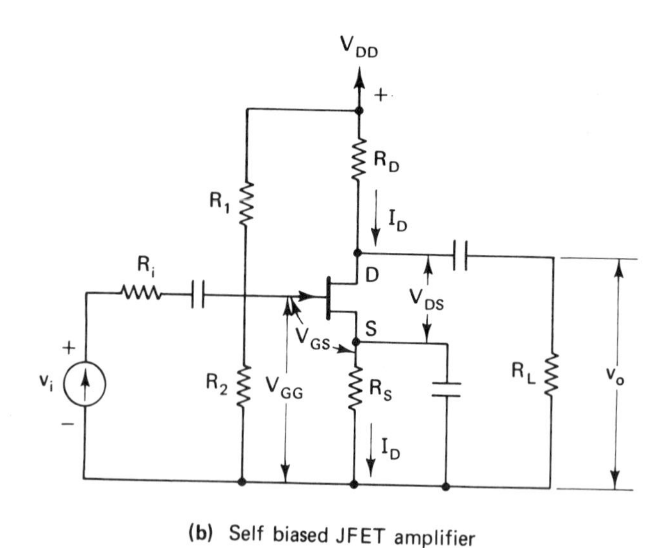

FIGURE 5.2 JFET and BJT amplifiers: (a) BJT amplifier; (b) self-biased JFET amplifier.

Finally, the output admittance (analogous to $1/h_{oe}$ for the BJT) is given as $|y_{os}|$ in the data sheet. This is also called r_D, the dynamic output resistance, and is defined by

$$r_D = \left.\frac{\text{variation in drain–source voltage}}{\text{variation in drain current}}\right|_{V_{GS\ \text{constant}}}$$

$$= \left.\frac{\Delta V_{DS}}{\Delta I_D}\right|_{V_{GS}} \tag{5.3}$$

2N5457 (SILICON)
2N5458
2N5459

Silicon N-channel junction field-effect transistors depletion mode (Type A) designed for general-purpose audio and switching applications.

CASE 29 (5)
(TO-92)

Drain and source may be interchanged.

MAXIMUM RATINGS

Rating	Symbol	Value	Unit
Drain-Source Voltage	V_{DS}	25	Vdc
Drain-Gate Voltage	V_{DG}	25	Vdc
Reverse Gate-Source Voltage	$V_{GS(r)}$	25	Vdc
Gate Current	I_G	10	mAdc
Total Device Dissipation @ $T_A = 25°C$ Derate above 25°C	P_D (2)	310 2.82	mW mW/°C
Operating Junction Temperature	T_J (2)	135	°C
Storage Temperature Range	T_{stg} (2)	-65 to +150	°C

ELECTRICAL CHARACTERISTICS (T_A = 25°C unless otherwise noted)

Characteristic		Symbol	Min	Typ	Max	Unit
OFF CHARACTERISTICS						
Gate-Source Breakdown Voltage (I_G = -10 μAdc, V_{DS} = 0)		BV_{GSS}	25	—	—	Vdc
Gate Reverse Current (V_{GS} = -15 Vdc, V_{DS} = 0) (V_{GS} = -15 Vdc, V_{DS} = 0, T_A = 100°C)		I_{GSS}	— —	— —	1.0 200	nAdc
Gate-Source Cutoff Voltage (V_{DS} = 15 Vdc, I_D = 10 nAdc)	2N5457 2N5458 2N5459	$V_{GS(off)}$	0.5 1.0 2.0	— — —	6.0 7.0 8.0	Vdc
Gate-Source Voltage (V_{DS} = 15 Vdc, I_D = 100 μAdc) (V_{DS} = 15 Vdc, I_D = 200 μAdc) (V_{DS} = 15 Vdc, I_D = 400 μAdc)	 2N5457 2N5458 2N5459	V_{GS}	— — —	2.5 3.5 4.5	— — —	Vdc
ON CHARACTERISTICS						
Zero-Gate-Voltage Drain Current (1) (V_{DS} = 15 Vdc, V_{GS} = 0)	2N5457 2N5458 2N5459	I_{DSS}	1.0 2.0 4.0	3.0 6.0 9.0	5.0 9.0 16	mAdc
DYNAMIC CHARACTERISTICS						
Forward Transfer Admittance (1) (V_{DS} = 15 Vdc, V_{GS} = 0, f = 1 kHz)	2N5457 2N5458 2N5459	$\|y_{fs}\|$	1000 1500 2000	3000 4000 4500	5000 5500 6000	μmhos
Output Admittance (1) (V_{DS} = 15 Vdc, V_{GS} = 0, f = 1 kHz)		$\|y_{os}\|$	—	10	50	μmhos
Input Capacitance (V_{DS} = 15 Vdc, V_{GS} = 0, f = 1 MHz)		C_{iss}	—	4.5	7.0	pF
Reverse Transfer Capacitance (V_{DS} = 15 Vdc, V_{GS} = 0, f = 1 MHz)		C_{rss}	—	1.5	3.0	pF

(1) Pulse Test: Pulse Width ≤ 630 ms; Duty Cycle ≤ 10%

(2) Continuous package improvements have enhanced these guaranteed Maximum Ratings as follows: P_D = 1.0 W @ T_C = 25°C, Derate above 25°C – 8.0 mW/°C, T_J = -65 to +150°C, θ_{JC} = 125° C/W.

FIGURE 5.3 FET data sheet. Courtesy of Motorola, Inc.

These parameters allow us to make all bias and mid-frequency gain computations.

It is instructive to represent the JFET characteristics graphically. Figure 5.4 shows the curves of I_D versus V_{GS} and g_m versus V_{GS}. Notice the nonlinearity of the I_D–V_{GS} curve and the linear variation of g_m with V_{GS}.

The output (drain) characteristic I_D versus V_{DS} is also of interest, although it is not used in design procedures. This curve must be determined experimentally. A typical result is given in Figure 5.5. Notice that above V_{po} (for $V_{GS} = 0$) the curve is almost flat, meaning that r_D is large (between 20 and 100 kΩ for the 2N5457 family). At some value of V_{DS} determined by the gate bias, the gate–channel diode breaks down. These breakdown points are given by the vertical lines at the right. The region between V_{po} and the beginning of the breakdown region is called the *active* or *pinch-off region*. The JFET as a linear amplifier is restricted to this region. Below V_{po}, the $V_{GS} = 0$ curve bends downward until it becomes a straight line. Along this line the JFET functions as a relatively low resistance. This region is called the *ohmic* or *saturation region*. JFETs operate in this region when they are used in switching circuits or when used as resistors in ICs. Also notice that the pinch-off voltage (V_p) for V_{GS}'s other than zero decreases as V_{GS} becomes more negative. V_p is given by

$$V_p = V_{po} - |V_{GS}| \tag{5.4}$$

5.2.2 JFET Biasing

Given the values of I_{DSS} and $V_{GS\,off}$, maximum and minimum, the quiescent operating biases (I_{DQ} and V_{DSQ}) can be found. Because of the nonlinearity of the JFET's I_D–V_{GS} characteristic, this problem is most easily solved graphically. Because this characteristic varies widely within a JFET type, it is necessary to deal with two characteristics, a maximum and a minimum. The maximum characteristic is plotted using $I_{DSS\,max}$ and $V_{GS(off)\,max}$, the minimum, with $I_{DSS\,min}$ and $V_{GS(off)\,min}$. Plotting is simplified by using the following points. (Manufacturers will also supply these curves.)

$$V_{GS} = 0, \qquad I_D = I_{DSS} \tag{5.5}$$

$$V_{GS} = \frac{V_{GS(off)}}{4} \qquad I_D = I_{DSS} \times (1 - \tfrac{1}{4})^2 = 0.5625 I_{DSS} \tag{5.6}$$

$$V_{GS} = \frac{V_{GS(off)}}{2} \qquad I_D = I_{DSS} \times (1 - \tfrac{1}{2})^2 = \frac{I_{DSS}}{4} \tag{5.7}$$

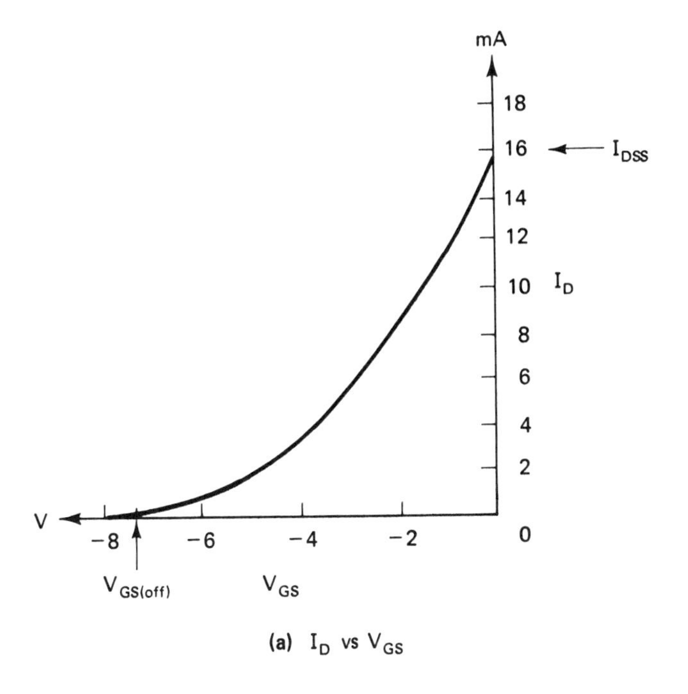

(a) I_D vs V_{GS}

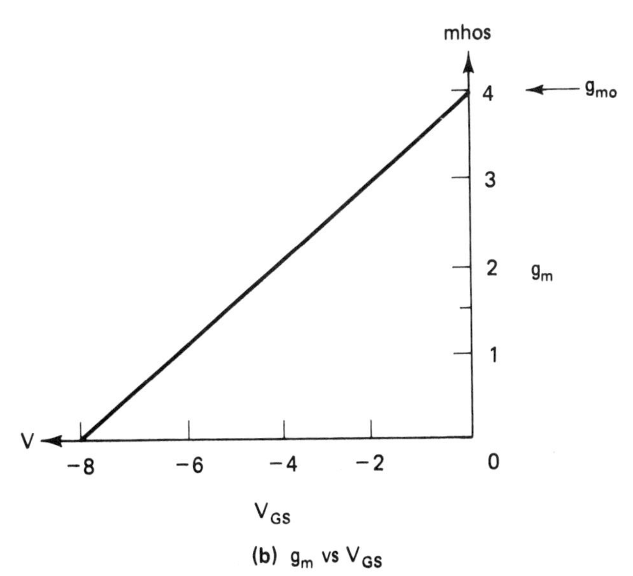

(b) g_m vs V_{GS}

FIGURE 5.4 I_D and g_m versus V_{GS} for the JFET: (a) I_D versus V_{GS}; (b) g_m versus V_{GS}.

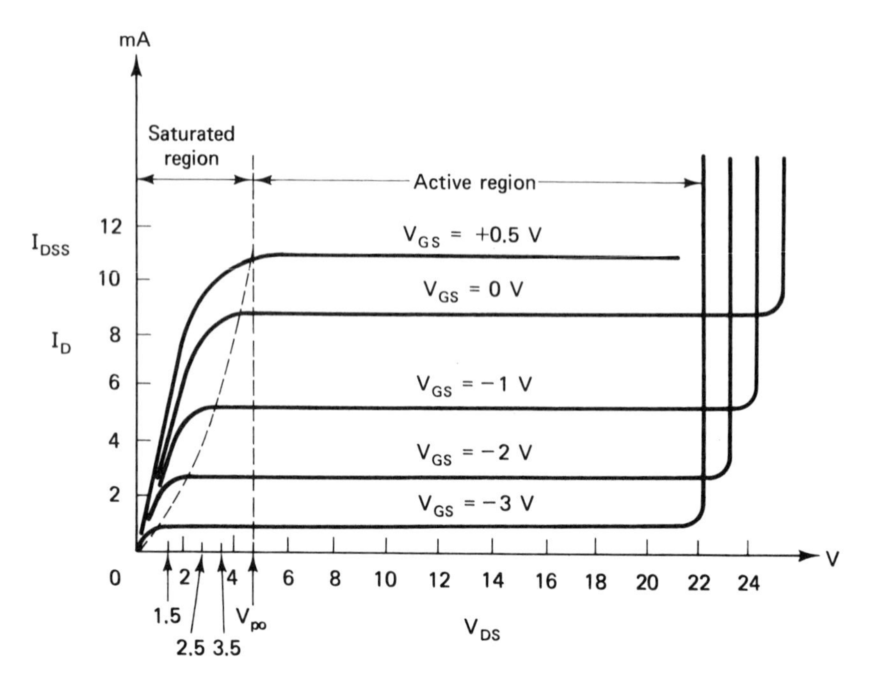

FIGURE 5.5 N-channel JFET drain characteristics.

$$V_{GS} = \frac{V_{GS(\text{off})}}{\frac{4}{3}} \qquad I_D = I_{DSS} \times (1 - \tfrac{3}{4})^2 = \frac{I_{DSS}}{16} \tag{5.8}$$

Figure 5.6 illustrates the construction of the maximum and minimum I_D–V_{GS} characteristics for the 2N5459 JFET.

To find the quiescent operating point (I_{DQ} and V_{DSQ}), we draw the curve derived from the equation relating V_{GS} and I_D,

$$V_{GS} = V_{GG} - I_D R_S \tag{5.9}$$

on the I_D–V_{GS} characteristics. (V_{GG} is the voltage from gate to ground as shown in Figure 5.2b.) The quiescent operating points are then given by the intersections of the $V_{GS} = V_{GG} - I_D R_S$ curve with the two JFET characteristics. Example 5.1 illustrates this procedure.

Example 5.1 *Finding the Quiescent Operating Points of a Self-Biased JFET*

Given the circuit of Figure 5.2b, using a 2N5459 JFET with $V_{DD} = 24$ V, $R_1 = 2.15$ MΩ, $R_2 = 0.25$ MΩ, $R_S = 1.5$ kΩ, and $R_D = 1.5$ kΩ, find I_{DQ} and V_{DSQ}, maximum and minimum.

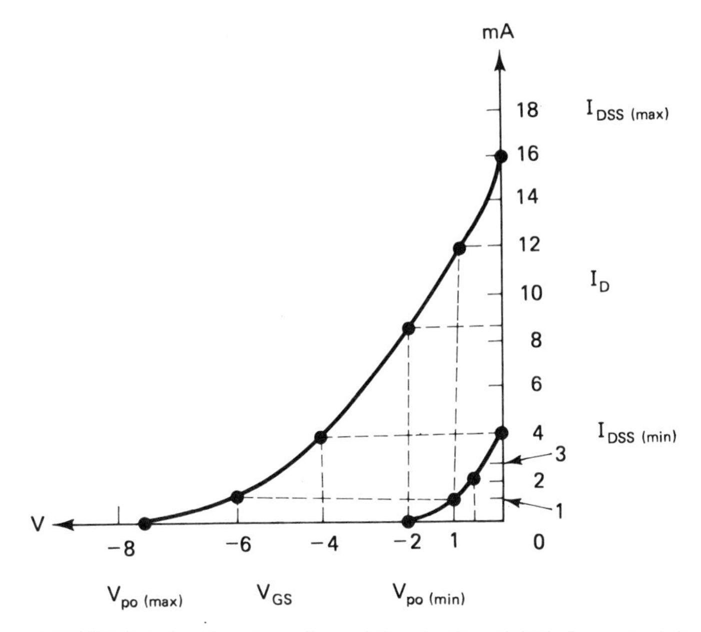

FIGURE 5.6 Construction of the I_D–V_{GS} JFET characteristics.

(a) Construct the max–min I_D–V_{GS} characteristics of Figure 5.6 (repeated in Figure 5.7).

(b) *Compute* V_{GG}:

$$V_{GG} = \frac{V_{DD}R_2}{R_1 + R_2} = \frac{24\text{ V} \times 0.25\text{ M}\Omega}{2.75\text{ M}\Omega + 0.25\text{ M}\Omega} = 2\text{ V}$$

(Note that this computation is accurate only if R_2 is much smaller than the JFET input resistance, which is typically many megohms.)

(c) Plot the $V_{GS} = V_{GG} - I_DR_S$ line on the JFET I_D–V_{GS} characteristics (Figure 5.7).

$$I_D = 0, \qquad V_{GS} = 2 - 0 = 2\text{ V}$$

$$I_D = 4\text{ mA}, \qquad V_{GS} = 2 - 4\text{ mA} \times 1.5\text{ k}\Omega = -4\text{ V}$$

(d) Read off I_{DQ} at the intersections (Figure 5.7):

$$I_{DQ\,\text{max}} = 4\text{ mA}$$

$$I_{DQ\,\text{min}} = 2\text{ mA}$$

(e) Compute V_{DSQ}, max and min:

$$V_{DSQ} = V_{DD} - I_{DQ}(R_S + R_D)$$

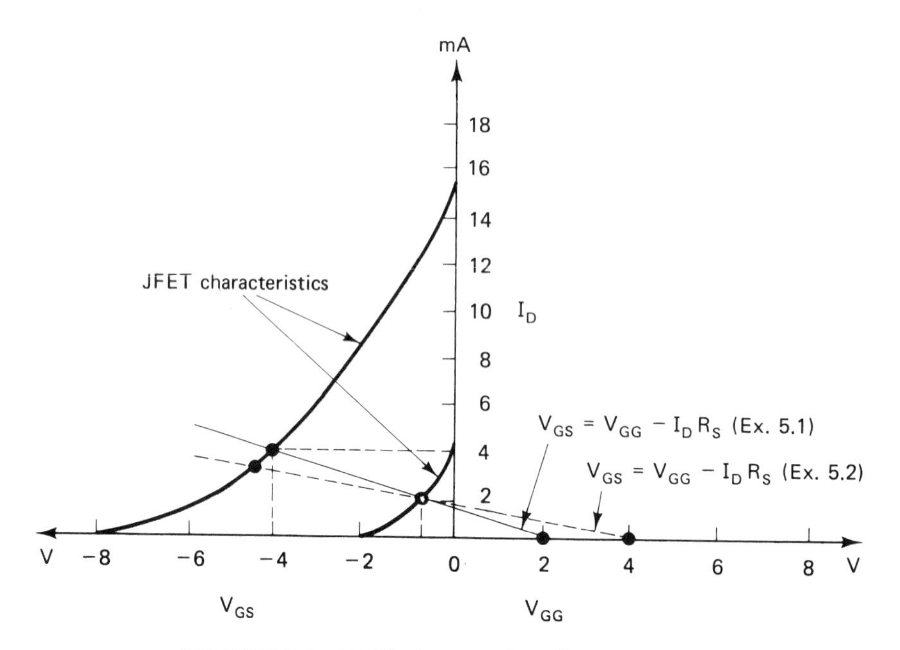

FIGURE 5.7 JFET characteristic for Example 5.1.

$$V_{DSQ\,\min} = 24\,\text{V} - 4\,\text{mA}\,(1.5\,\text{k}\Omega + 1.5\,\text{k}\Omega) = 12\,\text{V}$$

$$V_{DSQ\,\max} = 24\,\text{V} - 2\,\text{mA}\,(1.5\,\text{k}\Omega + 1.5\,\text{k}\Omega) = 18\,\text{V}$$

The design procedure for setting the JFET amplifier within a specified I_{DQ}, V_{DSQ} range is largely the reverse of the analysis. The designer must determine R_1, R_2, R_S, R_D, and V_{DD}. The I_D–V_{GS} characteristics are obtained. $I_{DQ\,\max}$ and $I_{DQ\,\min}$ are located on the maximum and minimum characteristics, respectively. A straight line is drawn through these two points and extended to the V_{GS} axis. This gives V_{GG}, and the inverse slope of the line gives R_S. V_{DD} and R_D are then computed to give the V_{DSQ} range, and R_1 and R_2 are then computed to give V_{GG}. (The factors behind the specification of I_{DQ} and V_{DGS} are discussed later.) Example 5.2 illustrates this procedure.

Example 5.2 *Biasing the Self-Biased JFET*

Given the circuit of Figure 5.2b, using a 2N5459 JFET, determine V_{DD}, R_D, R_S, R_1, and R_2 for $I_{DQ} = 2.5\,\text{mA} \pm 0.5\,\text{mA}$ and $V_{DSQ} = 15$ to $20\,\text{V}$.

(a) *Find* V_{GG}: Locate $I_{DQ\,\max} = 3.0\,\text{mA}$ and $I_{DQ\,\min} = 2.0\,\text{mA}$ on their respective characteristics (points shown as circles in Figure 5.7). Draw a line through these points intersecting the V_{GS} axis (dashed line in Figure 5.7). This intersection gives V_{GG}.

$$V_{GG} = 4\,\text{V}$$

(b) *Compute R_S:* R_S is given by the slope of the $V_{GS} = V_G - I_D R_S$ line.

$$R_S = \frac{V_{GS}}{I_D} = \frac{8\,\text{V}}{2.9\,\text{mA}} = 2.76\,\text{k}\Omega$$

using the points $V_{GS} = 4$ V, $I_D = 0$, and $V_{GS} - 4$ V, $I_D = 2.9$ mA.

(c) *Compute V_{DD} and R_D:* Starting with the equations

$$\text{Eq. 1} \qquad V_{DSQ\max} = V_{DD} - I_{DQ\min}(R_S + R_D)$$

$$\text{Eq. 2} \qquad V_{DSQ\min} = V_{DD} - I_{DQ\max}(R_S + R_D)$$

which are the volt drops from V_{DD} to ground through the JFET, we derive by manipulation

$$V_{DD} = \frac{I_{DQ\max} V_{DSQ\max} - I_{DQ\min} V_{DSQ\min}}{I_{DQ\max} - I_{DQ\min}}$$

$$= \frac{3\,\text{mA} \times 20\,\text{V} - 2\,\text{mA} \times 15\,\text{V}}{3\,\text{mA} - 2\,\text{mA}} = 30\,\text{V}$$

Then using Eq. 1 and solving for R_D, we get

$$R_D = \frac{V_{DD} - V_{DSQ\max}}{I_{DQ\min}} - R_S = \frac{30\,\text{V} - 20\,\text{V}}{2\,\text{mA}} - 2.75\,\text{k}\Omega = 2.25\,\text{k}\Omega$$

(d) *Compute R_1 and R_2:* Starting with the voltage-divider equation,

$$V_{GG} = \frac{V_{DD} R_2}{R_1 + R_2}$$

we can derive

$$R_1 = \frac{V_{DD} - V_{GG}}{V_{GG}} R_2$$

Choosing an R_2 with much less than the typical input resistance of the JFET

$$R_2 = 0.25\,\text{M}\Omega$$

$$R_1 = \frac{30\,\text{V} - 4\,\text{V}}{4\,\text{V}} 0.25\,\text{M}\Omega = 1.63\,\text{M}\Omega$$

Figure 5.8 shows the resulting circuit. (In actual practice when composition resistors are used the values computed here can only be approximated.)

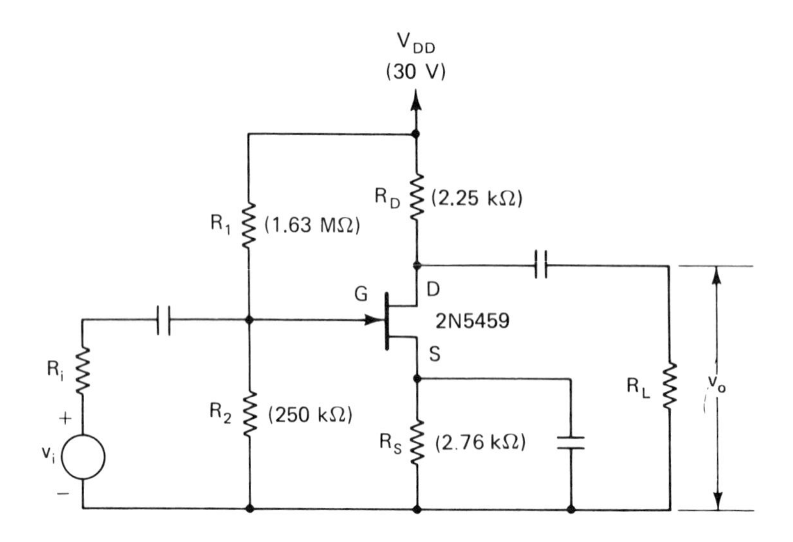

FIGURE 5.8 JFET amplifier for Example 5.2.

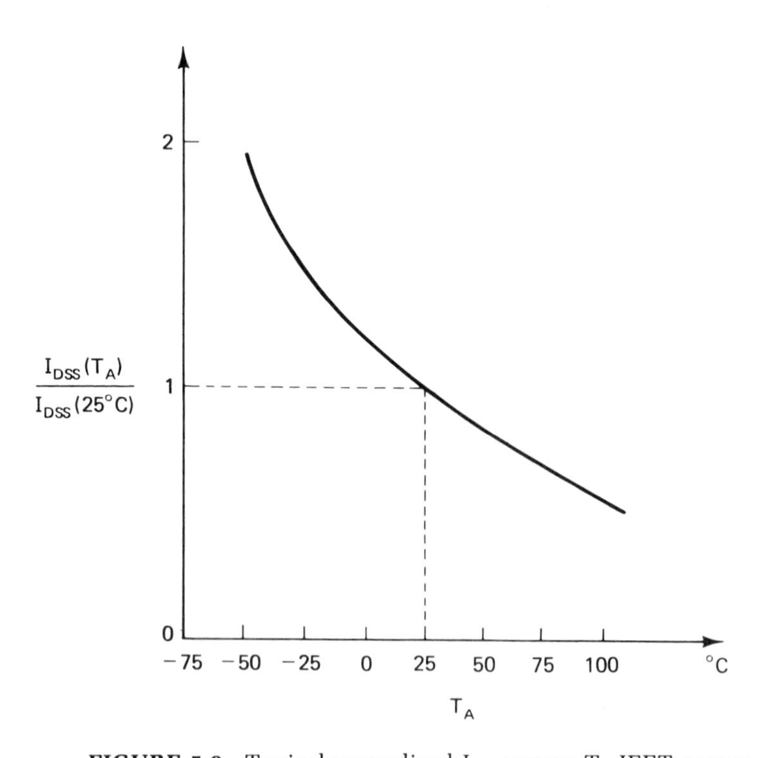

FIGURE 5.9 Typical normalized I_{DSS} versus T_A JFET curves.

5.2.3 Temperature Effects in the JFET

Temperature changes affect I_{DSS} and $V_{GS(off)}$ in the JFET. I_{DSS} varies approximately inversely to the 3/2 power of temperature. A typical curve relating $I_{DSS}(T_A)/I_{DSS}$ (25°) to the ambient temperature (T_A) is shown in Figure 5.9. Manufacturers provide similar curves for specific JFET types. The cutoff voltage $V_{GS(off)}$ and V_{po} consequently, are affected in the same way as the base–emitter voltage of the junction transistor; that is,

$$V_{GS(off)} = -2\,\text{mV}/°\text{C} \tag{5.10}$$

These temperature corrections are used to establish the maximum and minimum I_D–V_{GS} curves (highest T_A for the maximum, lowest for the minimum). Then the biasing design proceeds as described.

5.3 METAL OXIDE SEMICONDUCTOR FET (MOSFET)

Two typical MOSFET structures, a depletion type (Figure 5.10a) and an enhancement type (Figure 5.10c) are shown along with their circuit symbols. The depletion-type MOSFET contains a permanent physical *n*-channel. In the enhancement type the *n*-channel is induced by a positive V_{GS}. The operating principles of both are the same, however, and similar to that of the JFET. The *n*-channel MOSFETs of Figure 5.10 are etched (and diffused) into the surface of a supporting *p*-type substrate. Two highly doped (n^+) regions diffused into this substrate constitute the *source* and the *drain*. These regions are very close to each other, typically about 1 mil apart. For the depletion type a permanent *n*-channel is formed between the source and drain. A thin insulating layer of silicon dioxide (SiO_2) is deposited over the surface, and holes are cut into it to permit access to the source and drain. Finally, a conductive metal gate covering the entire channel region is deposited on the SiO_2 layer. The MOSFET structure is simpler than the junction transistor and particularly suited for use in IC chips.

5.3.1 Depletion-Enhancement Mode MOSFET

The depletion-mode MOSFET (or MOS) action is similar but not identical to that of the JFET. In the JFET, the depletion region is established between its gate and adjacent channel. In the depletion MOS, it is established primarily between its drain and substrate. In the JFET the gate controls I_D by regulating the extent of the depletion region. In the depletion MOS, the gate, while affecting the depletion region spread, also con-

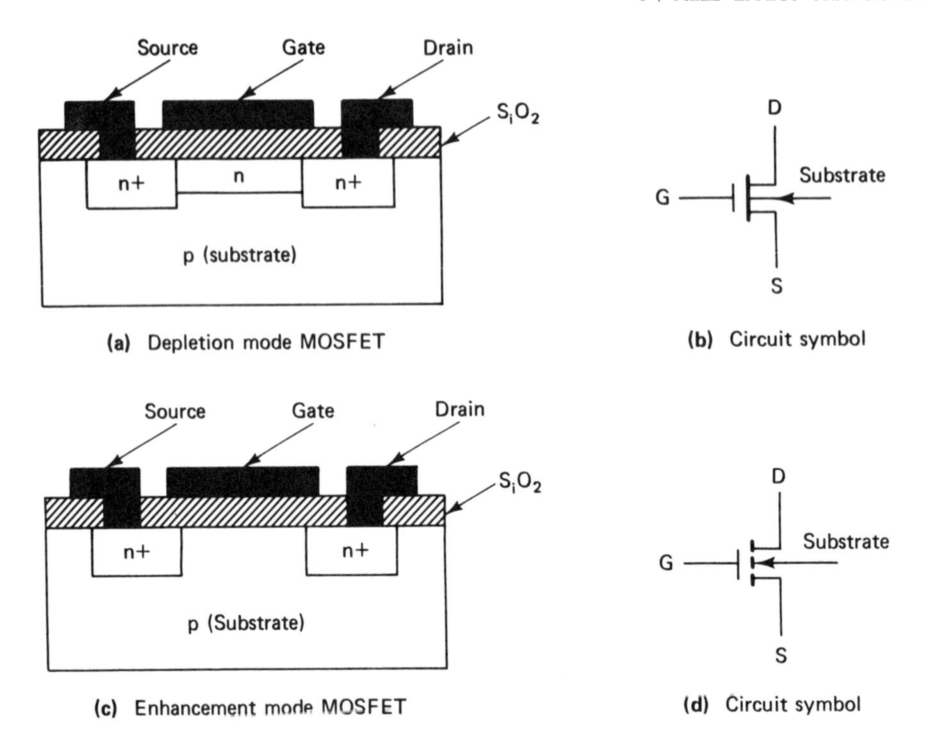

FIGURE 5.10 Depletion- and enhancement-mode MOSFETs: (a) depletion mode MOSFET; (b) circuit symbol; (c) enhancement mode MOSFET; (d) circuit symbol.

trols the conductivity of the channel by controlling the number of channel carriers.

The depletion MOS action takes place as follows. With V_{GS} at zero, an increase in V_{DS} (below V_{po}) will increase I_D as expected. As V_{DS} increases, the depletion region between the substrate and the drain grows and begins to intrude into the *n*-channel. At and beyond V_{po}, the tendency for I_D to increase with V_{DS} is counteracted by the reduction in the *n*-channel cross section due to the growth of the depletion region, and the I_D–V_{DS} (drain) characteristic levels off. At any fixed V_{DS} above or below V_{po}, a decrease in V_{GS} (more negative) will push negative carriers out of the *n*-channel into the substrate, thus reducing I_D.

The depletion MOS can also be operated with a positive V_{GS}. In this mode, called the enhancement mode, the positive V_{GS} attracts electrons out of the substrate into the *n*-channel, thus increasing I_D. A MOS operating in both the depletion and enhancement mode is called a depletion–enhancement MOS. The I_D–V_{GS} (transconductance) and the I_D–V_{DS} (drain characteristics) for a depletion–enhancement MOS are shown in Figure 5.11.

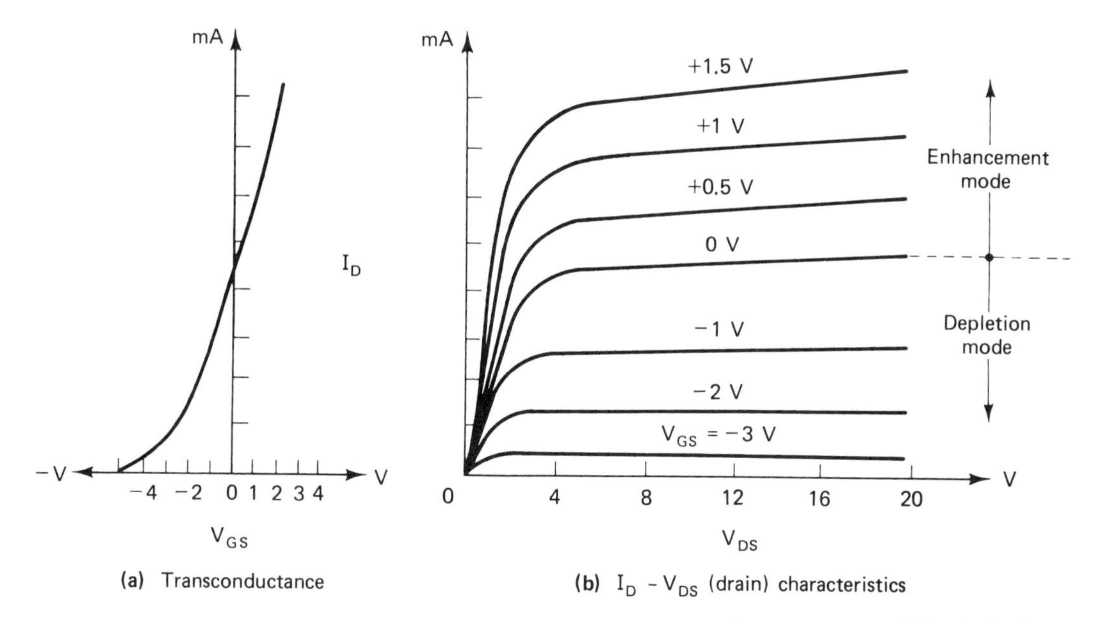

FIGURE 5.11 Transconductance and drain characteristics of the depletion–enhancement MOS: (a) transconductance; (b) I_D–V_{DS} (drain) characteristics.

Biasing the Depletion–Enhancement MOS. The simplest way to bias the depletion–enhancement MOS is to connect the gate to the source through a resistor, thus setting V_{GS} at zero. The resulting amplifier will then contain a minimum number of components (Figure 5.12a). The circuit, however, has the disadvantage of permitting wide variations in I_{DQ} ($I_{DQ} = I_{DSS}$), which are beyond the designer's control. This condition is shown in Figure 5.12b. If these swings can be tolerated, it is an economical and space-saving circuit. Example 5.3 illustrates the design procedure.

Example 5.3 *Design of a Fixed-Zero-Bias Depletion–Enhancement MOS Circuit*

Using a MFE825 *n*-channel depletion–enhancement MOS (MOS selection discussed later), design a fixed-zero-bias circuit to meet the following specifications:

$$V_{GSQ} = 0, \qquad V_{DD} = 20\text{ V}$$
$$V_{DSQ} \geq 8\text{ V}, \qquad R_i = 100\text{ k}\Omega$$

From the MFE825 data sheet,

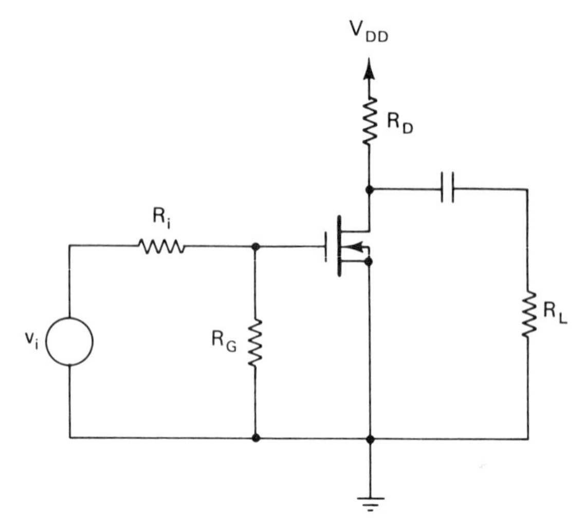

(a) Fixed-zero-bias circuit (depletion – enhancement MOS)

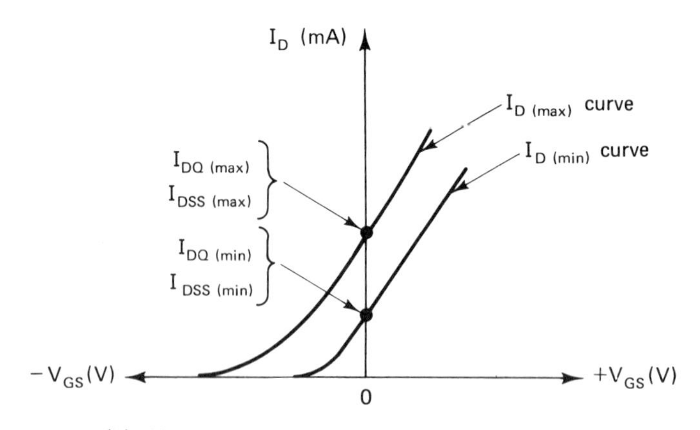

(b) Maximum and minimum transconductance curves

FIGURE 5.12 Fixed-zero-bias circuit for the depletion–enhancement MOS: (a) fixed-zero-bias circuit (depletion-enhancement MOS); (b) maximum and minimum transconductance curves.

$$I_{DSS} = 1.0/1.5 \text{ mA}$$

$$V_{GS(TN)} = V_{GS(\text{off})} = -2.0/-6.0 \text{ V}$$

$$V_{(BR)DSS} \text{ (maximum drain–source voltage)} = 20 \text{ V}$$

(a) Use the circuit of Figure 5.12a with R_L absent.

(b) *Choose* R_G: R_G must be very much smaller than the input resistance of the MOS (R_{in}) and much larger than the source resistance.

$$R_i \ll R_G \ll R_{in},\ 100 \text{ k}\Omega \ll R_G \ll 10^{14}\ \Omega,\ R_G = 2 \text{ M}\Omega$$

(c) *Find R_D:*

$$V_{DSQ} \geq 8\text{ V}$$
$$V_{DSQ\,\min} = V_{DD} - I_{DQ\,\max}R_D$$
$$I_{DQ\,\max} = I_{DSS\,\max} = 1.5\text{ mA}$$
$$8\text{ V} = 20\text{ V} - 1.5\text{ mA}\,R_D$$
$$R_D = 8\text{ k}\Omega$$

$V_{DSQ\,\max}$ must be checked for breakdown.

$$V_{DSQ\,\max} = V_{DD} - I_{DQ\,\min}R_D$$
$$I_{DQ\,\min} = I_{DSS\,\min} = 1\text{ mA}$$
$$V_{DSQ\,\max} = 20 - 1\text{ mA} \times 8\text{ k}\Omega = 12\text{ V}.$$

This is acceptable.

The self-bias scheme described for the JFET can be applied to the depletion–enhancement MOS. The design procedure is almost identical to the one used for the JFET except that $I_{DSS\,\max}$ and $I_{DSS\,\min}$ are usually chosen on either side of $V_{GS} = 0$ V. The circuit of Figure 5.2 is used with the JFET replaced by the depletion–enhancement MOS. The transconductance diagram for this design procedure is shown in Figure 5.13.

5.3.2 Enhancement MOS

FET action in the enhancement MOS is similar to that in the enhancement region of the enhancement–depletion MOS, but since no physical

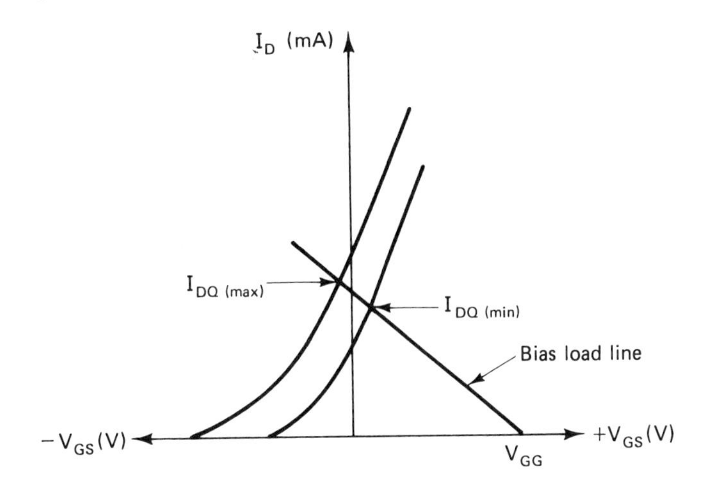

FIGURE 5.13 Transconductance diagram: self-biased depletion–enhancement MOS circuit.

n-channel exists, no significant I_D flows until the n-channel is created. This channel is induced by a positive V_{GS} (for the NMOS), which attracts electrons from the substrate into the region between the drain and source. At some V_{GS} called the *threshold voltage* (V_T), this channel is sufficiently established to permit significant conduction (typically about 10 μA) to begin. Therefore, the transconductance curve of the enhancement MOS begins at V_T. However, a small leakage current measured in nanoamperes does flow below V_T. The relationship between I_D and V_{GS} is given by

$$I_D = k(V_{GS} - V_T)^2 \tag{5.11}$$

where k is a constant proportional to the width–length ratio (W/L) of the channel. Values of k lie between 10^{-3} and 10^{-2} A/V^2 (amperes/volts^2).

Manufacturers specify the maximum and minimum transconduction curves in various ways. They will usually specify a range for V_T. They may give the maximum and minimum values of k. They may specify another point on the transconductance curve by giving an $I_{D\,on}$ for a V_{GS}. However done, enough information is provided to construct the maximum and minimum transconductance curves. Typical curves are shown in Figure 5.14.

Biasing the Enhancement MOS. The enhancement MOS can be biased within a maximum and minimum value of I_{DQ}, using the economical circuit of Figure 5.15, by the procedure illustrated in Figure 5.16. To begin, the maximum and minimum transconductance curves are drawn. $I_{DQ\,max}$

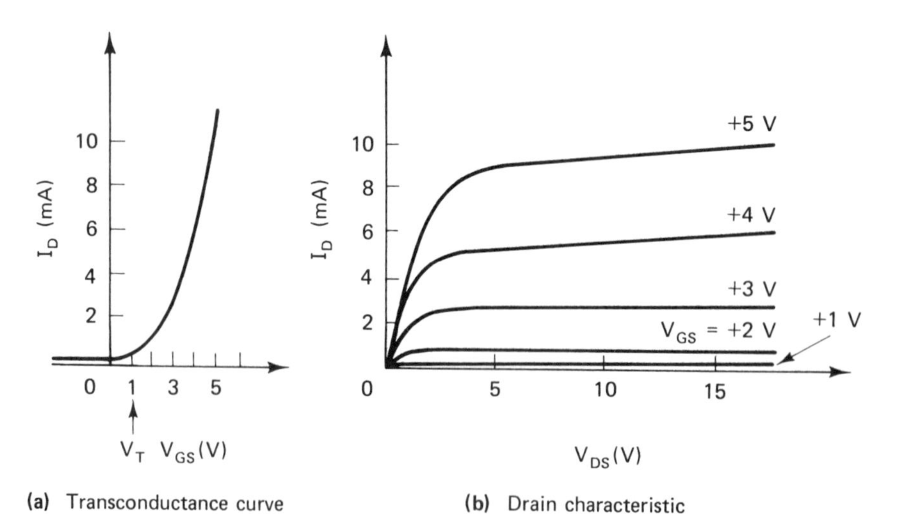

FIGURE 5.14 Characteristics of the enhancement MOS: (a) transconductance curve; (b) drain characteristic.

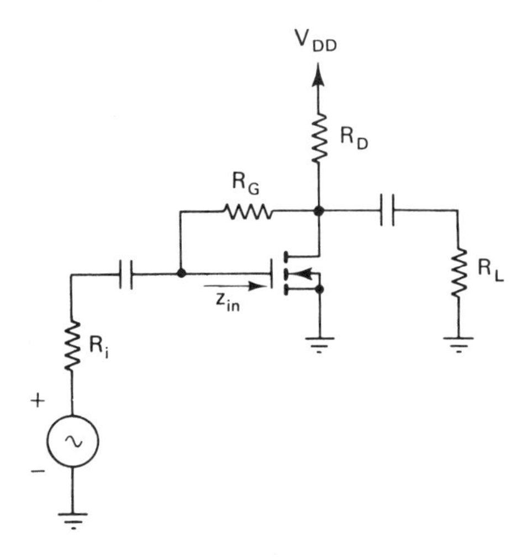

FIGURE 5.15 Bias circuit for the enhancement MOS.

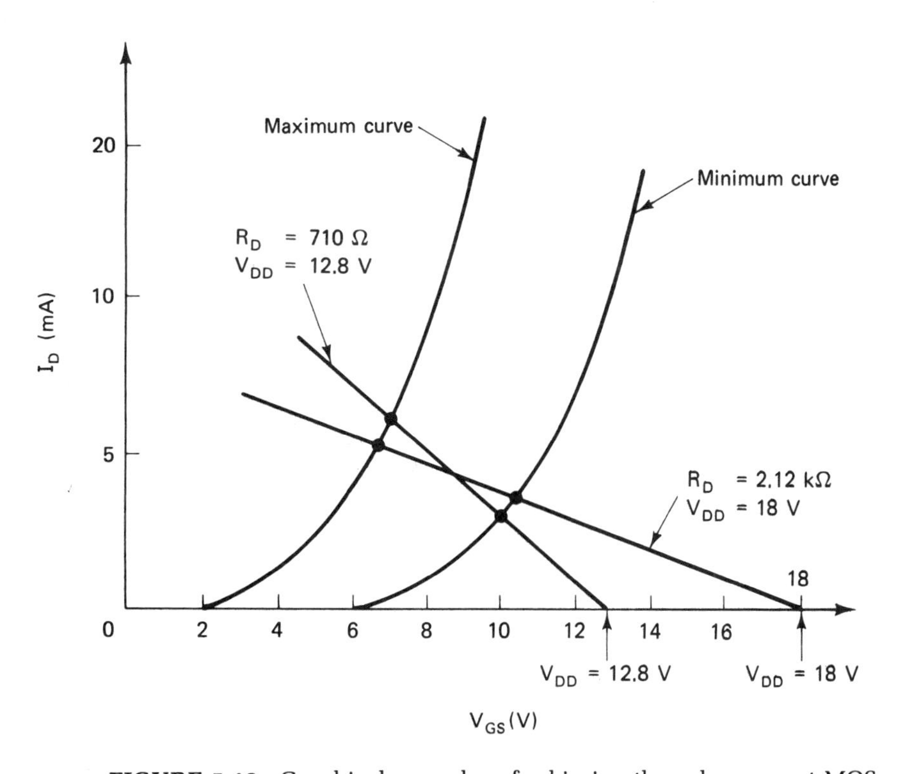

FIGURE 5.16 Graphical procedure for biasing the enhancement MOS.

and $I_{DQ\,min}$ are located on the curves, and a straight line is drawn through them to the V_{GS} axis. This line is the load line for the curve

$$V_{GS} = V_{DS} = V_{DD} - I_D R_D \tag{5.12}$$

describing the V_{GS}–I_D relationship for the circuit of Figure 5.15. R_D is equal to the reciprocal of the slope of this line, and V_{DD} is given by the intersection of the line and the V_{GS} axis.

Figure 5.16 illustrates this procedure for two I_{DQ} ranges. As can be seen, narrower ranges require higher values of R_D and, if V_{DSQ} is held constant, higher V_{DD}'s. However, there is a limit to this process, for unless precautions are taken to guarantee that the MOS will never be cut off, V_{DD} should never exceed the drain–source breakdown voltage, $V_{(BR)DSS}$.

To complete the design procedure, R_G must be chosen. Several considerations are involved in this choice. First, R_G must be large enough not to load R_D and to provide the desired input impedance. Second, to prevent a severe reduction in gain because of feedback (discussed later), R_G must be much larger than the source resistance.

Example 5.4 illustrates the entire procedure.

Example 5.4 *Design Procedure for Biasing the Enhancement MOS*

For the circuit of Figure 5.15, find R_D, R_G, and V_{DD} to meet the following specifications (use the transconductance curves of Figure 5.17):

$$I_{DQ\,max} = 6.5\text{ mA}, \qquad I_{DQ\,min} = 4.25\text{ mA}, R_i = 20\text{ k}\Omega$$

Using Figure 5.17, locate $I_{DQ\,max}$ on the $I_{D\,max}$ curve and $I_{DQ\,min}$ on the $I_{D\,min}$ curve. Draw a straight line through these points extended to meet both axes, as shown. The intersection of this load line and the V_{GS} axis gives $V_{DD} = 18$ V. R_D is the inverse slope of this line, given by

$$R_D = \frac{V_{DD}}{I_{DD}} = \frac{18}{10\text{ mA}} = 1.8\text{ k}\Omega$$

R_G is chosen to be $\gg$ than R_i and $\ll$ R_{in} of the MOS. Let $R_G = 100R_i = 100 \times 20\text{ k}\Omega = 2\text{ M}\Omega$. This is very much smaller than R_{in}, typically $10^{14}\ \Omega$.

5.3.3 The VFET

The VFET is an enhancement MOS with a unique geometry that considerably increases its power handling capacity and improves its linearity. Consequently, it is a formidable contender to the junction transistor for power amplification and switching, especially since FETs require much less input or control power than BJTs.

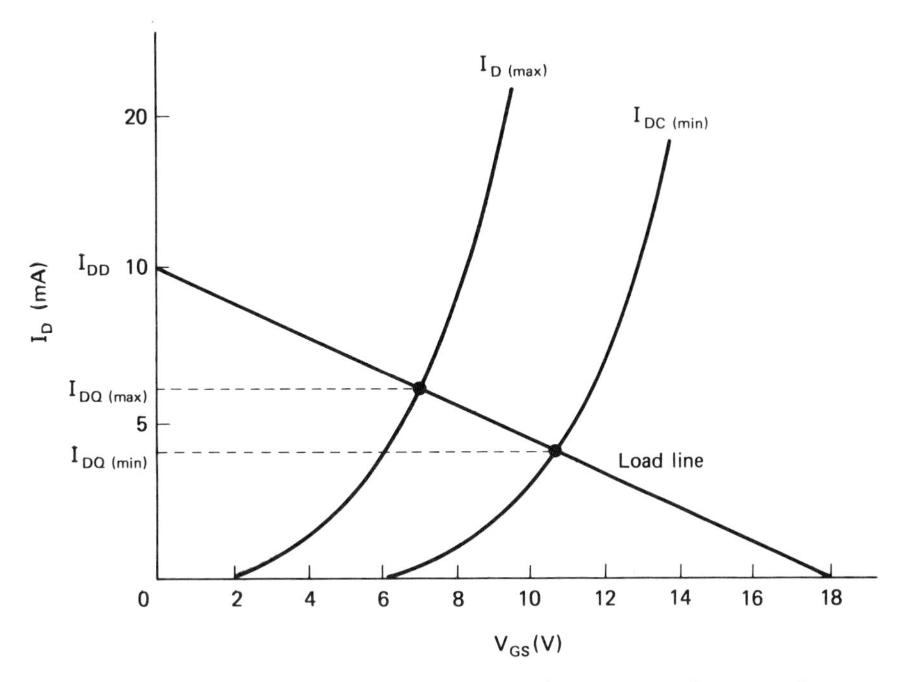

FIGURE 5.17 Enhancement MOS characteristic for Example 5.4.

The cross section of a VFET is shown above that of an enhancement MOS in Figure 5.18. Notice the differences in the channel geometries. The MOS contains one channel whose minimum length depends on the dimensions of a photographic mask. The VFET, on the other hand, contains two channels whose minimum length (upper n^+ to n^-) depends on a diffusion process that creates the p- and n-layers. Because of this difference in the manufacturing process, the VFET channel length can be made at least one-third that of the MOS with a corresponding decrease in channel resistance. The two parallel channels, possible in the VFET because of its vertical structure, also increase its current-handling capacity. (Multiple-channel VFETs are also available.)

The presence of a lightly doped n-layer between the drain and p-region increases the drain–source breakdown voltage (BV_{DSS}). With positive V_{DS} and $V_{GS} < V_T$, the n-to-p junction is reverse biased. Because of its light doping, the resulting depletion region penetrates deeply into the n-layer. This creates additional isolation between the drain and the source, avoiding breakdown (punch through) and increasing BV_{DSS}.

The short, paralleled channels of the VFET not only increase current flow and power-handling capacity but, as a direct consequence, increase transconductance (typically for the VFET, 250 mS, and for the MOS, 6 mS) and reduce switching times as well. Figure 5.19 is a section

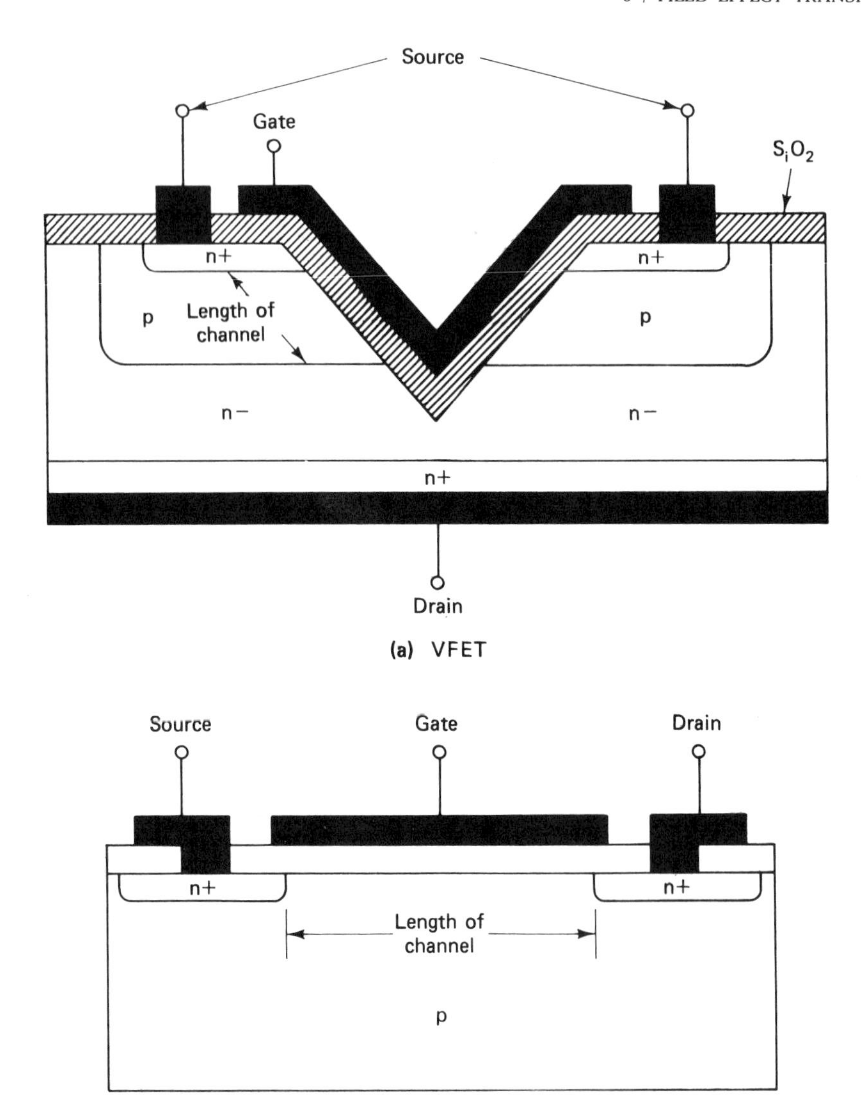

FIGURE 5.18 Geometry of the VFET and the enhancement MOS: (a) VFET; (b) MOSFET.

of a VFET data sheet showing the high BV_{DSS} (80 V), the low leakage current, I_{GSS} (100 nA maximum), the high transconductance (g_m) (170 mS), the high *on* current ($I_{D\text{ on}} = 1.5$ A), and the short rise, fall, turn-on, turn-off, and delay times (10 ns maximum) typical of the VFET. Figure 5.20 shows a typical transconductance curve.

VN88AD ■ VN89AD

ELECTRICAL CHARACTERISTICS ($T_C = 25°C$ unless otherwise noted)

	Parameter	Part Number	Min	Max	Unit	Test Conditions
Static						
BV_{DSS}	Drain-Source Breakdown	All	80		V	$V_{GS} = 0$, $I_D = 10 \mu A$
$V_{GS(th)}$	Gate Threshold Voltage	All	0.8	2.5	V	$V_{DS} = V_{GS}$, $I_D = 1 mA$
I_{GSS}	Gate Body Leakage	All		100	nA	$V_{GS} = 15 V$, $V_{DS} = 0$
I_{DSS}	Zero Gate Voltage Drain Current	All		10	μA	V_{DS} = Max Ratings, $V_{GS} = 0$
		All		500	μA	V_{DS} = 0.8 Max Ratings, $V_{GS} = 0$, $T_A = 125°C$
$V_{DS(on)}$	Drain-Source Saturation Voltage[1]	All		1.5	V	$V_{GS} = 5 V$, $I_D = 300 mA$
		VN88AD		4.0	V	$V_{GS} = 10 V$, $I_D = 1 mA$
		VN89AD		4.5	V	$V_{GS} = 10 V$, $I_D = 1 mA$
$r_{DS(on)}$	Drain-Source On Resistance[1]	All		5.0	Ω	$V_{GS} = 5 V$, $I_D = 300 mA$
		VN88AD		4.0	Ω	$V_{GS} = 10 V$, $I_D = 1 A$
		VN89AD		4.5	Ω	$V_{GS} = 10 V$, $I_D = 1 A$
$I_{D(on)}$	On-State Drain Current[1]	All	1.5		A	$V_{DS} = 25 V$, $V_{GS} = 10 V$
Dynamic						
g_{fs}	Forward Transconductance[1]	All	170		mS	$V_{DS} = 25 V$, $I_D = 0.5 A$
C_{iss}	Input Capacitance	All		50	pF	$V_{DS} = 25 V$, $V_{GS} = 0$, $f = 1 MHz$
C_{rss}	Reverse Transfer Capacitance	All		10	pF	$V_{DS} = 25 V$, $V_{GS} = 0$, $f = 1 MHz$
C_{oss}	Common-Source Output Capacitance	All		50	pF	$V_{DS} = 25 V$, $V_{GS} = 0$, $f = 1 MHz$
t_{ON}	Turn-ON Time	All		10	ns	$V_{DD} = 25V$, $I_D \approx 1A$, $R_L = 23 Ω$, $R_g = 25 Ω$
t_{OFF}	Turn-OFF Time	All		10	ns	$V_{DD} = 25V$, $I_D \approx 1A$, $R_L = 23 Ω$, $R_g = 25 Ω$
Drain-Source Diode Characteristics						
			Typ			
V_{SD}	Forward ON Voltage[1]	All	−0.9		V	$I_S = -1A$, $V_{GS} = 0$
t_{rr}	Reverse Recovery Time	All	35		ns	$V_{GS} = 0$, $I_F = I_R = 1 A$

Note 1: Pulse test — 80μs to 300μs, 1% duty cycle

Output characteristics

80 μs 1% Duty cycle pulse test
V_{GS} = 10 V
9 V
8 V
7 V
6 V
5 V
4 V
3 V
2 V
1 V
I_D — drain current (amps)
V_{DS} — drain to source voltage (volts)

Transfer characteristic

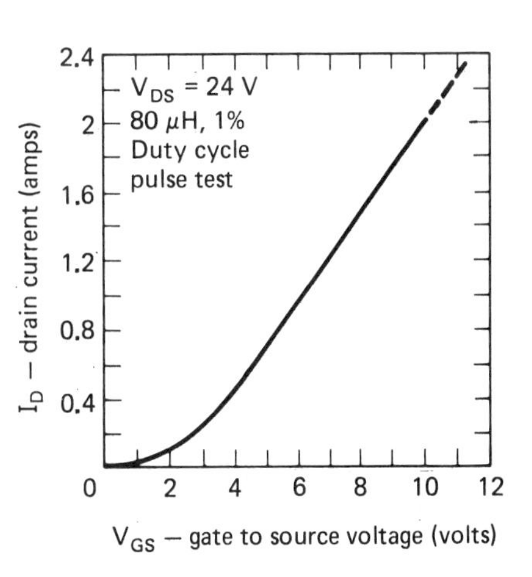

FIGURE 5.19 Portion of data sheet for VFET. Courtesy of Siliconix, Inc.

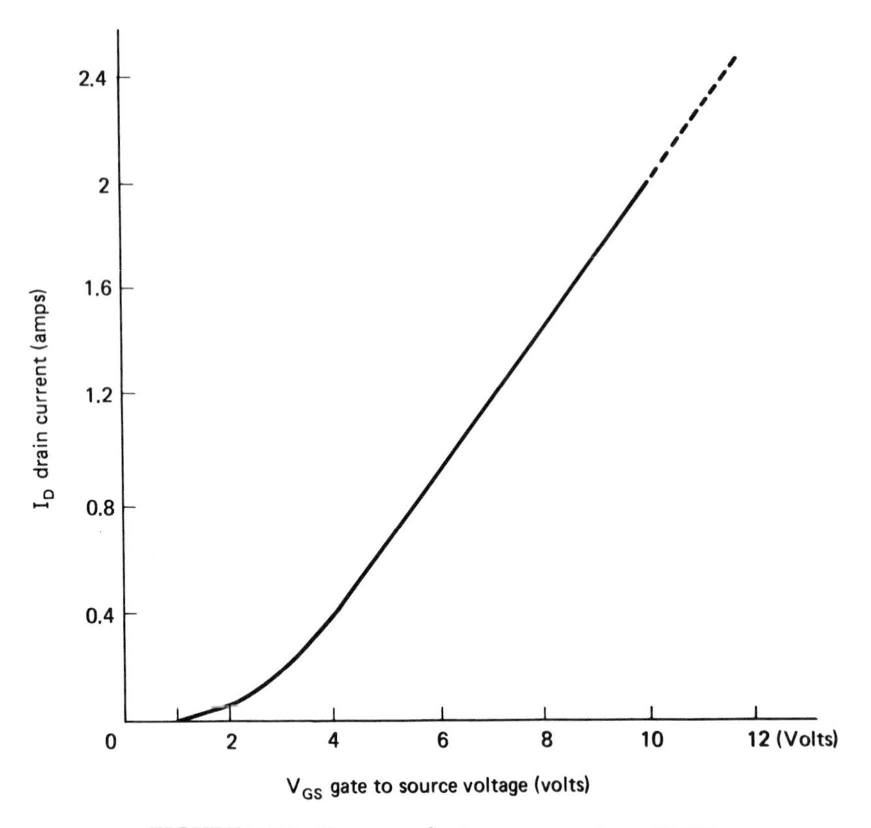

FIGURE 5.20 Transconductance curve for a VFET.

The design of the bias circuit of the VFET depends on its application. If it were to be used as a single FET low-level amplifier, it would be biased as any other enhancement MOS. However, since it is used primarily in power circuits requiring other configurations, it is biased differently.

5.3.4 Biasing Other Amplifier Configurations

As is the case for the junction transistor amplifier, the JFET and the depletion MOSFET are usable in the common-drain and common-gate configurations as well. The common-source configuration, which we have covered, is the one most used for amplifiers, the common drain is used when a low output impedance is required, and the common gate is rarely used at all.

Figure 5.21 shows the three configurations. For the DC bias case, all

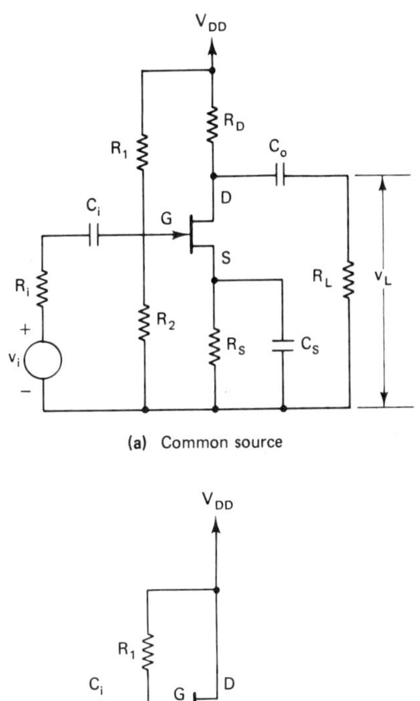

(a) Common source

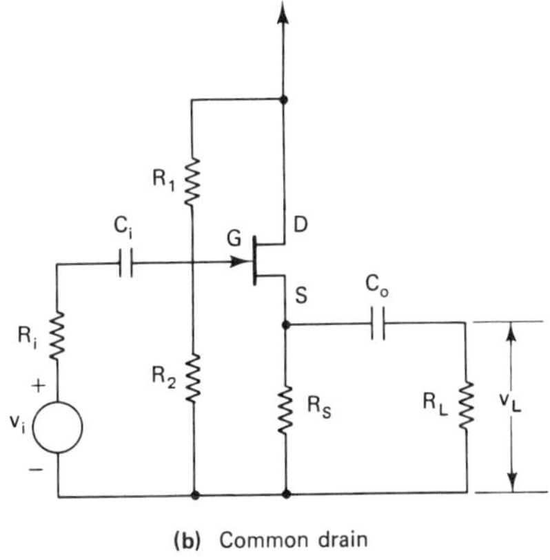

(b) Common drain

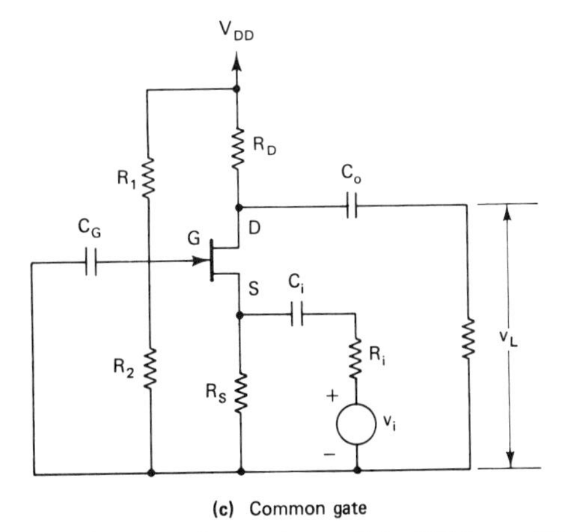

(c) Common gate

FIGURE 5.21 Three configurations of the FET amplifier: (a) common source; (b) common drain; (c) common gate.

capacitors are open circuits so that the R_L's and the signal source circuits are effectively absent. The remaining circuits are virtually identical: the common gate identical to the common source, and the common drain identical to the common source with $R_D = 0$. Therefore, the biasing procedures previously described apply.

5.4 SMALL-SIGNAL FET CIRCUIT

The small-signal FET circuit is used to find gain and input–output impedances. The small-signal circuit depends on the circuit configuration, not on the kind of FET used. Of the three possible configurations, only the common source and the common drain are important. For these two configurations, only the voltage gain is significant, since when gate current is zero, current and power gain are infinite. Of the two, only the common-source configuration can have a voltage gain greater than 1. The common drain A_V, it will be shown, is always less than 1. However, the common drain has an output impedance considerably lower than the common source while retaining its high input impedance. Therefore, it is useful as an impedance-matching device. Because of its low output impedance and because it offers power gain as well, the common drain is also widely used as a power amplifier.

5.4.1 Common-Source Circuit

The small-signal circuit for the common-source FET amplifier is shown in Figure 5.22. The input impedance of the FET, being extremely high, may be disregarded, so R_G, the physical resistor connected between gate and source, constitutes the circuit input resistance. The current source in the output circuit is given by $-g_m v_{GS}$, and the internal, output resistance of the FET, by r_D $(1/y_{os})$.

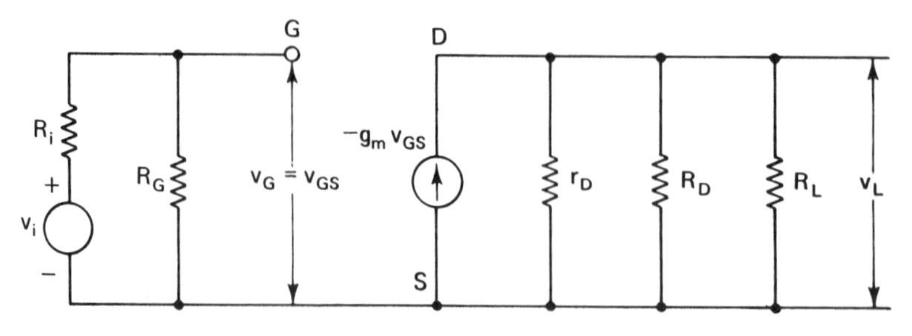

FIGURE 5.22 Common-source small-signal circuit of the FET amplifier.

The transconductance at $v_{GS} = 0$ (g_{mo}) is given by the manufacturer for the JFET and the depletion–enhancement MOS. The g_m for a specific quiescent operating point is then given by

$$g_m = g_{mo}\left(1 - \frac{V_{GSQ}}{|V_{GS(\text{off})}|}\right) \tag{5.13}$$

This transconductance is only slightly affected by V_{DS} above pinch off. The transconductance of the enhancement MOS is given by

$$g_m = 2k(V_{GSQ} - V_T) \tag{5.14}$$

where k (or the data needed to compute k) is supplied by the manufacturer.

The dynamic output resistance, r_D (or $y_{os} = 1/r_D$) is also supplied by the manufacturer for some $V_{DS} > V_{po}$ for all FETs. This resistance reflects the slope of the drain characteristic and for all FETs is almost independent of I_{DQ} or V_{DSQ} above pinch off. Therefore, the r_D supplied may be used as is for the small-signal circuit.

Once the circuit parameters, g_m, r_D, R_L, R_D, R_i, and R_G are known, the voltage gain (A_V) is found as follows:

$$A_V = \frac{v_L}{v_i} \tag{5.15}$$

$$v_L = -g_m v_{GS} R_o \tag{5.16}$$

where $R_o = r_o||R_D||R_L$ and

$$v_i = \frac{v_{GS}(R_i + R_G)}{R_G} \tag{5.17}$$

(v_i is derived from the voltage-divider equation). A_V is then given by

$$A_V = \frac{v_L}{v_i} = \frac{-g_m v_{GS} R_o}{v_{GS}(R_i + R_G)/R_G} = \frac{-g_m R_o}{(R_i + R_G)/R_G} \tag{5.18}$$

If, as is usually the case, $R_G \gg R_i$, A_V simplifies to

$$A_V = -g_m R_o \tag{5.19}$$

Since the gate current is assumed to be zero, the current gain (A_I) is infinite. The output impedance of the FET is given by

$$R_{\text{out}} = r_D||R_D \tag{5.20}$$

Gain specifications are usually given as minimums, so in our computations we will use those values that give us the minimum gain, $g_{mo\,\min}$ and $r_{D\,\min}$. An example follows.

Example 5.5 *Finding the Voltage Gain of a FET Amplifier*

Using the circuit of Figure 5.8 and the data sheet of Figure 5.3, find A_V. Let $R_i = R_L = 10\ \text{k}\Omega$. From the data sheet (Figure 5.3) for the 2N5459,

$$g_{mo}(y_{FS}) = 2000 \text{ to } 6000\,\mu\text{S}$$

$$V_{GS\,\text{off}} = -2 \text{ to } -8\ \text{V}$$

(a) *Find the lowest* g_m: Use the highest $V_{GSQ}/V_{GS\,\text{off}}$. From Figure 5.7,

$$\frac{V_{GSQ\ \text{max}}}{V_{GS(\text{off})\ \text{max}}} = \frac{-5\ \text{V}}{-8\ \text{V}} = 0.66, \qquad \frac{V_{GSQ\ \text{min}}}{V_{GS(\text{off})\ \text{min}}} = \frac{-1.2}{-2} = 0.6$$

$$g_m = g_{mo\,\text{min}}\left(1 - \frac{V_{GSQ\ \text{max}}}{V_{GS\ (\text{off})\ \text{max}}}\right) = 2000\,\mu\text{S}\,(1 - 0.66) = 680\,\mu\text{S}$$

(b) *Find* r_D: Use the lowest r_D.

$$r_D = 1/y_{os\,\text{max}} = \frac{1}{50\,\mu\text{S}} = 20\,\text{k}\Omega$$

(c) *Find* A_V:

$$A_V = -g_m R_o \qquad \text{when } R_G \gg R_i$$

$$R_G = R_1 || R_2 = 1.63\,\text{M}\Omega || 250\,\text{k}\Omega = 217\,\text{k}\Omega \qquad (R_i = 10\,\text{k}\Omega)$$

$$R_o = r_D || R_D || R_L = 20\ \text{k}\Omega || 2.25\ \text{k}\Omega || 10\ \text{k}\Omega = 1.68\ \text{k}\Omega$$

$$A_v = -680\,\mu S \times 1.68\ \text{k}\Omega = -1.14$$

Notice that the A_V is low for this example. This is a result of the low minimum g_{mo} of the JFET. Somewhat higher gains can be obtained by increasing $I_{DQ\ \text{min}}$. Still higher gains can be obtained from a higher $g_{mo\ \text{min}}$ FET. A 2N4416 with a $g_{mo\ \text{min}}$ of 4500 μS gives an A_V of 2.5. Yet higher gains require cascading. Three cascaded stages using the 2N4416 give an A_V of 16.

5.4.2 Common-Drain Small-Signal Circuit

The small-signal circuit for the common-drain amplifier of Figure 5.21b is shown in Figure 5.23. The input circuit is identical to the common-source amplifier. Two versions of the output circuit are shown. Version 1 follows the circuit diagram of Figure 5.21b closely except that the capacitors and the supply voltage are replaced by shorts. Version 2 is obtained by redrawing version 1.

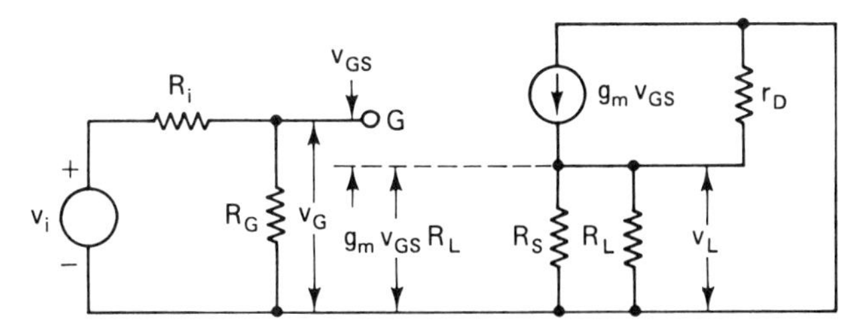

(a) Common drain small signal circuit (version 1)

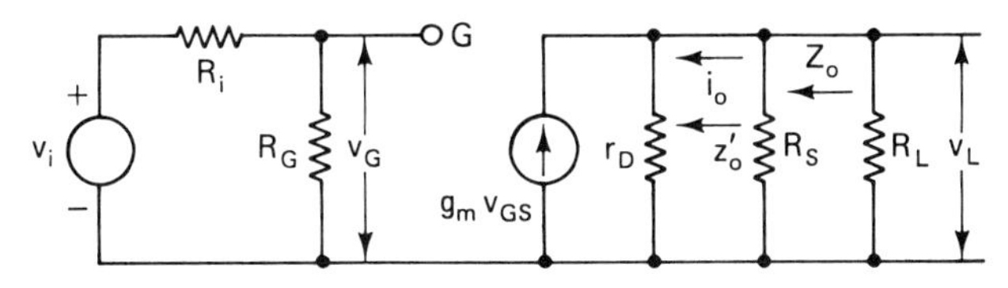

(b) Common drain small sginal circuit (version 2)

FIGURE 5.23 Common-drain small-signal circuits: (a) common-drain small-signal circuit version 1; (b) version 2.

The common-drain output circuit differs from the common source in two ways. First, $g_m v_{GS}$ is positive instead of negative so that v_L is not 180° out of phase with v_i as it is for the common source. Second, v_{GS} is not equal to v_G as it is for the common source but is given by

$$v_{GS} = v_G - g_m v_{GS} R_o \tag{5.21}$$

$$v_{GS} + g_m v_{GS} R_o = v_G \tag{5.22}$$

$$v_{GS} = \frac{v_G}{1 + g_m R_o} \tag{5.23}$$

where $R_o = R_s || R_L || r_D$.

The voltage gain (A_V) can be computed from Figure 5.23 as follows:

$$A_V = \frac{v_L}{v_i} = \frac{g_m v_{GS} R_o}{v_i} = \frac{g_m v_G R_o}{v_i (1 + g_m R_o)} \tag{5.24}$$

Since $v_G = v_i R_G/(R_i + R_D)$,

$$A_V = \frac{R_G}{R_i + R_G} \times \frac{g_m(R_o)}{1 + g_m R_o} \tag{5.25}$$

Maximum gain occurs when $g_m R_o \gg 1$ and $R_G \gg R_i$. For this case,

$$A_V \text{ (approaches) } \frac{g_m R_o}{g_m R_o} = 1 \tag{5.26}$$

The output impedance (z_o') can be determined as follows. By definition, z_o' is given by

$$z_o' = \frac{v_o}{i_o}\bigg|_{\text{with } v_G = 0} \tag{5.27}$$

where v_o is an external voltage applied across the FET output (R_L absent) and i_o is the resulting current. The circuit for the z_o computation is shown in Figure 5.24b. Since $v_G = 0$ and since any voltage applied from source to ground elevates the source above ground (see Figure 5.24a),

$$v_{GS} = -v_o \tag{5.28}$$

Referring to Figure 5.24, we see

$$i_o = \frac{v_o}{r_D} - g_m v_{GS} = \frac{v_o}{r_D} + g_m v_o \tag{5.29}$$

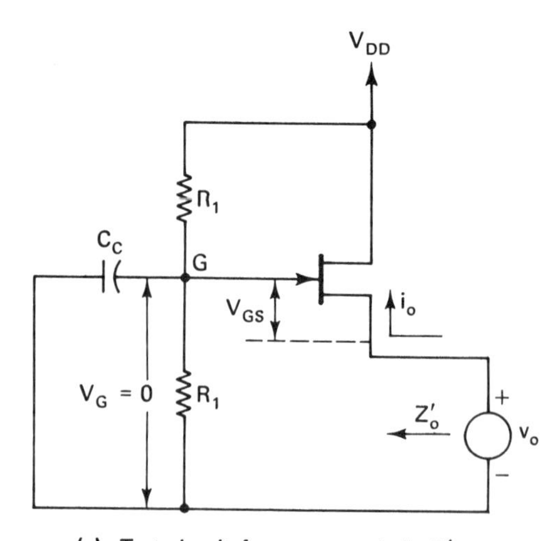

(a) Test circuit for common drain Z_o'

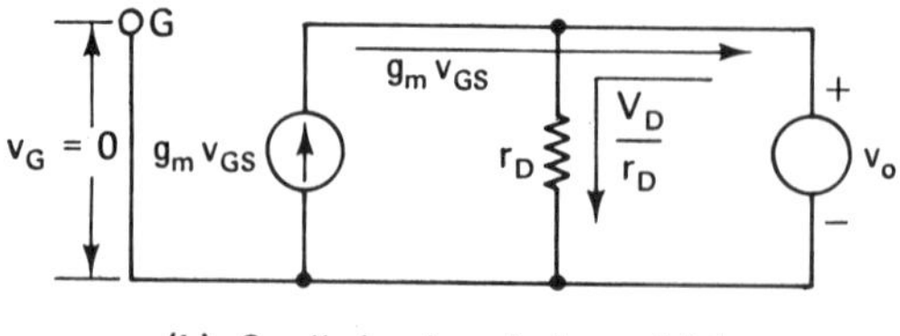

(b) Small signal equivalent of (a)

FIGURE 5.24 Common-drain circuits for finding z'_o: (a) test circuit for common-drain z'_o; (b) small-signal equivalent of (a).

Substituting Eq. (5.29) in Eq. (5.27),

$$z_o' = \frac{v_o}{v_o/r_D + g_m v_o} = \frac{1}{(1/r_D) + g_m} \tag{5.30}$$

With $1/r_D \ll g_m$, which is always the case,

$$z_o' \approx \frac{1}{g_m} \tag{5.31}$$

Since R_S is across z_o', the total output impedance across $R_S(z_o)$ is given by

$$z_o = R_S || \frac{1}{g_m} = \frac{R_S}{1 + g_m R_S} \tag{5.32}$$

Example 5.6 *Finding A_V and Z_o for the Common-Drain Amplifier*

For the common-drain circuit of Figure 5.21b with $I_{DQ} = 3$ mA, $R_s = R_L = 5$ kΩ, $R_i = 10$ kΩ, $R_G = 250$ kΩ, $r_D = 20$ kΩ, and $g_m = 2000\ \mu$S, find A_V and Z_o.

(a) *Find A_V:*

$$R_o = r_D || R_S || R_L = 20\,\text{k}\Omega || 5\,\text{k}\Omega || 5\,\text{k}\Omega = 2.22\,\text{k}\Omega$$

Since $R_G \gg R_i$,

$$A_V = \frac{g_m R_o}{1 + g_m R_o} = \frac{2000\,\mu\text{S} \times 2.22\,\text{k}\Omega}{1 + 2000\,\mu\text{S} \times 2.22\,\text{k}\Omega} = 0.8$$

(b) *Find Z_o:*

$$Z_o = \frac{R_S}{1 + g_m R_S} = \frac{5\,\text{k}\Omega}{1 + 2000\,\mu\text{S} \times 5\,\text{k}\Omega} = 455\,\Omega \quad \left(\text{true for } \frac{1}{r_D} \ll g_m\right)$$

5.5 FREQUENCY RESPONSE OF THE COMMON-SOURCE FET AMPLIFIER

The voltage gain of the common-source FET amplifier is reduced at low frequencies by the effects of three capacitors (C_i, C_o, and C_S in Figure 5.21a). At high frequencies, the gain is reduced by the intrinsic FET capacities and the stray circuit capacitance. Since the capacitances affecting the low end of the frequency response have no effect on the high end, and vice versa, the two ends can be studied separately.

5.5.1 Low-Frequency Response for the Depletion FET Amplifier

Figure 5.25 shows a common-source depletion FET amplifier and its low-frequency (LF), small-signal circuit. The LF voltage gain of this circuit is potentially limited by the three capacitors, C_i, C_o, and C_S, as noted previously. Since the input circuit is isolated from the output, the effect of C_i can be considered independently of C_o. Also, because C_S usually reduces the LF gain before C_i and C_o have significant effect, it too can be considered independently.

Effects of C_i and C_o. To study the effects of C_i and C_o, assume that C_s is open. The input and output circuits are seen to be series RC high-pass filters. As noted in Section 4.6.1, the low-end cutoff frequency for this filter type is given by

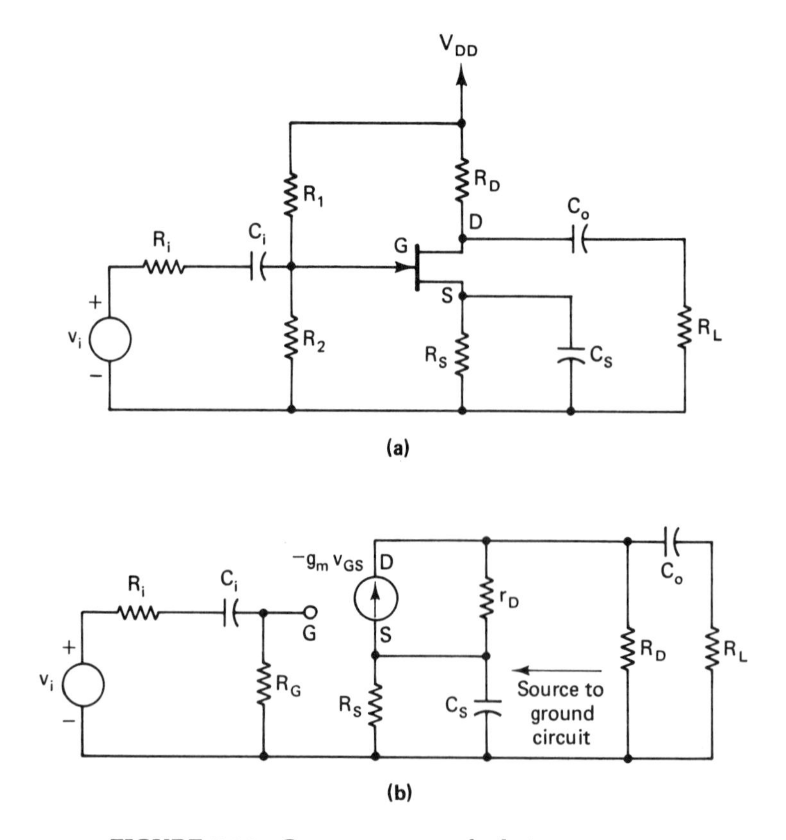

FIGURE 5.25 Common-source depletion FET LF circuit.

$$f_L = \frac{1}{2\pi R_T C} \tag{5.33}$$

For the input circuit, R_T is given by

$$R_T = R_G + R_i \tag{5.34}$$

and C is given by

$$C = C_i \tag{5.35}$$

For the output circuit, R_T to a close approximation is given by

$$R_T = r_D \,||\, R_D + R_L \tag{5.36}$$

and C is given by

$$C = C_o \tag{5.37}$$

In both cases, R_T is usually high enough so that reasonable values of capacitances suffice to meet extreme LF specificiations. Example 5.7 illustrates a typical case.

Example 5.7 *Finding C_i and C_o for a LF Specification*

For Figure 5.25b, $R_i = R_L = 10\ \text{k}\Omega$, $R_G = 300\ \text{k}\Omega$, $R_S = 1\ \text{k}\Omega$, $r_D = 20\ \text{k}\Omega$, and $R_D = 5\ \text{k}\Omega$. Find C_i and C_o for an f_L in each case of 10 Hz.

(a) *Input (C_i):*

$$C_i = \frac{1}{2\pi f_L R_T} = \frac{1}{2\pi f_L (R_i + R_G)} = \frac{1}{2\pi \times 10\ \text{Hz} \times 310\ \text{k}\Omega}$$

$$= 0.05\ \mu\text{F}$$

(b) *Output (C_o):*

$$C_o = \frac{1}{2\pi f_L R_T} = \frac{1}{2\pi f_L (r_D \,||\, R_D + R_L)}$$

$$= \frac{1}{2\pi \times 10\ \text{Hz} \times (20\ \text{k}\Omega \,||\, 15\ \text{k}\Omega + 10\ \text{k}\Omega)} = 0.856\ \mu\text{F}$$

Effect of the Bypass Capacitor (C_S). To study the effect of the bypass capacitor (C_S) on LF gain, assume that C_i and C_o are AC shorts. (This is a valid assumption since C_i and C_o can usually be made large enough to have little effect at the f_L due to C_S.) Under these conditions the LF gain is controlled by the impedance of the source-to-ground circuit; the higher this impedance, the lower the gain. This circuit is shown in Figure 5.25. It consists primarily of the source-to-ground output impedance (z_{os}) given by

$$z_{os} = \frac{R_S}{1 + g_m R_S} \tag{5.38}$$

in parallel with C_s. (Actually, $R_D||R_L + r_D$ is also across z_{os}, but if this is $\gg R_S$ it can be neglected.) As frequency falls, the total impedance $(z_{os}||Z_{CS})$ increases, thus reducing the gain of the amplifier. The resulting low-end cutoff frequency (f_L') is given by

$$f_L' = \frac{1}{\dfrac{2\pi R_S C_S}{1 + g_m R_S}} = \frac{1 + g_m R_S}{2\pi R_S C_S} \tag{5.39}$$

Equation (5.39) is only precisely true when

$$f_l' \gg f_l \tag{5.40}$$

where
$$f_l = \frac{1}{2\pi R_S C_S} \tag{5.41}$$

If Eq. (5.40) is not true, then the true low-end cutoff frequency f_{co} is given by

$$f_{co} = \sqrt{f_L'^2 - 2f_l^2} \tag{5.42}$$

Example 5.8 *Finding C_S for a Given f_L'*

In Figure 5.25, $R_S = 1\ \text{k}\Omega$, $r_D = 40\ \text{k}\Omega$, $R_D = 2\ \text{k}\Omega$, $R_L = 20\ \text{k}\Omega$, and $g_m = 3000\ \mu\text{S}$. Find C_S for $f_L' = 60$ Hz.

(a) *Find C_S:* Since $R_D||R_L + r_D = 2\ \text{k}\Omega||20\ \text{k}\Omega + 40\ \text{k}\Omega = 41.8\ \text{k}\Omega \gg R_S$, Eq. (5.39) may be used.

$$f_L' = \frac{1 + g_m R_S}{2\pi R_S C_S}$$

and
$$C_S = \frac{1 + g_m R_S}{2\pi R_S f_L'} = \frac{1 + 3 \times 10^{-3} \times 1\ \text{k}\Omega}{2\pi \times 1\ \text{k}\Omega \times 60} = 10.6\,\mu\text{F}$$

(b) *Check validity of work:*

$$f_l = \frac{1}{2\pi R_S C_S} = \frac{1}{2\pi \times 1\ \text{k}\Omega \times 10.6\ \mu\text{F}} = 15\ \text{Hz}$$

Sixty hertz is not very much larger than 15 Hz; therefore, the actual cutoff frequency is lower than 60.

$$f_{co} = \sqrt{f_L'^2 - 2f_l^2} = \sqrt{60^2 - 2 \times 15^2} = 56\ \text{Hz}$$

5.5.2 Low-Frequency Response of the Enhancement FET Amplifier

Figure 5.26 shows the actual and LF small-signal circuit for an enhancement FET amplifier. In this circuit the FET source is grounded, and the output circuit is effectively identical to the depletion FET case. Therefore, we need only consider the input circuit, which does differ from the depletion case. There R_G is returned to ground (Figure 5.25), but in Figure 5.26 R_G is connected between gate and drain. The voltage across R_G is then $V_{GS} + |A_V| V_{GS}$, and the Miller effect, discussed in Section 4.62, occurs. The apparent resistance of R_G (R_G') seen from the input circuit is given by

$$R_G' = \frac{R_G}{1 + |A_V|} \tag{5.43}$$

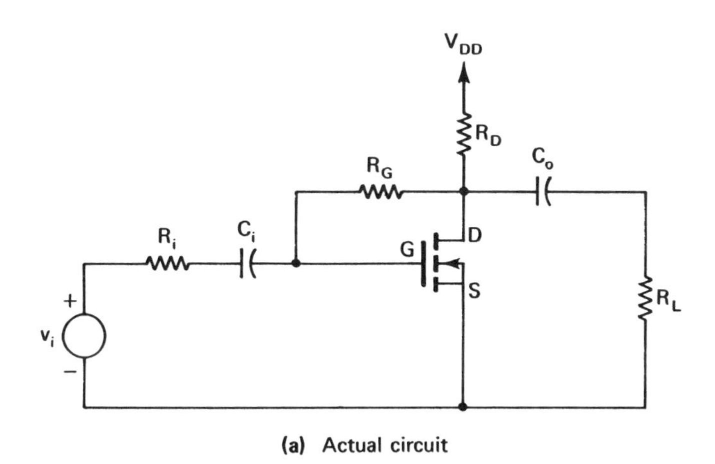

(a) Actual circuit

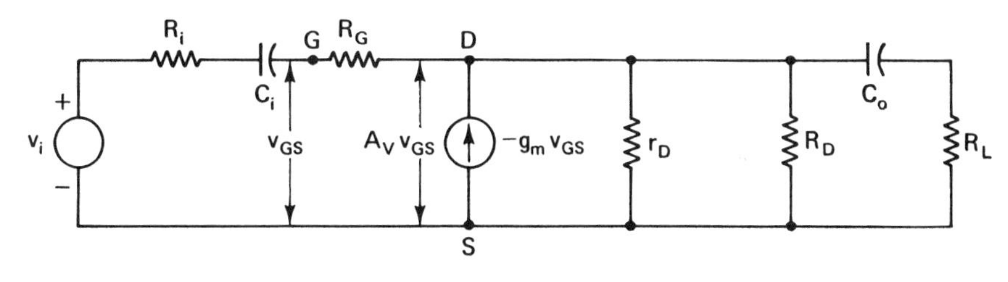

(b) Small signal circuit

FIGURE 5.26 Low-frequency circuit, enhancement FET amplifier: (a) actual circuit; (b) small-signal circuit.

and f_L is given by

$$f_L = \frac{1}{2\pi(R_i + R'_G)} \tag{5.44}$$

For high gain, R'_G may be much smaller than R_G, the actual resistance connected from gate to drain, necessitating a larger capacitance than an equivalent depletion FET circuit. Example 5.9 illustrates this effect.

Example 5.9 *Finding C_i for the Enhancement FET Amplifier*

In Figure 5.26, let $R_G = 5\ \text{M}\Omega$; R_L, R_D, r_D, and $R_i = 12\ \text{k}\Omega$; and $g_m = 5$ mS. Choose C_i for $f_L = 30$ Hz.

$$A_V = -g_m R_L||R_D||r_D = -5 \times 10^{-3}\ \text{mS} \times 4\ \text{k}\Omega = -20$$

$$R'_G = \frac{R_G}{1 + |A_V|} = \frac{5\ \text{M}\Omega}{1 + 20} = 238\ \text{k}\Omega$$

$$C_i = \frac{1}{2\pi \times f_L(R_i + R'_G)} = \frac{1}{2\pi \times 30\ \text{Hz}(12\ \text{k}\Omega + 238\ \text{k}\Omega)} = 0.021\ \mu\text{F}$$

5.5.3 High-Frequency Response of the Common-Source FET Amplifier

Two versions of a high-frequency (HF), small-signal circuit for a depletion FET amplifier are shown in Figure 5.27. This circuit is derived from the actual circuit in Figure 5.25a by shorting C_i, C_o, and C_S and adding C_{GS} and C_{GD}, the gate-to-source and gate-to-drain capacitances. Also included is the stray capacitance C_w, representing the capacitance to ground of all elements in the output. When R_G is replaced by $R'_G = R_G(1 + |A_V|)$, Figure 27b serves for the enhancement FET as well.

In Figure 5.27a, C_{GD} is seen to be connected from gate to drain. As a result, the Miller effect causes the apparent capacitor seen from the gate (C_M) to be

$$C_M = C_{GD}(1 - A_V) \tag{5.45}$$

where $A_V = -g_m R_o = -g_m r_D||R_D||R_L$, C_M is shown in Figure 5.27b in parallel with C_{GD}. Although C_{GS} is usually much larger than C_{GD}, Miller effect multiplication makes C_M significant.

The input circuit of Figure 5.27b is a parallel RC low-pass filter with a high-frequency cutoff (f_H) given by

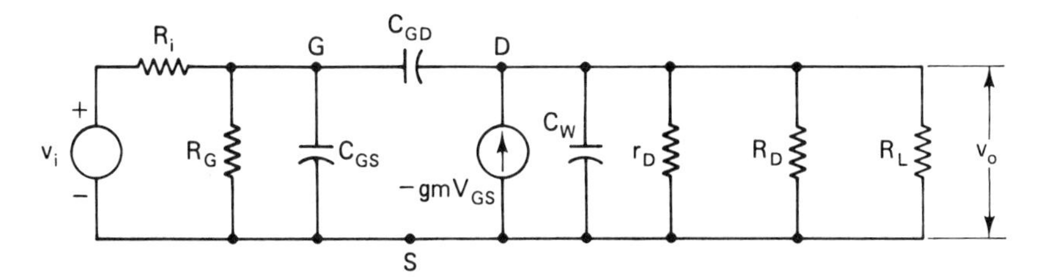

(a) Small signal HF circuit

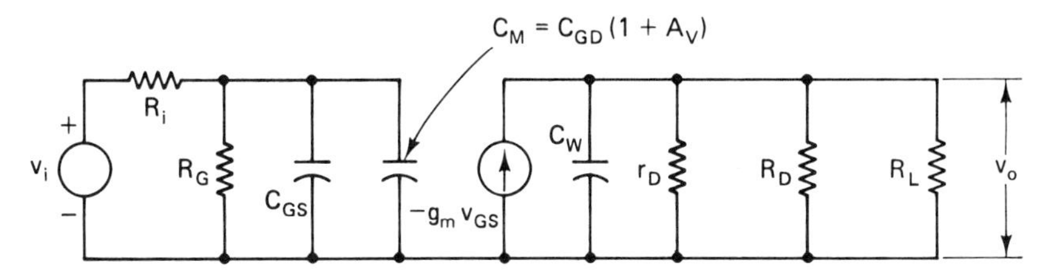

(b) Small signal HF circuit, C_{GD} replaced by equivalent capacitance C_M

FIGURE 5.27 Small-signal HF circuit for the common-source depletion FET amplifier: (a) small-signal HF circuit; (b) small-signal HF circuit, C_{GD} replaced by equivalent capacitance C_M.

$$f_H = \frac{1}{2\pi R_T C_T} \tag{5.46}$$

where $R_T = R_i || R_G$ and $C_T = C_{GS} + C_M$.

The output circuit of Figure 5.27a (or b) also resolves into a parallel RC high-pass filter. Its f_H is also given by Eq. (5.46) except that $R_T = r_D || R_D || R_L$ and $C_T = C_W$. The circuit f_H is usually determined by the input circuit with its large C_T, but if C_W is very large the output may take control.

Manufacturers sometimes give the intrinsic FET capacitances as C_{rss} (the reverse transfer capacitance) and C_{iss} (the common-source input capacitance). C_{rss} is equal to C_{GD}, and C_{iss} is the total input capacitance with $A_V = 0$ (no Miller effect). The two ways of presenting capacitance data are related by

$$C_{rss} = C_{GD} \tag{5.47}$$

$$C_{iss} = C_{GS} + C_{rss} \tag{5.48}$$

An example illustrating the computation of the HF response of an FET amplifier follows.

Example 5.10 *Computing the HF Response of the FET Amplifier*

For the amplifier of Example 5.5 (circuit of Figure 5.8), find the HF cutoff. From Example 5.5 and Figure 5.8, $R_i = R_L = 10\ \text{k}\Omega$, $R_1 = 1.63\ \text{M}\Omega$, $R_2 = 250\ \text{k}\Omega$, $R_D = 2.76\ \text{k}\Omega$, $r_D = 20\ \text{k}\Omega$, and $A_V = -1.12$. In addition, assume $C_w = 10\ \text{pF}$.

From the data sheet of Figure 5.3 for the 2N5459 JFET,

$$C_{iss\,max} = 7.0\ \text{pF}$$

$$C_{rss\,max} = 3.0\ \text{pF}$$

(a) *Find C_T and R_T for the input circuit:*

$$C_{GD} = C_{rss} = 3.0\ \text{pF}$$

$$C_{GS} = C_{iss} - C_{rss} = 7\ \text{pF} - 3\ \text{pF} = 4\ \text{pF}$$

$$C_M = C_{GD}(1 - A_V) = 3\ \text{pF}\,(1 + 1.12) = 6.36\ \text{pF}$$

$$C_T = C_{GS} + C_M = 4\ \text{pF} + 6.36\ \text{pF} = 10.36\ \text{pF}$$

$$R_T = R_i||R_G = R_i||R_1||R_2 = 10\ \text{k}\Omega||1.63\ \text{M}\Omega||250\ \text{k}\Omega$$

$$= 9.56\ \text{k}\Omega$$

(b) *Find $f_{H\,input}$:*

$$f_{H\,input} = \frac{1}{2\pi R_T C_T} = \frac{1}{2\pi \times 9.56\ \text{k}\Omega \times 10.36\ \text{pF}} = 1.61\ \text{MHz}$$

(c) *Find f_H for the output circuit:*

$$R_T = r_D||R_D||R_L = 20\ \text{k}\Omega||2.75\ \text{k}\Omega||10\ \text{k}\Omega = 1.95\ \text{k}\Omega$$

$$f_{H\,output} = \frac{1}{2\pi \times 1.95\ \text{k}\Omega \times 10.36\ \text{pF}} = 7.88\ \text{MHz}$$

The input $f_H = 1.61\ \text{MHz}$, being lowest, determines the cutoff frequency.

Problems

1. For the 2N5457 JFET (Figure 5.3), determine the following parameters: maximum drain–source voltage (V_{DS}), total device dissipation (P_D), gate–source breakdown voltage (BV_{GSS}), gate–source cutoff voltage ($V_{GS(off)}$), leakage current at cutoff, g_{mo}, r_D, and I_{DSS} (typical values only).
2. For the 2N5457, find I_D and g_m for $V_{GS}/V_{po} = \frac{1}{4}$. (Use typical values.)
3. Plot V_p versus V_{GS} as V_{GS} varies from $V_{GS(off)}$ to 0 V for the 2N5457.

4. Plot I_D versus V_{GS} for the 2N5457. (Plot maximum and minimum curves.)

5. For $R_S = 1$ kΩ, $V_{DD} = 20$ V, $R_1 = 3$ MΩ, $R_2 = 150$ kΩ, and $R_D = 5$ kΩ, find I_{DQ} for the 2N5457. (Use the curves of problem 4.)

6. Using the values of problem 5, except for R_S, find R_S for $I_{DQ} = I_{DSS}/4$ (2N5457 JFET).

7. Repeat problem 6 for $I_{DQ} = I_{DSS}/8$ and also find V_{GSQ} (the quiescent V_{GS}) and V_{DSQ}. (Use typical values.)

8. Repeat problem 7 and find maximum and minimum values.

9. Using the circuit of Figure 5.2b and the I_D versus V_{GS} curves of problem 4, determine V_{DD}, R_D, R_S, R_1, and R_2 for $I_{DQ} = 0.75$ mA ± 0.15 mA.

10. Determine the bias circuit for a 2N5458 JFET for $I_c = 1$ mA ∓ 0.5 mA.

11. Using the curve of Figure 5.9, determine the $I_{DSS\ max}$ and $I_{DSS\ min}$ for a 2N5458 JFET operating in an ambient temperature of 75°C.

12. Using Figure 5.12a with $V_{DD} = 24$ V, $R_D = 8$ kΩ, $I_{DSS} = 1.0/2.0$ mA, and $g_m = 5$ mS, find $V_{DS(\text{max and min})}$ and $g_{mo(\text{max and min})}$.

13. Design a fixed-zero-bias circuit for a depletion-enhancement MOS to meet the following specifications: $V_{GSQ} = 0$ and $V_{DSQ} \geq 5$ V. Let $V_{DD} = 20$ V, $R_i = 10$ kΩ, $I_{DSS} = 2/3$ mA, and $V_{BR} = 20$ V. Find R_G, R_D and check the MOS for reverse breakdown. Draw the circuit.

14. Design a self-bias circuit for the MFE825 MOS to meet the following specifications: $I_{DQ} = 0.5$ mA ± 20%, $V_{DSQ} \geq 8$ V, $R_i = 10$ kΩ, and $V_{(BR)DSS} = 20$ V. Find R_1, R_2, R_D, V_{DD}, and R_S. Check for drain–source breakdown.

15. Using Figure 5.28, compute R_B and V_{DD} for the following specifications:
(a) $I_{DQ} = 10$ mA ± 30%
(b) $I_{DQ} = 5$ mA ± 10%
(c) $I_{DQ} = 5$ mA ± 20%

16. For the circuit of Figure 5.15, using the curves of Figure 5.28, find R_D, R_G, and V_{DD} to meet the following specifications: $I_{DQ} = 10$ mA ± 30%.

17. Given $I_{DSS} = 10$ mA, $g_{mo} = 5$ mS, and $V_{GS(\text{off})} = -5$ V, find I_D and g_m for a FET operating at $V_{GS} = -2$ V.

18. Using the circuit of Figure 5.8 with $R_D = 5$ kΩ, $R_S = 1$ kΩ, $R_1 = 800$ kΩ, $R_2 = 125$ kΩ, $R_i = R_L = 10$ kΩ, $g_{mo} = 3000$ to $7000\ \mu S$, $V_{GS(\text{off})} = -1.2$ to -2 V, and $r_D = 40$ kΩ, find $A_{V(\text{max and min})}$. What is the current gain?

19. For the common drain circuit of Figure 5.21b, with $R_1 = 2$ MΩ, $R_2 = 300$ kΩ, $R_i = 47$ kΩ, $R_L = 1$ kΩ, $R_s = 10$ kΩ, $V_{DD} = 20$ V, $V_{GS(\text{off})} = -5$ V, and $g_{mo} = 5$ mS, find A_V.

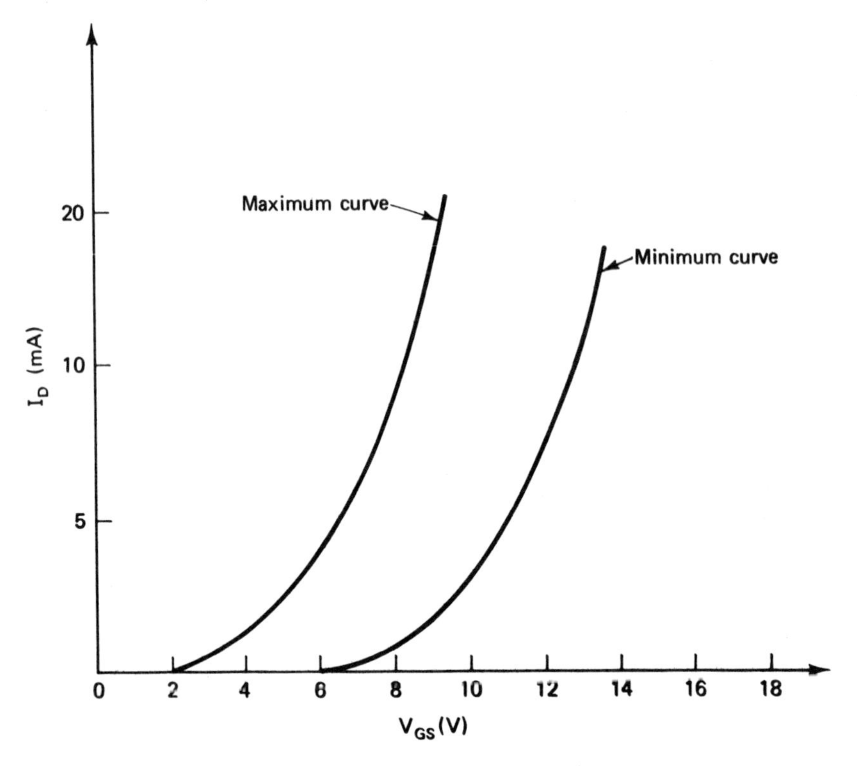

FIGURE 5.28 Problem 15.

20. For Figure 5.25b, $R_i = R_L = 22\ \text{k}\Omega$, $R_G = 250\ \text{k}\Omega$, $R_S = 1\ \text{k}\Omega$, $r_D = 20\ \text{k}\Omega$, and $R_D = 10\ \text{k}\Omega$. Find C_i and C_o for (a) $f_L = 20$ Hz, and (b) $f_L = 5$ Hz.

21. For Figure 5.25, $R_S = 2\ \text{k}\Omega$, $r_D = 20\ \text{k}\Omega$, $R_D = 5\ \text{k}\Omega$, $R_L = 30\ \text{k}\Omega$, and $g_m = 2000\ \mu\text{S}$. Find C_S for $f_L' = 30$ Hz. Repeat for $g_m = 1000\ \mu\text{S}$ and $g_m = 5000\ \mu\text{S}$.

22. An enhancement FET amplifier (Figure 5.26) has an $R_G = 2\ \text{M}\Omega$, but in operation it has an $R_G' = 250\ \text{k}\Omega$. What is the A_V of the amplifier?

23. In Figure 5.26, $R_G = 2\ \text{M}\Omega$, $R_L = 10\ \text{k}\Omega$, $R_D = 5\ \text{k}\Omega$, $r_D = 20\ \text{k}\Omega$, $R_i = 10$ kΩ, and $g_m = 2$ mS. Choose C_i to give $f_L = 15$ Hz. If R_G were returned to V_{DD} instead of the drain, what C_i would be needed to give the same f_L?

24. For a JFET, the manufacturer supplies the following data: $C_{iss\,max} = 10$ pF and $C_{rss\,max} = 4$ pF. If this JFET is used in an amplifier having $A_V = 10$, what is the total input capacity?

25. For the amplifier of Figure 5.8 with $R_i = R_L = 25\ \text{k}\Omega$, $R_1 = 2\ \text{M}\Omega$, $R_2 = 250\ \text{k}\Omega$, $R_D = 20\ \text{k}\Omega$, $r_D = 20\ \text{k}\Omega$, $g_m = 10\ \text{mS}$, $C_{iss} = 12\ \text{pF}$, and $C_{rss} = 3\ \text{pF}$, find f_H (assume $C_w = 10\ \text{pF}$).

References

Bell, David A. *Electronic Devices and Circuits,* 2nd ed., Reston Publishing Co., Reston, Virginia, 1980.

Nashelsky, Louis, and Robert Boylestad. *Devices: Discrete and Integrated,* Prentice-Hall, Inc., Englewood Cliffs, New Jersey, 1981.

Schilling, D. L., and C. Belove. *Electronic Circuits: Discrete and Integrated,* 2nd ed., McGraw-Hill Book Co., New York, 1979.

6
SPECIAL CIRCUITS

A number of important special amplifier circuits, other than the small-signal amplifiers, are widely used in electronic systems. These include power amplifiers, dc amplifiers, and differential amplifiers. These circuits introduce new considerations, require new design procedures, reflect new operating principles, and in general illustrate the potential for diversity and ingenuity in amplifier design.

6.1 POWER AMPLIFIERS: GENERAL

Power amplifiers are used to drive devices requiring substantial power, such as speakers, transmitting antennas, and servomotors. Gain is not usually an important consideration since the required driving signal can usually be provided by preceding low-level stages. Instead, the design aims at getting the maximum output power despite supply voltage and transistor limitations while maintaining extremely low distortion levels. The power amplifier stage acts as an interface between low-level stages primarily concerned with voltage or current gain and an output device requiring power.

Power amplifiers fall into three broad classifications; each is based on the percentage of time the transistor conducts within a cycle. This percentage is called the *duty cycle* and is given by

$$\text{duty cycle} = \frac{t_{on}}{T} \times 100 \tag{6.1}$$

where t_{ON} is the conduction time and T is the duration of the cycle. Table 6.1 summarizes these classifications.

6.1.1 Class A Power Amplifiers

Class A power amplifiers provide low-distortion output but at the cost of low efficiency and high transistor dissipation. The small-signal amplifiers we have studied previously were all class A. However, these configurations make poor power amplifiers because too much power is lost in their resistances, particularly R_C. A better configuration is shown in Figure 6.1. This circuit has the disadvantage of requiring an output transformer, but it provides the maximum possible efficiency for a class A amplifier. Moreover, the transformer, if tapped, permits matching the amplifier to a variety of loads while maintaining maximum efficiency.

Power relationships for the class A power amplifier are easily calculated by reference to the output characteristic of Figure 6.2. In the construction of this characteristic, R_E is assumed negligible and the transformer, ideal. These assumptions are close to reality and simplify the discussion without affecting the validity of the results. Because $R_E = 0$ and the transformer is ideal, the dc load line is erected vertically from V_{CC}. (If R_E is not zero, this load line slants as shown by the dashed line.) For maximum output (voltage, current, and power), the ac load is drawn from $2V_{CC}$ to $i_{c\,max}$, intersecting the dc load line at

$$V_{CEQ} = V_{CC} \tag{6.2}$$

$$I_{CQ} = \frac{i_{c\,max}}{2} \tag{6.3}$$

The power supplied by the power supply (P_{SS}) is constant, independent of signal, and is given by

$$P_{SS} = V_{DC}I_{DC} = V_{CC}I_{CQ} \tag{6.4}$$

TABLE 6.1 *Power Amplifiers*

Class	*Duty Cycle*	*Applications*	*Maximum Efficiency*
A	100%	High-quality audio	50%
B	50%	High-power audio	78.5%
C	< 50%	Broadcasting	Almost 100%

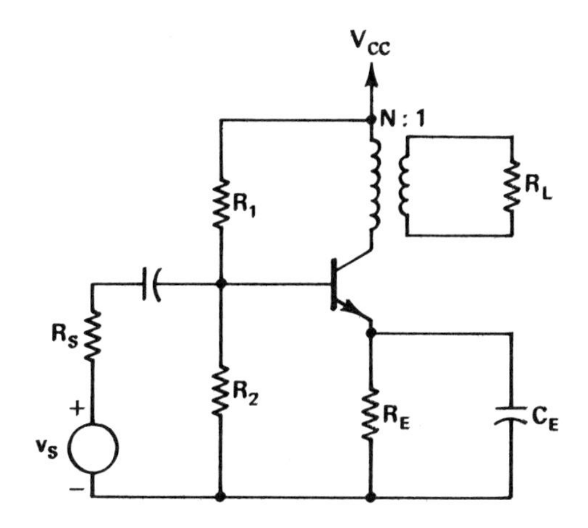

FIGURE 6.1 Class A power amplifier.

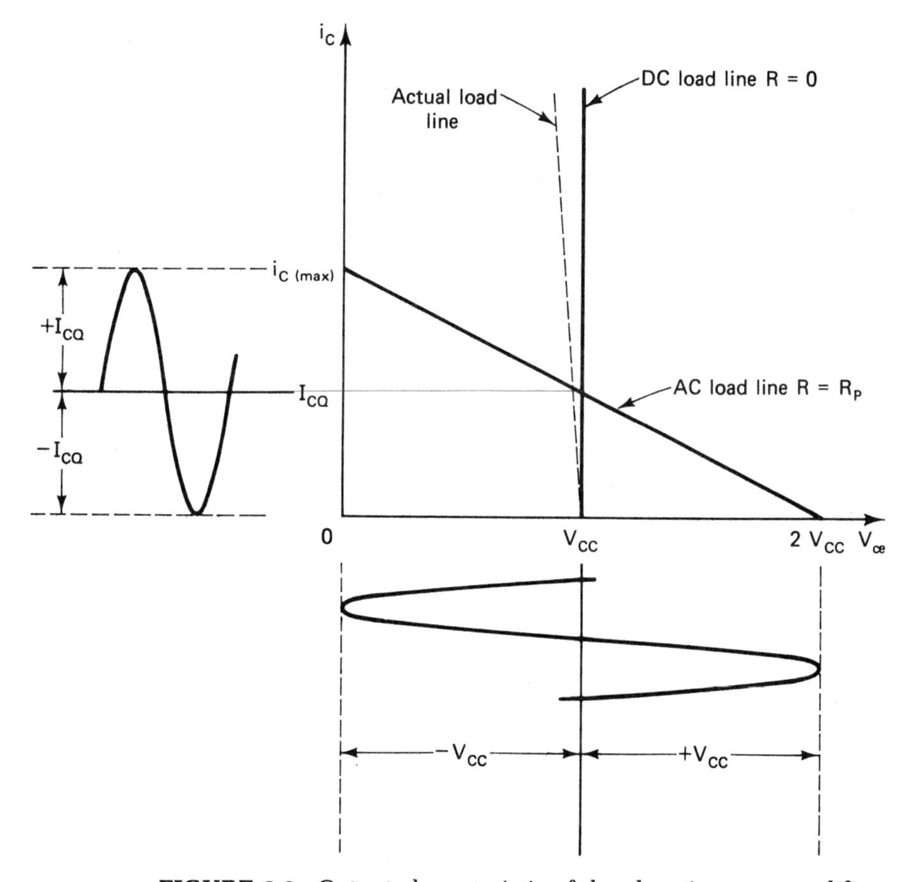

FIGURE 6.2 Output characteristic of the class A power amplifier.

In the absence of signal, P_{SS} is entirely dissipated by the transistor. Therefore, the maximum power dissipated by the transistor, $P_{D\,max}$, occurs in the absence of signal and is given by

$$P_{D\,max} = P_{SS} = V_{CC}I_{CQ} \tag{6.5}$$

The maximum load power, $P_{L\,max}$, is generated when the transistor swings $\pm V_{CC}$ and $\pm I_{CQ}$ (Figure 6.2). $P_{L\,max}$ is then given by the product of the rms voltage and current.

$$P_{L\,max} = \frac{V_{CC}}{\sqrt{2}} \times \frac{I_{CQ}}{\sqrt{2}} = \frac{V_{CC}I_{CQ}}{2} \tag{6.6}$$

The efficiency (η) of the amplifier is given by

$$\eta = \frac{P_{L\,max}}{P_{SS}} \times 100 = \frac{V_{CC}I_{CQ}}{2V_{CC}I_{CQ}} \times 100 = 50\% \tag{6.7}$$

A useful relationship, the figure of merit (F_M), is given by

$$F_M = \frac{P_{D\,max}}{P_{L\,max}} = \frac{V_{CC}I_{CQ}}{V_{CC}I_{CQ}/2} = 2 \tag{6.8}$$

The ac resistance (R_p) seen by the output circuit is given by the inverse slope of the ac load line:

$$R_p = \frac{1}{m} = \frac{V_{CC}}{I_{CQ}} \tag{6.9}$$

The ideal transformer relationships depend on the fact that it is a lossless device. Therefore,

$$P_p = P_s \tag{6.10}$$

where P_p is the power into the primary and P_s is the power out of the secondary. Also, transformer action gives

$$\frac{V_p}{V_s} = \frac{N_p}{N_s} = N:1 \tag{6.11}$$

where V_p and V_s are, respectively, the voltages (either peak or rms) across the primary and secondary windings and N_p/N_s or $N:1$ is the primary-to-secondary turns ratio. Starting with

$$P_p = \frac{V_p^2}{R_p} = P_s = \frac{V_s^2}{R_L} \tag{6.12}$$

where R_p is the ac resistance reflected into the primary because of R_L across the secondary (this is the R_p of the collector circuit of the power amplifier), we can show that the turns ratio ($N : 1$) required to match the power amplifier to R_L is given by

$$\frac{V_p^2}{V_s^2} = N^2 : 1 = \frac{R_p}{R_L} \tag{6.13}$$

and

$$N : 1 = \sqrt{\frac{R_p}{R_L}} \tag{6.14}$$

The design of the class A power amplifier starts from performance specifications, usually a P_L into an R_L using a given V_{CC}. The designer must specify P_{Dmax}, I_{CQ} and the maximum collector–emitter voltage (BV_{CEO}) for the transistor and the turns ratio ($N : 1$) of the transformer. Also, R_E must be chosen using some criteria and the bias resistors R_1 and R_2 computed. Example 6.1 illustrates this procedure.

Example 6.1 *Designing a Class A Power Amplifier*

Performance specifications:

$$P_L = 10\,\text{W}, R_L = 3\,\Omega, V_{CC} = 25\,\text{V}$$

(a) *Transistor specifications:*

$$P_{D\max} = 2P_{L\max} = 2 \times 10\,\text{W} = 20\,\text{W} = P_{SS}$$

$$P_{SS} = V_{CC}I_{CQ} = 20\ \text{W}$$

so

$$I_{CQ} = \frac{P_{SS}}{V_{CQ}} = \frac{20\ \text{W}}{25\ \text{V}} = 0.8\ \text{A}$$

$$R_p = \frac{V_{CC}}{I_{CQ}} = \frac{25\ \text{V}}{0.8\ \text{A}} = 31.25\,\Omega$$

$$BV_{CEO} = 2 \times V_{CC} = 50\ \text{V}$$

(At $i_c = 0$, $2V_{CC}$ is across the transistor.)

(b) *Transformer specifications:*

$$N : 1 = \sqrt{\frac{R_p}{R_s}} = \sqrt{\frac{31.25\,\Omega}{3\,\Omega}} = 3.23$$

(c) *Choosing the transistor:* Referring to a General Electric transistor catalog, we choose a D44CB silicon power transistor with the following specifications: $I_{C(\text{cont.})} = 4$ A, $V_{CEO} = 60$ V, and $h_{fe} = 20$. At $I_C = 2$ A, $P_D = 30$ W at 25°C. This provides safety factors and power derating above 25°C.

(d) *The bias circuit:* To satisfy the initial assumption that $R_E = 0$, choose $R_E = 0.1R_p$.

$$R_E = 0.1R_p = 0.1 \times 31.25\,\Omega = 3.13\,\Omega$$

Using the technique of Section 4.4.1,

$$R_B = 0.1\beta R_E, \qquad \beta = 20$$

$$R_B = 0.1 \times 20 \times 3.13\,\Omega = 6.26\,\Omega$$

$$V_{BB} = 1.1 \times I_{CQ} \times R_E + 0.7\,\text{V} = 1.1 \times 0.8\,\text{A} \times 3.13\,\Omega + 0.7\,\text{V}$$
$$= 3.45\,\text{V}$$

$$R_1 = \frac{V_{CC}R_B}{V_{BB}} = \frac{25\,\text{V} \times 6.26\,\Omega}{3.45\,\text{V}} = 45.36\,\Omega$$

$$R_2 = \frac{V_{CC}R_B}{V_{CC} - V_{BB}} = \frac{25\,\text{V} \times 6.26\,\Omega}{25\,\text{V} - 3.45\,\text{V}} = 7.26\,\Omega$$

Because the power equations used in Example 6.1 assumed an R_E of zero, a condition approached but not realized, it is necessary to introduce a safety factor. An R_E equal to $0.1R_p$ introduces about a 10% power loss. As a result, P_L falls short of the design goal. To make up for this loss and for additional losses (transformer loss, transistor saturation loss), the performance specification for P_L should be increased about 20%.

6.1.2 Class B Push-Pull Power Amplifier

The class B push–pull power amplifier supplies an output inherently of higher distortion than the class A, but with a large gain in efficiency and a reduction in transistor dissipation. Figure 6.3 shows a popular configuration, which has the advantage of few components, a single-ended power supply, and no transformer. However, for acceptably low distortion it requires matched complementary transistors and a bias network containing diodes matched to the transistors.

To understand the operation of the push–pull class B amplifier, consider Figure 6.4, which shows the circuit without a bias network. Also, to simplify this explanation, assume that the transistor T_1 turns on into linear operation when $v_{BE1} > 0$ and T_2 when $v_{BE2} > 0$. In the absence of signal, leakage currents through the transistor charge C_o and C_i to $V_{CC}/2$. The positive half of the input signal will now turn T_1 on (T_2 will be off). AC current will flow through T_1, C_o, and R_L, giving the output voltage v_{L1} of Figure 6.4c. The negative half of the signal will turn T_1 off and T_2 on. Although V_{CC} is now disconnected from T_2, C_o, charged to

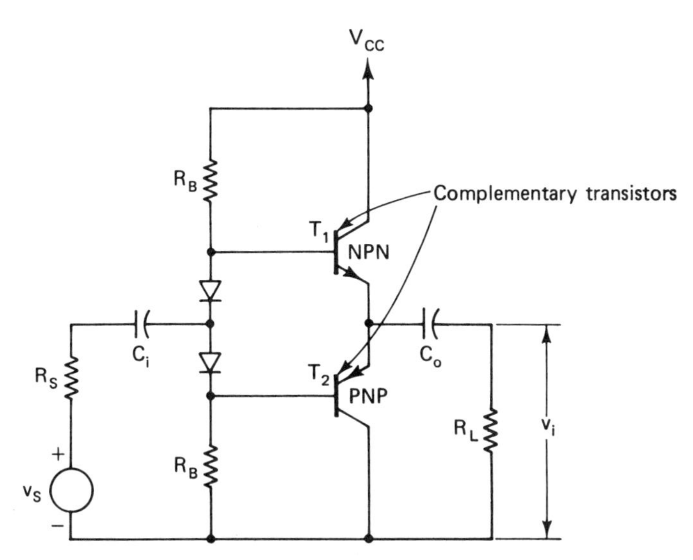

FIGURE 6.3 Class B push–pull power amplifier.

$V_{CC}/2$, is of the proper polarity to act as the power supply for T_2. AC current now flows downward from the collector of T_2, upward through R_2 and through C_o, producing v_{L2} of Figure 6.4c. Adding the two v_L's gives the desired output v_L of Figure 6.4d.

Crossover Distortion. The ideal operation described in the preceding section is marred by the fact that the turn-on region of the transistor is rounded and nonlinear, substantial linearity not being achieved until v_{BE} exceeds 0.8 V. Therefore, the lower regions of v_{L1} and v_{L2} are distorted by compression. This distortion is called *crossover distortion.* Its cause and shape are illustrated by Figure 6.5. In this figure the transconductance curve (i_c versus v_{BE}) of a transistor is shown. An input sinusoidal voltage (v_{BE}) produces an output current wave form with a compressed lower region (Figure 6.5a). Since both transistors act in this way, the resultant output is unacceptably distorted (Figure 6.4b).

Crossover distortion could be eliminated if v_{BE} could immediately cause the transistors to begin conduction in their linear regions. One way to accomplish this is to forward bias the transistors (+0.8 V for T_1, −0.8 V for T_2). As a result, amplification of the input signal will start above the lower nonlinear section of the transconductance curve and be restricted to its linear region. Each transistor will then output a fairly clean output perched on top of a small DC bias. This bias will be eliminated by C_o and the desired output obtained.

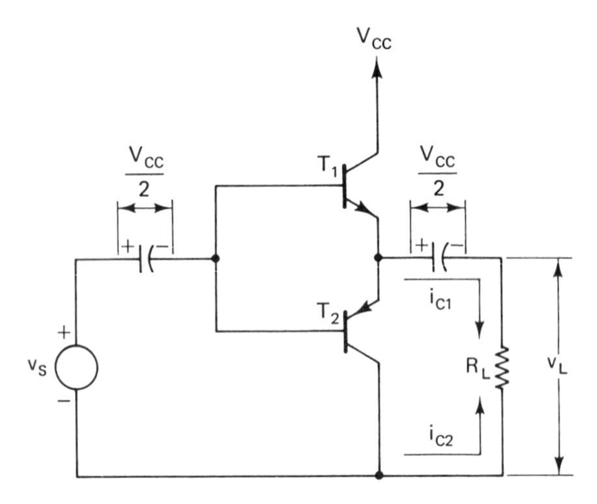

(a) Class B push pull amplifier - - unbiased

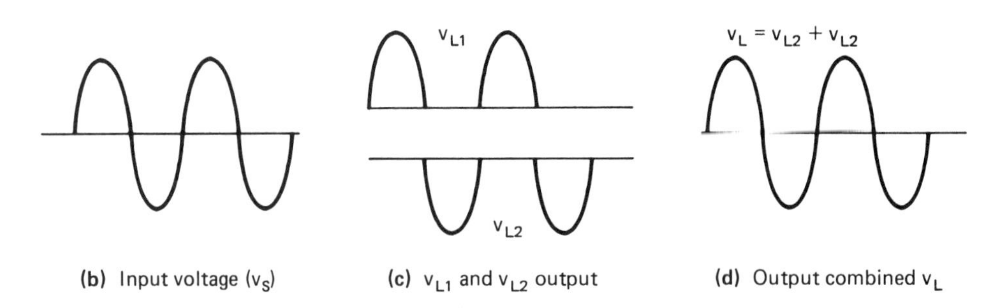

(b) Input voltage (v_S) (c) v_{L1} and v_{L2} output (d) Output combined v_L

FIGURE 6.4 Push–pull operation: (a) class B push–pull amplifier, unbiased; (b) input voltage (v_S); (c)v_{L1} and v_{L2} output; (d) output combined v_L.

Figure 6.6 shows two ways this bias could be established and shows the effect of temperature on the Q points arising from each procedure. Constant voltage bias established by the purely resistive circuit of Figure 6.6b produces considerable Q-point shifts with temperature. The constant-current circuit of Figure 6.6c is preferable.

For the circuit of Figure 6.6c to furnish a constant bias current, the two diodes must have transconductance curves (i_D versus v_D) similar to the transistors'. Also, they should be so positioned that they share the same ambient temperature as the transistors. Then, with reference to Figure 6.6c, we can show that $I_D = I_{CQ}$ and that I_D is almost independent of temperature.

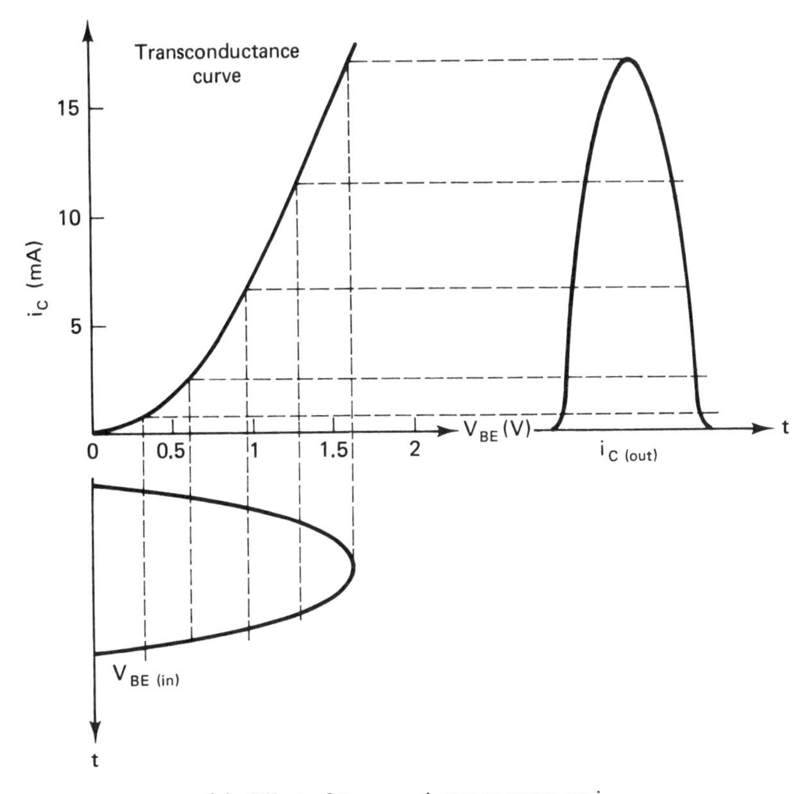

(a) Effect of transconductance curve on $i_{C\ (out)}$

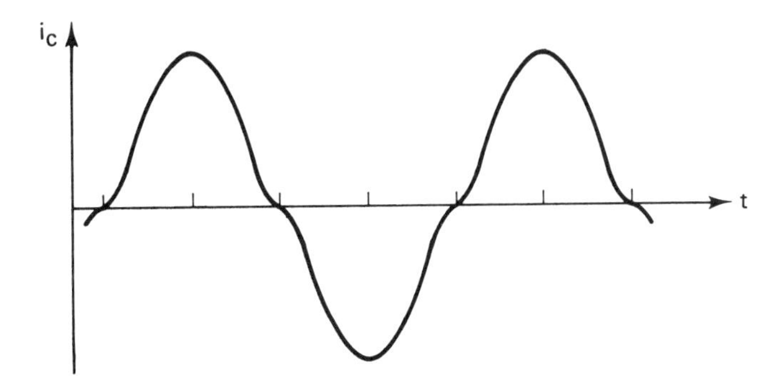

(b) $i_{C\ (out)}$ for push pull class B amplifier without bias

FIGURE 6.5 Crossover distortion in the class B push–pull amplifier: (a) effect of transconductance curve on i_C (out); (b) i_C (out) for push-pull Class B amplifier without bias.

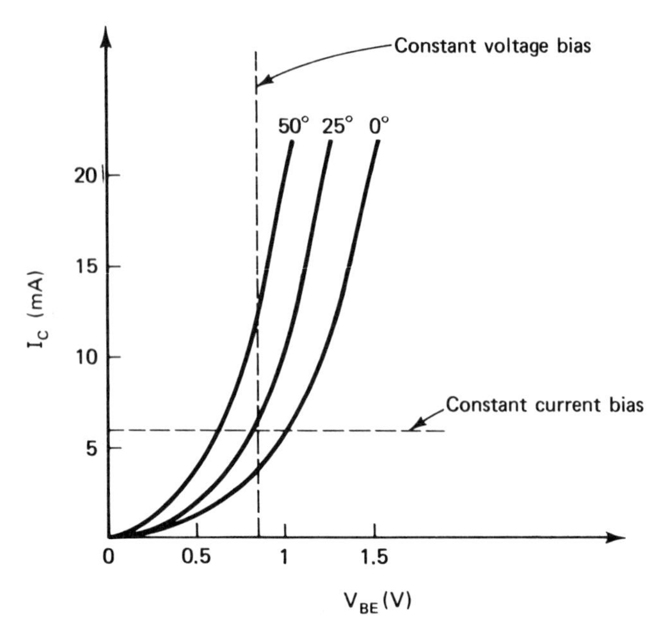

(a) Temperature effect on transconductance

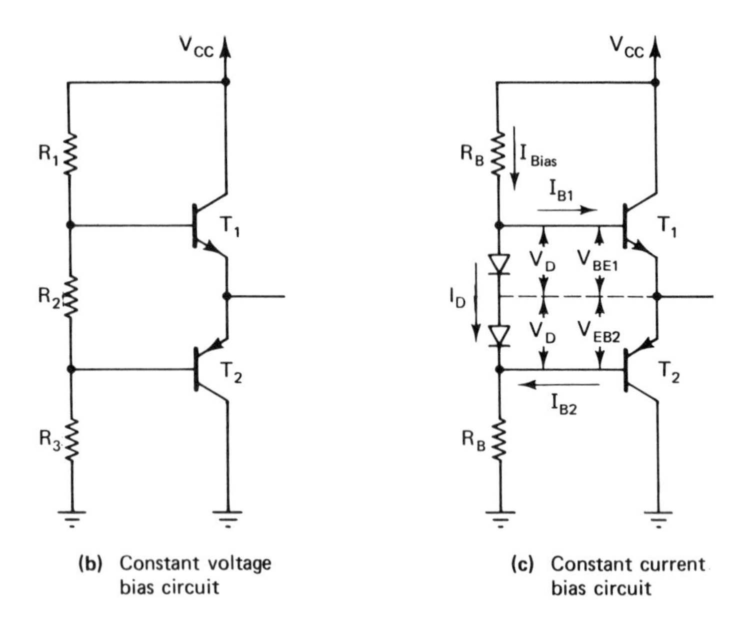

(b) Constant voltage bias circuit

(c) Constant current bias circuit

FIGURE 6.6 Bias circuits for the class B push–pull amplifier: (a) temperature effect on transconductance; (b) constant voltage bias circuit; (c) constant current bias circuit.

Figure 6.6c shows that the voltage drop across both diodes ($2V_D$) is equal to the drop across both base–emitter circuits of T_1 and T_2 ($V_{BE1} + V_{EB2}$). Therefore,

$$V_D = V_{BE1} = V_{EB2} \tag{6.15}$$

These voltages are of the correct polarity to forward bias both transistors. Also, because the diode and transistor transconductance curves are identical and the same voltage exists across both, I_D must equal I_{CQ}. For this reason the circuit is sometimes called a *current mirror*. Finally, because most of the supply voltage exists across the bias resistances ($2V_D = 1.4$ V as compared to a typical $V_{CC} = 25$ V), the current (I_{Bias}) through R_B is almost constant as well. Ignoring the small base currents (I_{B1} and I_{B2}), $I_{\text{Bias}} = I_D$. Therefore, I_D and I_{CQ} are also constant.

Power Relations. Power relations for the class B are readily derived with the aid of its output characteristic, which is shown in Figure 6.7. Here the output characteristic of each transistor is positioned to reflect push–pull action. Bias current is zero, and the output current and voltage wave shapes are assumed distortionless. As was true for the class A, we must determine $P_{L\,\max}$ (maximum supplied power), F_M (figure of merit), and η (efficiency).

$P_{L\,\max}$ is supplied when the amplifier is driven to maximum undistorted output voltage and current. Figure 6.7 shows that for sinusoidal input this condition is reached when $V_{\text{peak}} = V_{CC}/2$ and $I_{\text{peak}} = I_{C\,\text{sat}}$. Therefore,

$$P_{L\,\max} = V_{\text{rms}} I_{\text{rms}} = \frac{V_{CC}}{2\sqrt{2}} \times \frac{I_{C\,\text{sat}}}{\sqrt{2}} = \frac{V_{CC} I_{C\,\text{sat}}}{4} \tag{6.16}$$

Since R_L, given by the inverse of the slope of the load line, is given by

$$R_L = \frac{V_{CC}}{2I_{C\,\text{sat}}} \tag{6.17}$$

replacing $I_{C\,\text{sat}}$ by $V_{CC}/2R_L$ gives another useful form for $P_{L\,\max}$:

$$P_{L\,\max} = \frac{V_{CC}^2}{8R_L} \tag{6.18}$$

To find $P_{SS\,\max}$, note that the power supply supplies current only during the half-cycle when T_1 is on. This current (I_{C1}) has the wave shape shown in Figure 6.8 for sinusoidal input. Its average or dc value (I_{DC}) is given by

$$I_{DC} = \frac{I_{C1\,\text{peak}}}{\pi} \tag{6.19}$$

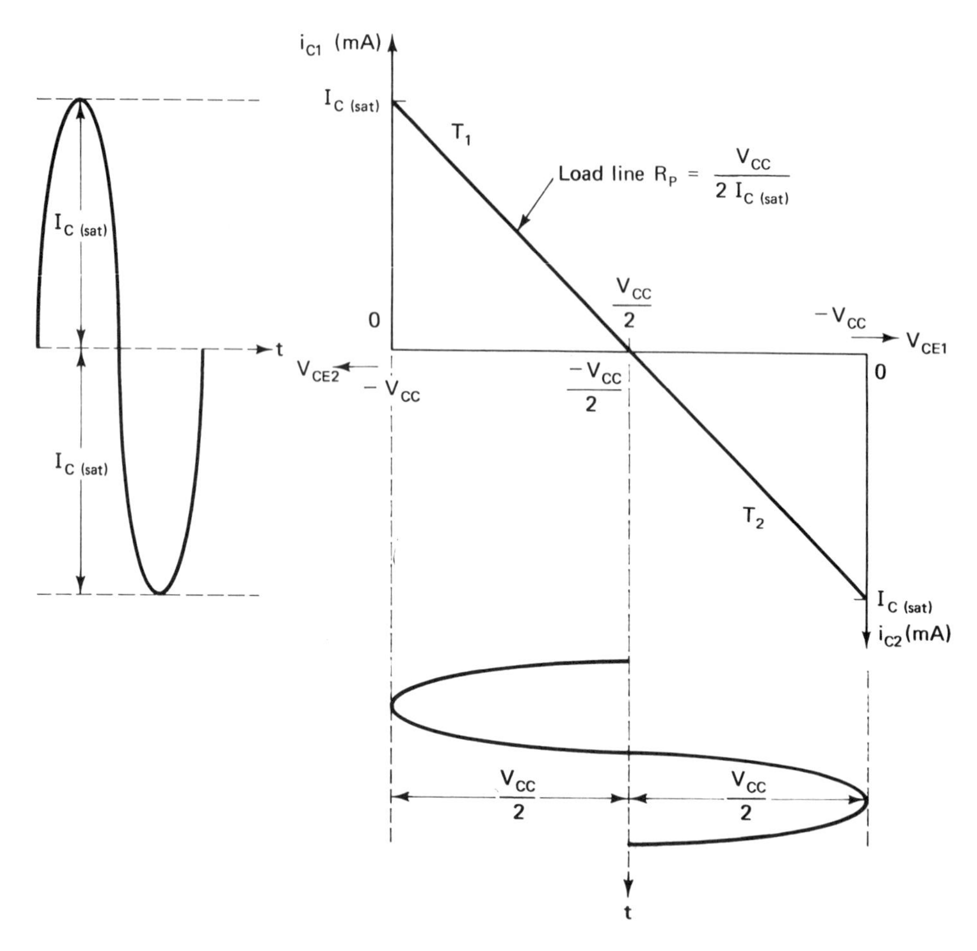

FIGURE 6.7 Class B output characteristic.

For zero input, $I_{C_1\,peak}$ equals zero and P_{SS} equals zero as well. As the input increases, $I_{C_1\,peak}$ increases to its maximum, $I_{C_1\,sat}$. Therefore,

$$P_{SS\,max} = V_{DC}I_{DC\,max} = \frac{V_{CC}I_{C_1\,sat}}{\pi} \tag{6.20}$$

P_D, the power dissipated by each transistor, is a nonsimple function of the input signal (v_s). With v_s zero, both P_L and P_{SS} are also zero. As v_s increases, both P_L and P_{SS} increase, but at different rates. $2P_D$ is the difference between P_{SS} and P_L and is given by

$$2P_D = P_{SS} - P_L = \frac{V_{CC}I_{C\,peak}}{\pi} - \frac{I_{C\,peak}^2 R_L}{2} \tag{6.21}$$

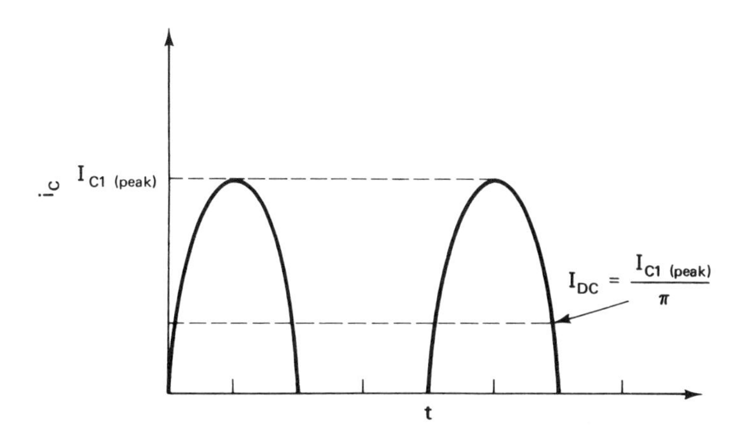

FIGURE 6.8 Power supply current.

where $I_{C\,peak}$ is the peak value of I_C for any v_s, and $I^2_{C\,peak}R_L/2$ is an alternative form of P_L at any v_s.

$P_{D\,max}$ can be found graphically from Eq. (6.20) by plotting $2P_D$ versus $I_{C\,peak}$ for $0 \leq I_{C\,peak} \leq I_{C\,sat}$. A normalized version derived by substituting $R_L = V_{CC}/(2I_{C\,sat})$ and dividing both sides by $V_{CC}I_{C\,sat}$,

$$\frac{2P_D}{V_{CC}I_{C\,sat}} = \frac{P_{SS}}{V_{CC}I_{C\,sat}} - \frac{P_L}{V_{CC}I_{C\,sat}} = \frac{I_{C\,peak}}{I_{C\,sat}\,\pi} - \frac{I^2_{C\,peak}}{4I^2_{C\,sat}}$$

is shown in Figure 6.9. $2P_D/V_{CC}I_{C\,sat}$ maximizes at 0.1. Therefore,

$$P_D = 0.05V_{CC}I_{C\,sat} \tag{6.23}$$

The figure of merit (F_M) is given by

$$F_M = \frac{P_{D\,max}}{P_{L\,max}} = \frac{0.05V_{CC}I_{C\,sat}}{V_{CC}I_{C\,sat}/4} = 0.2 \tag{6.24}$$

Efficiency (η) is given by

$$\eta = \frac{P_{L\,max}}{P_{SS\,max}} \times 100 = \frac{V_{CC}I_{C\,sat}/4}{V_{CC}I_{C\,sat}/\pi} \times 100 = 78.5\% \tag{6.25}$$

Table 6.2 compares the power relations of the class A and class B power amplifiers. Class B has the higher efficiency, but of particular importance is its lower F_M. For example, for a P_L of 20 W, the class A requires a P_D of 20 W while a class B requires only 4 W per transistor.

Design of a Class B Power Amplifier. The design of a class B power amplifier starts from performance specifications giving P_L and R_L. The designer must specify V_{CC} and R_B for the circuit and $P_{D\,max}$, $I_{C\,sat}$ or $I_{C(dC)}$,

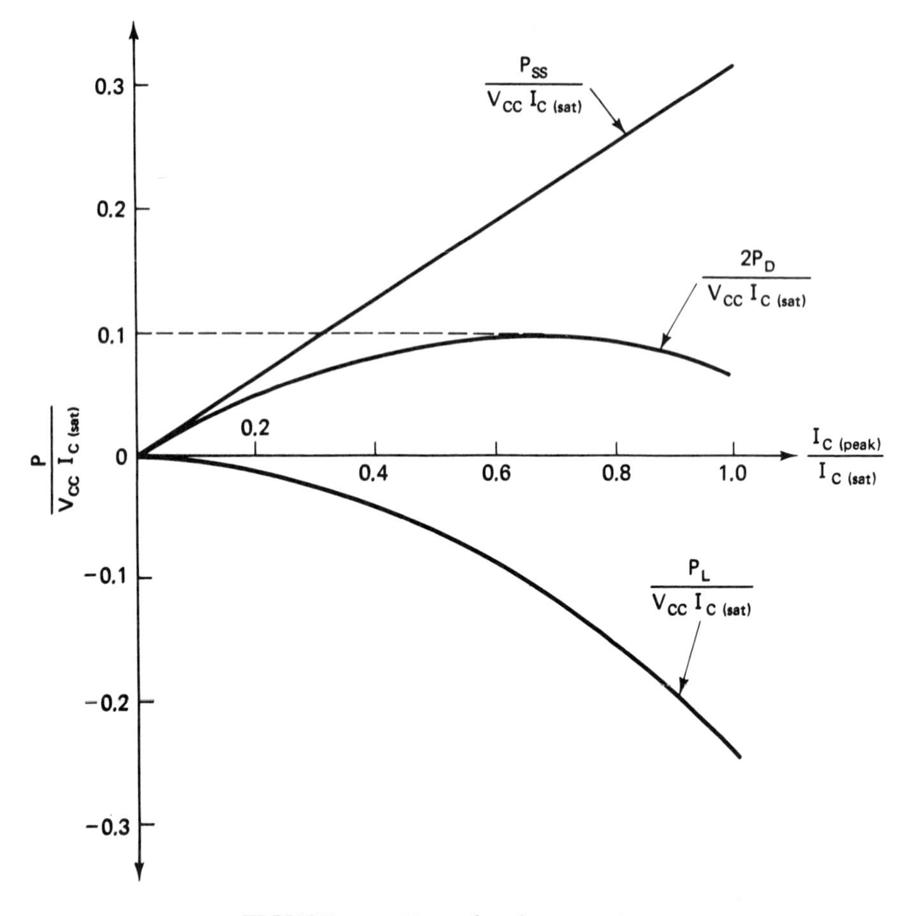

FIGURE 6.9 Normalized curve of P_D versus $I_{c\ peak}$.

and BV_{CEO} for the transistor. A bias current (I_{Bias}) of about 5% of $I_{C\ sat}$ is usually enough to reduce crossover distortion to satisfactory levels, but reference to actual transconductance curves may be in order. Either R_L or V_{CC} may be specified, but not both, since one depends on the other for a given P_L. An example of class B design follows.

TABLE 6.2 *Power Relations: Classes A and B*

Relation	*Class A*	*Class B*
$P_{L\ max}$	$V_{CC}I_{CQ}/2$	$V_{CC}I_{C\ sat}/4$
$P_{SS\ max}$	$V_{CC}I_{CQ}$	$V_{CC}I_{C\ sat}/\pi$
$P_{D\ max}$	$V_{CC}I_{CQ}$	$0.05V_{CC}I_{C\ sat}$
F_M	2	0.2
η	50%	78.5%

Example 6.2 *Design of a Class B Power Amplifier*

Performance specifications:

$$P_L = 40\,\text{W}, \qquad R_L = 5\,\Omega$$

(a) *Circuit values:* From $P_L = V_{CC}^2/8R_L$

$$V_{CC} = \sqrt{8R_L P_L} = \sqrt{8 \times 5 \times 40} = 40\text{ V}$$

$$I_{C\,\text{sat}} = \frac{V_{CC}}{2R_L} = \frac{40}{2 \times 5} = 4\text{ A}$$

From Figure 6.5c, ignoring I_{B1} and I_{B2},

$$V_{CC} = I_{\text{Bias}} \times 2R_B + 2V_D$$

where V_D is the drop across each diode. Then

$$R_B = \frac{V_{CC} - 2V_D}{2 \times I_{\text{Bias}}}$$

Letting $I_{\text{Bias}} = 0.05 \times I_{C\,\text{sat}} = 0.05 \times 4\text{ A} = 0.2\text{ A}$,

$$R_B = \frac{40\text{ V} - 1.4\text{ V}}{2 \times 0.2\text{ A}} = 96.5\,\Omega$$

(b) *Transistor specifications:*

$$P_D = F_M R_L = 0.2 \times 40 = 8\text{ W}$$

$$I_{\text{DC(each transistor)}} = \frac{I_{C\,\text{sat}}}{\pi} = \frac{4\text{ A}}{\pi} = 1.27\text{ A}$$

$$BC_{CEO} = V_{CC} \quad \text{(from Figure 6.6, where max. voltage } V_{CE} = V_{CC}) = 40\text{ V}$$

Referring to a GE transistor catalog and adding a safety factor to P_D, we select the D42C4 and its complement the D43C4:

$$P_D = 12.5\text{ W at } 25\,^\circ\text{C}, \qquad BV_{CEO} = \pm 45\text{ V}, \qquad I_{\text{(cont.)}} = \pm 3\text{ A}$$

Integrated Class B Power Amplifiers. Because transistors are inexpensive when etched into chips, monolithic (integrated) class B power amplifiers contain more special features than the discrete design. A commercially popular monolithic class B incorporating a short-circuit-proof output is shown in Figure 6.10. Instead of bias resistors, the circuit uses an integrated constant-current stage to provide bias current. An input transistor (T_1) with an unbypassed emitter resistor provides gain and an input impedance independent of R_L. A double-ended power supply eliminates the need for a dc blocking capacitor in the output. This

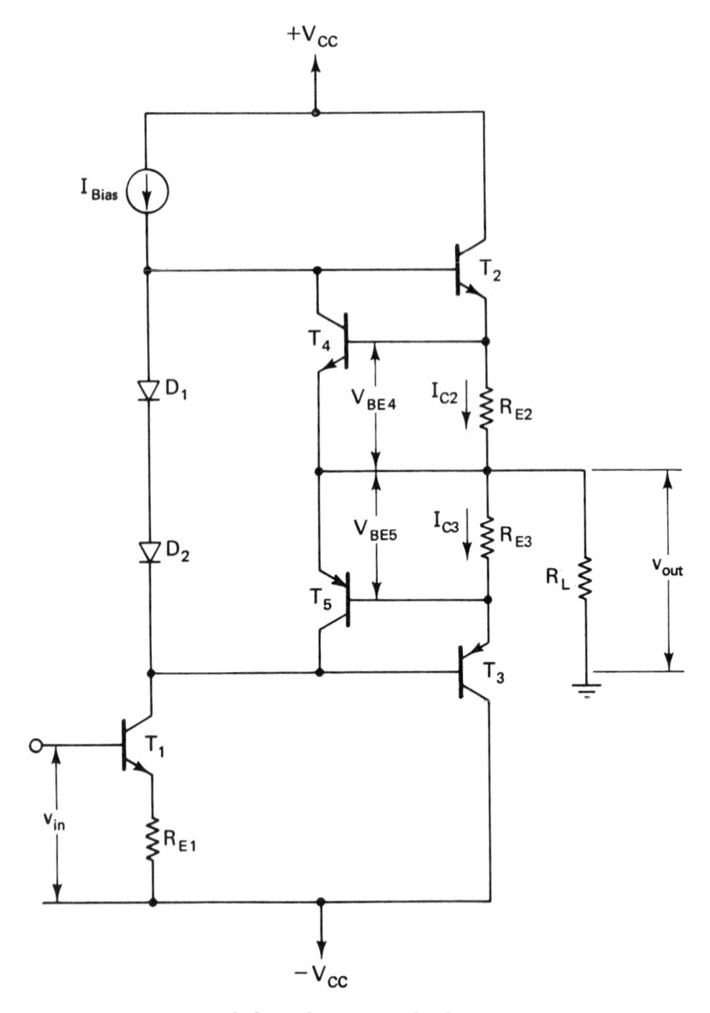

FIGURE 6.10 Monolithic class B with short-circuit protection.

capacitor is often undesirable because it limits the low-frequency response. The transistors T_4 and T_5 provide the short-circuit protection. The volt drops across R_{E2} and R_{E3} under normal operating conditions are not large enough to turn T_4 and T_5 on. Therefore, they do not affect T_2 and T_3. Suppose, however, that the top of R_L were shorted to $-V_{CC}$. If T_4 were absent, i_{c2} would increase dangerously. However, with T_4 in, i_{c2} can increase only to the point where T_4 turns on. T_4 now diverts the base drive (i_{B2}) of T_2 and limits the voltage across R_{E2} to $V_{BE4\ on}$, about 0.7 V. Consequently, i_{c2} is limited to a safe current ($V_{BE4\ on}/R_{E2}$). T_5 protects T_3 in the same way, from a short to $+V_{CC}$.

6.1.3 Class C Power Amplifier

The class C amplifier is basically a pulsed amplifier operating at duty cycles of less than 50%. As such, the transistor operates as a switch passing high currents for short times at very low collector–emitter voltages. Consequently, the average power dissipated in the transistor is small relative to the load power it controls. Being a pulsed amplifier, it cannot process the kind of signals handled by class A and B amplifiers. Its use is limited instead to signals whose essential information can be carried by pulses and whose original wave shapes can subsequently be restored. This is exactly what happens when a class C with a tuned collector is used to amplify a constant-amplitude radio-frequency (RF) signal. Because class C amplifiers are mainly used for this purpose (RF power amplification), we restrict this section to this application.

Figure 6.11 shows a typical class C amplifier with a tuned collector circuit and a clamped bias input. The collector circuit is tuned to the signal frequency. The time constant of the input circuit (T_i) is made long relative to the period (T_o) of the signal, and the signal magnitude is made large enough to drive the transistor from cutoff to saturation. Under these conditions, class C action occurs as follows, producing the wave shapes shown in Figure 6.12. The input circuit of the transistor constitutes a clamper circuit with the base–emitter circuit forming the diode (Section 3.4). Therefore, the input signal v_s becomes the clamped signal v_{Base}, as shown. Because C_i discharges slightly through R_B and R_S between

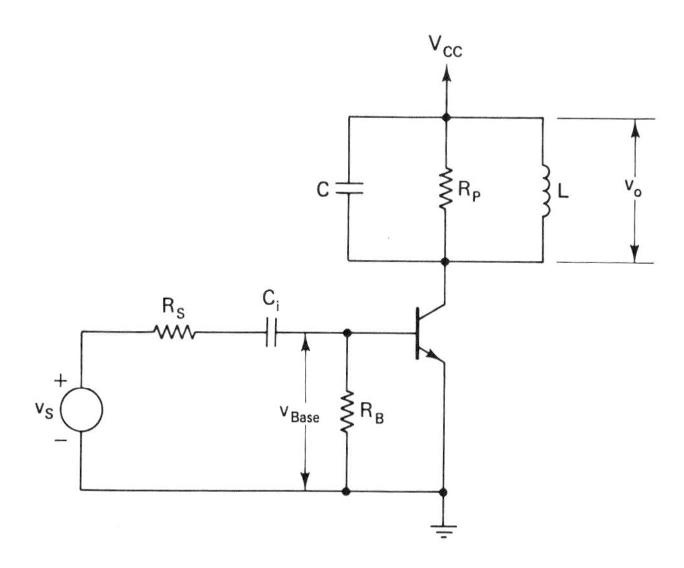

FIGURE 6.11 Tuned collector class C power amplifier.

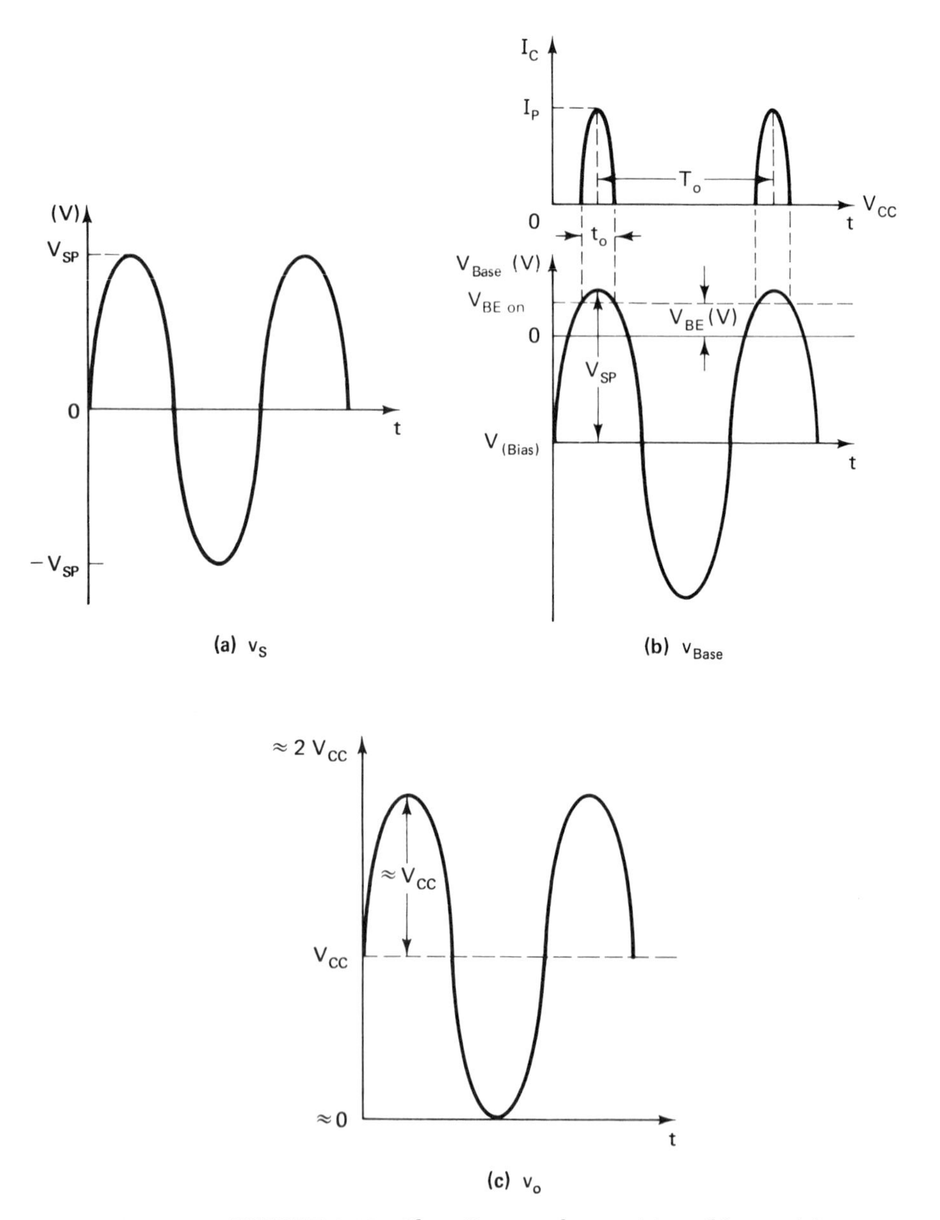

FIGURE 6.12 Class C wave shapes: (a) v_S; (b) v_{base}; (c) v_o.

cycles of the carrier, a small portion of v_{Base} will project above $v_{BE\,on}$, producing the low duty cycle current pulses of Figure 6.12b. These current pulses excite the tuned collector circuit and cause it to resonate at f_o, which restores the original sinusoidal wave, but at much higher power levels.

Tuned Circuit. The parallel tuned circuit in the collector of the transistor (Figure 6.11) transforms the current pulses into a sinusoidal wave of the same frequency as the input signal. Its action is analogous to that of a clock pendulum, which, when activated by a brief push at the appropriate time, swings steadily with precise timing. In a similar way, the class C current pulse charges the tuned circuit capacitor (C) to some voltage (v_C) at the appropriate time. At the end of the current pulse, C discharges, sending current through R_p and L. The current through R_p produces I^2R power losses, but the current through L stores energy in a magnetic field; this energy reaches maximum when $v_C = 0$. When this condition occurs, the magnetic field surrounding L begins to collapse, which causes a current flow that charges C to $-v_C$. When i_L drops to zero, C is fully charged and a half-cycle of opposite polarity begins. At the end of a full cycle, another current pulse initiates the next cycle, replenishing the energy drained from C and lost in the resistance. Because of constraints on the rate of change of v_C and i_L introduced by C and L, the voltage and currents must be sinusoidal and of a fixed frequency.

The parallel tuned RLC circuit is specified by its center frequency (f_o) and its bandwidth (B). Also useful is its quality factor (Q_c). These parameters can be derived by straightforward network analysis. They can be found in most network texts and so will not be repeated here. For the high Q case ($Q_c \geq 10$), they are defined as follows:

$$f_o = \frac{1}{2\pi\sqrt{LC}} \tag{6.26}$$

$$B = \frac{1}{2\pi R_p C} \tag{6.27}$$

$$Q_c = \frac{f_o}{B} = 2\pi f_o RC \tag{6.28}$$

They are illustrated in Figure 6.13, which shows a test circuit and a graph of the voltage across the tuned circuit, $V_{o(\omega)}$, versus the frequency of an applied constant magnitude sinusoidal current.

The specifications that the tuned circuit must satisfy depend on whether it is amplifying a single frequency or a band of frequencies. For the single-frequency case (amplifying the output of a low-level oscillator for example), the important specification is the center frequency (f_o) and R_p, the impedance at f_o. For this application, Q_c is kept as high as is feasible, consistent with other practical factors. For the second case (amplifying a frequency-modulated carrier for example), the tuned circuit must also meet the bandwidth specification so that the extreme frequencies of the signal are not unacceptably attenuated at output.

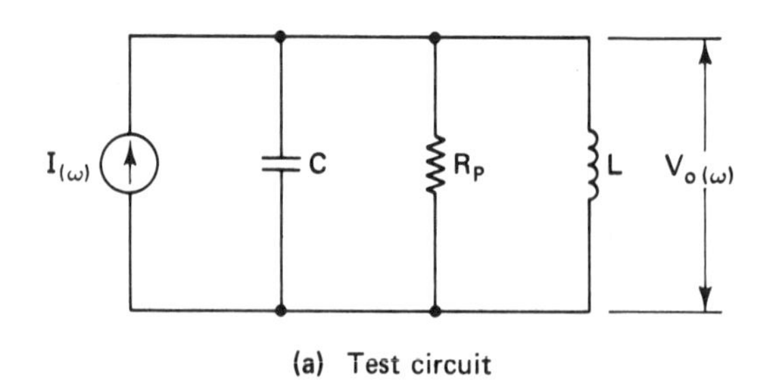

(a) Test circuit

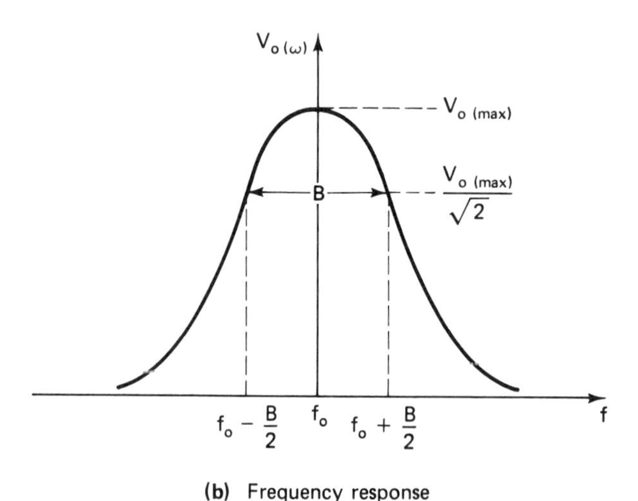

(b) Frequency response

FIGURE 6.13 Frequency response of the tuned circuit: (a) test circuit; (b) frequency response.

Real tuned circuits differ in some respects from the ideal circuit of Figure 6.13a. In that circuit, C and L are lossless so that all the circuit resistance is embodied in R_p. In real circuits, L also introduces a significant resistance (representable as either a series or parallel resistance), which dissipates power. This resistance (R_s' for the series case, R_p' for the parallel representation) is indirectly given by the inductor quality factor (Q_L) given by

$$Q_L = \frac{2\pi f_o L}{R_s'} = \frac{R_p'}{2\pi f_o L} \tag{6.29}$$

To avoid undue power loss in the inductor, Q_L should be at least $10Q_c$.

The design of a tuned circuit usually starts with an R_p derived from some previous calculation, the center frequency, and the bandwidth.

The designer must determine L, C, and the Q_L of the inductor. An example follows.

Example 6.3 *Tuned Circuit Design*

Performance specifications:

$$f_o = 1\,\text{MHz}, \qquad B = 10\,\text{kHz}, \qquad R_p = 10\,\text{k}\Omega$$

Find L, C, and Q_L. Since $B = 1/2\pi R_p C$

$$C = \frac{1}{2\pi B R_p} = \frac{1}{2\pi \times 10\,\text{kHz} \times 10\,\text{k}\Omega} = 1591\,\text{pF}$$

Since $f_o = \dfrac{1}{2\pi\sqrt{LC}}$,

$$L = \frac{1}{(2\pi f_o)^2 C} = \frac{1}{(2\pi \times 1\,\text{MHz})^2 \times 1591\,\text{pF}} = 15.9\,\mu\text{H}$$

$$Q_L \geq 10\,Q_c \geq \frac{10 \times f_o}{B} \geq \frac{10 \times 1\,\text{MHz}}{10\,\text{kHz}} \geq 1000$$

Power Relations. Power relations for the class C amplifier can be determined with the aid of the output characteristic shown in Figure 6.14. The load line in this figure is appropriate only for signals at or very close to f_o, where the impedance of the tuned circuit is effectively R_p. The current pulses at the left reach $I_{C\,sat}$ and drive v_{CE} down to $V_{CE\,sat}$. Between current pulses, clamping bias keeps the transistor at cutoff, and the tuned circuit generates the sinusoidal voltage shown.

Since the class C is usually operated at maximum drive, it is desirable to define the power relations at that level. Consequently, their definitions differ from those for the class A and B. For example, $P_{D\,max}$, which defines the maximum transistor power dissipation regardless of signal level, is replaced by $P_{D\,at\,max}$, which defines the P_D of the transistor at maximum drive. Similar alterations apply to the figure of merit (F_M) and the efficiency (η).

$P_{L\,at\,max}$ equals $P_{L\,max}$ and with reference to Figure 6.14 can be seen to be given by

$$P_{L\,max} = \frac{V_{rms}^2}{R_p} = \frac{(V_{CC} - V_{CE\,sat})^2}{2R_p} \approx \frac{V_{CC}^{\;2}}{2R_p} \tag{6.30}$$

since V_{CE} is ≤ 0.3 V, typically.

Because the class C collector current is pulsed, computation of $P_{D\,at\,max}$ involves some complications. If the current pulse were rectangular and if V_{CE} were assumed to be constant at $V_{CE\,sat}$ during the entire *on* interval, $P_{D\,at\,max}$ would be given by

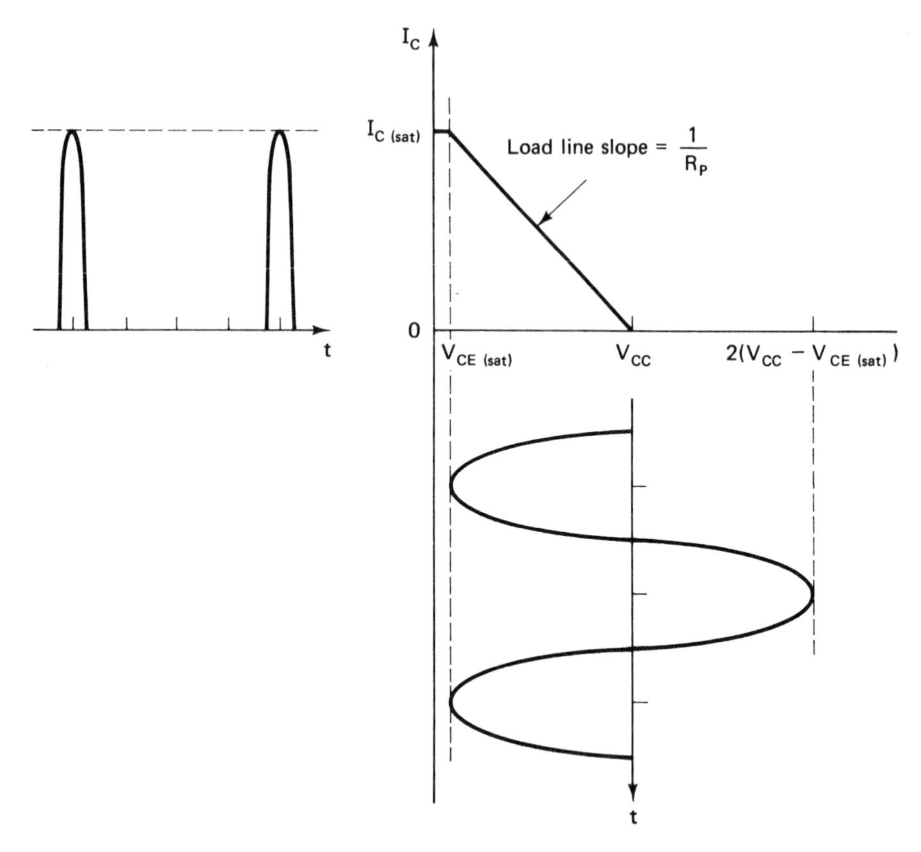

FIGURE 6.14 Output characteristic of the class C.

$$P_{D\,\mathrm{at\,max}}\,(\text{special case}) = kV_{CE\,\mathrm{sat}}I_{C\,\mathrm{sat}} \tag{6.31}$$

where k equals the duty cycle. However, because i_c and v_{CE} are sinusoidal, v_{CE} in particular being greater than $V_{CE\,\mathrm{sat}}$ during some of the *on* interval, a more conservative value of $P_{D\,\mathrm{at\,max}}$ when the duty cycle is $\leq 10\%$ is given by

$$P_{D\,\mathrm{at\,max}} = \frac{V_{CE\,\mathrm{sat}}I_{C\,\mathrm{sat}}}{2} = \frac{V_{CE\,\mathrm{sat}}V_{CC}}{2R_p} \tag{6.32}$$

Knowing $P_{L\,\mathrm{max}}$ and $P_{D\,\mathrm{at\,max}}$, we can define the figure of merit at maximum drive, $F_{M\,\mathrm{at\,max}}$, by

$$F_{M\,\mathrm{at\,max}} = \frac{P_{D\,\mathrm{at\,max}}}{P_{L\,\mathrm{max}}} = \frac{V_{CE\,\mathrm{sat}}V_{CC}}{2R_p\;V_{CC}^2/2R_p} \tag{6.33}$$

$$= \frac{V_{CE\,\mathrm{sat}}}{V_{CC}}$$

The efficiency at maximum drive, η at max, given by

$$\eta \quad \text{at max} = \frac{P_{L\,\text{max}}}{P_{SS\,\text{at max}}} \tag{6.34}$$

can be computed by letting

$$P_{SS\,\text{at max}} = P_{L\,\text{max}} + P_{D\,\text{at max}} \tag{6.35}$$

Then

$$\eta \quad \text{at max} = \frac{P_{L\,\text{max}}}{P_{L\,\text{max}} + P_{D\,\text{at max}}} = \frac{1}{1 + (P_{D\,\text{at max}}/P_{L\,\text{max}})} = \frac{1}{1 + F_{M\,\text{at max}}} \tag{6.36}$$

which, since $F_{M\,\text{at max}}$ can be very small, can approach 100%.

Driving the Class C. For maximum efficiency and lowest figure of merit, it is necessary to drive the class C amplifier into saturation. Saturation occurs when v_{ce} is ≤ 0.3 V. If, for example, v_{CE} fell only to a minimum of $V_{CC}/2$, then P_D would increase and efficiency would fall. For the case where the duty cycle is near 10%, P_D at $v_{ce} = V_{CC}/2$ is given by

$$P_{D(V_{CC/2})} = \frac{V_{CC/2} \times I_{C\,\text{sat}/2}}{2} = \frac{V_{CC}^2}{8R_p} \tag{6.37}$$

Efficiency and output power are affected by the duty cycle. Low duty cycles produce high efficiencies. However, unless the base driving signal is increased, output power also falls. The designer must opt either for high efficiency, high output power, or some compromise between the two.

Class C Design. The design of a tuned class C power amplifier starts from performance specifications establishing P_L, f_o, and B (bandwidth). Often V_{CC} is also specified. The designer must determine the transistor specifications P_D, $I_{C\,\text{sat}}$, and BV_{CEO} and the circuit parameters L, C, R_p, and Q_L. If the input time (T_i) constant is known, C_i is calculated; otherwise, T_i can be found experimentally. An example follows.

Example 6.4 *Design of a Class C Power Amplifier*

Performance specifications:

$$P_L = 100\,\text{W}, \qquad f_o = 500\,\text{kHz}, \qquad B = 20\,\text{kHz}$$

$$V_{CC} = 100\,\text{V}, \qquad T_i = 10T_o \quad (T_o = 1/f_o)$$

(a) *Finding transistor specifications:* From $P_D/P_L = F_M$,

$$P_D = F_M P_L = \frac{V_{ce\,sat}}{V_{CC}} P_L = \frac{0.3\text{ V}}{100\text{ V}} \times 100 = 0.3\text{ W}$$

From

$$P_L = \frac{V_{CC}^2}{2R_p} = \frac{V_{CC}}{2} \times \frac{V_{CC}}{R_p} = \frac{V_{CC} I_{C\,sat}}{2}$$

we obtain

$$I_{C\,sat} = \frac{2P_L}{V_{CC}} = \frac{2 \times 100\text{ W}}{100\text{ V}} = 2\text{ A}$$

From Figure 6.14,

$$BV_{CEO} = 2V_{CC} = 2 \times 100\text{ V} = 200\text{ V}$$

Provided the circuit is protected against failure of input signal, a 2-or 3-W power transistor could be safely used here.

(b) *Finding the tuned circuit parameters:* From $P_L = V_{CC}^2/2R_p$,

$$R_p = \frac{V_{CC}^2}{2P_L} = \frac{(100\text{ V})^2}{2 \times 100\text{ W}} = 50\,\Omega$$

R_p is also given by

$$R_p = \frac{V_{CC}}{I_{c\,sat}} = \frac{100\text{ V}}{2\text{ A}} = 50\,\Omega$$

From $B = 1/(2\pi R_p C)$,

$$C = \frac{1}{2\pi R_p B} = \frac{1}{2\pi \times 50\,\Omega \times 20\text{ kHz}} = 0.159\,\mu\text{F}$$

From $f_o = 1/(2\pi\sqrt{LC})$,

$$L = \frac{1}{(2\pi f_o)^2 C} = \frac{1}{(2\pi \times 500\text{ kHz})^2 \times 0.159\,\mu\text{F}} = 0.64\,\mu\text{H}$$

$$Q_c\text{ (circuit Q)} = \frac{f_o}{B} = \frac{500\text{ kHz}}{20\text{ kHz}} = 25$$

$$Q_L\text{ (inductor Q)} \geq 10Q_c \geq 250$$

(c) *Finding C_i:* Assuming the reverse-bias base–emitter resistance (r_{BE}) to be about 1 MΩ, set

$$R_B = 0.1 r_{BE} = 100\text{ k}\Omega$$

This effectively eliminates r_{BE}, which is variable and not precisely known, from the input time constant. From $T_i = 10T_o = R_B C_i$,

$$C_i = \frac{10}{R_B} T_o = \frac{10}{R_B f_o} = \frac{10}{100\text{ k}\Omega \times 500\text{ kHz}} = 200\text{ pF}$$

6.2 DIFFERENCE AMPLIFIER

The difference amplifier (sometimes called the differential amplifier) is an extremely versatile and widely used small-signal amplifier. A typical configuration is shown in Figure 6.15. As shown, the circuit has two input ports and one output, which can be taken either between collectors or from one to ground, the collector-to-ground output being the most common. The difference amplifier usually forms the major part of a monolithic amplifier etched into a chip and called an operational amplifier (op-amp). Because etched transistors can be virtually identical and are exposed to the same environment, their parameters can be considered identical. This improves operation and simplifies design.

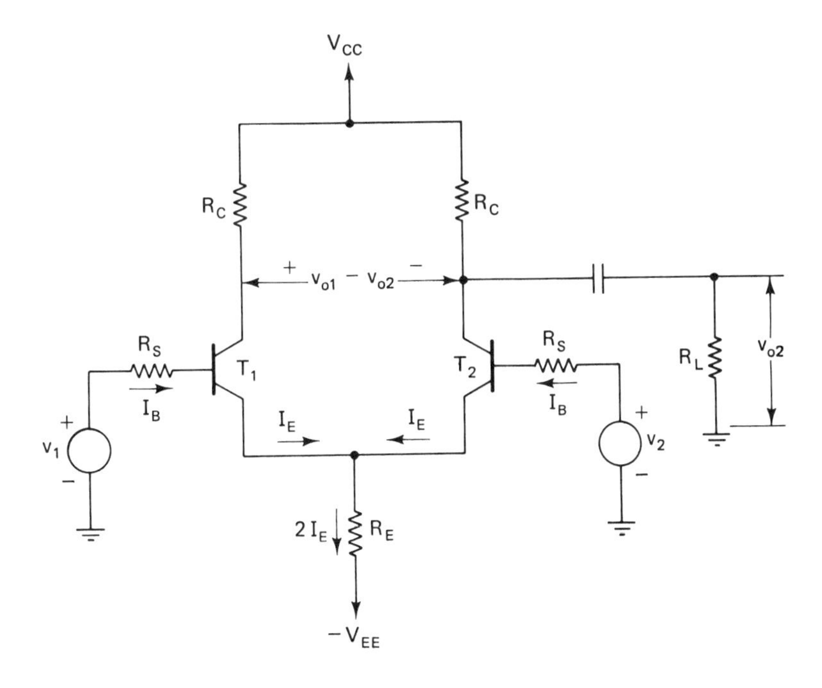

FIGURE 6.15 Difference amplifier.

An understanding of the difference amplifier is facilitated by analyzing the input signals into common-mode (v_c) and difference-mode (v_d) components. The common mode refers to the identical components of the signals, the difference mode to the unlike components. For example, two dc signals $V_{s1} = 10$ V and $V_{s2} = 14$ V can be analyzed into a common-mode component given by their average value

$$V_c = \frac{V_{s1} + V_{s2}}{2} = \frac{10\text{ V} + 14\text{ V}}{2} = 12\text{ V} \tag{6.38}$$

and into two difference mode components given by their actual minus their average value.

$$V_{d1}/2 = V_{s1} - V_c = 10\text{ V} - 12\text{ V} = -2\text{ V} \tag{6.39}$$

$$V_{d2}/2 = V_{s2} - V_c = 14\text{ V} - 12\text{ V} = +2\text{ V} \tag{6.40}$$

This analysis is by no means limited to DC signals. Any two signals, DC, AC, or otherwise, can be expressed in terms of their common- and difference-mode components and given by

$$v_{s1} = v_c + \frac{v_d}{2} \tag{6.41}$$

$$v_{s2} = v_c - \frac{v_d}{2} \tag{6.42}$$

The difference amplifier exhibits an extremely low gain to the common-mode components and a relatively high gain to the difference components. This action can be demonstrated with the aid of Figure 6.15. Consider first the case where the output (v_o) is taken between collectors. With perfectly matched transistors, both collector voltages will vary in step when the input signals are identical (common mode), and v_o will equal zero. For input signals of equal magnitude but opposite polarities (difference mode), collector voltages will vary in equal but opposite directions, and the gain will be twice that of each stage. (If the output is taken from one collector to ground, the gain will be halved.) When both components are present, both actions will occur simultaneously and suppress the common-mode components and amplify the difference mode.

The difference amplifier has several advantages over the conventional small-signal amplifier. It can distinguish differences between signals. When used as an ordinary amplifier (by feeding in only one signal and grounding the other port), an inverted or noninverted output can be obtained by choosing input ports. Output drift is very low because temperature effects act like common-mode signals. However, its chief advantage arises from the fact that positive or negative feedback (discussed later) is easily obtained. This property is exploited in op-amps.

6.2.1 Small-Signal Analysis of the Difference Amplifer

For a deeper and more useful understanding of the difference amplifier, its small-signal circuit is required. Figure 6.16 shows the small-signal circuit of the difference amplifier with a single-ended output from the collector of T_2. Because there are two transistors sharing the emitter resistor (R_E), it was necessary to refer each input circuit to the emitter. This was done by writing the base–emitter equations in terms of i_e rather than i_b and subsequently reconstructing the circuit, as follows.

$$v_s = i_b(R_s + h_{ie}) + i_e R_E \quad \text{(base–emitter equation)} \tag{6.43}$$

Substituting $i_b = i_e/(h_{fe} + 1)$ into Eq. (6.43) gives

$$v_s = \frac{i_e(R_s + h_{ie})}{h_{fe} + 1} + i_e R_E = i_e \left(\frac{R_s}{h_{fe} + 1} + h_{ib}\right) + i_e R_E \tag{6.44}$$

Interpreting Eq. (6.44) to be a circuit in which i_e flows through $R_s/(h_{fe} + 1) + h_{ib}$ and R_E, which by definition is the emitter circuit, gives the input circuit of Figure 6.16. The output circuit is the conventional one for a common-emitter amplifier except that it is driven by $i_{c2} \approx i_{e2}$. The input signals are expressed in terms of their common- and difference-mode components.

Using Figure 6.16, the operation of the difference amplifier can be described analytically. For the common-mode input component, $i_{e1} = i_{e2}$ and the volt drop across R_E is equal to $2i_{e2}R_E$. Therefore, the common-mode emitter current, $i_{c(e2)}$, is given by

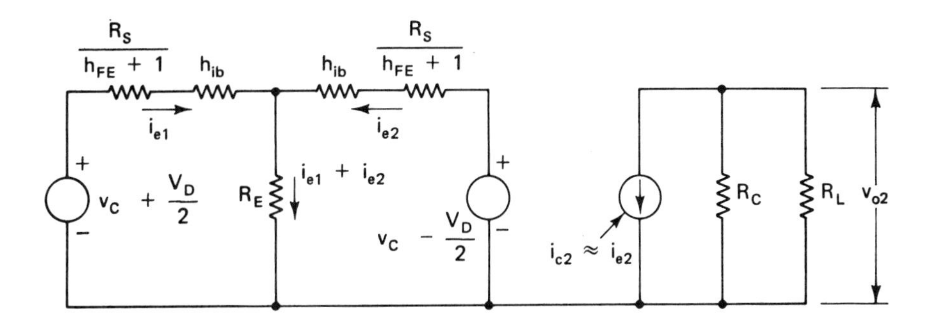

FIGURE 6.16 Small-signal circuit of the difference amplifier.

$$i_{c(e2)} = \frac{v_c}{h_{ib} + [R_s/(h_{fe} + 1)] + 2R_E} \tag{6.45}$$

In normal operation,

$$2R_E \gg h_{ib} + \frac{R_s}{h_{fe} + 1}$$

so Eq. (6.45) simplifies to

$$i_{c(e2)} \approx \frac{v_c}{2R_E} \tag{6.46}$$

In response to the difference-mode components, $i_{e1} = -i_{e2}$ (signal inputs are $+v_d/2$ and $-v_d/2$) and the volt drop across R_E equals zero. Therefore, the difference-mode emitter current, $i_{d(e2)}$, is given by

$$i_{d(e2)} = \frac{v_d/2}{h_{ib} + [R_s/(h_{fe} + 1)]} \tag{6.47}$$

which, when $R_s/(h_{fe} + 1)$ is made very much smaller than h_{ib}, simplifies to

$$i_{d(e2)} \approx \frac{v_d}{2h_{ib}} \tag{6.48}$$

Assuming $R_C \ll R_L$ (usually true), the difference-mode voltage gain (A_d) and the common-mode voltage gain (A_c) are given by

$$A_d = \frac{v_{d(o2)}}{v_d} = \frac{i_{d(e2)}R_C}{v_d} = \frac{R_C}{2h_{ib}} \tag{6.49}$$

$$A_c = \frac{v_{c(o2)}}{v_c} = \frac{i_{c(e2)}R_C}{v_c} = \frac{R_C}{2R_E} \tag{6.50}$$

where the simplified equations for $i_{d(e2)}$ and $i_{c(e2)}$ are used.

Equations (6.45) through (6.50) provide further insights into the workings of the difference amplifier. We see that R_E affects the common-mode circuit but is effectively shorted in the difference-mode circuit. Since a large unbypassed emitter resistor seriously diminishes transistor gain, this accounts for the difference in gain between the two modes.

A linear difference mode gain is an important requirement for the difference amplifier. Otherwise, distortion occurs. Since v_{o2} is given by

$$v_{o2} = A_d v_d + A_c v_c \tag{6.51}$$

we see that the last term on the right ($A_c v_c$) must be much smaller than $A_d v_d$ to achieve a v_{o2} proportional to v_d. How much less depends upon the design requirements, but a ratio given by

$$A_d v_d = 10 A_c v_c \tag{6.52}$$

is often satisfactory. Equation (6.52) can be transformed into

$$\frac{A_d}{A_c} = \frac{10 v_c}{v_d} \tag{6.53}$$

where the left term (A_d/A_c) is called the *common-mode rejection ratio* (CMRR). A sufficient condition for linearity then is given by

$$\text{CMRR} = \frac{A_d}{A_c} = \frac{R_C/2h_{ib}}{R_C/2R_E} = \frac{R_E}{h_{ib}} = \frac{10 v_c}{v_d} \tag{6.54}$$

6.2.2 Design of the Difference Amplifier

The goal of difference amplifier design is a device capable of linear amplification of small difference signals. Therefore, the primary performance specification is the ratio v_c/v_d (or CMRR). Voltage gain and input and output impedances are secondary and are obtainable by additional circuitry. The designer starting from this ratio must specify V_{CC}, V_{EE}, R_E, R_C and pertinent transistor specifications.

The design presented here is a straightforward and useful one using the approximate equations and based on certain conditions. These include the following:

$$R_C \ll R_L$$

$$2R_E \gg \begin{cases} h_{ib} + R_s/(h_{fe} + 1) \\ R_C \end{cases}$$

$$h_{ib} \gg R_s/(h_{fe} + 1)$$

$$V_{CC} = -V_{EE}$$

These conditions are not restrictive and are easily met.

The design starts with a determination of the required CMRR by

$$\text{CMRR} = \frac{10 v_c}{v_d} \tag{6.55}$$

$V_{CC} = |V_{EE}|$ is found from the CMRR by using the following relations:

$$I_{EQ} = \frac{25 \times 10^{-3}}{h_{ib}} \qquad \text{(see Section 4.7.3)} \tag{6.56}$$

and since $2R_E \gg R_s/(h_{fe} + 1)$, the emitter current I_{EQ} is given by

$$V_{CC} = 2R_E I_{EQ} = \frac{50 \times 10^{-3} R_E}{h_{ib}} = 50 \times 10^{-3}\,\text{CMRR} \tag{6.57}$$

Next a reasonable I_{EQ} is chosen, and h_{ib} is computed from Eq. (6.56). (I_{EQ} may be fixed a priori by the maximum input current swing.) R_E is then computed from

$$\text{CMRR} = \frac{R_E}{h_{ib}}, \qquad R_E = h_{ib}\,\text{CMRR} \tag{6.58}$$

R_C is computed from the differential gain (A_d):

$$A_d = \frac{R_C}{2h_{ib}}, \qquad R_C = 2h_{ib}A_d \tag{6.59}$$

However, if the restrictions $R_C \ll R_L$ and $R_E \gg R_C$ cannot be satisfied, a lower A_d must be accepted and an amplifer stage added. Finally, R_S is chosen to satisfy the restriction

$$h_{ib} \gg \frac{R_S}{h_{fe} + 1} \tag{6.60}$$

Example 6.5 *Difference Amplifier Design*

Performance specifications:

$$\frac{v_c}{v_d} = 10, \qquad A_d \geq 5, \qquad R_L = 10\,\text{k}\Omega$$

Design:

$$\text{CMRR} = \frac{10v_c}{v_d} = 100$$

$$V_{CC} = |V_{EE}| = 50 \times 10^{-3}\,\text{CMRR} = 50 \times 10^{-3} \times 100 = 5\,\text{V}$$

Letting $I_{EQ} = 2.5$ mA,

$$h_{ib} = \frac{25 \times 10^{-3}}{I_{EQ}} = \frac{25 \times 10^{-3}}{2.5\,\text{mA}} = 10\,\Omega$$

$$R_E = h_{ib}\,\text{CMRR} = 10\,\Omega \times 100 = 1000\,\Omega$$

$$R_C = 2h_{ib}A_d = 2\,\Omega \times 10\,\Omega \times 5 = 100\,\Omega$$

which satisfies the restrictions $R_C \ll R_L$ and $2R_E \gg R_C$. To satisfy $h_{ib} \gg R_S/(h_{fe} + 1)$ ($h_{fe} = 100$), let $R_S = 0.1 \times h_{ib} \times (h_{fe} + 1) = 0.1 \times 10 \times 101 = 101$. The condition $2R_E \gg h_{ib} + R_S/(h_{fe} + 1)$ is also satisfied.

6.2.3 Difference Amplifier Using a Constant Current Source

Because CMRR depends upon R_E, very large CMRRs cannot be realized with feasible V_{EE}'s. If CMRR were 10,000 in Example 6.4, a V_{EE} of -500 V would be required. Other less restrictive designs would require similar nonfeasible V_{EE}'s. The basic problem arises from the need for a large dynamic R_E together with a reasonable I_{EQ} at a feasible V_{EE}.

One solution is shown in Figure 6.17. Here R_E is replaced by a common-base transistor configuration that provides a high dynamic collector resistance ($1/h_{ob}$ typically $> \frac{1}{2}$ MΩ) and the needed I_{CQ} at feasible V_{EE}'s. I_{CQ} for T_1 and T_2 is supplied through T_3 and is given by

$$I_{CQ} = \frac{V_{EE} - 0.7}{2R_E'}$$

For $V_{EE} = 10$ V, an I_{CQ} equal to 2.5 mA is supplied by making R_E' equal to 1.86 kΩ. For these reasonable values, $1/h_{ob}$ might be 1 mΩ, giving a CMRR near 100,000, a value typical of commercially available difference amplifiers.

6.2.4 Darlington Difference Amplifier

Often a difference amplifier with a higher input impedance than the configuration of Figure 6.17 is needed. (This is only h_{ie} in the difference mode.) Also, the low R_S required by the design procedure may be difficult to provide. The Darlington-input difference amplifier of Figure

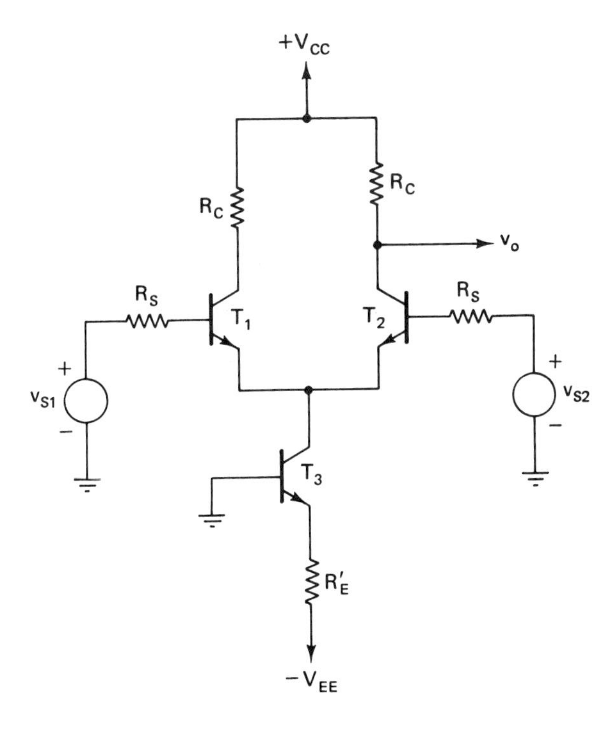

FIGURE 6.17 Difference amplifier with constant-current source.

6.18 solves both problems by giving a high input impedance and permitting a higher R_S.

The input impedance (z_i) of the Darlington difference amplifier can be derived with the aid of Figure 6.18. In the difference mode, R_E appears as a short circuit so that, looking into the base of T_1, we see h_{ie_1}. This is reflected into the base of T_4 as $(h_{fe_4} + 1)h_{ie_1}$ and is in series with h_{ie_4}. Therefore, z_i is given by

$$z_i = h_{ie_4} + (h_{fe_4} + 1)h_{ie_1}$$

Since h_{ie} varies inversely with I_{EQ} and since $I_{EQ1} = (h_{fe} + 1)I_{EQ4}$, then

$$h_{ie_1} = \frac{h_{ie4}}{(h_{fe_1} + 1)} \tag{6.62}$$

and assuming $h_{fe_1} = h_{fe_4}$,

$$z_i = h_{ie_4} + \frac{(h_{fe_1} + 1)h_{ie_4}}{(h_{fe_1} + 1)} = 2h_{ie_4} \tag{6.63}$$

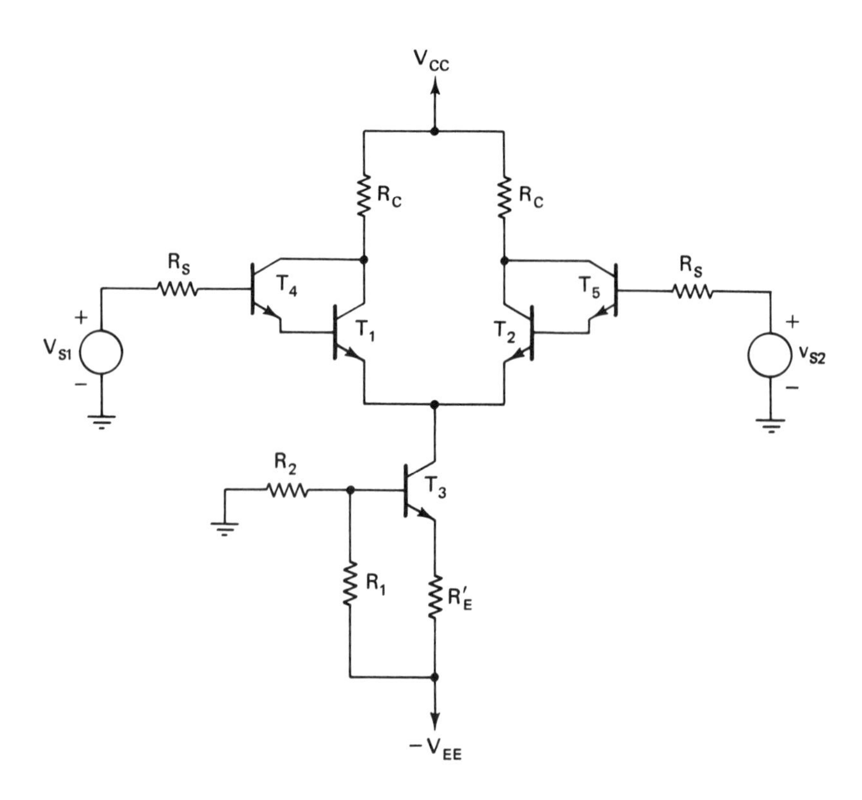

FIGURE 6.18 Darlington-input difference amplifier.

Because I_{EQ4} must be very small to prevent saturating T_1, h_{ie_4} can be very large. For example, with $h_{fe_4} = h_{fe_1} = 100$, $I_{EQ1} = 1$ mA, h_{ie_4} is given by

$$h_{ie4} = \frac{(h_{fe} + 1)25 \times 10^{-3}}{I_{EQ4}} = \frac{(h_{fe} + 1) \times 25 \times 10^{-3}}{I_{EQ1}/(h_{fe} + 1)}$$

$$= \frac{(h_{fe} + 1)^2 \times 25 \times 10^{-3}}{I_{EQ2}} = \frac{(101)^2 \times 25 \times 10^{-3}}{1 \text{ mA}} = 255 \text{ k}\Omega$$

and

$$z_i = 2h_{ie_4} = 510 \text{ k}\Omega$$

The small-signal circuit of the Darlington difference amplifier is shown in Figure 6.19. We saw earlier that reflecting the base circuit into the emitter divided R_S by $(h_{fe} + 1)$. In the Darlington circuit this occurs twice, reducing the reflecting source resistance to

$$\frac{R_S}{(h_{fe} + 1)^2} \tag{6.64}$$

Therefore, R_S for the Darlington configuration can be $(h_{fe} + 1)$ times larger than the non-Darlington R_S and still meet the design restrictions. Also, h_{ie_4}, which equals $(h_{fe} + 1)h_{ie_1}$ is reduced to h_{ib_1} when reflected twice. Otherwise, difference amplifier operation is unchanged.

The difference amplifier may use FETs advantageously in its input stages (T_1 and T_2 of Figure 6.17). Using FETs requires changes in biasing circuits, of course, but high input impedances are obtained without the use of the Darlington configuration. Circuit operation is otherwise unchanged.

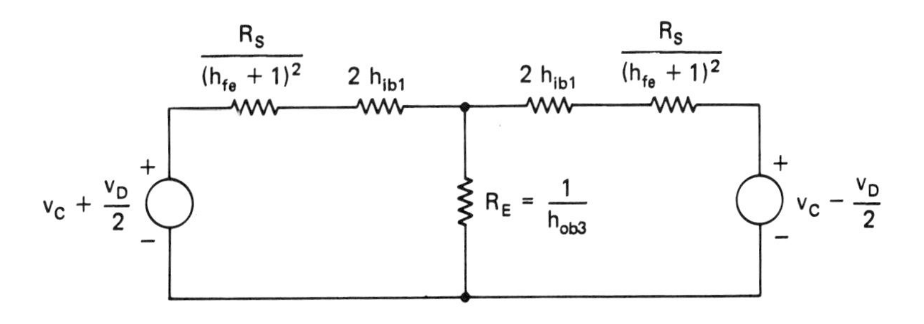

FIGURE 6.19 Small-signal circuit of the Darlington difference amplifier.

6.3 DC AMPLIFIERS

For many amplifier applications, input and output coupling capacitors are troublesome or impractical. This is true for applications requiring extremely low cutoff frequencies down to DC. For integrated circuits (ICs) this is particularly true, since large coupling capacitors cannot be etched onto a chip.

Figure 6.20 illustrates the difficulties associated with DC coupling the base of a transistor to a preceding collector. The base voltage (V_{B2}) is the collector voltage (V_{C_1}), so to maintain a reasonable emitter current, the bottom of the emitter circuit of T_2 must be lifted above ground. This is a poor solution. It requires a second supply voltage and reduces the net supply voltage of T_2. In Figure 6.20, R_{E2} is lifted 12 V, leaving 2.3 V across R_{E2} but only an 8 V supply for T_2.

One solution to these difficulties is shown in Figure 6.21. A zener diode interposed between C_1 and B_2 reduces V_{B2} as required. As long as the zener current (I_z) is maintained above $I_{z\,min}$, the zener acts as a very low impedance voltage dropping device and has no effect on overall gain. Fortunately, zeners are feasible for ICs.

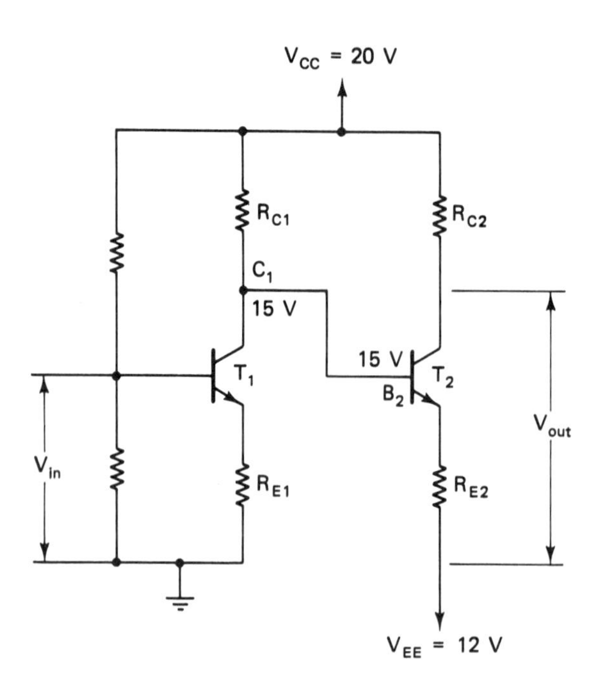

FIGURE 6.20 DC cascaded transistor stages.

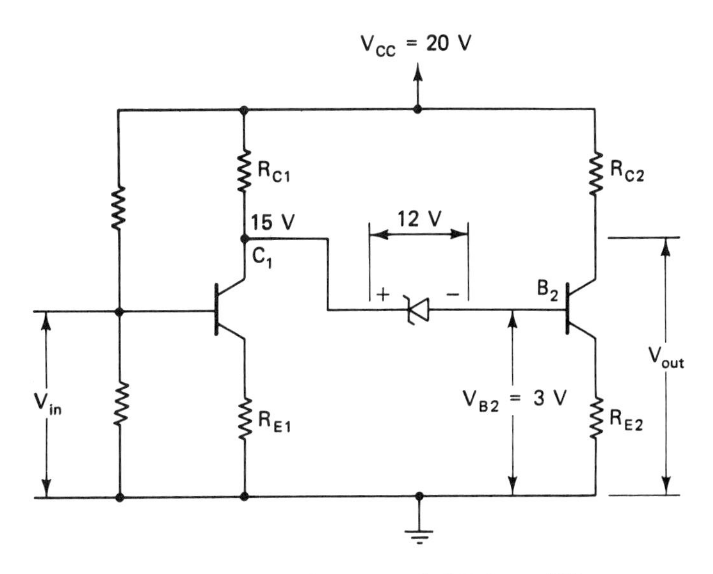

FIGURE 6.21 Zener-coupled DC amplifier.

Another level shifting circuit applicable to ICs is shown in Figure 6.22. This circuit drops a high DC voltage V_{in} to a level suitable for the base of a common-emitter amplifier. Circuit action can be verified as follows. Since I_E is approximately equal to I_C for all BJTs and since

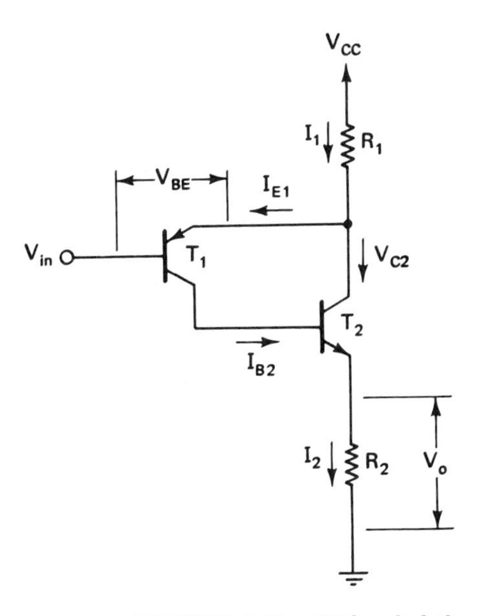

FIGURE 6.22 DC level shifter.

$$I_2 = I_{C_2} + I_{E_1} = I_{C_2} + I_{C_1} \tag{6.65}$$

and

$$I_1 = I_{C_2} + I_{C_1} \tag{6.66}$$

Then

$$I_2 = I_1 \tag{6.67}$$

V_{C_2} is constrained by

$$V_{C_2} = V_{in} + V_{BE_1} \tag{6.68}$$

Therefore,

$$I_2 = \frac{V_{CC} - (V_{in} + V_{BE_1})}{R_1} \tag{6.69}$$

and

$$V_{out} = I_1 R_2 = I_2 R_2 = \frac{R_2}{R_1}(V_{CC} - (V_{in} + V_{BE})) \tag{6.70}$$

The voltage gain (A_V) of this circuit can be derived from Eq. (6.70) by differentiation and is given by

$$A_v = \frac{-R_2}{R_1} \tag{6.71}$$

Example 6.6 *Level Shifting*

Performance specifications:

$$V_{in} = 15\text{ V}, \qquad V_{out} = 2\text{ V}, \qquad V_{CC} = 20\text{ V}$$

Assume that $R_1 = 1\text{ k}\Omega$ and $V_{BE} = 0.7$ V. Using Figure 6.22 and Eq. (6.70),

$$V_{out} = \frac{R_2(V_{CC} - (V_{in} + V_{BE}))}{R_1}$$

$$2\text{ V} = \frac{R_2}{1\text{ k}\Omega} \times (20 - 15 - 0.7) = \frac{4.3R_2}{1\text{ k}\Omega}$$

$$R_2 = \frac{1\text{ k}\Omega \times 2\text{ V}}{4.3} = 465\,\Omega$$

$$A_v = \frac{-R_2}{R_1} = \frac{465\,\Omega}{1\text{ k}\Omega} = 0.465$$

Problems

1. In a class A power amplifier (Figure 6.1), $V_{CC} = 20$ V and $I_{CQ} = 1$ A. Compute R_p, P_{SS}, $P_{D\,max}$, $P_{L\,max}$, η, $I_{C\,max}$, and F_M.

2. In a class A power amplifier (Figure 6.1), $I_{C\,max} = 4$ A and $R_p = 20\ \Omega$. Find I_{CQ}, P_{SS}, and $P_{L\,max}$.

3. Derive the relation between I_p and I_s as a function of $N:1$ for the ideal transformer.

4. An ideal transformer has an $N:1 = 5$. A sinusoidal voltage $V_{peak} = 25$ V is across its primary and a 5-Ω resistor across its secondary. Find $I_{p\,peak}$, $I_{s\,peak}$, $V_{s\,peak}$, P_p, P_s, and R_p.

5. Design a class A power amplifier to meet the following specifications: $P_{L\,max} = 5$ W, $R_L = 3\ \Omega$, and $V_{CC} = 30$ V. Choose the transistor from any available catalog. Find $P_{D\,max}$, $I_{C\,max}$, P_{SS}, R_E, $N:1$, R_1, R_2, and the power lost in the bias circuit (P_B).

6. Repeat problem 5 for $P_{SS} = 12$ W, $R_L = 3\ \Omega$, and $N:1 = 2$.

7. The text derives the relation between I_D and I_{CQ} (Figure 6.6), ignoring the effect of I_{B1} and I_{B2}. Derive the I_{Bias} versus I_{CQ} relation but include I_{B1} and I_{B2}.

8. For a class B power amplifier with $V_{CC} = 40$ V and $R_L = 20\ \Omega$, find $P_{L\,max}$, $P_{SS\,max}$, $P_{D\,max}$, and $I_{c\,sat}$.

9. For a class B power amplifier with $V_{CC} = 40$ V and $I_{C\,sat} = 2$ A, find $P_{L\,max}$ and R_L.

10. For a class B power amplifier with $P_{D\,max} = 10$ W and $R_L = 10\ \Omega$, find V_{CC} and $I_{c\,sat}$.

11. Design a class B power amplifier to supply 20 W across 10 Ω. Compute the power dissipated in its bias network. Select transistors.

12. Repeat problem 11 for 40 W across 10 Ω.

13. In Figure 6.10, let $R_{E2} = R_{E3} = 10\ \Omega$. Assuming $V_{BE\,max} = 0.75$ V, find $I_{C2\,max}$ if the top of R_L is shorted to $-V_{CC}$.

14. Find f_o, B, and Q_c for a parallel tuned resonant circuit with $L = 1$ mH, $C = 0.001\ \mu$F, and $R_p = 20$ kΩ.

15. Design a parallel tuned circuit to meet the following specifications: $f_o = 2$ MHz, $B = 20$ kHz, and $R_p = 10$ kΩ. Find L, C, Q_c, and Q_L.

16. A parallel tuned circuit has an $f_o = 500$ kHz and a bandwidth $= 25$ kHz. Find R_p if $C = 0.001\ \mu$F.

17. A class C power amplifier has the following parameters: $V_{CC} = 20$ V and $R_p = 500\ \Omega$. Assuming $V_{CE\,sat} = 0.3$ V, find $P_{L\,max}$, $P_{D\,at\,max}$, $I_{C\,sat}$, F_M (at max), and η (at max). Compare η and F_M against the class B amplifier.

18. A class C power amplifier supplies 50 W at $\eta = 98\%$. Find F_M, $P_{D\,at\,max}$, R_p, $I_{C\,sat}$, V_{CC}, and C_i for $T_i = 10T_o$. Assume $V_{CE\,sat} = 0.3$ V.

19. Design a class C amplifier to meet the following specifications: $P_L = 50$ W, $f_o = 1$ MHz, $B = 20$ kHz, $V_{CC} = 100$ V, and $T_i = 10T_o$. Let $r_{BE} = 1\ \mathrm{M}\Omega$.

20. Find the common- and difference-mode components if $v_{s_1} = 10$ V dc and $v_{s_2} = 20$ V dc.

21. Repeat problem 20 with $v_{s_1} = 10 \sin 2t$ V and $v_{s_2} = 20 \sin 2t$ V.

22. Repeat problem 20 with $v_{s_1} = 10\ V_{DC}$ and $v_{s_2} = 10\ V + 10 \sin 2tV$.

23. In Figure 6.16, let $R_S = 100\ \Omega$, $h_{fe} = 100$, $h_{ib} = 10\ \Omega$, $R_C = 200\ \Omega$, $R_L = 10\ \mathrm{k}\Omega$, $v_c = 5$ V, $v_{\alpha 1}/2 = +2$ V, and $v_{\alpha 1}/2 = -2$ V. For $R_E = 500\ \Omega$, $1\ \mathrm{k}\Omega$, $10\ \mathrm{k}\Omega$, and $50\ \mathrm{k}\Omega$, find the common- and difference-mode emitter current I_{e2}.

24. Using the data of problem 23, compute A_d, A_c, and CMRR. Use both exact and simplified equations and compare results.

25. Repeat problem 24, letting $R_S = 2\ \mathrm{k}\Omega$. Explain the results.

26. Design a difference amplifier (Figure 6.15) to meet the following specifications: $v_c/v_d = 200$, $A_d \geq 3$ and $R_L = 20\ \mathrm{k}\Omega$. Specify V_{CC}, V_{EE}, CMRR, R_E, R_S, and R_C (h_{fe} for all transistors $= 100$).

27. Design a difference amplifier (Figure 6.17) to meet the following specifications: CMRR $= 50$k, $A_d \geq 3$, $R_L = 20\ \mathrm{k}\Omega$, h_{fe} for all T's $= 100$, and $h_{ob3} = \frac{1}{2}\ \mu$S. Specify V_{EE}, V_{CC}, R_E', and R_S.

28. In Figure 6.18, all h_{fe}'s $= 100$ and $I_{C3} = 4$ mA. Find z_i looking into base T_4 or T_5.

29. If, in problem 27 (Figure 6.17), the input circuits were replaced by Darlingtons, how large could R_S be and still satisfy the design restrictions?

30. In Figure 6.18, let $R_2 = R_1 = 1\ \mathrm{k}\Omega$, $I_{CQ1} = I_{CQ2} = 2$ mA, and all h_{fe}'s $= 100$. Find R_E'. $V_{EE} = 10$ V.

31. Design a Darlington difference amplifier (Figure 6.18) to meet the following specifications: $v_c/v_d = 20$ k, $A_d \geq 5$ ($h_{ob3} = \frac{1}{2}\mu$S and all h_{fe}'s $= 50$). Specify R_S, V_{CC}, V_{EE}, R_E', R_1, R_2, and R_C.

32. Design a level shifting circuit to meet the following specifications: $V_{in} = 10$, $V_{CC} = 15$ V, and $V_{out} = 3$ V. Assume $R_1 = 2\ \mathrm{k}\Omega$ and $V_{BE} = 0.7$ V.

References

Bell, David A. *Electronic Devices and Circuits*, 2nd ed., Reston Publishing Co., Reston, Virginia, 1980.

Comer, David J. *Modern Electronic Circuit Design*, Addison-Wesley Publishing Co., Reading, Massachusetts, 1968.

General Electric Semiconductors: Short Form Catalog, General Electric Co., Auburn, N.Y., 1980.

Schilling, D. L., and C. Belove. *Electronic Circuits: Discrete and Integrated*, 2nd ed., McGraw-Hill Book Co., New York, 1979.

7

OPERATIONAL AMPLIFIER

7.1 INTRODUCTION

A typical operational amplifier (op-amp) consists of a difference amplifier combined with a high-gain linear amplifier, a level shifting circuit, and an output amplifier. The difference amplifier may use Darlington inputs or FETs and will usually use the constant-current configuration of Figure 6.17. The difference amplifier is followed by a high-gain linear amplifier, which is often another difference amplifier. Very high differential gains are attained (typically 10^4 to 10^5), while common-mode gains are held to about 1. CMRRs, therefore, are typically 10^4 to 10^5. Since dc coupling is desirable, capacitors other than those required to prevent oscillation or for frequency compensation are absent, and level shifting circuits are used when needed, for example, between the high-gain amplifier and the output stage. The output stage is designed for a low output impedance and power gain and is often a complementary amplifier.

Since op-amps are usually monolithic etched circuits, they can afford to be very sophisticated. For example, the 741, a general-purpose op-amp, contains 20 transistors. The increased circuit complexity is used to improve stability and to provide overload protection. Figure 7.1 shows a simpler version of a commercial op-amp, which nevertheless contains many of the features of more complex circuits. T_1 through T_5

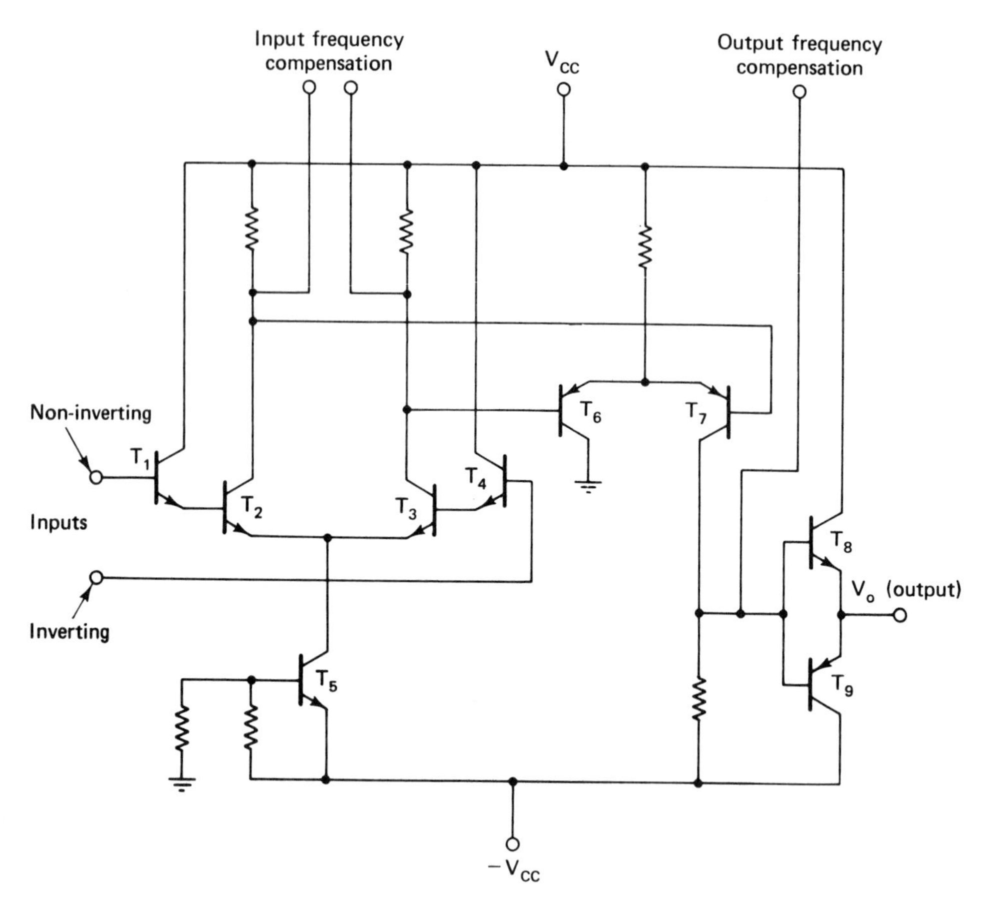

FIGURE 7.1 Typical op-amp.

comprise a Darlington input, constant-emitter-current difference amplifier. T_6 and T_7 act as a high-gain linear amplifier, an additional difference amplifier stage, and a level shifter. T_8 and T_9 comprise the output stage. External ports for frequency compensation and the prevention of oscillation are also provided.

The op-amp is an extremely versatile and widely used device, which will satisfy many circuit needs. Essentially, with appropriate external circuitry, the op-amp performs arithmetic operations such as summing, exponentiation, differentiating, and integrating. It can also be used as an oscillator, active filter, impedance transformer, active capacitor, active inductor, stable amplifier, and more. Because manufacturers supply such a wide range of op-amps, much circuit design can be reduced to interfacing external circuitry and op-amps.

7.2 OP-AMP CHARACTERISTICS

Rarely are technologists who are not working for IC manufacturers called upon to design op-amps. Instead, the op-amp is treated as a discrete device such as a BJT or a FET. The circuit designer's task is limited to designing required external circuitry and interfacing the op-amp. To do this requires, first, a familiarity with op-amp characteristics and specifications.

Specifications for a general-purpose op-amp (741) are given in Table 7.1. Unfamiliar terms are defined in the following.

TABLE 7.1 *Typical Specifications for a 741 Op-Amp*

Electrical Characteristics

Parameter	Conditions*	Min	Typ	Max	Units
Input offset voltage (V_{IO})	$R_s \leq 10$ kΩ†			6.0	mV
Input offset current (I_{IO})				500	nA
Input bias current (I_{IB})				1.5	μA
Input resistance	$T_A = 25°C$	0.3	1		MΩ
Supply current	$T_A = 25°C$ $V_{CC} = \pm 15$ V		1.7	2.8	mA
Large-signal voltage gain	$V_{CC} = \pm 15$ V $V_{out} = \pm 10$ V $R_L \geq 2$ kΩ	25×10^3			V/V
Output voltage swing	$V_{CC} = \pm 15$ V $R_L = 10$ kΩ $R_L = 2$ kΩ	±12 ±10	±14 ±13		V
Input voltage range	$V_{CC} = \pm 15$ V	±12			V
Common-mode rejection ratio (CMRR)	$R_S \leq 10$ kΩ	70	90		dB
Supply voltage rejection ratio	$R_s \leq 10$ kΩ	77	96		dB

Absolute maximum ratings

Supply voltage (V_{CC})	±22 V
Power dissipation	500 mV
Differential input voltage	±30 V
Input voltage for supply voltage	±15 V = supply voltage
Output short-circuit duration	Indefinite
Operating temperature range	−55° to 125°C
Storage temperature range	−65° to 150°C
Lead temperature (soldering 10 spc.)	300°C

*Specifications apply for $V_{CC} = \pm 15$ V and $-55°C \leq T_A \leq 125°C$ *unless otherwise noted.*

†R_s is source resistance. R_L is load resistance.

The *input offset currents* and *voltage* specifications (I_{IO}, V_{IO}) describe op-amp imbalances. The input-offset current is the difference between the two input currents flowing through mandatory external dc returns. This imbalance produces a false differential mode signal, which must be tolerated. FET input op-amps with extremely low offset currents (10 pA) are available. The input offset voltage specification gives the input differential voltage required to zero the quiescent output voltage. The 741 provides terminals to which corrective dc voltages may be applied.

The *input bias current* (I_{IB}) gives the average value of the two input currents flowing through external dc returns. The lower this value, the lower I_{IO} will be. FET input op-amps with 50-pA I_{IB}'s are available.

The 741, having Darlington inputs, has a high input resistance between 0.3 and 1 MΩ. However, FET input op-amps have much higher input resistances.

The *output voltage swing* gives the maximum output. It is usually a volt or two less than the supply voltages because of internal voltage drops. The 741 requires positive and negative supply voltages, but single-ended op-amps (single supply voltage) are available.

The *supply voltage rejection ratio* defines the effects of supply voltage changes on the quiescent output voltage. Since changes in supply voltages show up as common-mode signals (attenuated somewhat), the supply rejection ratio is similar to the CMRR.

7.2.1 Slew Rate

Another op-amp characteristic, important in high-speed operation, is called *slew rate* (SR). Slewing is displayed in its simplest form when a difference-mode step voltage large enough to drive the output voltage to maximum is input (Figure 7.2a). The modified step output shown in Figure 7.2b is produced. The SR is equal to the slope of the oblique segment of the output and is given by

$$SR = \frac{V_{max}}{T} \quad \text{(Figure 7.2b)} \tag{7.1}$$

Available op-amps have SRs between 0.2 and 75 V/μs.

The transformation of an abrupt input transition into an oblique output occurs only in circuits containing capacitance or inductance. Inductors are not normally used in op-amps, but capacitors are introduced to prevent oscillation. In Figure 7.1 a capacitor would usually be connected across the input frequency compensation terminals (or across the output frequency terminal and ground). In the 741 op-amp, compensa-

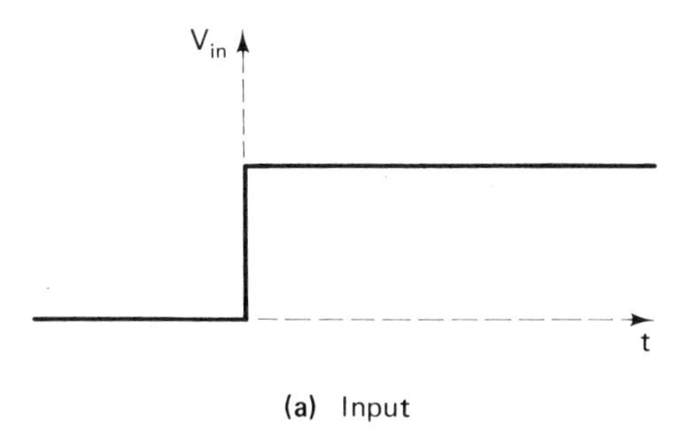

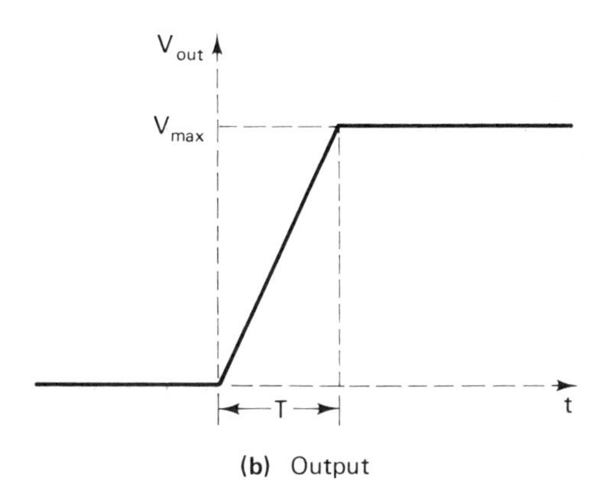

FIGURE 7.2 Slewing in op-amps: (a) input; (b) output.

tion is provided internally by a 30-pF capacitor connected, effectively, from the output stage of the difference amplifier to ground.

Whenever a voltage is taken across (or depends on one taken across) a capacitance being charged by a constant current, slewing takes place. The current through a capacitance depends on the rate of change of the voltage across it and is given by

$$i = C\frac{\Delta V}{\Delta t} \tag{7.2}$$

The slew rate $(\Delta V/\Delta t)$ is therefore given by

$$\text{SR} = \frac{\Delta V}{\Delta t} = \frac{i}{C} = \frac{I_{\max}}{C} \tag{7.3}$$

where $I_{\max}$ is the maximum current into capacitor C. Consequently, SR depends markedly on the effective compensating capacitor, which obviously should be held to a minimum.

Slew rate also places an upper limit on maximum swing sinusoidal operation. Slew rate distortion begins when the zero crossing slope of the sinusoid just exceeds the slew rate, this being the point where the sinusoid's slope is steepest. This distortion is illustrated in Figure 7.3.

The highest undistorted frequency that can be output at maximum swing (maximum power) is given by

$$f_{max} = \frac{SR}{2\pi V_{max}} \tag{7.4}$$

where V_{max} is the maximum output voltage. This frequency is often called the *power bandwidth* since low-frequency cutoff is zero. Higher undistorted frequencies can be output, but at lower levels.

Example 7.1 *Finding the Power Bandwidth of an Op-Amp*

A 741 op-amp with $R_L = 2\ \text{k}\Omega$ and $V_{CC} = \pm 15$ V, has an SR $= 0.5$ V/μs. Find the power bandwidth.

From Table 7.1 for $R_L = 2\ \text{k}\Omega$ and $V_{CC} = \pm 15$ V, $V_{max} = 13$ V (use typical value). From Eq. (7.4),

$$f_{max} = \frac{SR}{2\pi V_{max}} = \frac{0.5\ \text{V}/\mu\text{s}}{2\pi \times 13} = 6.12\ \text{kHz}$$

7.2.2 Op-Amp Packaging and Representation

Op-amps are available typically in three package types, the top-hat metal can, the flat in-line package, and the rectangular molded package.

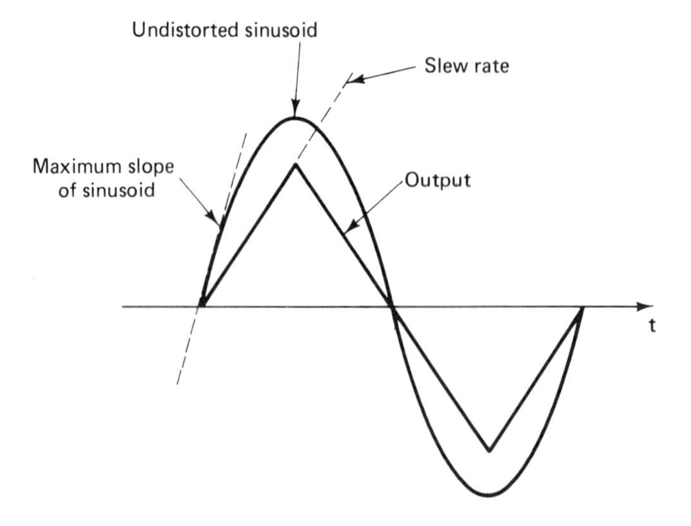

FIGURE 7.3 Slew-rate distortion in sinusoids.

Each package may have different numbers of leads, depending on the op-amp type and the number of op-amps in each package. (Packages containing one to four op-amps are available.) For example, the single package 741 op-amp will have 8 leads on the metal can (all used), 10 on

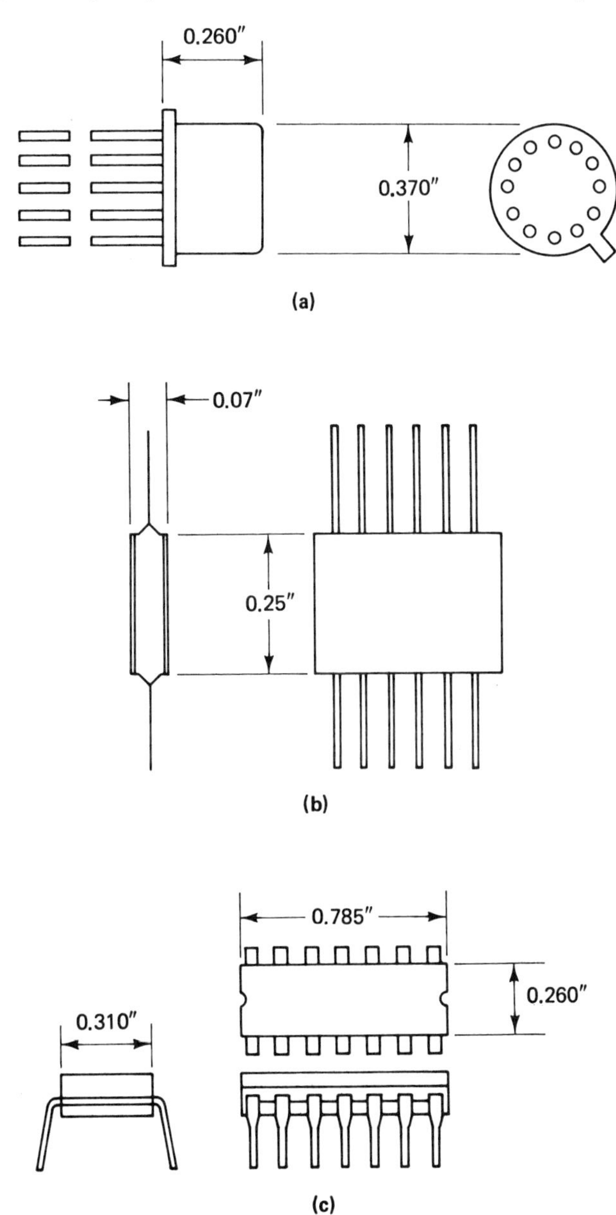

FIGURE 7.4 Op-amp packages: (a) typical metal can (12 leads); (b) typical flat package (10 leads); (c) typical cavity DIP (14 leads).

the flat package, and 14 on the rectangular package (not all used). The dual 747 op-amp (two op-amps in one package) has a 10-lead can and 14-lead flat and rectangular packages.

These packages are all small. Metal cans have typical diameters of about 0.375 in. and are about 0.260 in. thick. Flat packages may be about 0.25 in. square and 0.070 in. thick; rectangular molded packages, 0.75 in. long, 0.29 in. wide, and 0.18 in. thick. The metal can and the rectangular molded packages are convenient for mounting on prepunched printed circuit boards. Figure 7.4 illustrates these packages.

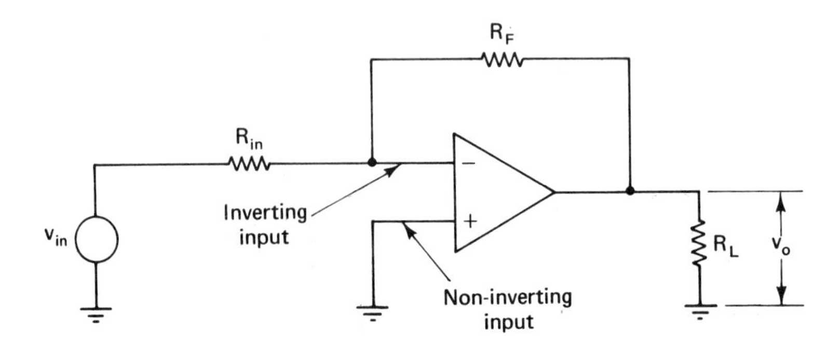

(a) Simplified op-amp circuit

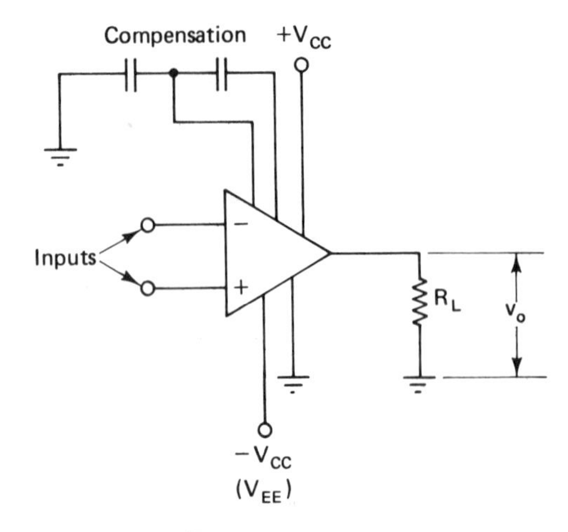

(b) Complete op-amp representation

FIGURE 7.5 Op-amp representations: (a) simplified op-amp circuit; (b) complete op-amp representation.

Symbolically, the op-amp is represented as an amplifier with two inputs, an inverting input (−) and a noninverting input (+), and one output. Supply voltages, compensation, and ground terminals are included only when necessary. Figure 7.5 shows two op-amp representations.

7.3 FEEDBACK

When a part of the output signal of an amplifier is reintroduced into its input (feedback), remarkable changes in system performance occur. Gain is reduced and made independent of transistor parameters with one type of feedback or the amplifier becomes an oscillator with another. Input and output impedances are made very large or very small depending on the feedback mode. When the feedback network is frequency dependent, the output frequency response is affected. Bandwidths can be increased, decreased, or shaped.

Four types of feedback are used in op-amp circuits. They are defined in terms of their input and output configurations. They are illustrated in Figure 7.6. If the feedback is taken across the load resistor (R_L), it is called *voltage ratio feedback*. If then it is introduced in series with the signal input, it is called either *voltage-series* or *series–parallel* (SP) feedback (Figure 7.6a). If instead it is introduced in parallel with the signal input (Figure 7.6b), it is called *voltage-shunt* or *parallel–parallel* (PP) feedback. If the feedback is dependent on the current flowing through R_L, it is called *current feedback*, which, again, can be introduced either in series (current-series or series–series, SS) or in parallel (current-shunt or parallel–series, PS) with the input signal. In all cases, if the feedback reduces the signal input (returns to the inverting input), it is negative feedback; if it increases the signal input (returns to the noninverting input), it is positive feedback.

7.3.1 Characteristics of Negative SP Feedback

Negative SP feedback decreases and stabilizes gain, increases bandwidth, decreases distortion, increases input impedance, and decreases output impedance. These results are easily verified analytically.

Closed Loop Gain. Op-amps amplify the voltage difference between the two inputs (difference mode). This difference is called the *error voltage* (v_e). The specified (open loop) gain of the op-amp operates on v_e. A signal across the noninverting input to ground ($v_{in(+)}$) produces an inphase v_o. In the SP amplifier a portion of v_o is fed back to the inverting in-

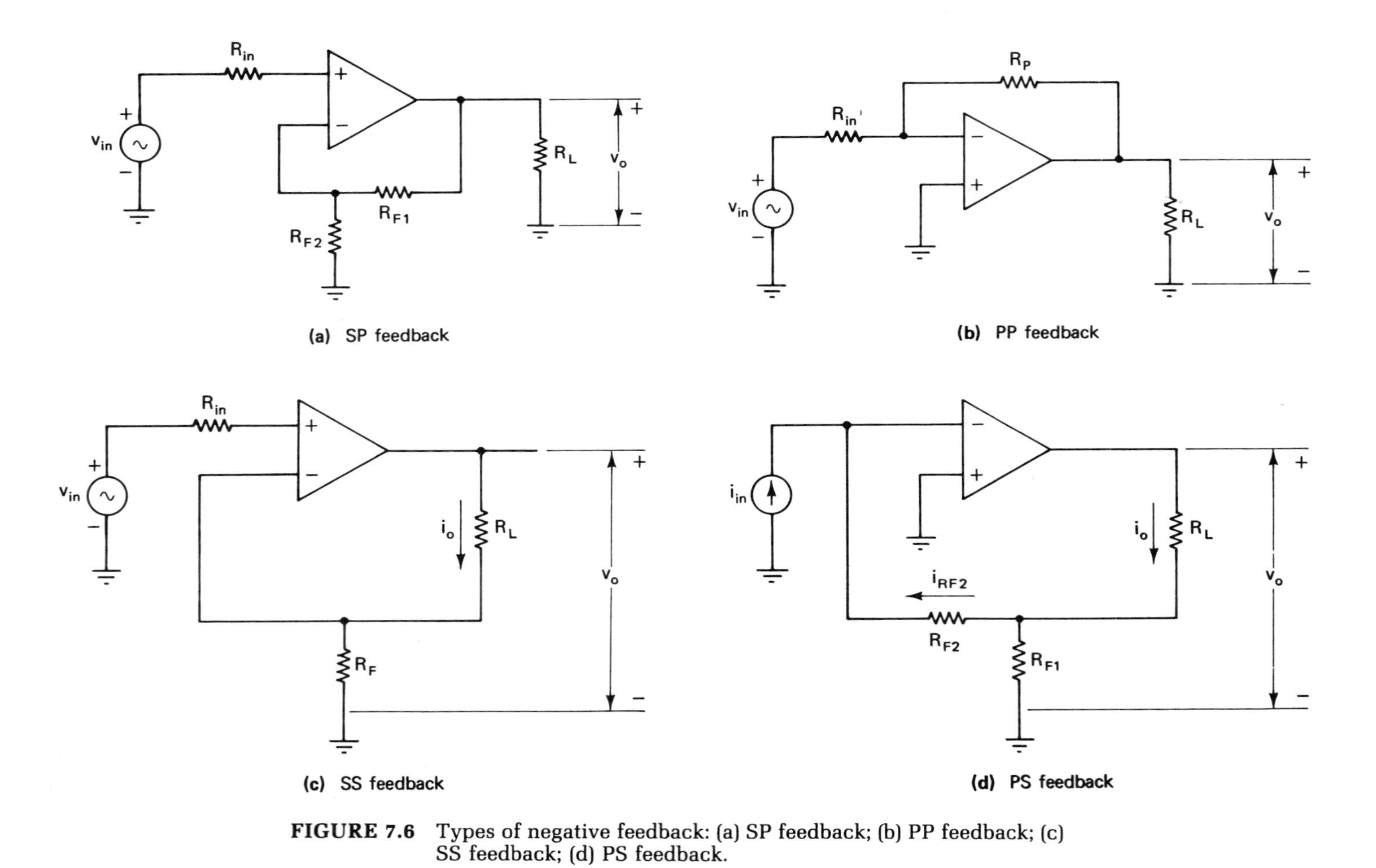

FIGURE 7.6 Types of negative feedback: (a) SP feedback; (b) PP feedback; (c) SS feedback; (d) PS feedback.

put and produces a signal across $v_{in(-)}$ in phase with $v_{in(+)}$. The error voltage is then given by

$$v_e = v_{in(+)} - v_{in(-)} \quad (7.5)$$

Although the full open-loop gain operates on v_e, the error voltage is smaller than $v_{in(+)}$, and the actual gain (closed-loop gain) is reduced.

Analytically, the closed-loop gain (A_{Vf}) and $v_{in(-)}$ are given by

$$A_{Vf} = \frac{v_o}{v_{in(+)}} \quad (7.6)$$

$$v_{in(-)} = Bv_o \quad (7.7)$$

where B is the feedback ratio. The output voltage (v_o) is always given by

$$v_o = A_V v_e \quad (7.8)$$

where A_V is the open-loop gain. Combining Eqs. (7.5) to (7.8), we get

$$A_{Vf} = \frac{A_V}{1 + A_V B} \quad (7.9)$$

which, when $AB \gg 1$ (typically true), gives

$$A_{Vf} = \frac{A_V}{A_V B} = \frac{1}{B} \quad (7.10)$$

Since the feedback ratio B is determined by stable components and is independent of A_V, A_{Vf} is extremely stable.

Example 7.2 *Finding SP Closed-Loop Gain*

In the amplifier of Figure 7.6a, $R_L = 1\ \text{k}\Omega$, $R_{F1} = 10\ \text{k}\Omega$, and $R_{F2} = 1\ \text{k}\Omega$. If the open-loop gain $= 1 \times 10^5$, find the closed-loop gain.

The feedback resistors are large enough not to load v_o, so the feedback ratio B is given by

$$B = \frac{v_{in(-)}}{v_o} = \frac{v_o R_{F2}}{v_o(R_{F1} + R_{F2})} = \frac{R_{F2}}{R_{F1} + R_{F2}} = \frac{1\ \text{k}\Omega}{11\ \text{k}\Omega} = \frac{1}{11}$$

Since $A_V B = 1 \times 10^5 \times 0.5 \gg 1$,

$$A_{Vf} = \frac{1}{B} = \frac{1}{1/11} = 11$$

Nonlinear distortion in amplifiers occurs because their gain changes for large swings. Since SP feedback stabilizes gain, distortion is reduced. By an analysis similar to the preceding gain analysis, distortion with feedback (D_f) compared to distortion without (D) can be shown to be given by

$$D_f = \frac{D}{1 + A_V B} \tag{7.11}$$

Input and Output Impedances. Input and output impedances for any device or circuit are found by computing the ratio v/i at the input and output ports respectively. The device impedance is then given by this ratio.

$$z = \frac{v}{i} \tag{7.12}$$

Any circuit action that alters this ratio changes the apparent. In the SP feedback op-amp, feedback increases the input impedance and reduces the output impedance.

The effect of SP feedback on the input impedance can be explained with the aid of Figure 7.7. Without feedback and with the inverting (−)

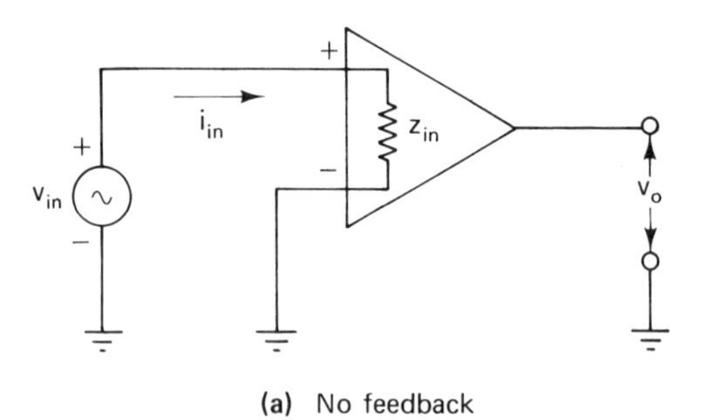

(a) No feedback

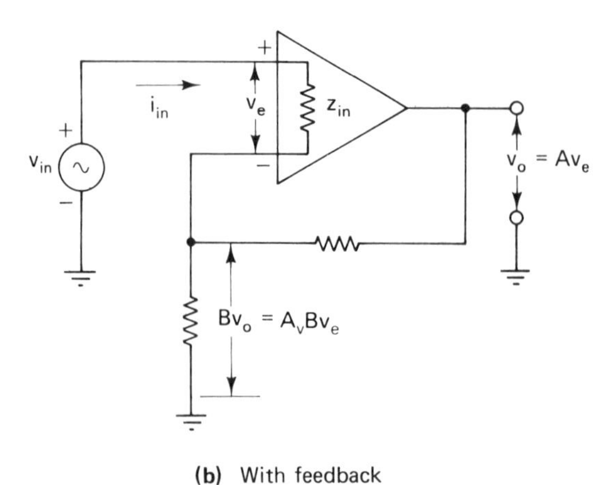

(b) With feedback

FIGURE 7.7 Effect of SP feedback on input impedance: (a) no feedback; (b) with feedback.

input grounded (Figure 7.7a), all of v_{in} appears across the internal (open-loop) input impedance (z_{in}). Therefore, the apparent input impedance equals the open-loop input impedance. (In op-amps with Darlington inputs, this ranges from a few hundred kilohms to several megohms.) With SP feedback (Figure 7.7b), only the error voltage (v_e) appears across z_{in}. Since v_e is always less than v_{in} for SP feedback, i_{in} is reduced, and the apparent input impedance with feedback (z_{if}), called the closed-loop input impedance, is increased.

Analytically,

$$z_{if} = \frac{v_{in}}{i_{in}} \tag{7.13}$$

With SP feedback

$$i_{in} = \frac{v_e}{z_{in}} \tag{7.14}$$

where z_{in} is the open-loop input impedance. From Figure 7.7b, we see that

$$v_{in} = v_e + A_V B v_e = v_e(1 + A_V B) \tag{7.15}$$

Combining Eqs. (7.12) to (7.14), we get

$$z_{if} = \frac{v_{in}}{i_{in}} = \frac{v_{in} z_{in}}{v_e} = \frac{v_e(1 + A_V B) z_{in}}{v_e} = (1 + A_V B) z_{in} \tag{7.16}$$

The output impedance (z_{of}) of the SP feedback op-amp is found by positing a voltage source across the output, shorting the noninverting input, and computing the ratio v_o/i_o. This configuration is illustrated in Figure 7.8. Also, in this figure, the output circuit is shown schematically as a Thevenin's circuit with z_o as the open-loop (no feedback) output impedance. We see that i_o with i_B assumed to be 0 (note assumed direction) is given by

$$i_o = \frac{v_o - A_V v_e}{z_o} \tag{7.17}$$

and

$$v_e = -B v_o \tag{7.18}$$

Combining Eqs. (7.17) and (7.18), we get

$$i_o = \frac{v_o + A_V B v_o}{z_o} = \frac{v_o(1 + A_V B)}{z_o} \tag{7.19}$$

Then

$$z_{of} = \frac{v_o}{i_o} = \frac{v_o z_o}{v_o(1 + A_V B)} = \frac{z_o}{1 + A_V B} \tag{7.20}$$

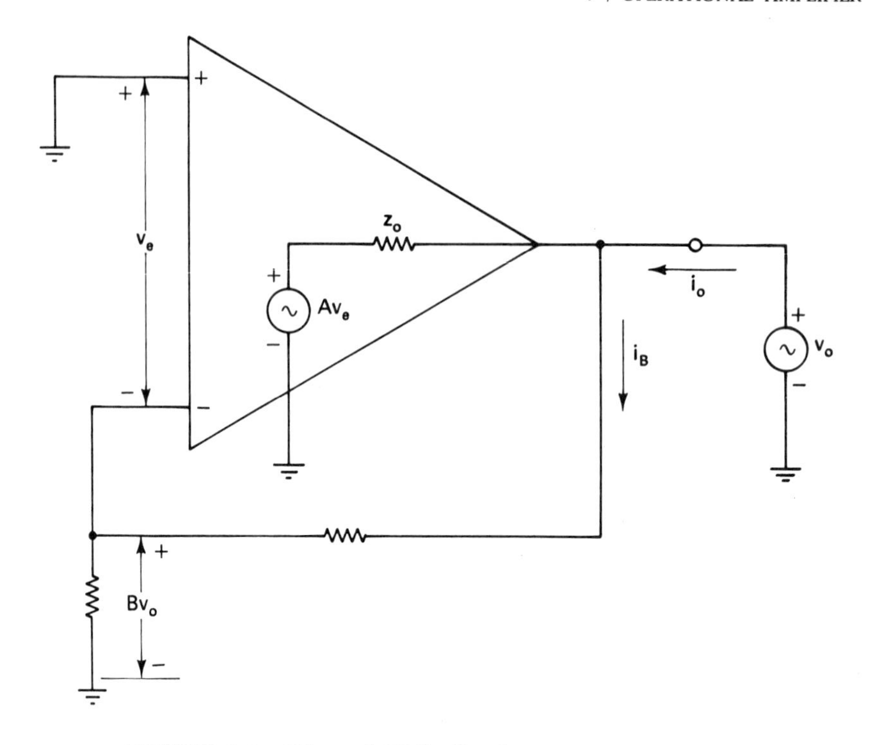

FIGURE 7.8 Effect of SP feedback on output impedance.

Example 7.3 *Finding z_{if} and z_{of} for an SP Op-Amp*

For the op-amp circuit of Example 7.2, the open-loop input and output impedances are 300 kΩ and 75 Ω, respectively. Find the closed-loop input and output impedances, z_{if} and z_{of}.

From Example 7.2,

$$A_V = 1 \times 10^5, \qquad B = 0.091, \qquad A_V B = 0.091 \times 10^5$$

Therefore,

$$z_{if} = (1 + A_V B) z_{in} \approx A_V B z_{in} = 0.091 \times 10^5 \times 300\ \text{k}\Omega$$
$$= 2.73 \times 10^9\ \Omega$$

$$z_{of} = \frac{z_o}{1 + A_V B} \approx \frac{z_o}{A_V B} = \frac{75}{0.091 \times 10^5} = 824 \times 10^{-5}\ \Omega$$

Bandwidth. The product of the voltage gain and the bandwidth of the op-amp is a constant called the *gain–bandwidth product.* Because of this, gain can be reduced for more bandwidth, and vice versa. Since SP feedback reduces voltage gain by the factor $(1 + A_V B)$, it increases bandwidth by the same factor. Because op-amps are dc amplifiers, bandwidth is established only by the high-frequency cutoff (f_H). Therefore, the closed-loop high-frequency cutoff (f_{Hf}) is related to the open-loop high-frequency cutoff (f_{Ho}) by

$$f_{Hf} = (1 + A_V B) f_{Ho} \tag{7.21}$$

For Eq. (7.21) to be strictly true, the decrease in A_V with frequency (roll-off) between f_{Ho} and f_{Hf} must be 20 dB per decade (A_{Vf} decreases by 10 when frequency increases by 10). This condition is normally true for all op-amps, but special compensation can change this relationship. The normal case is displayed in Figure 7.9. Notice that roll-off is unaffected by feedback and that closed-loop bandwidth is increased because, when gain is decreased, roll-off is reached at higher frequencies.

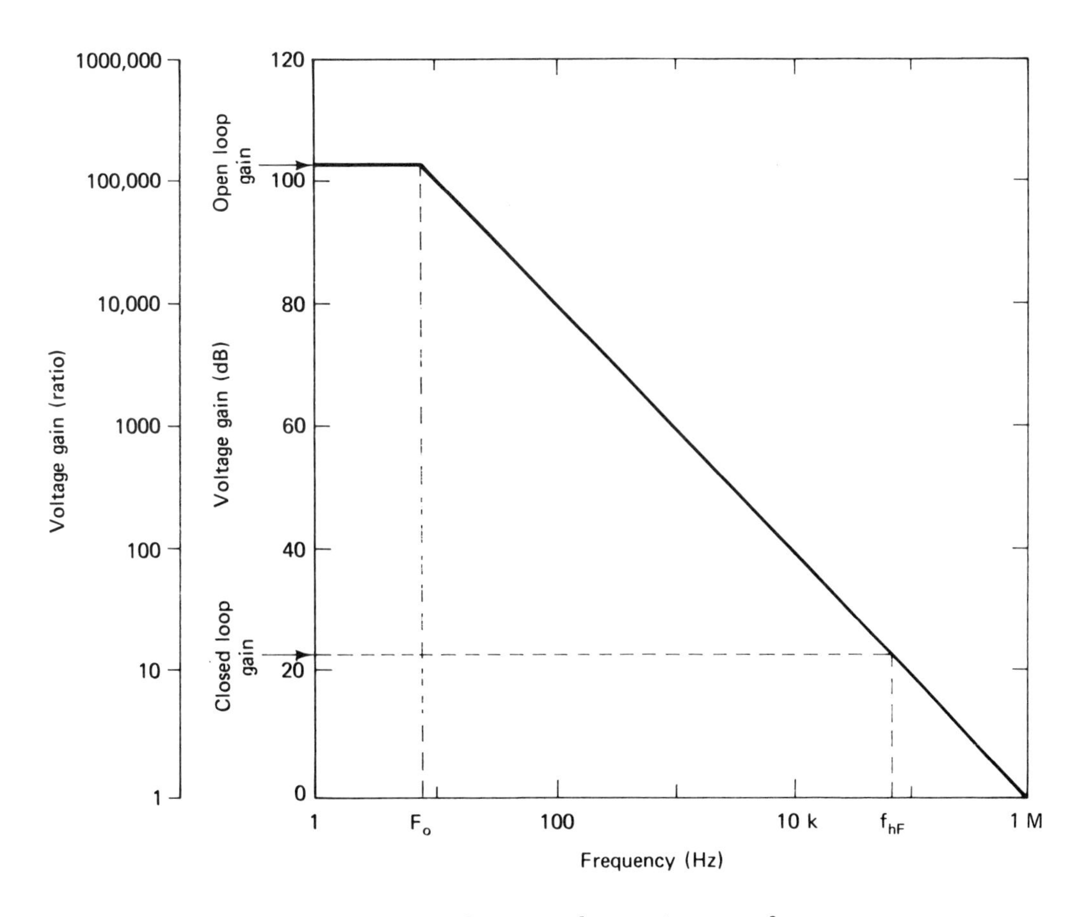

FIGURE 7.9 Op-amp voltage gain versus frequency.

7.3.2 Applications

SP feedback op-amps are primarily used in circuits requiring extremely high input impedances and extremely low output impedances. Applications that exploit these characteristics include noninverting amplifiers, voltage followers, instrumentation amplifiers, and sample and hold circuits.

Noninverting Amplifier. Figure 7.10 shows an SP op-amp used as a noninverting amplifier. Since in this case

$$B = \frac{R_{F2}}{R_{F1} + R_{F2}} \tag{7.22}$$

Eqs. (7.10), (7.16), and (7.20) with $A_V B \gg 1$ become

$$A_{VF} = \frac{1}{B} = 1 + \frac{R_{F1}}{R_{F2}} \tag{7.23}$$

$$z_{if} = A_V B z_{in} = \frac{A_V z_{in}}{(1 + R_{F1}/R_{F2})} \tag{7.24}$$

$$z_{of} = \frac{z_o}{A_V B} = \frac{z_o}{A_V}\left(1 + \frac{R_{F1}}{R_{F2}}\right) \tag{7.25}$$

This amplifier provides high input impedances and low output impedances at very stable voltage gains.

The design of a noninverting amplifier must take into account the effects of the input bias currents (I_{IB}) and the offset input current (I_{IO}). To prevent I_{IB} from developing an excessive false voltage across the inputs, R_S should equal $R_{F1}||R_{F2}$ and be small enough so that I_{IO} develops a voltage difference much smaller than the lowest signal voltage. Also, for

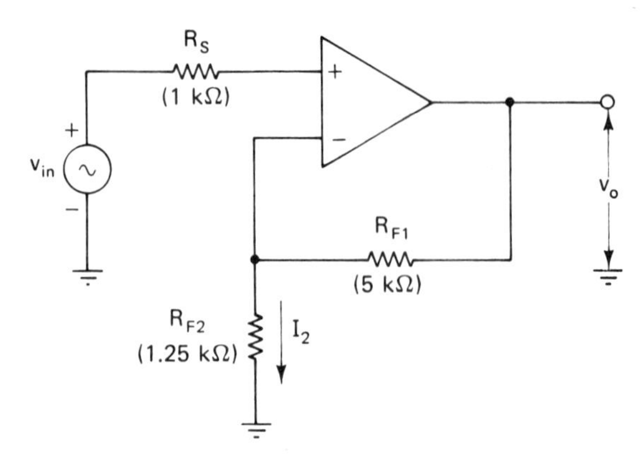

FIGURE 7.10 Noninverting amplifier.

ease of computation the current flowing through R_{F2} (I_2) should be at least 10 times larger than I_{IB}.

Performance specifications may include minimum values of A_{Vf} and z_{if} and maximum values of z_{of}. The design procedure calls for selecting a B that satisfies all specifications and then, in light of the preceding discussion, selecting an appropriate R_S, R_{F1}, and R_{F2}. (If R_S is too high and is beyond the control of the designer, a buffer stage may be required between the source and the amplifier.)

Example 7.4 *Design of a Noninverting Amplifier*

Performance specifications: Op-amp 741 with $A_V = 25 \times 10^3$ min., $z_i = 0.3\ \text{M}\Omega$ min., $z_O = 75\ \Omega$ max., $I_{IB} = 1.5\ \mu\text{A}$ max., and $I_{IO} = 500$ nA max.

Amplifier specifications: $A_{Vf} \geq 5$, $z_{if} \geq 10^8\ \Omega$, $z_{of} \leq 1\ \Omega$

Signal specifications: $v_{in} \rightarrow$ 0.05 to 1 V

(a) *Choose B:*

$$z_{of} = \frac{z_o}{A_V B_1}, \qquad B_1 = \frac{z_o}{z_{of} A_V} = \frac{75\ \Omega}{1\ \Omega \times 25 \times 10^3} = 3 \times 10^{-3}$$

$$z_{if} = A_V B_2 z_{in}, \qquad B_2 = \frac{z_{if}}{A_V z_i} = \frac{1 \times 10^8\ \Omega}{25 \times 10^3 \times 0.3\ \text{M}\Omega} = 1.33 \times 10^{-2}$$

$$A_{Vf} = \frac{1}{B_3}, \qquad B_3 = \frac{1}{A_{Vf}} = \frac{1}{5} = 0.2$$

Selecting maximum B (B_3) ensures that all specifications are met in this case.

(b) *Compute R_S, R_{F1}, and R_{F2}:* Since $v_{in\,min} = 0.05$ V, we accept an unbalanced (false) signal of $0.05/100 = 500\ \mu\text{V}$. R_S is then given by

$$R_S = \frac{500\ \mu\text{V}}{500\ \text{nA}} = 1\ \text{k}\Omega$$

Next

$$\frac{1}{B} = 1 + \frac{R_{F1}}{R_{F2}}, \qquad \frac{R_{F1}}{R_{F2}} = 5 - 1 = 4, \qquad R_{F1} = 4R_{F2}$$

Using $R_{F1} || R_{F2} = R_S$ and $R_{F1} = 4R_{F2}$,

$$\frac{R_{F1} R_{F2}}{R_{F1} + R_{F2}} = \frac{4R_{F2}}{5} = 1\ \text{k}\Omega, \qquad R_{F2} = 1.25\ \text{k}\Omega,\ R_{F1} = 5\ \text{k}\Omega$$

(c) *Check for $I_2 \gg I_{IB}$:*

$$I_2 = \frac{V_{o\,min}}{R_{F1} + R_{F2}} = \frac{V_{in\,min} A_{Vf}}{R_{F1} + R_{F2}} = \frac{0.05 \times 5}{6.25\ \text{k}\Omega} = 40\ \mu\text{A}, \qquad I_2 \gg I_{IB}$$

The circuit values are enclosed in parentheses in Figure 7.10.

Voltage Follower. The ideal voltage follower would have a voltage gain of 1, an infinite input impedance, and zero output impedance. It would be a perfect buffer, providing total isolation between input and output with no change in signal level.

The voltage follower of Figure 7.11 closely approximates this ideal. With $B = 1$ and $A_V B \gg 1$, the gain and impedance equations become

$$A_{Vf} = 1 \tag{7.26}$$

$$z_{if} = A_V z_{in} \tag{7.27}$$

$$z_{of} = \frac{z_o}{A_V} \tag{7.28}$$

For an op-amp with $A_V = 50{,}000$, $z_{in} = 300\ \text{k}\Omega$, and $z_o = 75\ \Omega$ (typical values), the circuit of Figure 7.11 gives $A_{Vf} = 1$, $z_{if} = 15 \times 10^9\ \Omega$, and $z_{of} = 1.5 \times 10^{-3}\ \Omega$. Physical limitations (internal voltage drops and contact resistance) alter these results slightly, but commercial voltage followers (feedback connected internally), such as the 102, 202, and 302 series, give $A_{Vf} = 0.999$, $z_{if} = 10^{12}\ \Omega$, and $z_{of} = 0.8\ \Omega$.

Instrumentation Amplifier. Some measurement environments require an amplifier with an extremely high gain and a high input impedance capable of distinguishing a small signal in the presence of relatively high background noise. Measuring electromyographic signals produced by heart contractions or other biological actions is an example of such an environment. Figure 7.12 shows an instrumentation amplifier composed of op-amps suitable for this environment. The two voltage followers O_1 and O_2 provide the required high input impedances, typically $10^{12}\ \Omega$, and the op-amp O_3 functioning as a difference amplifier provides the CMRR (often > 110 dB) and the high difference gain required. When used to measure biological signals, for example, the ground electrode is located at a position that makes the background noise (actually unwanted signals) common mode and the desired signal, difference mode.

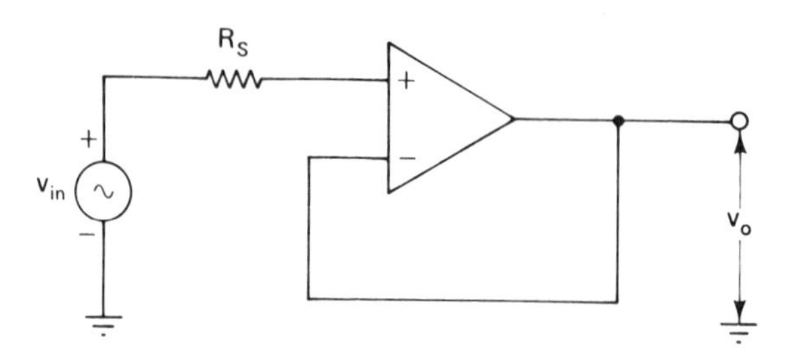

FIGURE 7.11 Voltage follower.

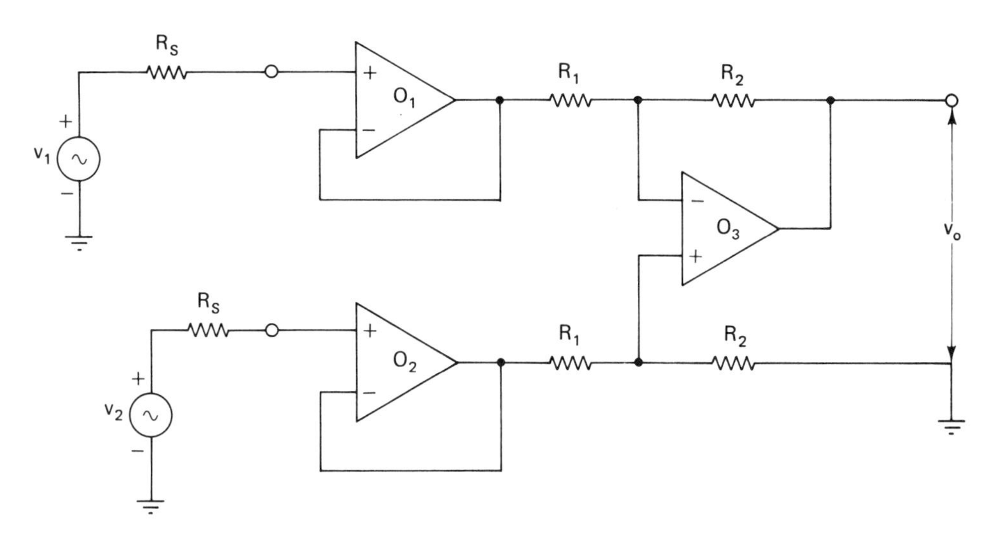

FIGURE 7.12 Instrumentation amplifier.

Instrumentation amplifiers are available as monolithic chips, often with FET inputs replacing the voltage followers. The Bifet LF 352 with JFET input is an example. It has an input impedance of about 2×10^{12} Ω and a minimum CMRR of 110 dB (316,000).

Sample and Hold Circuits. Sample and hold circuits are used in analog-to-digital converters to sample the analog signal and to retain the sampled voltage until the next sampling period. Essentially, a capacitor connected across a high input impedance amplifier is charged to the sampled analog voltage through a low-impedance source (to ensure rapid charging) and then disconnected from the source. A high-quality capacitor will retain its charge a relatively long time under these conditions. In the sample and hold circuit of Figure 7.13, a voltage follower is the low-impedance charging source, a logic gating circuit, the connect–disconnect switch, and another voltage follower is the high input impedance amplifier. This configuration permits rapid capacitor charging and allows the capacitor voltage to be sensed with negligible current drain.

Sample and hold circuits are available as monolithic chips, such as the LH0023 and LH0043 series. Their specifications are similar to op-amps with the addition of those relating to switch response time (aperture time), the interval required to charge the capacitor to the signal voltage (acquisition time), and to accuracy. The sample and hold circuit is important and will be studied later.

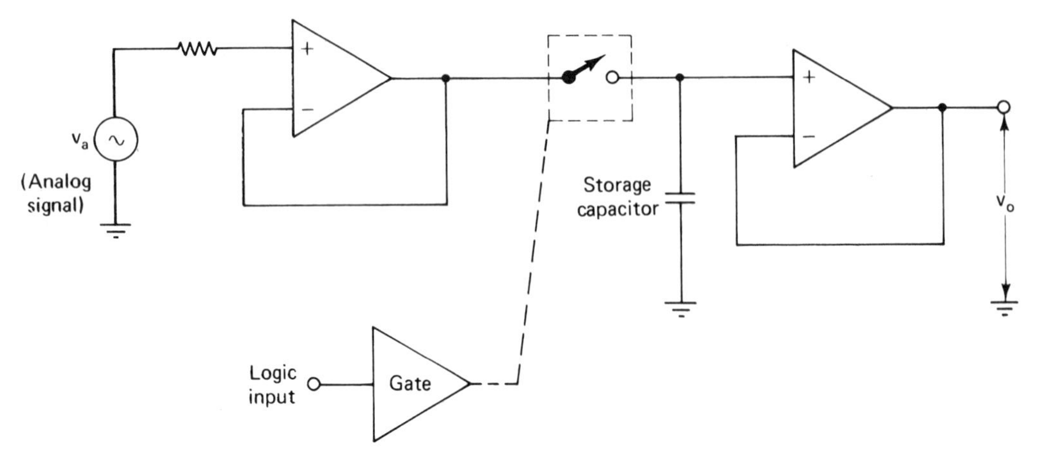

FIGURE 7.13 Sample and hold circuit.

7.3.3 Characteristics of Negative PP Feedback

Negative PP feedback has the same effect on voltage gain, bandwidth, distortion and output impedance as negative SP feedback. However, it reduces input impedance drastically, transforming the op-amp into a virtually ideal current-in to voltage-out amplifier. This configuration, in turn, lends itself to highly versatile and useful applications.

Input Impedance and the Virtual Ground. The input impedance of the negative PP feedback op-amp can be determined with the aid of Figure 7.14, which shows the op-amp with a feedback resistance (R_F) connected from output to input. The current flowing through R_F (I_F) as a result of an input voltage is given by

$$i_F = \frac{v_e + |A_V v_e|}{R_F} \tag{7.29}$$

where A_V is the open-loop voltage gain and v_e is the voltage across the inputs. The apparent resistance of R_F (R_{FM}) seen at the input is given by

$$R_{FM} = \frac{v_e}{i_F} = \frac{v_e R_F}{v_e(1 + A_V)} = \frac{R_F}{1 + A_V} \tag{7.30}$$

The total input impedance is given by $R_{FM} || z_i$, but for a high-gain op-amp R_{FM} is usually so small relative to z_{in} that current taken by z_{in} (inverting to noninverting input) is insignificant and z_{in} itself may be ignored or considered to be infinite. Note that this means

$$i_{in} = i_F \tag{7.31}$$

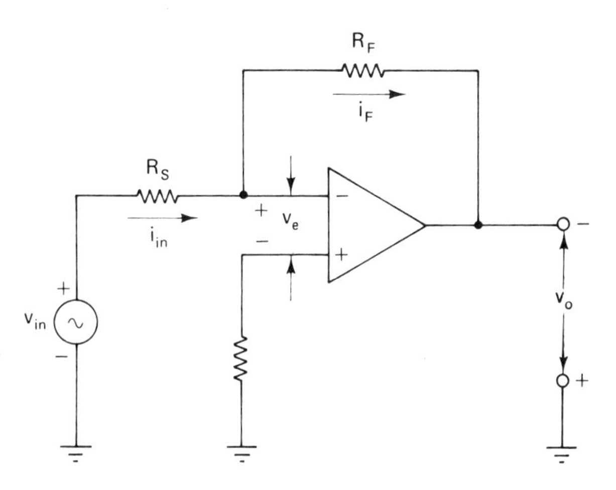

FIGURE 7.14 Negative PP feedback amplifier.

For a high-gain op-amp with a typical R_S, v_e usually approximates zero despite the fact that z_{in} is relatively large. This condition occurs because virtually no current flows through z_{in} but is diverted instead through R_F by Miller action. If the noninverting input is at ground potential, then so is the inverting input. The inverting input may therefore be considered an open circuit held at ground. For this reason it is called a *virtual ground.*

7.3.4 Applications

Applications of the negative feedback PP amplifier primarily exploit the virtual ground at its input, the extremely small input impedance, and the fact than $i_{in} = i_F$.

Inverting Amplifier. Figure 7.14 is actually an op-amp used as an inverting amplifier, so called because v_o goes positive when v_{in} goes negative. Because of its virtually zero input impedance, z_{if} is given by R_S only. The ability to fix z_{if} is often advantageous. Also, the circuit functions as a high-gain, very stable, low-output-impedance amplifier.

Equations for voltage gain, bandwidth, and output inpedance are derived as follows:

$$B = \frac{V_F}{v_o} = \frac{i_F R_S}{i_F R_F} = \frac{R_S}{R_F} \tag{7.32}$$

where V_F is feedback signal.

$$A_{VF} = \frac{v_o}{v_{in}} = \frac{i_F R_F}{i_S R_S} = \frac{R_F}{R_S} \qquad (\text{since } i_F = i_S) \tag{7.33}$$

$$f_{Hf} = \text{bandwidth} = (1 + A_V B) f_{Ho} \approx \frac{A_V R_S}{R_F} f_{Ho} \tag{7.34}$$

when $A_V B \gg 1$.

$$z_{of} = \text{closed-loop output impedance}$$

$$= \frac{z_o}{1 + A_V B} \approx \frac{z_o R_F}{A_V R_S} \tag{7.35}$$

when $A_V B \gg 1$.

Example 7.5 *Design of an Inverting Amplifier*

Performance specifications: Using a 740 op-amp, $A_V = 40 \times 10^3$, $z_o = 75\ \Omega$, and $f_{Ho} = 40$ Hz, design an amplifier to meet the following specifications: $A_{VF} \geq 30$, $z_{if} = 1\ \text{k}\Omega$, bandwidth $= 40$ kHz, and $z_{of} \leq 1\ \Omega$.

(a) Since $z_{if} = 1\ \text{k}\Omega$, $R_S = 1\ \text{k}\Omega$.

(b) Since $f_{Hf} = \dfrac{A_V R_S}{R_F} f_{Ho}$,

$$R_F = \frac{A_V R_S f_{Ho}}{f_{Hf}} = \frac{40 \times 10^3 \times 1\ \text{k}\Omega \times 40\ \text{Hz}}{40\ \text{kHz}} = 40\ \text{k}\Omega$$

(c) *Check A_{Vf} and z_{of}:*

$$A_{Vf} = \frac{R_F}{R_S} = \frac{40\ \text{k}\Omega}{1\ \text{k}\Omega} = 40$$

(If A_{Vf} were too low, it would have been necessary to cascade 2 amplifiers of slightly wider bandwidth.)

$$z_{of} = \frac{z_o R_F}{A_V R_S} = \frac{75 \times 40\ \text{k}\Omega}{40 \times 10^3 \times 1\ \text{k}\Omega} = 75 \times 10^{-3}$$

Summing Amplifier. Adding multiple inputs to the inverting amplifier converts it into a *summing amplifier*, or if $A_{Vf} = 1$, into a *summer* (Figure 7.15). Here the virtual ground is used to isolate each input so that i_{in} is the sum of $i_1 + i_2 + i_3$, and each i is proportional to each v. Therefore,

$$i_{in} = i_1 + i_2 + i_3 = \frac{(v_1 + v_2 + v_3)}{R_S} \tag{7.36}$$

and since

$$i_{in} = i_F$$

and because of the virtual ground, all of v_o is across R_F so that

$$v_o = -i_F R_F \tag{7.37}$$

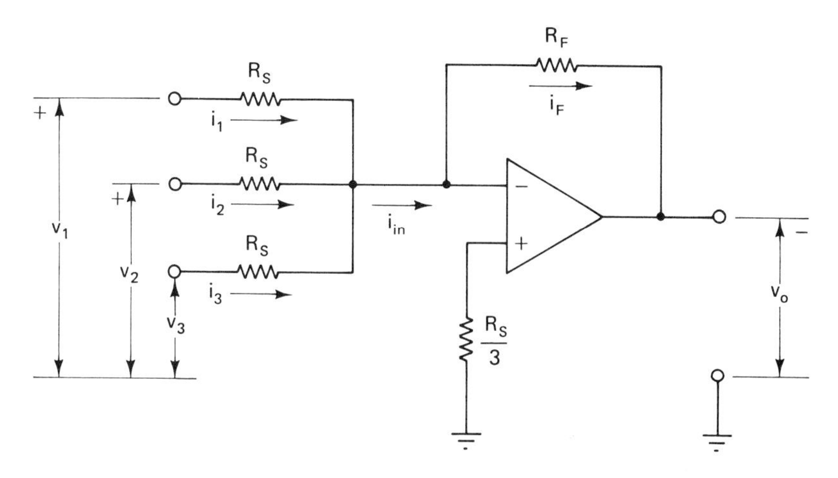

FIGURE 7.15 Summing amplifier.

$$v_o = -(v_1 + v_2 + v_3)\frac{R_F}{R_S} \tag{7.38}$$

If R_F is set equal to R_S, the summing amplifier becomes a summer, and

$$v_o = -(v_1 + v_2 + v_3) \tag{7.39}$$

Nonlinear and Frequency-Sensitive Feedback. Figure 7.16 shows a negative feedback PP amplifier where the complex (frequency sensitive and possibly nonlinear) impedance Z_S and Z_F replace R_S and R_F. The previous op-amp equations apply, but in the following generalized form:

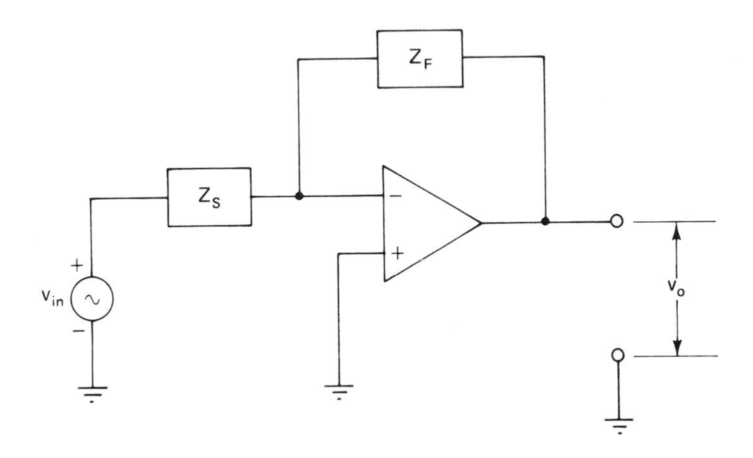

FIGURE 7.16 Negative PP feedback amplifier with complex feedback.

$$I_{in} = I_F \tag{7.40}$$

$$V_{in} = I_{in}Z_S \tag{7.41}$$

$$V_o = -I_F Z_F = -I_{in}Z_F \tag{7.42}$$

where all V's, I's, and Z's are complex (possibly nonlinear).

Equations (7.40) through (7.42) can be combined to yield

$$V_o = \frac{Z_F}{Z_S} \times V_{in} \tag{7.43}$$

Equation (7.43) is important because it extends the range of op-amp applications to nonlinear and frequency-sensitive areas. We offer two examples.

Logarithmic Amplifier

Logarithmic amplification ($v_o = k \ln v_{in}$) results when Z_F is supplied by the emitter–collector circuit of a grounded base transistor (Figure 7.17). This result is easily derived after noting that the transistor collector current i_c is given by

$$i_c = I_o e^{V_{BE}/V_T} \tag{7.44}$$

where I_o and v_T are constants. Since the base is grounded and because of the virtual ground,

$$v_{BE} = v_{CE} = -v_o \tag{7.45}$$

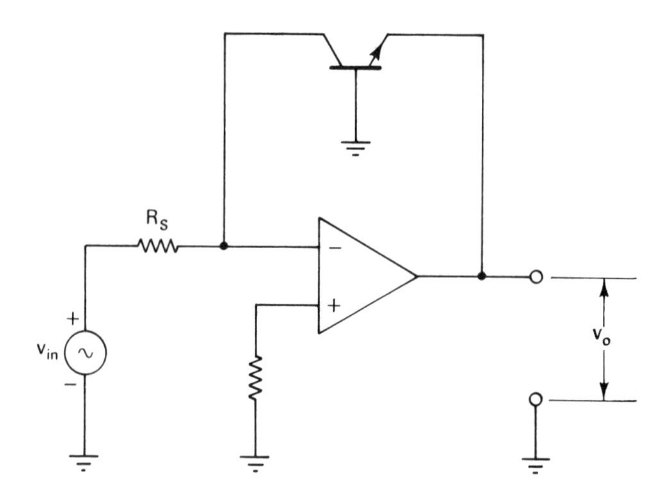

FIGURE 7.17 Logarithmic amplifier.

and

$$i_c = I_o e^{-v_o/V_T} \tag{7.46}$$

$$Z_F = \frac{v_o}{i_c} = \frac{v_o}{I_o e^{-v_o/V_T}} \tag{7.47}$$

Therefore,

$$A_{Vf} = \frac{v_o}{v_{in}} = \frac{Z_F}{R_S} = \frac{v_o}{I_o e^{-v_o/V_T}} \times \frac{1}{R_S} \tag{7.48}$$

and

$$v_{in} = I_o R_S e^{-v_o/V_T} \tag{7.49}$$

Dividing both sides of the equation by I_oR_S and taking the natural log of both sides yields

$$v_o = -V_T \ln \frac{v_{in}}{R_S I_o} \tag{7.50}$$

V_T, as noted, is a constant, typically 25 mV, and I_o is the reverse current of the transistor, typically small. Therefore, R_S acts as a scale factor, establishing gain. High-gain monolithic logarithmic amplifiers spanning several logarithmic decades are commercially available.

Active Filters

When Z_F decreases with increasing frequency, the negative PP feedback op-amp becomes an active low-pass filter. This circuit is shown in Figure 7.18. The magnitude of Z_F as a function of radian frequency (ω) is given by

$$|Z_F| = \left| \frac{R_F/J\omega C}{R_F + (1/J\omega C)} \right| = \frac{R_F}{\sqrt{1 + (\omega C_F R_F)^2}} \tag{7.51}$$

Closed-loop gain is given by

$$|A_F| = \frac{|Z_F|}{R_S} = \frac{R_F}{R_S\sqrt{1 + (\omega C_F R_F)^2}} \tag{7.52}$$

Low-frequency gain (A_{FL}) for $\omega = 0$ is given by

$$|A_{FL}| = \frac{R_F}{R_S} \tag{7.53}$$

The high-end cutoff frequency (f_H), as usual for RC circuits, is given by

$$f_H = \frac{1}{2\pi R_F C_F} \tag{7.54}$$

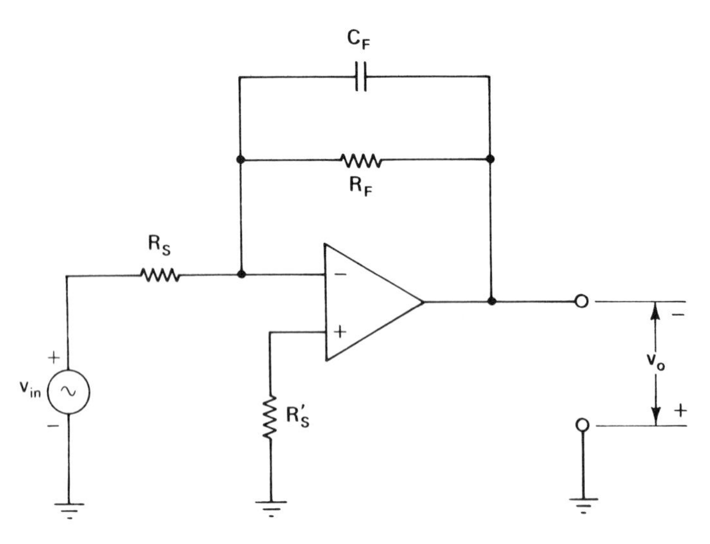

FIGURE 7.18 Active low-pass filter.

Control of gain and bandwidth is accomplished by varying R_F or C_S and R_S. Adjusting R_S controls low-frequency gain. Adjusting either R_F or C_F controls bandwidth (f_H). It is more desirable to control bandwidth through C_F than R_F because R_F also affects gain, but it is also more costly and bulky (especially for lower frequencies). However, if R_S and R_F are mounted on the same shaft (ganged) so that in any position their ratio remains constant, f_H may be controlled by R_F with gain constant. Example 7.6 illustrates the design of an active low-pass filter.

Example 7.6 *Design of an Active Low-Pass Filter*

Performance specifications: f_H from 20 kHz through 50 kHz, $A_{VFL} = 100$.

(a) *Ratio of* R_F/R_S*:*

$$A_{VFL} = \frac{R_F}{R_S} = 100$$

(b) *Find* R_F, R_S, *and* C_F*:* From $f_H = 1/(2\pi R_F C_R)$,

$$R_F = \frac{1}{2\pi f_H C_F}$$

Choosing $C_F = 100$ pF (capacitor should be small),

$$R_{F\,\max} = \frac{1}{2\pi \times 20\text{ kHz} \times 100\text{ pF}} = 79.6\text{ k}\Omega \approx 80\text{ k}\Omega$$

$$R_{F\,\text{min}} = \frac{1}{2\pi \times 50\text{ kHz} \times 100\text{ pF}} = 31.8\text{ k}\Omega \approx 32\text{ k}\Omega$$

$$R_S = \frac{R_F}{100}$$

Then

$$R_{S\,\text{max}} = 800\ \Omega$$

$$R_{S\,\text{min}} = 320\ \Omega$$

R_S' may be made the average of R_S:

$$R_S' = \frac{R_{S\,\text{max}} + R_{S\,\text{min}}}{2} = \frac{800\ \Omega + 320\ \Omega}{2} = 560\ \Omega$$

High-Pass Filter

When Z_F/Z_S increases with frequency, the op-amp becomes a high-pass filter. One possible configuration is shown in Figure 7.19. Here the magnitude of Z_F/Z_S ($|Z_FZ_S|$) is given by

$$|Z_F/Z_S| = \frac{R_F}{\sqrt{R_S^2 + 1/(\omega C_S)^2}} \tag{7.55}$$

When $\omega = 0$, Eq. (7.55) equals 0, and when ω is high Eq. (7.55) approaches R_F/R_S, the normal closed-loop gain. Therefore, the circuit has zero (or low) gain for low frequencies and high gain for higher frequencies. The low-end cutoff frequency (f_L) occurs where $R_S = 1/\omega C_S$ and is given by

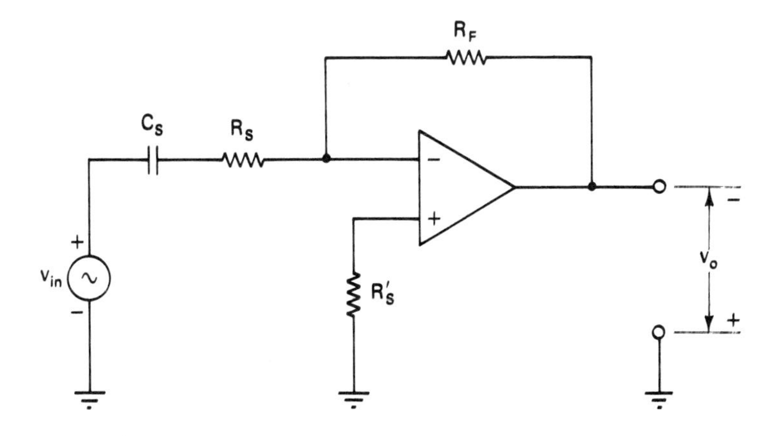

FIGURE 7.19 Active high-pass filter.

$$f_L = \frac{1}{2\pi R_S C_S}$$

High-frequency gain is controlled by R_F, and low-end cutoff frequency, by C_S or R_S. Ganging R_S *and* R_F permits control of f_L by R_S without affecting gain.

Band-pass Filters

Figure 7.20a shows an active band-pass filter called a Sallen and Key resonator. The action of this filter can be described qualitatively as follows. Without C_F, the op-amp is an open-loop inverting amplifier with a passive high-pass filter (R_S, C_S, R_1) input, giving a voltage gain increasing with frequency. With C_F present, however, feedback increases with

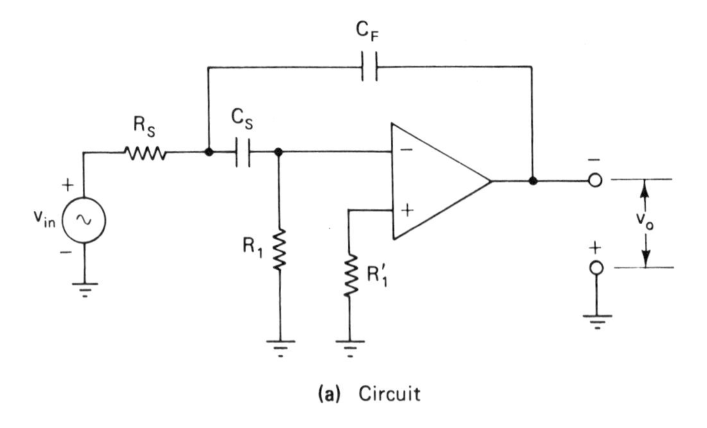

(a) Circuit

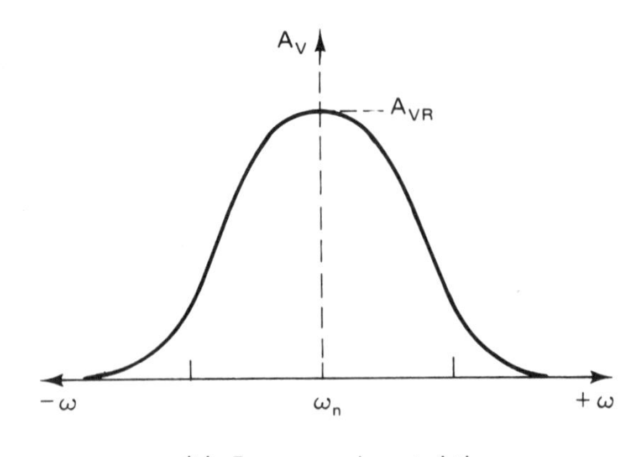

(b) Frequency characteristic

FIGURE 7.20 Active band-pass filter: (a) circuit; (b) frequency characteristic.

frequency, causing voltage gain to act inversely. At frequencies below resonance, the high-pass filter action of the input circuit predominates, and voltage gain rises with frequency. Above resonance, the feedback action of C_F predominates and voltage gain falls with frequency. The typical band-pass characteristic of Figure 7.20b results.

Mathematical analysis of this circuit is complicated by the frequency-sensitive impedances and feedback effects. However, it can be shown that with $R_F = R_S = R_1 = R$, $C_F = C_S = C$, and $A_V \gg 1$, then

$$\omega_n \text{ (center radian frequency)} = \frac{1}{\sqrt{A_V}RC} \tag{7.56}$$

$$Q \text{ (quality factor)} = \frac{\sqrt{A_V}}{3} \tag{7.57}$$

$$A_{VR} \text{ (voltage gain at resonance)} = \frac{A_V}{3} \tag{7.58}$$

These simplified equations allow only ω_n to be fixed, A_{VR} and Q being outside the designer's control. However, if the R's and C's are allowed to differ, ω_n, Q, and A_{VR} can be independently controlled. Typical active band-pass filters may have Q's of 60 and gains of 40,000.

7.3.5 Difference Amplifier: PP and SP Feedback

Figure 7.21 shows a PP plus SP feedback op-amp configured as a difference amplifier. Difference- and common-mode gains may be calculated using the approximate voltage gain equations, but these approximations must be kept in mind when drawing conclusions.

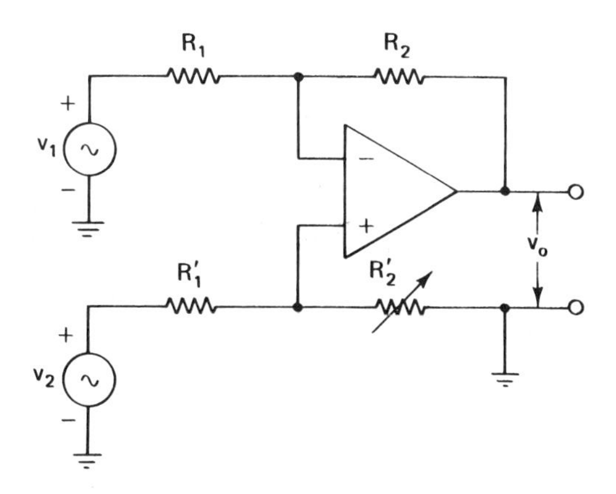

FIGURE 7.21 Difference amplifier.

The inverting input gain (A_{V_1}) is given by

$$A_{V_1} = \frac{R_2}{R_1} \tag{7.59}$$

The non-inverting input gain is given by

$$A_{V_2} = \frac{R_2'}{R_1' + R_2'}\left(\frac{R_2}{R_1} + 1\right) \tag{7.60}$$

When R_2/R_1 equals R_2'/R_1', $A_{V_2} = R_2/R_1$ and $A_{V_1} = -A_{V_2}$. For a common-mode signal ($v_1 = v_2$), v_o is the algebraic sum of A_{V_1} and A_{V_2}, which by adjusting R_2' can theoretically be made zero (CMRR $= \infty$). Actually, practical matters (including our two approximations) preclude this result, but the circuit does permit us to maximize CMRR.

7.3.6 Characteristics of Negative SS and PS Feedback

SS and PS negative feedback (Figure 7.6c and d) affect z_{if} in the same way as their respective counterparts, SP and PP feedback (z_{if} depends primarily on the way feedback is combined with signal). This correspondence holds true for distortion and bandwidth. Also, PS feedback creates a virtual ground at its input, as does PP feedback.

However, SS and PS negative feedback affect output impedance (z_{of}) differently than SP or PP. Both SS and PS feedback increase z_{of} considerably, making output current virtually independent of R_L. This makes the SS feedback amplifier an excellent voltage-in to current-out (i_o/v_{in}) amplifier and the PS feedback amplifier an equally excellent current amplifier.

Output Impedance. The closed-loop output impedance for both SS and PS cases can be derived from Figure 7.22. Both cases are representable by this single figure, since in each case feedback is delivered to the inverting input. Figure 7.22 is Figure 7.6c with the ground return of R_F repositioned and the + input grounded. It also is Figure 7.6d with i_{in} equal to zero (R_{F2} in series with the open-loop input resistance of the op-amp may be shorted or left in without affecting circuit operation). As usual, z_o is found by placing a voltage across R_L and determining the current (i_o) flowing into the output terminal of the op-amp as follows.

$$i_o = \frac{v_o - A_V v_e}{R_F + z_o} \tag{7.61}$$

where z_o is the open-loop output impedance of the op-amp.

$$v_e = i_o R_F \tag{7.62}$$

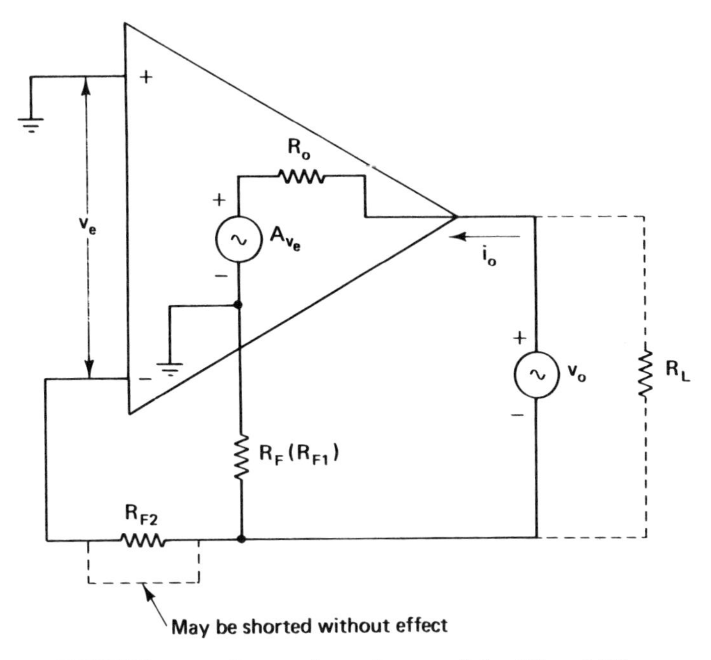

FIGURE 7.22 Output impedances of the SS and PS op-amp.

Combining Eqs. (7.60) and (7.61) and solving for i_o gives

$$i_o = \frac{v_o}{z_o + R_F(1 + A_V)} \tag{7.63}$$

z_{of} is given by

$$z_{of} = \frac{v_o}{i_o} = z_o + R_F(1 + A_V) \approx A_V R_F \tag{7.64}$$

when A_V is large.

SS Gain. Since the SS negative feedback amplifier transforms an input voltage to an output current, its gain is most usefully expressed as a transconductance G_{mf} defined by

$$G_{mf} = \frac{i_o}{v_{in}} \tag{7.65}$$

The closed-loop voltage gain A_{Vf} of the SS corresponds to the SP and is given by $A_{Vf} = A_V/(1 + A_V B)$. Since, in this case, $B = R_F/R_L$, A_{Vf} is given by

$$A_{Vf} = \frac{A_V}{1 + A_V R_F/R_L} \approx \frac{R_L}{R_F} \tag{7.66}$$

For resistive loads, G_{mf} is proportional to A_{Vf} and is given by

$$G_{mf} = \frac{i_o}{v_{in}} = \frac{v_o}{v_{in}R_L} = \frac{A_{Vf}}{R_L} = \frac{R_L}{R_F R_L} = \frac{1}{R_F} \tag{7.67}$$

Note that both gains are stabilized by feedback, depending only on R's.

PS Current Gain. For the PS negative feedback op-amp (Figure 7.6d), interest centers on the current gain (A_I). Current gain is most easily derived by noting that the general expression for closed-loop gain

$$A_{Vf} = \frac{A_V}{1 + A_V B} \tag{7.68}$$

applies to current as well as voltage gain, provided that the open-loop gain is A_I, and B is a current feedback ratio (B_I). Therefore, the closed-loop current gain A_{IF} is given by

$$A_{IF} = \frac{A_I}{1 + A_I B_I} \tag{7.69}$$

Referring to Figure 7.6d and recalling that a virtual ground exists at the inverting input, we see that A_{IF} is given by

$$A_{IF} = \frac{A_I}{1 + A_I[R_{F1}/(R_{F2} + R_{F1})]} \approx \frac{R_{F2}}{R_{F1}} + 1 \tag{7.70}$$

when $A_I[R_{F1}/(R_{F2} + R_{F1})] \gg 1$.

7.3.7 Electronic Voltmeter: An SS Application

A negative feedback SS op-amp makes an excellent electronic voltmeter, providing high input impedances and an output current proportional to input voltage. An electronic voltmeter circuit complete with range switching and meter zeroing is shown in Figure 7.23. The circuit operates in the following way. Since $G_{mf} = I_o/V_{in} = 1/R_F$, then

$$I_o = \frac{V_{in}}{R_F} \tag{7.71}$$

and I_o is proportional to V_{in}. I_o flows through a 100-μA full-scale ammeter with an internal resistance of 10 Ω. R_F is switched to change scale so that I_o never exceeds 100 μA. Because of meter sensitivity, offset voltage could prove troublesome. By using an op-amp that permits nulling the offset voltage by adjusting a potentiometer, as shown, meter zeroing is provided. Example 7.7 illustrates a design procedure for an op-amp electronic voltmeter using SS feedback.

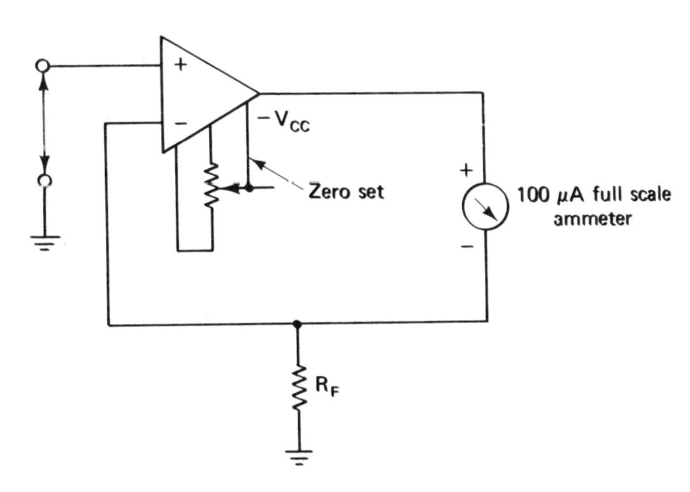

FIGURE 7.23 Electronic voltmeter.

Example 7.7 *Design of a DC Electronic Voltmeter*

Performance specifications:

Input impedance ≥ 20 MΩ
Voltage scales: 1, 10, and 100 mV, full scale, switchable.
Meter sensitivity: 100 μA per 1 mV, R_L = 1 mV/100 μA = 10 Ω
Open-loop op-amp voltage gain = 25 × 10^3, z_i (op-amp) = 300 kΩ

(a) *Use* $I_o = V_{in}/R_F$:

$$R_{F1} = \frac{V_{in}}{I_o} = \frac{1 \text{ mV}}{100\ \mu\text{A}} = 10\ \Omega \qquad \text{(1-mV scale)}$$

$$R_{F2} = \frac{10 \text{ mV}}{100\ \mu\text{A}} = 100\ \Omega \qquad \text{(10-mV scale)}$$

$$R_{F3} = \frac{100 \text{ mV}}{100\ \mu\text{A}} = 1\ \text{k}\Omega \qquad \text{(100-mV scale)}$$

R_F should be a switchable resistor equal to 10 Ω, 100 Ω, and 1 kΩ.

(b) *Input impedance:*

$$z_{if} = (1 + A_V B) z_{in} \approx A_V B z_{in} = \frac{A_V R_F}{R_L} z_{in}$$

$$= \frac{25 \times 10^3 \times 10\ \Omega}{10\ \Omega} \times 300\ \text{k}\Omega = 7{,}500\ \text{M}\Omega$$

on a 1-mV scale, 250 kΩ on a 10-mV scale, and 250 MΩ on a 100-mV scale.

7.3.8 Electronic Ammeter: A PS Application

PS negative feedback produces low input impedance and linear current gain. This makes the PS feedback op-amp suitable as an electronic ammeter. Figure 7.24 shows a typical configuration and Example 7.8 illustrates a design.

Example 7.8 *Design of a DC Electronic Ammeter*

Performance specifications:

Ranges: 1, 10, 100 μA
$z_{in} \leq 0.4\ \Omega$
Meter specifications: 100 μA/mV, $R_L = 10\ \Omega$
Op-amp specifications: $A_V = 50 \times 10^3$

(a) R_{F2}/R_{F1} *ratios:*

$$A_I = \frac{I_o}{I_{in}} = \frac{100\ \mu\text{A}}{1\ \mu\text{A}} = 100 = \frac{R_{F2}}{R_{F1}} + 1, \quad \frac{R_{F2}}{R_{F1}} = 99 \qquad (1\ \mu\text{A})$$

$$= \frac{100\ \mu\text{A}}{10\ \mu\text{A}} = 10 = \frac{R_{F2}}{R_{F1}} + 1, \quad \frac{R_{F2}}{R_{F1}} = 9 \qquad (10\ \mu\text{A})$$

$$= \frac{100\ \mu\text{A}}{100\ \mu\text{A}} = 1 = \frac{R_{F2}}{R_{F1}} + 1, \quad \frac{R_{F2}}{R_{F1}} = 0 \qquad (100\ \mu\text{A})$$

(b) *Establish* R_{F2}*:* When R_{F2} is $\gg R_L$ (usually the case), R_L may be ignored and z_{if} is given by

$$z_{if} = \frac{R_{F2}}{A_V} \qquad \text{(see Section 7.3.3)}$$

$$0.4 = \frac{R_{F2}}{50 \times 10^3}, \quad R_{F2} = 50 \times 10^3 \times 0.4\ \Omega = 20\ \text{k}\Omega \qquad (1\ \mu\text{A})$$

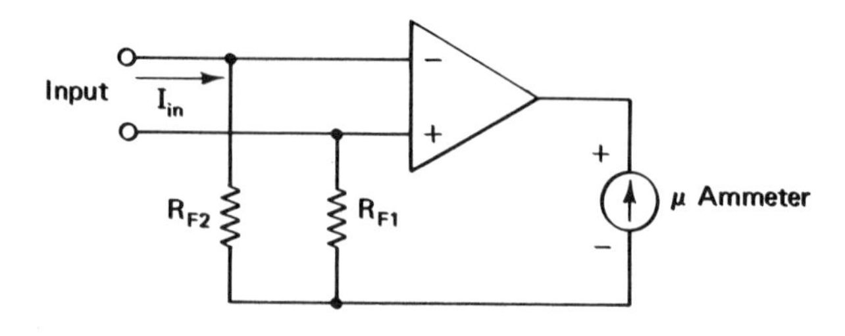

FIGURE 7.24 Electronic ammeter.

Then, since $R_{F2}/R_{F1} = 99$,

$$R_{F1} = \frac{R_{F2}}{99} = \frac{20\ \text{k}\Omega}{99} = 202\ \Omega$$

Then, using $R_{F1} = 202\ \Omega$ and $R_{F2} = R_{F1} \times (R_{F2}/R_{F1})$,

$$\begin{aligned} R_{F2} &= R_{F1}99 = 202\ \Omega \times 99 \approx 20\ \text{k}\Omega \qquad (1\ \mu\text{A}) \\ &= R_{F1}9 = 202\ \Omega \times 9 \approx 2\ \text{k}\Omega \qquad (10\ \mu\text{A}) \\ &= R_{F1}0 = 202\ \Omega \times 0 \approx 0\ \Omega \qquad (100\ \mu\text{A}) \end{aligned}$$

7.4 ACTIVE DIODE APPLICATIONS

Nonlinear PP and SP negative feedback can make an op-amp perform as a virtually ideal diode. Also, it can provide a buffering action that eliminates the effects of source resistance. Examples follow.

7.4.1 Active Diode Half-wave Rectifier

Figure 7.25 shows an op-amp half-wave rectifier that provides near perfect rectification. Break voltage is near zero because until v_o reaches

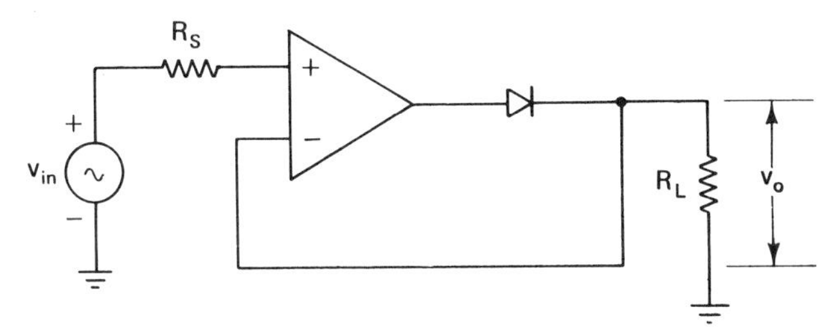

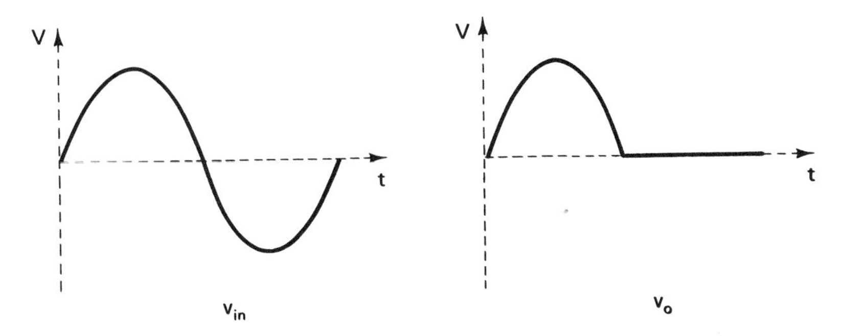

FIGURE 7.25 Active half-wave rectifier.

0.7 V the op-amp operates with open-loop gain (no feedback). Therefore, v_o reaches 0.7 V when

$$v_i = \frac{v_o}{A_V} = \frac{0.7}{A_V} \tag{7.72}$$

which is virtually zero for a high-gain op-amp. After the diode turns on, the circuit acts as a voltage follower, providing extremely low output resistance and eliminating source resistance (R_S) effects.

7.4.2 Active Clipper

Figure 7.26 shows an op-amp positive clipper in which the effects of R_S and the diode break voltage are eliminated. When v_{in} is less than V_R, the error voltage (v_e measured from − to + inputs) is negative. Therefore, v_o is driven positive and the diode is kept off. The input (v_{in}) appears across

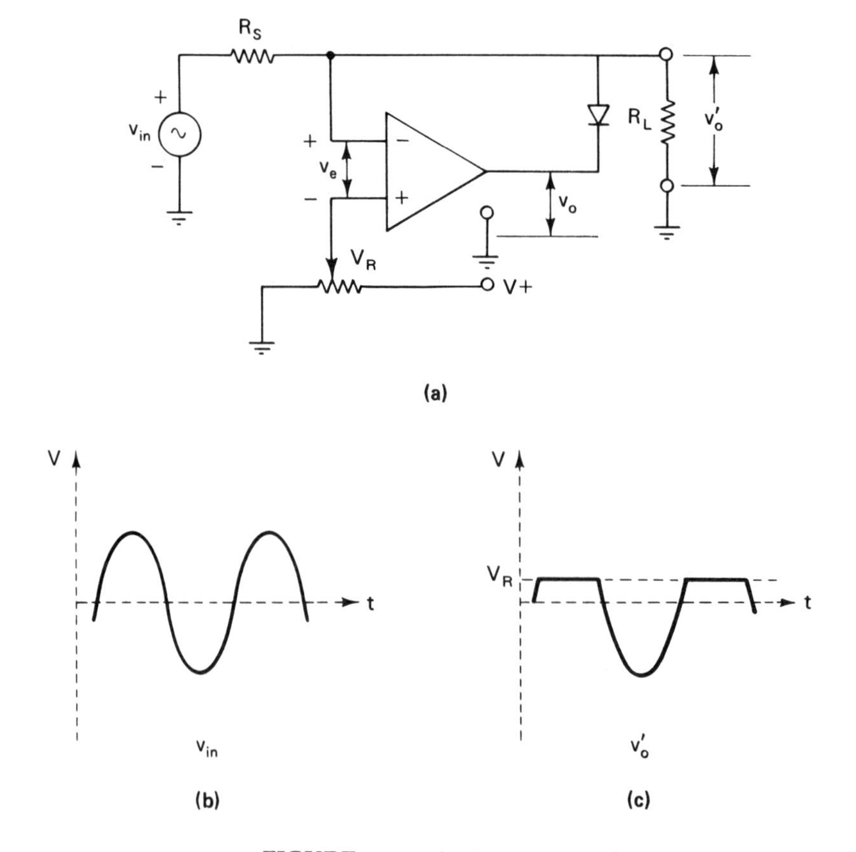

FIGURE 7.26 Active positive clipper.

R_L (diminished by the R_S, R_L voltage-division effect). When v_{in} exceeds V_R, v_e becomes positive, driving v_o negative and turning the diode on. In this state, PP feedback establishes a virtual ground between the op-amp inputs, tying v_o' to V_R.

7.4.3 Active Clamper

Figure 7.27 shows an active positive clamper in which the effect of the diode break voltage is eliminated. During the first negative excursion of v_{in}, v_e goes negative, driving v_o positive and turning on the diode. A virtual ground is established between the op-amp inputs, and C charges to the difference between $v_{in\ peak}$ and V_R. (For example, with $v_{in\ peak} = -5$ V and $V_R = -2$ V, V_C would charge to 3 V.) When C is completely charged, v_o' is at V_R and v_e equals 0 V. Subsequently, as v_{in} becomes less negative,

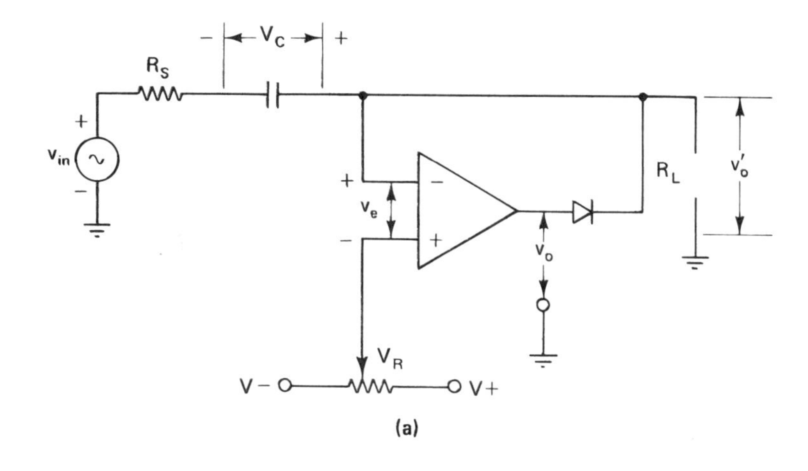

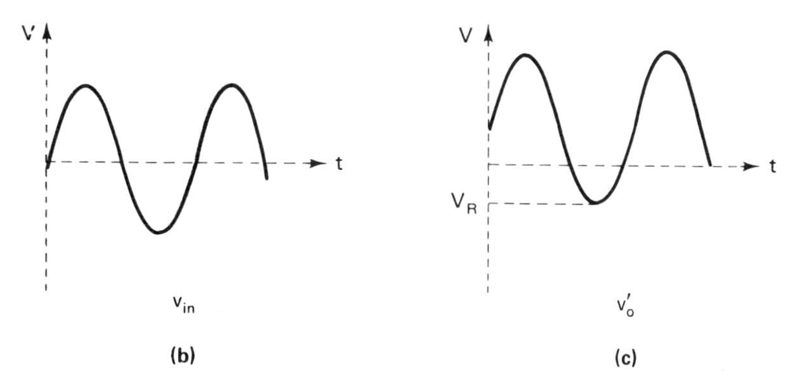

FIGURE 7.27 Active positive clamper.

v_e goes positive and the diode turns off. The output voltage v_o' is then equal to the algebraic sum of v_{in} and V_C (Figure 7.27c). In effect, the average value of v_{in} is boosted positively by V_C clamping its lower edge at V_R. This action continues through the cycle to the next negative peak of v_{in} when the diode goes on for a brief interval to recharge C. As in the previous example, the open-loop op-amp gain acts to virtually eliminate the diode break voltage.

7.5 POSITIVE FEEDBACK: SINUSOIDAL OSCILLATORS

Positive feedback obtains when feedback increases its originating signal. The most important result of positive feedback is oscillation, a sustained output in the absence of input. Sustained oscillation will occur in an amplifier when its closed-loop gain is just enough to make the fed-back signal identical in shape, magnitude, and polarity to its originating signal. Under these conditions, output will continue without external input.

As a practical matter, it is not possible to simply set the closed-loop gain of a linear amplifier to this condition and achieve stable oscillation. Small parameter changes induced by temperature and supply voltage shifts would either decrease or increase v_o, causing the oscillation to die down or to run away.

To achieve stability, the oscillator must reach a condition where the fed-back voltage generates the originating output voltage ($A_{Vf}v_f = v_o$) and also where a decrease in v_o causes an increase in $A_{Vf}v_f$, and vice versa.

This condition is illustrated schematically in Figure 7.28. Initially, below a, v_o increases faster than $A_{Vf}v_f$. This means the fed-back voltage is larger than its originating signal and v_o increases. Past a, v_o decreases as $A_{Vf}v_f$ increases, but until b, v_o is still greater than $A_{Vf}v_f$. At b the condition for stable oscillation is achieved. If, at this point, $A_{Vf}v_f$ attempts to increase, v_o is reduced, reducing $A_{Vf}v_f$ and restoring operation to b. If at this point $A_{Vf}v_f$ is reduced, v_o will be increased, restoring b.

Conditions for stable operation are most easily satisfied by oscillators generating sinusoids. The sinusoid is the only periodic wave form that will preserve its shape in linear reactive circuits (circuits containing L or C as well as R). The conditions for stable sinusoidal oscillation are given by

$$A_{Vf}v_f = v_o \tag{7.73}$$

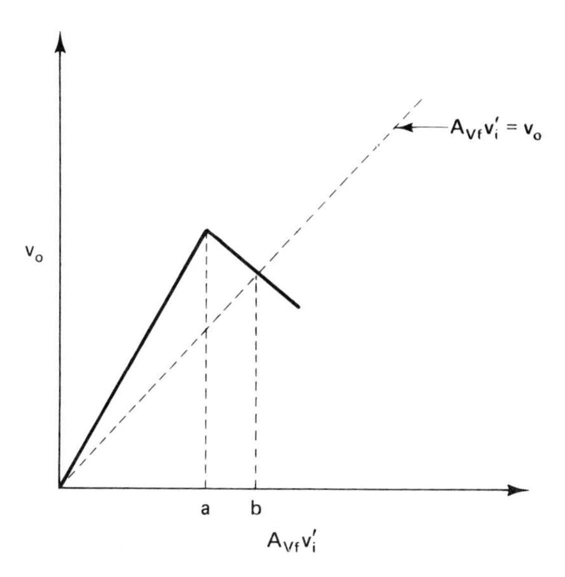

FIGURE 7.28 Stability condition in oscillators.

where v_f is the signal fed back to the input by positive feedback; and since

$$v_f = B_p v_o \tag{7.74}$$

where B_p is the ratio (v_f/v_o) fed back to the input by positive feedback, then

$$A_{Vf}B_p v_o \quad \text{must equal} \quad v_o \tag{7.75}$$

and

$$A_{Vf}B_p = 1 \tag{7.76}$$

Also,

$$\theta_f = n \times 360^\circ \tag{7.77}$$

where θ_F is the total closed-loop phase shift (measured from input to output back to input). The criteria for stability must also apply.

7.5.1 Wien-Bridge Oscillator

Figure 7.29 shows a circuit that satisfies the required conditions for stable sinusoidal operation. This circuit is called a Wien-bridge oscillator and is widely used to generate 5-Hz to 1-MHz sinusoids.

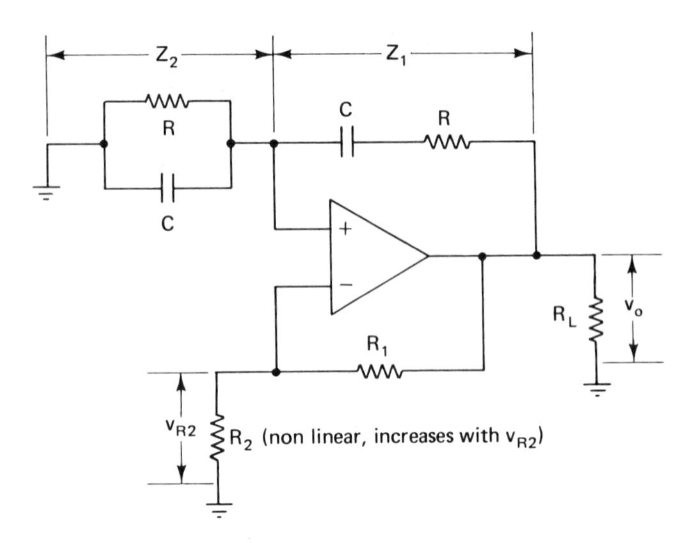

FIGURE 7.29 Wien-bridge oscillator.

Oscillation occurs at a frequency where the phase shift across Z_1 (θ_1) cancels the phase shift across Z_2, giving $\theta_f = 0$. Since

$$Z_1 = R + \frac{1}{J\omega C} = \sqrt{R^2 + \left(\frac{1}{\omega C}\right)^2}^{\;\tan^{-1}(-1/\omega CR)} \tag{7.78}$$

and

$$Z_2 = \frac{R}{J\omega C\left(R + \dfrac{1}{J\omega C}\right)} = \frac{R}{1 + J\omega C R} = \frac{R}{\sqrt{1 + (\omega CR)^2}}^{\;-\tan^{-1}(\omega CR)} \tag{7.79}$$

setting $\theta_1 - \theta_2 = 0$ and solving for ω_o gives

$$\tan^{-1}\left(-\frac{1}{\omega_o CR}\right) - (-\tan^{-1}\omega_o CR) = 0 \tag{7.80}$$

$$-\frac{1}{\omega_o CR} + \omega_o CR = 0 \tag{7.81}$$

$$\omega_o = \frac{1}{CR} \quad \text{or} \quad f_o = \frac{1}{2\pi CR} \tag{7.82}$$

B_p at f_o can be found from $B_p = Z_2/(Z_1 + Z_2)$ and Eq. (7.81).

$$B_p = \frac{Z_2}{Z_2 + Z_1} = \frac{R}{1 + J\omega_o CR} \times \frac{1}{\left(\dfrac{R}{1 + J\omega_o CR} + R + \dfrac{1}{J\omega_o C}\right)} \tag{7.83}$$

Setting $\omega_o = 1/CR$ in Eq. (7.82) gives

$$B_p = \tfrac{1}{3}\angle 0^\circ \tag{7.84}$$

Therefore, for the necessary condition $A_{Vf}B_p = 1$, A_{Vf} must be equal to 3.

Stabilization is provided by the action of the nonlinear resistance (R_2) in the SP feedback network as follows. Closed-loop gain A_{Vf} is given by

$$A_{Vf} = \frac{R_1}{R_2} + 1 \tag{7.85}$$

R_2 is selected so that at $v_{R2} = 0$ V it is $< R_1/2$, so A_{Vf} is > 3 and oscillation builds up. As v_o increases, v_{R2} follows and R_2 increases as well. When $R_2 = R_1/2$,

$$A_{Vf} = \frac{R_1}{R_1/2} + 1 = 3 \tag{7.86}$$

This is the lower limit on A_{Vf} for sustained oscillation. An increase in v_o beyond this point will decrease A_{Vf}; a decrease in v_o will increase A_{Vf}. Hence stability is established at $A_{Vf} = 3$.

The peak output voltage (V_{op}) depends on the value of R_2 when

$$R_2 = \frac{R_1}{2} \tag{7.87}$$

and is given by

$$V_{op} = 3V_{R2} \tag{7.88}$$

since $A_{Vf} = 3$. Therefore, we can vary V_{op} by making R_1 adjustable.

Example 7.9 *Design of a Wien-Bridge Oscillator*

Performance specifications: $f_o = 100$ Hz (sinusoid), $R_L = 2\ \text{k}\Omega$, $V_{op} = 6$ V, op-amp 741. Assume that $R_2 = 10\ \text{k}\Omega$ at 2 V.

(a) *Find R and C:* From $f_o = 1/2\pi RC$,

$$RC = \frac{1}{2\pi f_o} = \frac{1}{2\pi \times 100\ \text{Hz}} = 1.6 \times 10^{-3}$$

Choose $R > 2\ \text{k}\Omega$ to prevent circuit loading; $R = 20\ \text{k}\Omega$:

$$C = \frac{1.6 \times 10^{-3}}{R} = \frac{1.6 \times 10^{-3}}{20\ \text{k}\Omega} = 0.08\ \mu\text{F}$$

(b) *Choose R_1:* Since $R_2 = 10\ \text{k}\Omega$ at 2 V and $V_{op} = 3V_{R2} = 3 \times 2 = 6$ V, set $R_1 = 2R_2 = 2 \times 10\ \text{k}\Omega = 20\ \text{k}\Omega$.

7.6 VOLTAGE-CONTROLLED RELAXATION OSCILLATOR

Relaxation oscillators generate nonsinusoidal (sawtooth) wave forms by alternately charging and discharging a capacitor through a resistor. A widely used sawtooth generator, the voltage-controlled oscillator (VCo), is shown in Figure 7.30.

The circuit of Figure 7.30a is a Miller integrator with a four-layer diode (D_4) across C. Without D_4, The circuit generates a ramp wave form slewing from 0 V to maximum output voltage (Figure 7.30b). Its parameters can be derived as follows:

$$i_{\text{in}} = i_f = \frac{V_C}{R} \tag{7.89}$$

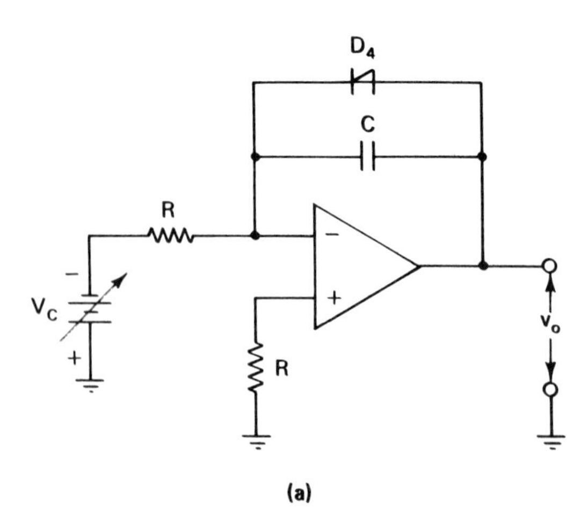

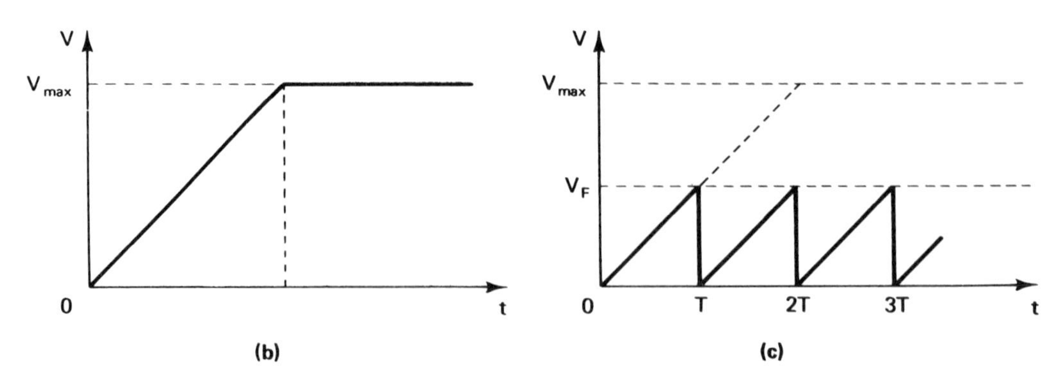

FIGURE 7.30 Voltage-controlled oscillator.

$$\frac{dv_o}{dt} = \frac{-i_{in}}{C} = \frac{-i_F}{C} = \frac{-v_C}{RC} \tag{7.90}$$

$$v_0 = -\int_0^t \frac{v_C}{RC}\, d\tau \tag{7.91}$$

When v_C is constant (V_C) then

$$v_o = \frac{-V_C}{RC} \times t \tag{7.92}$$

The four-layer diode (discussed later) becomes a very low resistance at a fixed voltage (V_F), discharging the capacitor and reinitiating the slewing action. The sawtooth wave shown in Figure 7.30c is generated.

The period (T) required for the capacitor voltage to reach V_F depends on R, C, and V_C. T and its inverse f are found from Eq. (7.91) by letting $t = T$ and $v_o = -V_F$.

$$T = \frac{V_F RC}{V_C} \tag{7.93}$$

$$f = \frac{1}{T} = \frac{V_C}{V_F RC} \tag{7.94}$$

Example 7.10 *Design of a VCO*

Performance specifications: f = 100 to 1000 Hz, V_F (firing voltage of D_4) = 6 V, $V_{C\,max}$ = 10 V. Find R, C, and $V_{C\,min}$.

(a) *Find R and C:* Using $V_{C\,max}$ and f(max),

$$RC = \frac{V_C}{V_F f} = \frac{10\text{ V}}{6\text{ V} \times 1000\text{ Hz}} = 1.67 \times 10^{-3}$$

Letting $R = 1\text{ k}\Omega$, $C = 1.67\ \mu\text{F}$.

(b) *Find* $V_{C\,min}$*:* Using RC, V_F, and f(min),

$$V_{C\,min} = RC \times V_F \times f = 1.67 \times 10^{-3} \times 6\text{ V} \times 100\text{ Hz} = 1\text{ V}$$

Problems

1. Figure 7.31 shows an input circuit for an op-amp. If $R_{S1} = R_{S2} = 1\text{ k}\Omega$ and I_{IB} and I_{IO} are as per Table 7.1, find the maximum difference-mode voltage appearing across the two inputs.
2. Repeat problem 1 with $R_{S1} = R_{S2} = 10\text{ k}\Omega$.
3. Repeat problem 2 with $R_{S1} = 1\text{ k}\Omega$ and $R_{S2} = 5\text{ k}\Omega$.

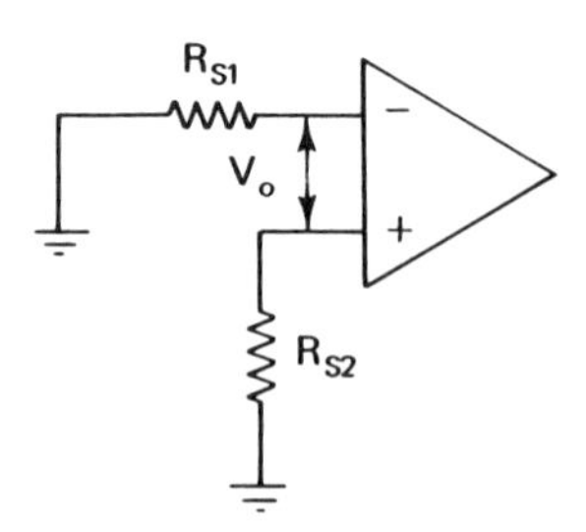

FIGURE 7.31 Input circuit of an op-amp.

4. An op-amp has an SR = 0.05 V/μs. Find f_{max} (a) if V_{CC} = 15 V; (b) if V_{CC} = 10 V; (c) if V_{max} = 10 V.
5. Describe the characteristics of the following feedbacks in terms of how they are derived from the output and inserted into the input: (a) PP, (b) voltage-series, (c) current-series, (d) PS, (e) series–series.
6. In the amplifier of Figure 7.6a, R_L = 2 kΩ, R_{F1} = 5 kΩ, R_{F2} = 10 kΩ, and A_V = 25 k. Find A_{Vf}.
7. For Figure 7.6a, A_v = 25 k, R_{F1} = 10 kΩ, and A_{Vf} = 4. Find R_{F2}.
8. Design the feedback circuit for an SP op-amp with R_L = 5 kΩ so that A_{Vf} = 10.
9. For Figure 7.6a, A_V = 25 k, z_{in} = 1 MΩ, z_o = 100 Ω, R_{F1} = 10 kΩ, and R_{F2} = 20 kΩ. Find A_{vf}, z_{if}, z_{of}, and f_{hf}. F_{Ho} = 100 Hz.
10. Design the amplifier of Figure 7.6a for z_{if} = 100 MΩ.
11. Repeat problem 10 for $z_{if} \geq$ 100 MΩ and $z_{of} \leq$ 1 Ω.
12. Design a noninverting amplifier to meet the following specifications: op-amp 741, z_o = 75 Ω, A_{Vf} = 10, $z_{if} \geq$ 10 MΩ, $z_{of} \leq$ 0.5 Ω; signal range (v_{in}) is 0.01 to 2 V.
13. Repeat problem 12 with $A_{Vf} \geq$ 30.
14. The 741 op-amp is used in a voltage-follower circuit. Compute z_{if}, z_{of}, and A_{Vf}. (Assume z_o = 75 Ω.)
15. Design an inverting amplifier to meet the following specifications: op-amp 740, z_o = 75 Ω, f_{Ho} = 40 Hz, R_{in} = 2 kΩ, $A_{Vf} \geq$ 100.
16. Using the 740, design an inverting amplifier for A_{Vf} = 200, f_{Hf} = 25 kHz, and $z_{of} \leq$ 1 Ω.
17. Design an op-amp circuit that will multiply an input by 1, 10, and 20.
18. Design an op-amp circuit that will divide an input by 1, 10, and 20.

19. Design an active low-pass filter to meet the following specifications: f_{Hf} is 10 kHz through 100 kHz and $A_{Vf} = 50$.

20. Design an active high-pass filter to meet the following specifications: $f_L = 100$ Hz and $A_{Vf} = 100$.

21. An active band-pass filter is required to have the following specifications: $A_V = 20$ k and f_o (center frequency) = 100 kHz. Design a Sallen and Key resonator to meet these specifications. Find Q and A_R.

22. What are the input impedances seen by v_1 and v_2 in Figure 7.21? Sketch a multi-op-amp circuit that will function as a difference amp but have higher input impedances.

23. Design the op-amp circuit of problem 22 for a difference-mode gain equal to 100.

24. Using Norton's transformation theorem, demonstrate that an op-amp with a very high output impedance ($z_{of} = \infty$) constitutes an ideal current source.

25. An SS negative feedback amplifier has a $G_m = 0.1$, $A_V = 25$ k, and $R_F = 1$ kΩ. For $v_{in} = 0.01$ V, find i_o for (a) $R_L = 1$ kΩ, (b) $R_L = 10$ kΩ, and (c) $R_L = 10$ MΩ.

26. A PS negative feedback op-amp has $A_{Vf} = 100$. Find A_{If}.

27. Design an electronic voltmeter to meet the following specifications:
Input impedance ≥ 200 MΩ
Scales 10-, 100-, and 1000-mV full scale
Op-amp $A_V = 50 \times 10^3$
Meter resistance = 10 Ω
Meter sensitivity 100 μA/mV.

28. For problem 27, determine the lowest scale for an input impedance ≥ 2 MΩ.

29. Design a dc electronic ammeter to meet the following specifications:
Ranges: 5, 50, 500 μA
$z_{in} \leq 0.002$ Ω
Meter specifications: 1000 μA/mV
Op-amp $A_V = 50 \times 10^3$

30. For the circuit of Figure 7.25, find the v_{in} required to turn on the diode in the output circuit. Assume that the open-loop gain of the op-amp is 50 k and the break voltage of the diode is 0.7 V.

31. Using Figure 7.26, design a positive clipper that will clip input wave forms at precisely +2 V. How do R_L and the diode break voltage (V_B) affect clipper operation?

32. Compare the effects of R_L and V_B in the active clamper of Figure 7.27 with their effects on the diode clamper of Chapter 3.

33. Design a Wien-bridge oscillator to meet the following specifications: $f_o = 50$ Hz (sinusoid), $R_L = 2$ kΩ, $V_{o\,peak} = 10$ V, op-amp 741. Assume $R_2 = 5$ kΩ at 3.33 V.

34. Design a VCO to meet the following specifications: $f = 500$ Hz, V_F (of D_4 in Figure 7.30a) = 5 V, and $V_{C\,max} = 12$ V.

35. Extend the design of problem 34 to permit f to vary from 500 to 1000 Hz.

References

Bell, David. *Electronic Devices and Circuits,* 2nd ed., Reston Publishing Co., Reston, Virginia, 1980.

Comer, David J. *Modern Electronic Circuit Design,* Addison-Wesley Publishing Co., Reading, Massachusetts, 1968.

Malvino, Albert P. *Electronic Principles,* 2nd ed., McGraw-Hill Book Co., 1979.

Schilling, Donald L., and Charles Belove. *Electronic Circuits: Discrete and Integrated,* McGraw-Hill Book Co., 1979.

8
SPECIAL DEVICES

8.1 INTRODUCTION

The physics of the four-layer solid-state device (thyristors) was discussed in Section 1.7. There it was shown that these devices functioned as current- or voltage-actuated latching switches capable of carrying and controlling large currents. Consequently, they are widely used in high-power switching. Four-layer switches include the Schockley (four-layer) diode, the silicon unilateral switch (SUS), the gate-turn-off switch (GTO), the silicon bilateral switch (SBS), the programmable unijunction transistor (PUT), the silicon-controlled switch (SCS), the light-activated silicon controlled rectifier (LASCR), and the silicon-controlled rectifier (SCR). Also, we have devices with more than four layers (actually combinations of four layers) such as the DIAC and TRIAC. The unijunction transistor (UJT) discussed in Section 8.3, although not a four-layer device acts as one and is included. The chapter concludes with a discussion of some special light-activated and light-producing devices.

8.2 THE TRIGGERING PROBLEM

The devices listed in the preceding section (light-producing devices excepted) fall into roughly two categories, triggers and controllers. The SCR is the primary controller. Its ability to switch from the nonconducting to conducting state in response to a small control signal, plus its ability to handle high power, account for its wide utility. However, precision triggering of the SCR calls for circuits that can supply adequate gate current and voltage at precise times. This is the function of the trigger devices: the four-layer diode, the SBS, the SCS, the PUT, the UJT, and others. Light-activated devices also function as triggers.

An SCR may be triggered by applying a slowly rising voltage to its gate, but the initiation of the *on* state becomes too dependent on the specific triggering characteristic of the individual SCR. Pulse triggering, however, can accommodate wide tolerances in SCR gate characteristics by overdriving the gate. Also the initiation of the *on* state is made more precise. In addition, low-powered trigger circuits lend themselves to automatic, self-programmed, and feedback control systems using low-power electronic circuitry.

8.2.1 Unilateral Trigger-Pulse Generators

Figure 8.1 shows a basic relaxation oscillator used for trigger-pulse generation. Also shown is a typical characteristic of a unilateral trigger device. Before (or between) pulses, capacitor C is charged from V_1 through R_1, the trigger device being in the *open* or nonconducting state. As V_C (capacitor voltage) increases, the operating point of the trigger device moves to the right along the lowest curve of the characteristic. When V_S is reached, the trigger device switches into the *on* or conducting state, and C discharges through R_2, generating the trigger pulse. The cycle then resumes.

Circuit oscillation depends critically on R_1 (assuming V_1 constant). For oscillation, the trigger device must turn off between capacitor discharges. For this to happen, the current i_p, given by the intersection of the R_1 load line and the device characteristic, must be less than I_H (holding current). Figure 8.1 shows a minimum R_1 that just satisfies this condition. Also, the trigger device cannot turn on unless the R_1 load line intersects the characteristic along its negative resistance portion (above I_S). Therefore, R_1 max is the largest value that will permit turn on.

The magnitude of the pulse voltage e_p and the available pulse current i_p depends on R_2 (for a given characteristic). When the device switching time is very fast compared with the R_2C time constant ($R_2C >$ $10 \times$ switching time), the peak pulse current is given by the intersection of the R_2 load line and the device characteristic. Then e_p is given by

$$e_p = V_S - V_F \tag{8.1}$$

However, when switching time is slower, approaching R_2C, e_p and i_p are reduced by the increased device resistance during switching. Also, a device that slowly switches from point 1 to point 2 (Figure 8.1) may never

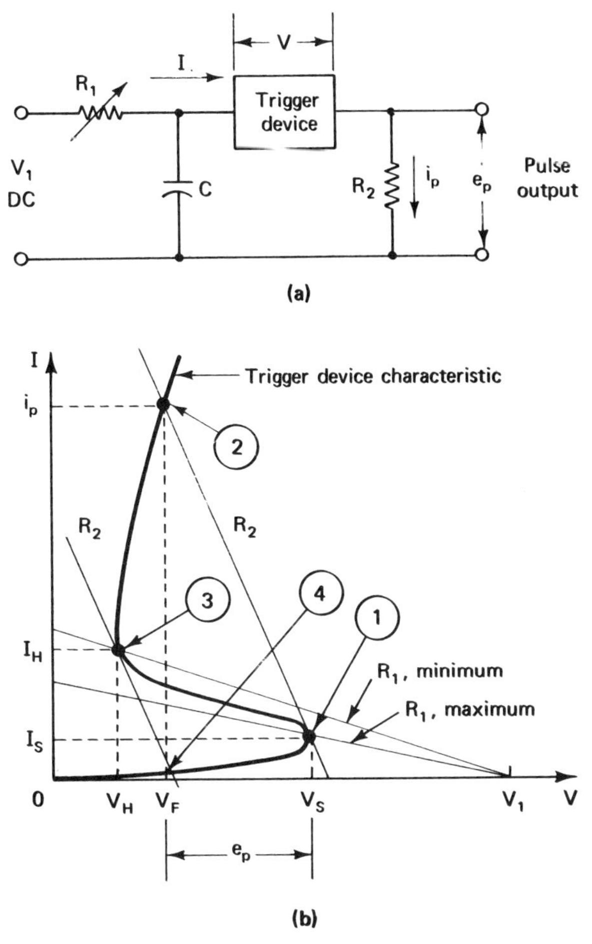

FIGURE 8.1 Basic relaxation oscillator circuit and characteristics. Courtesy of General Electric.

get there, since the capacitor is discharging during transition and the intersection of R_2 and the characteristics moving from point 2 to point 3.

A figure of merit for triggering devices is provided by e_p for the device in a typical test circuit. The higher the e_p under these conditions, the better the device (ignoring other considerations).

8.3 UNIJUNCTION TRANSISTOR

The unijunction transistor (UJT) is a single p–n junction device, which nevertheless functions as a latching switch. The physics underlying this action is discussed in Section 1.9. Here we view the UJT purely as a circuit element with the characteristic shown in Figure 8.2. Although traditionally UJT characteristics are shown with V_E as ordinate versus I_E as abscissa, here we have interchanged the axes to emphasize the UJTs relation to other trigger devices. A commercial input characteristic (GE 2N2646) is shown in Figure 8.3.

Comparing the UJT characteristics against Figure 8.1b, we note that V_p, the firing voltage, corresponds to V_S, and I_p' to i_p. The valley point, the lowest value of the characteristic voltage values (V_V), corresponds to V_H of Figure 8.1b, and the holding current I_V, to I_H. V_p is seen to be a function of V_{BB}, the voltage across the UJT.

The UJT (insert in Figure 8.3) is a three-terminal device that functions as a thyristor. The terminals are labeled B_1 (base 1), B_2 (base 2), and E (emitter). Between B_2 and B_1 the UJT looks like a resistance. This resistance is called the *interbase resistance* (R_{BB}) and ranges between 4 and 12 kΩ at 25°C with the emitter current (I_E) equal to zero. Internally, R_{BB} divides into a resistance between E and B_2 called r_{B2} and one between E and B_1 (r_{B1}). The ratio $r_{B1}/(r_{B1} + r_{B2})$, together with the supply voltage, determines the firing voltage (V_p), but thyristor action occurs between E and B_1, r_{B1} switching to a very low value when the UJT fires. Short-form UJT specifications usually give R_{BB}, the parameter η (discussed later), the minimum valley current (I_V), the peak point emitter current (I_F), the reverse emitter current I_{EO}, and the peak pulse voltage across B_1

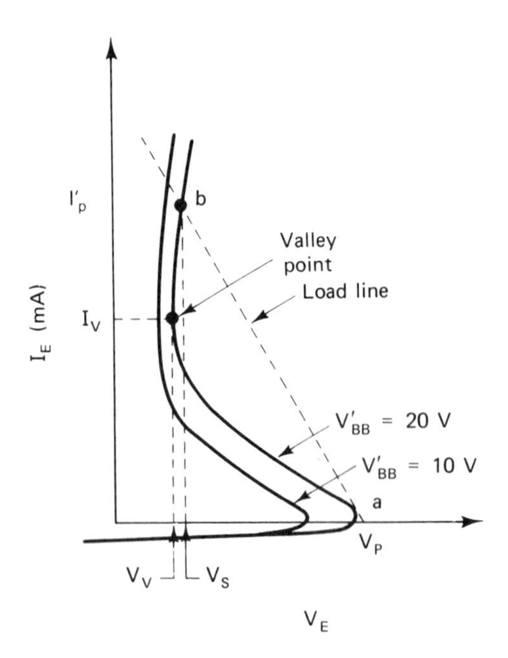

FIGURE 8.2 UJT characteristics.

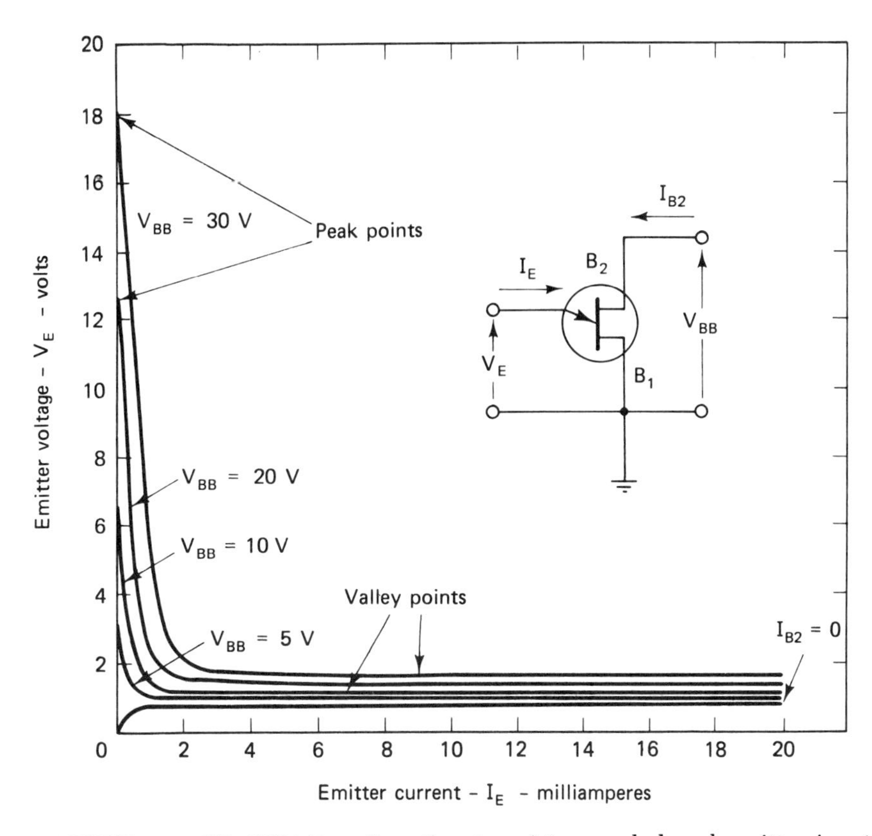

FIGURE 8.3 GE 2N2646 unijunction transistor symbol and emitter input characteristics. Courtesy of General Electric.

measured in a specified test circuit. For GE UJTs, this pulse voltage lies between 3 and 6 V.

The UJT operates like other thyristors, with firing and latching taking place in the emitter–base 1 circuit. Refer to Figure 8.3, referring all voltages to B_1. When V_E is below V_p, the emitter junction is reverse biased and the UJT is *off*. When V_E reaches V_p, the emitter–base 1 junction is forward biased and the UJT turns *on*. In this condition, r_{B1} becomes small, and the emitter current is limited mainly by the resistance in series with the emitter–base 1 circuit. V_p is determined by V_{BB}, the emitter-diode voltage drop (V_D), and the intrinsic stand-off ratio (η). V_p is given by

$$V_p = \eta V_{BB} + V_D \tag{8.2}$$

The parameter η is a function of the division of the interbase resistance by the emitter junction (see Section 1.9) and is determined by UJT construction. It typically lies between 0.51 and 0.82.

V_D, the emitter diode voltage drop, acts like the typical cross-diode voltage. At 25 °C it is approximately 0.5 to 0.7 V and has a temperature coefficient of −3 mV/°C. This variability can be compensated for by external resistances connected to B_2 (R_{B2}) given by empirical equations supplied by the manufacturer. For example, for the 2N2646,7 UJTs,

$$R_{B2} = \frac{1000}{\eta V_1} \tag{8.3}$$

where V_1 is the external supply voltage.

8.3.1 Design of a UJT Trigger Circuit

The goal of UJT trigger circuit design is to produce a train of pulses at a specified rate that will dependably fire an SCR type. Figure 8.4 shows a typical UJT relaxation oscillator trigger circuit, and Figure 8.5 shows manufacturer's curves that expedite design.

The circuit of Figure 8.4 operates as follows. Between pulses, C_1 charges through R_1, increasing V_E as shown. When V_E reaches V_p the UJT turns on, discharging C_1 through R_{B1}. When V_E drops (to about 2 V), the UJT turns off and the cycle repeats. The output pulses (V_{B1}) are delivered to the gate of an SCR. Example 8.1 illustrates the design of a UJT trigger circuit (using a 2N2646 UJT) for driving members of a group of SCRs. The design includes temperature compensation and drives the SCR through a pulse transformer that replaces the external base 1 resistance.

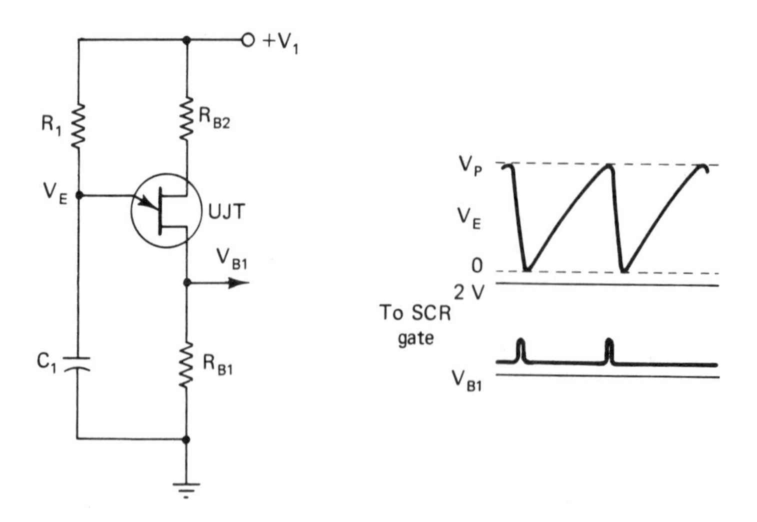

FIGURE 8.4 Basic unijunction transistor relaxation oscillator-trigger circuit with typical wave forms. Courtesy of General Electric.

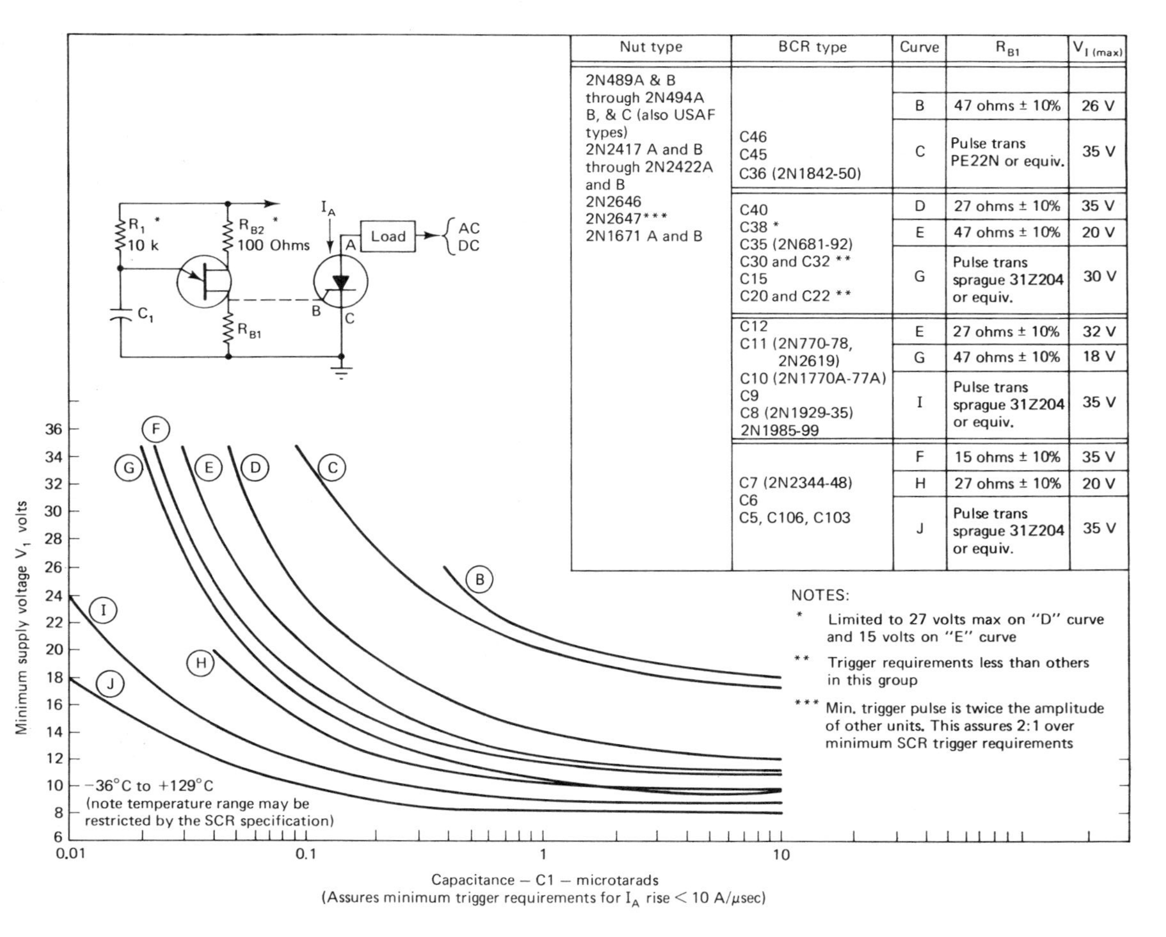

Nut type	BCR type	Curve	R_{B1}	$V_{1\,(max)}$
2N489A & B through 2N494A B, & C (also USAF types) 2N2417 A and B through 2N2422A and B 2N2646 2N2647*** 2N1671 A and B	C46 C45 C36 (2N1842-50)	B	47 ohms ± 10%	26 V
		C	Pulse trans PE22N or equiv.	35 V
	C40 C38 * C35 (2N681-92) C30 and C32 ** C15 C20 and C22 **	D	27 ohms ± 10%	35 V
		E	47 ohms ± 10%	20 V
		G	Pulse trans sprague 31Z204 or equiv.	30 V
	C12 C11 (2N770-78, 2N2619) C10 (2N1770A-77A) C9 C8 (2N1929-35) 2N1985-99	E	27 ohms ± 10%	32 V
		G	47 ohms ± 10%	18 V
		I	Pulse trans sprague 31Z204 or equiv.	35 V
	C7 (2N2344-48) C6 C5, C106, C103	F	15 ohms ± 10%	35 V
		H	27 ohms ± 10%	20 V
		J	Pulse trans sprague 31Z204 or equiv.	35 V

FIGURE 8.5 UJT trigger circuit design curves. Courtesy of General Electric.

Example 8.1 *UJT Trigger Circuit*

Circuit specifications and requirements:
Use a 2N2646 UJT with a Sprague 31Z204 pulse transformer coupling. Assume $\eta = 0.66$.
Must trigger a C11 type SCR at the lowest possible supply voltage.
Pulse frequency = 928 Hz.

(a) *Find C:* A straightforward derivation shows that the operating period (T) (Figure 8.4) is given by

$$\frac{1}{f} = T = 2.3R_1C_1 \log \frac{1}{1-\eta} \tag{8.4}$$

Using the circuit of Figure 8.4 with $R_1 = 10\ \text{k}\Omega$, C_1 is computed to be

$$C_1 \cong 0.1\ \mu\text{F}$$

(In practice, R_1 is limited to values between 3 kΩ and 3 MΩ.)

(b) *Find R_{B2} for temperature compensation:* From Figure 8.5, we see that for a 2N2646 driving a C11 with pulse transformer coupling, curve I applies. Entering the abscissa at $C = 0.1\ \mu\text{F}$, we read the minimum supply voltage (ordinate of intersection of 0.1 μF and curve I) as 12 V. From Eq. (8.3),

$$R_{B2} = \frac{1000}{\eta V_1} = \frac{1000}{0.66 \times 12} = 126\ \Omega \tag{8.5}$$

(20% resistor). When R_{B2} is greater than 100 Ω, the minimum supply voltage (V_1) used in Eq. (8.3) must be corrected by

$$V_1' = \frac{(2200 + R_{B2})V_1}{2300} \tag{8.6}$$

Therefore,

$$V_1' = \frac{(2200 + 126)12}{2300} = 12.14\ \text{V}$$

The completed design for this example is therefore given by the circuit of Figure 8.5, with $C_1 = 0.1\ \mu\text{F}$, $R_{B2} = 126$, and $V_1'(V_1) = 12.14$ V. (In practice, an R_{B2} of 120 or even 100 Ω would be used.)

Temperature compensation being the controlling element of the preceding design, the computed R_{B2} must be used. If, as may happen, this results in an impractical V_1, then the design must be repeated with another UJT (a 2N1671B for example). Alternatively, temperature compensation at extreme values may be sacrificed by using a lower R_{B2}. If an SCR larger than the C45 is to be triggered, the PUT trigger circuit, discussed next, is recommended.

8.4 PROGRAMMABLE UNIJUNCTION TRANSISTOR (PUT)

The PUT is a small four-layer device with an anode gate that controls the peak voltage at which the device turns on. The PUT turns on whenever its anode voltage exceeds the gate voltage by 0.5 to 0.7 V (the voltage drop across a forward-biased diode). Since the anode gate voltage is controllable, the PUT acts like a UJT with an adjustable η. Hence its name. However, because the PUT has a much lower forward resistance in its on state than the UJT (typically 3 Ω), it delivers higher peak pulse currents and can drive larger SCRs.

A PUT relaxation oscillator is shown in Figure 8.6. In this circuit, E_S, R_1, and R_2 establish the gate voltage, while E_S, R_T, and C_T determine the oscillation frequency at this gate voltage. It can be shown by combining these factors that the period of oscillation (T) is approximately given by

$$T \cong R_T C_T \ln\left(\frac{E_S}{E_S - V_p}\right) \tag{8.7}$$

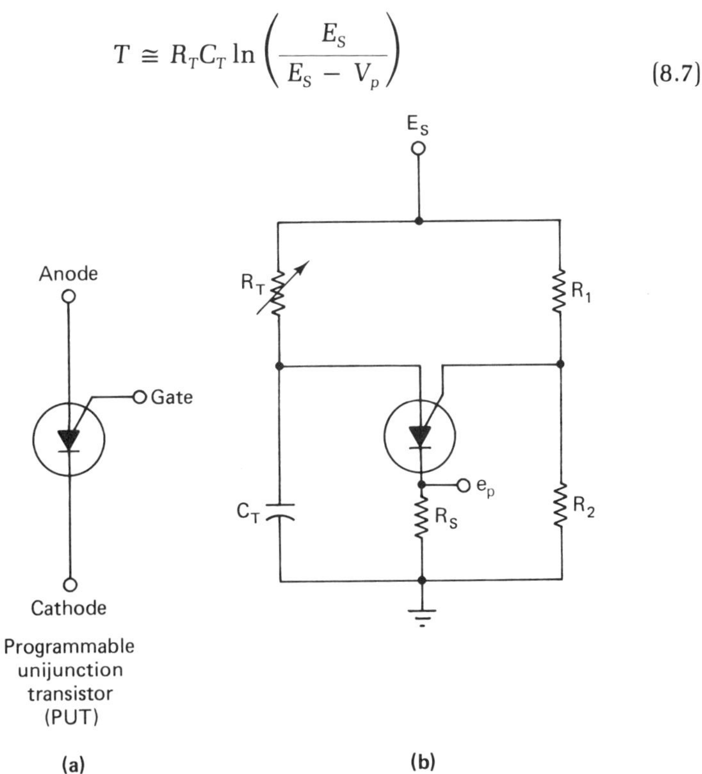

FIGURE 8.6 PUT relaxation oscillator. Courtesy of General Electric.

or

$$T \cong R_T C_T \ln\left(1 + \frac{R_2}{R_1}\right) \tag{8.8}$$

where ln is the natural log.

8.4.1 Design of a PUT Trigger Circuit

Triggering large SCRs, the major application of PUT trigger circuits, calls for the consideration of factors that can be ignored in UJT applications. It will be shown later that SCRs are triggered by charge accumulation in their gate regions. Therefore, the PUT trigger must be designed to deliver sufficient charge, rather than limiting consideration to peak pulse voltage alone, as for the UJT. The charge required to trigger an SCR as a function of time is given by (or can be computed from) manufacturer's data. The charge delivered by the PUT trigger circuit's exponentially decaying pulse is a function of R_S, C_T, and the firing voltage V_p. To successfully fire the SCR, the trigger circuit charge must exceed the charge required by the SCR. This problem can be solved graphically, giving an R_S, C_T, and V_p from which R_T, R_1, and R_2 can be found. These computations are illustrated by Example 8.2.

Example 8.2 *Designing a PUT Trigger Circuit**

Design a free-running relaxation oscillator to trigger a C20 flasher from a 12-V supply. The operating frequency must be adjustable from 5 to 50 pulses per second. Use a 2N6027 PUT.

(a) *Selection of C_T:* Compute or secure from the manufacturer a plot of charge versus time required to fire a C20 SCR. This is shown in Figure 8.7. Assuming an exponentially decaying current pulse with a peak of 80 mA and an RC time constant of 8 μs as a trial and error test, plot charge delivered versus time on the same graph (Figure 8.7). If this curve intersects and rises above the C20 curve, it will fire the SCR. If it does not, increase peak current and repeat. Figure 8.7 shows that the 80-mA peak, 8-μs pulse will fire the C20 at 3.6 μs. (In practice, a reasonable safety factor should be included in the design to provide for variability in SCR characteristics.)

Select a value for R_S. This should be low enough so that the volt drop across R_S (e_p) plus the volt drop across the PUT in its on state is much less than E_S. Let R_S = 39 Ω. Then the peak triggering voltage (V_p) is given by

$$V_p = I_p R_S + V_{PUT}$$
$$= (80 \text{ mA})(39\ \Omega) + 1 \text{ V} = 4.1 \text{ V}$$

*Courtesy of General Electric.

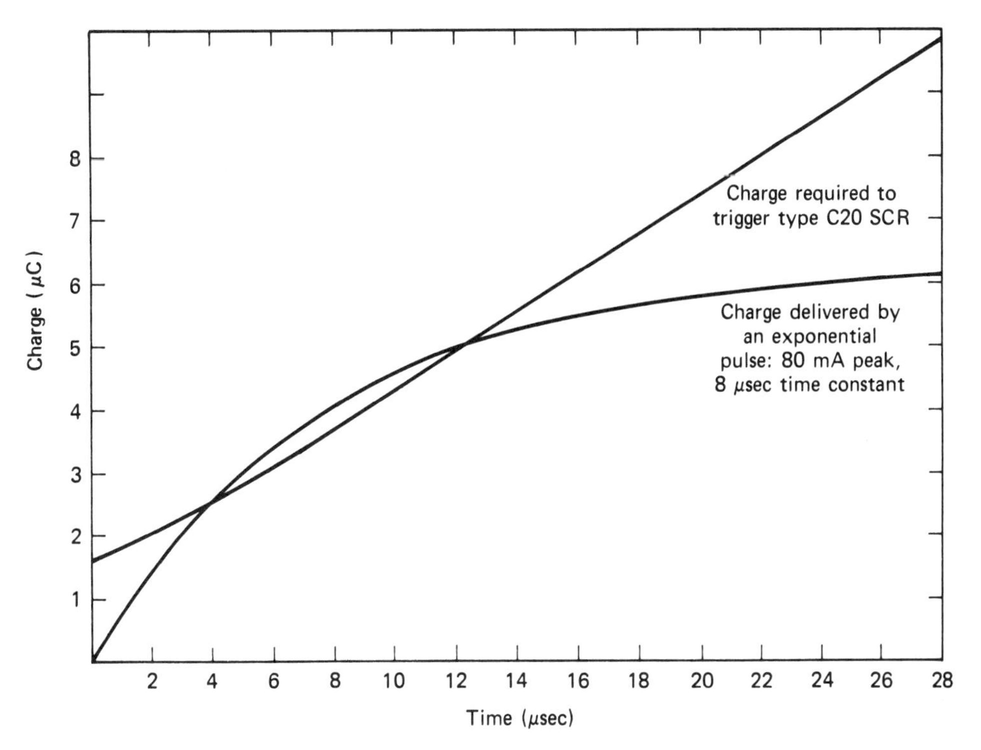

FIGURE 8.7 Charge to trigger an exponential pulse. Courtesy of General Electric.

The ratio V_p/E_S is called η. This is closely related to the η of the UJT and is given by

$$\eta = \frac{V_p}{E_S} = \frac{4.1 \text{ V}}{12 \text{ V}} \approx \frac{1}{3}$$

C_T can now be computed.

$$R_S C_T = 8\ \mu\text{s}$$

$$C_T = \frac{8\ \mu\text{s}}{R_S} = \frac{8\ \mu\text{s}}{39\ \Omega} \cong 0.2\ \mu\text{F}$$

(b) *Computing R_T(max) and R_T(min)*: Using Eq. (8.7),

$$R_T(\text{max}) = \frac{1}{C_T\left(\ln \dfrac{E_s}{E_s - V_p}\right) f_{\text{min}}} = \frac{1}{0.2\ \mu\text{F} \times \left(\ln \dfrac{12}{12 - 4.1}\right) 5} = 2.4\ \text{M}\Omega$$

and

$$R_T(\text{min}) = \frac{1}{C_T\left(\ln\frac{E_s}{E_s - V_p}\right)f_{\text{max}}} = \frac{1}{0.2\,\mu\text{F} \times \left(\ln\frac{12}{12 - 4.1}\right)50} = 240\text{ k}\Omega$$

Before continuing, $R_T(\text{min})$ must be tested to see if it permits the current through the PUT to drop below I_H (holding current). This is a necessary condition for oscillation. If not satisfied, computations must be repeated. The maximum value of the minimum PUT current (I_L) is given by

$$I_L = \frac{E_S}{R_T(\text{min})} = \frac{12\text{ V}}{250\text{ k}\Omega} = 48\,\mu\text{A}$$

I_H for the 2N6027 is 70 μA when the gate current (I_G) is 1 mA. Therefore, $R_T(\text{min})$ will be satisfactory provided I_G is held to 1 mA.

(c) *Computing R_1 and R_2:* The restriction on I_G permits us to compute R_1 and R_2. Transform the gate circuit into its Thevenin's equivalent. Since $\eta = 1/3$, 2/3 of E_S is across the Thevenin's resistance (R_G). Therefore

$$I_G\text{ (gate current)} = \frac{2E_S}{3R_G}$$

And

$$R_G = \frac{2 \times 12}{3 \times 1\text{ mA}} = 8\text{ k}\Omega$$

$$R_G = \frac{R_1R_2}{R_1 + R_2} = R_1\eta = R_2(1-\eta)$$

Therefore

$$R_1 = \frac{R_G}{\eta} = \frac{8\text{ k}\Omega}{1/3} = 24\text{ k}\Omega$$

$$R_2 = \frac{R_G}{1-\eta} = \frac{8\text{ k}\Omega}{2/3} = 12\text{ k}\Omega$$

8.5 OTHER TRIGGER DEVICES

Other devices are available that can serve as the trigger device in Figure 8.1a. These devices have basically the same characteristic curve shape and operating principles as those already covered. They differ in cost, response time, current output, peak pulse voltage, and other factors that may be important to a particular design.

8.5.1 Four-Layer (Schockley) Diode

The Schockley diode is a simple four-layer device possessing only anode and cathode terminals. Its characteristic along with a test circuit are shown in Figure 8.8. This characteristic exhibits the necessary firing voltage point V_p and the negative resistance section to satisfy the requirements of a trigger device. However, V_p is fixed, ranging from approximately 8 to 15 V, with switching currents of 50 to 100 μA. V_p is not

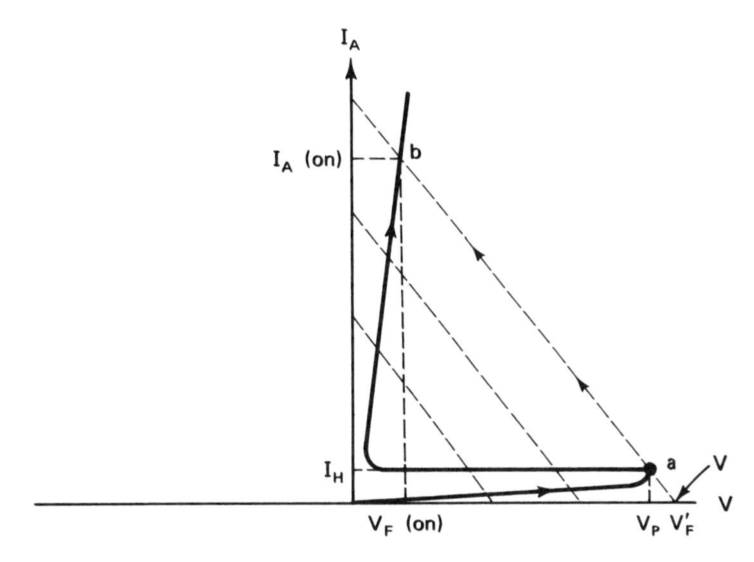

(a) Characteristic

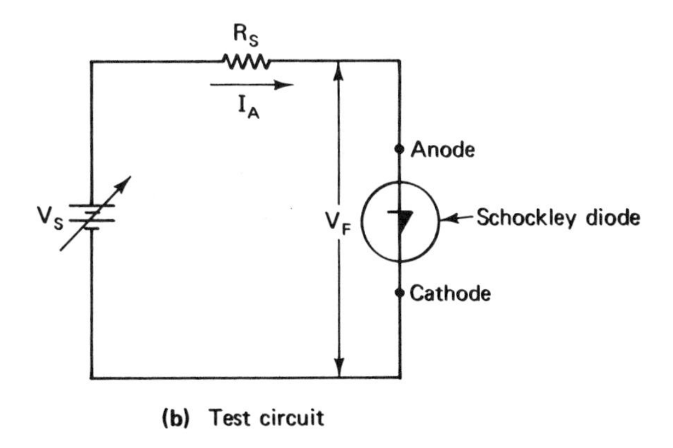

(b) Test circuit

FIGURE 8.8 Schockley diode: (a) characteristic; (b) test circuit.

held precisely, variations of ± 2 V being typical. Typical peak pulse currents are high, 10 A for 50-μs pulses. Holding currents vary widely within a given type. For example, the 1N5780 has a minimum I_H specification of 1 mA and a maximum of 20 mA.

The test circuit of Figure 8.8b can be solved and circuit action analyzed by conventional graphical diode analysis. Figure 8.8a shows I_A increasing along the lower portion of the characteristic as V_S increases from zero. When V_S reaches V_F', the circuit is at its unstable point *a*. Operation immediately switches to the stable point *b*. At *b*, I_A can be read from the graph and is given analytically by

$$I_A(\text{on}) = \frac{V_F' - V_F(\text{on})}{R_S} \tag{8.9}$$

Once turned on, the diode remains on (latches) until I_A falls below I_H.

8.5.2 Silicon Unilateral Switch (SUS)

The SUS is a four-layer diode with an anode gate. Typically, it is used as a Schockley diode with the anode gate used for synchronization, disabling, or forced switching. The circuit symbol, equivalent circuit, and characteristic curve are shown in Figure 8.9. Note the similarity of this curve to that of the Schockley diode.

Table 8.1 gives manufacturer's data for the 2N4987 SUS. When used as a trigger device, the peak pulse voltage (V_o) specification is used as a figure of merit. V_o is the only measure of the ability of the device to transfer charge from the trigger circuit capacitor to the SCR gate. V_o is

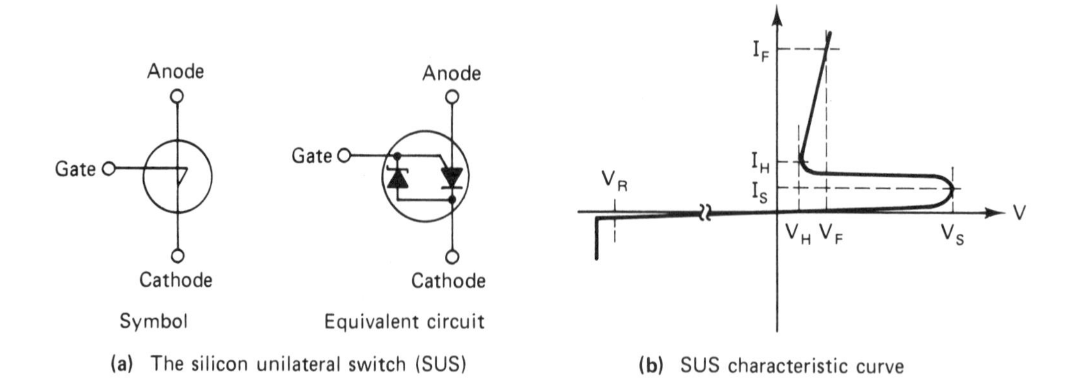

FIGURE 8.9 Silicon unilateral switch: (a) the silicon unilateral switch (SUS); (b) SUS characteristic curve. Courtesy of General Electric.

TABLE 8.1 *Specifications for the 2N4987*

Switching voltage, V_S	6 to 10 V
Switching current, I_S	0.5 mA, max
Holding voltage, V_H	Not specified ($\cong$ 0.7 V at 25°C)
Holding current, I_H	1.5 mA
Forward voltage	
V_F (at I_F = 175 mA)	1.5 V
Reverse voltage rating, V_R	30 V
Peak pulse voltage, V_o	3.5 V min

Courtesy of General Electric.

typically measured in the circuit of Figure 8.1a with $V_1 = 15$ V, $R_1 = 10$ kΩ, $C = 0.1$ μF, and $R_2 = 20$ Ω. V_o is of course measured across R_2.

Compared to the UJT and Schockley diode, the SUS has the highest switching current of the three and lower switching voltages than the Schockley diode.

8.5.3 Silicon Control Switch (SCS)

The SCS is a four-layer device with external access to all four layers. Its schematic construction and symbol are shown in Figure 8.10. The gate closest to the cathode is called the cathode gate; the one closest to the anode is the anode gate.

Both gates can be used to turn on the SCS at an anode voltage below

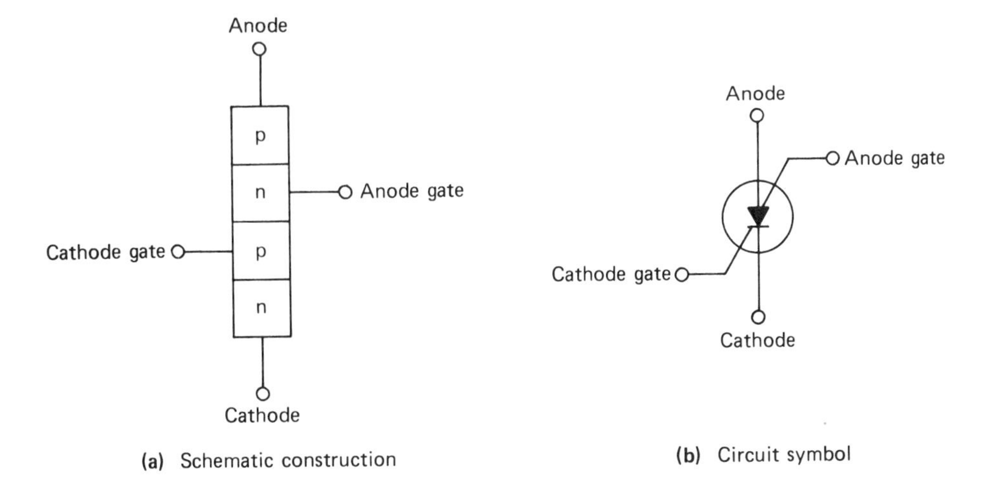

FIGURE 8.10 Silicon-controlled switch: (a) schematic construction; (b) circuit symbol.

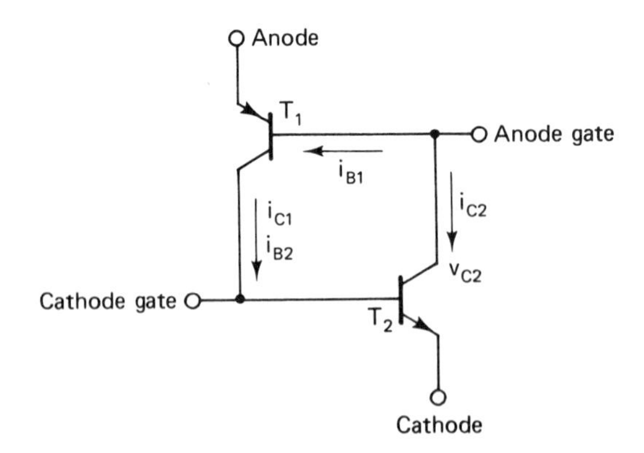

FIGURE 8.11 Equivalent circuit of SCS.

V_p, but, in addition, the anode gate can be used to turn off the SCS as well. The ability to turn off a device by a gate signal is lacking in the devices studied so far and in the SCR as well.

Figure 8.11 shows an equivalent circuit, which can be used to explain this action. In this figure, the SCS is represented as two transistors (see Section 1.7) tied base to collector. When in the *on* condition, a positive voltage pulse of sufficient magnitude applied to the anode gate will reduce i_{C1}. Since i_{C1} is the base current of T_2, i_{C2} is reduced and V_{C2} is increased. This action further increases the anode gate voltage, and regenerative action occurs, turning the SCS off. Aside from this unique capability, the device functions in the normal thyristor manner. Each

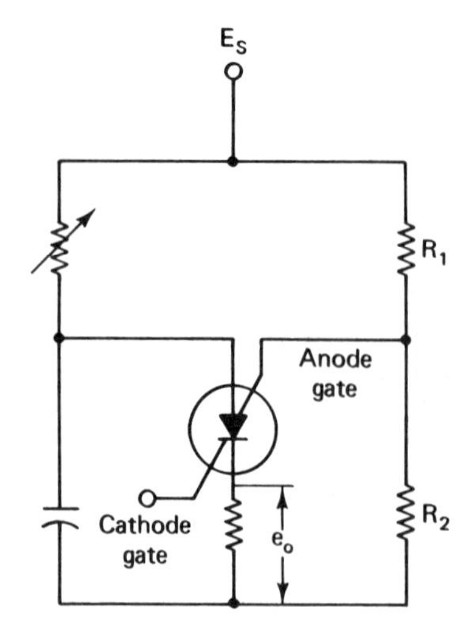

FIGURE 8.12 SCS relaxation oscillator.

gate turns the device on when an appropriate voltage pulse is applied. However turn-on anode gate current is always much greater than turn-on cathode gate currents, for example, 1.5 mA against 1 μA for the 3N81.

An SCS relaxation oscillator is shown in Figure 8.12. The firing voltage V_p is controlled by the dc voltage at the anode gate and is given approximately by

$$V_p \cong \frac{E_S R_2}{R_1 + R_2} \tag{8.10}$$

The peak point current (firing current, I_p) and the holding current (I_V) are also functions of R_1 and R_2. The cathode gate is available for synchronization or forced firing.

8.6 GENERAL CHARACTERISTICS OF TRIGGERS

The characteristics of trigger devices of concern to the circuit designer include V_p (peak firing or peak voltage), I_p, the current required at V_p to fire the device, I_V(min), the lowest device current in the *on* state, peak pulse voltage (e_p), turn-on time, and ease of synchronization. The choice of a particular device hinges on these characteristics. V_p, I_p, and e_p have already been discussed. I_V(min) determines the holding current I_H below which the device turns off. Turn-on time and ease of synchronization require further discussion.

Turn-on time (or rise time) is a measure of the time required for a device to switch from the unstable point (point 1 in Figure 8.1) to the stable *on* position (point 2 in Figure 8.1). Since firing an SCR depends on the rate at which charge is delivered to its gate, this is an important parameter governing maximum firing rates.

In practice, relaxation oscillators are synchronized to precise oscillators instead of running free. This ensures more precise triggering and better control. Therefore, the ease with which a trigger circuit may be synchronized is an important parameter. Because the UJT firing point depends on the supply voltage, its trigger circuit is easily synchronized by reducing either its supply or interbase voltage. Figure 8.13 illustrates synchronization by means of a negative pulse at base 2. The PUT trigger circuit may be synchronized in an analogous way by introducing a negative voltage pulse into the anode gate. The SUS circuit may also be synchronized using its anode gate. The SCS is best synchronized via its sensitive cathode gate. To synchronize the four-layer diode, positive voltage pulses must be introduced across the device (anode to cathode). Figure 8.14 summarizes the major characteristics of the circuits discussed here.

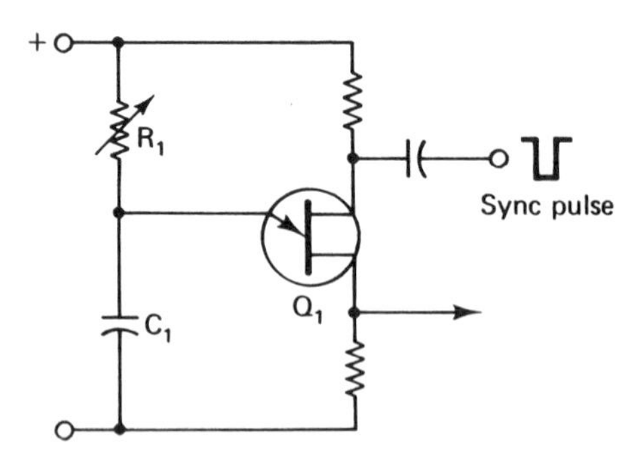

FIGURE 8.13 Pulse synchronization of UJT relaxation oscillator. Courtesy of General Electric.

GATE TRIGGER CHARACTERISTICS, RATINGS, AND METHODS

	CLASS	E-I CHARACTERISTICS	BASIC CIRCUIT	MAJOR TYPES	V_p PEAK POINT VOLTAGE	I_p (max) PEAK POINT CURRENT	I_v (min) VALLEY CURRENT	TON TURN ON TIME	SPECIFICATION NUMBER
Unidirectional	UJT Unijunction Transistor			TO-5: 2N489A, 2N489B, 2N1671A, 2N1671B, 2N1671C TO-18: 2N2417A, 2N2417B, 5G515, 5G516, 2N2646, 2N2647	Fixed Fraction of Interbase Voltage	12 μa 6 μa 25 μa 6 μa 2 μa 5 μa 2 μa	8 ma 8 ma 8 ma 8 ma 8 ma 4 ma 8 ma	1-2 μsec Typ	60.10 60.10 60.50 60.50 60.50 60.62 60.62
	PUT Programable Unijunction Transistor			2N6027 2N6028	$\frac{R_2 E_s}{R_1 + R_2}$	As low as 2 μa As low as .15 μa (A function of R_1, R_2)	70 μa 25 μa (A function of R_1, R_2)	80 nsec Max	60.20
	SUS Silicon Unilateral Switch			TO-18: 2N4983, 2N4984, 2N4985, 2N4986 TO-98: 2N4987, 2N4988, 2N4989, 2N4990	6-10 v 7.5-9 v 7.5-8.2 v 7-9 v	500 μa 150 μa 300 μa 200 μa	1.5 ma .5 ma 1.0 ma .75 ma	1.0 μsec Max	65.25, 65.28 65.27, 65.28 65.27, 65.28 65.26, 65.28
	SCS Silicon Control Switch			3N84	$\frac{R_2 E_s}{R_1 + R_2}$ (40v max)	A function of R_1 and R_2	10 ma Max (A function of R_1 and R_2)	1.5 μsec Max	65.18
Bidirectional	SBS Silicon Bilateral Switch			TO-18: 2N4993 TO-98: 2N4991, 2N4992	6-10 v 7.5-9 v	500 μa 120 μa	1.5 ma .5 ma	1.0 μsec Max	65.30, 65.31 65.32
	DIAC			ST2	28 v-36 v	200 μa	Very high	1 μsec Typ	175.30
	ATS Assymetrical AC Trigger Switch (ST4)			ST4	14-18V 7-9 V	80 μa 80 μa		1 μsec	175.32

FIGURE 8.14 Summary of trigger devices. Courtesy of General Electric.

8.7 SILICON-CONTROLLED RECTIFIER (SCR)

Conceptually, the SCR is a four-layer device with a cathode gate connected to the lower *p*-type layer. The cathode gate provides a sensitive means of reducing the firing voltage, V_p. At zero gate current ($I_G = 0$), the SCR is essentially a four-layer diode with V_p dependent on construction. As I_G is increased, V_p is reduced. Therefore, the SCR can function as a latching switch controlled by a sensitive gate. Figure 8.15 shows this schematic construction along with the circuit symbol, and Figure 8.16 shows the characteristic curves with V_p labeled V_{BR}. Note how the curve is changed and V_{BR} reduced with increased I_G.

The SCR has proved to be a device with widespread utility in the field of power control. Extensive development to meet the needs of this industry has resulted in SCRs with current-handling ranges from 0.5 A (low-current SCRs) to 2500 A (high-current SCRs), operating at anode voltages up to 3600 V. They are used to control motors of all sizes and to rectify and control heavy power. Figure 8.17 summarizes the ranges of SCRs available.

The characteristics of a 2N681–92 SCR are given in Table 8.2 to illustrate the high level of performance available from SCRs.

TABLE 8.2 *Short-Form Specifications for the 2N681–92 SCR*

Type		2N681–92
Voltage range		25–800 V
Forward conduction		
I_T(rms)	Max. rms on-state current	25 A
I_T(av)	Max. average on-state current at 180° conduction	16 A at 65 °C
I_{TSM}	Max. peak one cycle nonrepetitive surge current	150 A
V_{TM}	Peak on-state voltage at 25 °C, 180° conduction, rated I_T(av)	2.0 V
I_H	Max. holding current at 25 °C	100 mA
t_d, t_r	Typical turn-on time	1.6 μs
Firing		
I_{GT}	Max. required gate current to trigger	
	At −65 °C	80 mA
	At 25 °C	40 mA
V_{GT}	Max. required gate voltage to trigger	
	At −65 °C	3 V
	At 25 °C	3 V

Courtesy of General Electric.

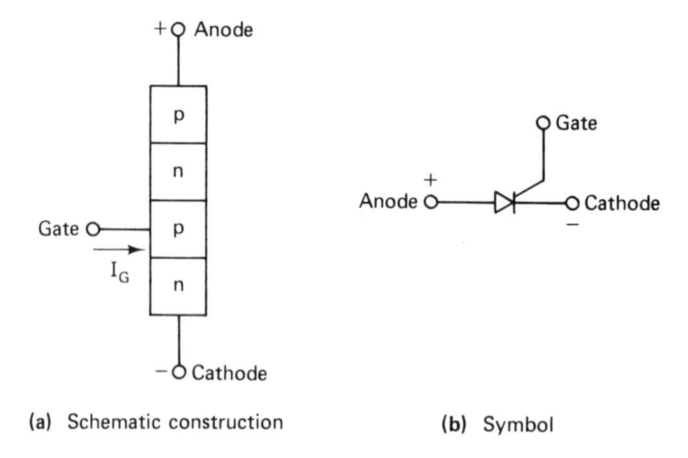

FIGURE 8.15 SCR construction and symbol: (a) schematic construction; (b) symbol.

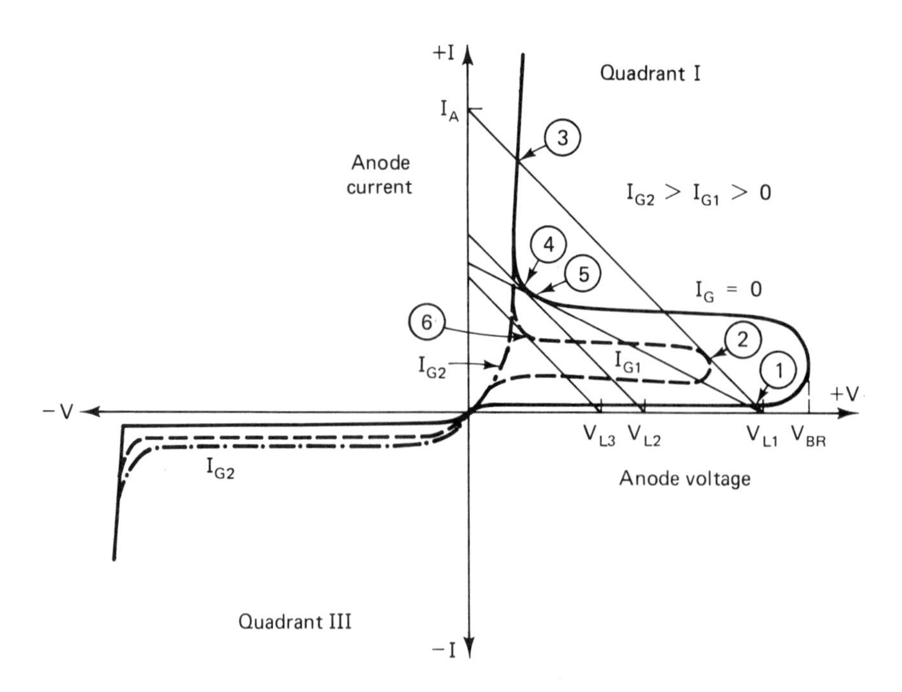

FIGURE 8.16 SCR anode-cathode characteristics with gate current. Courtesy of General Electric.

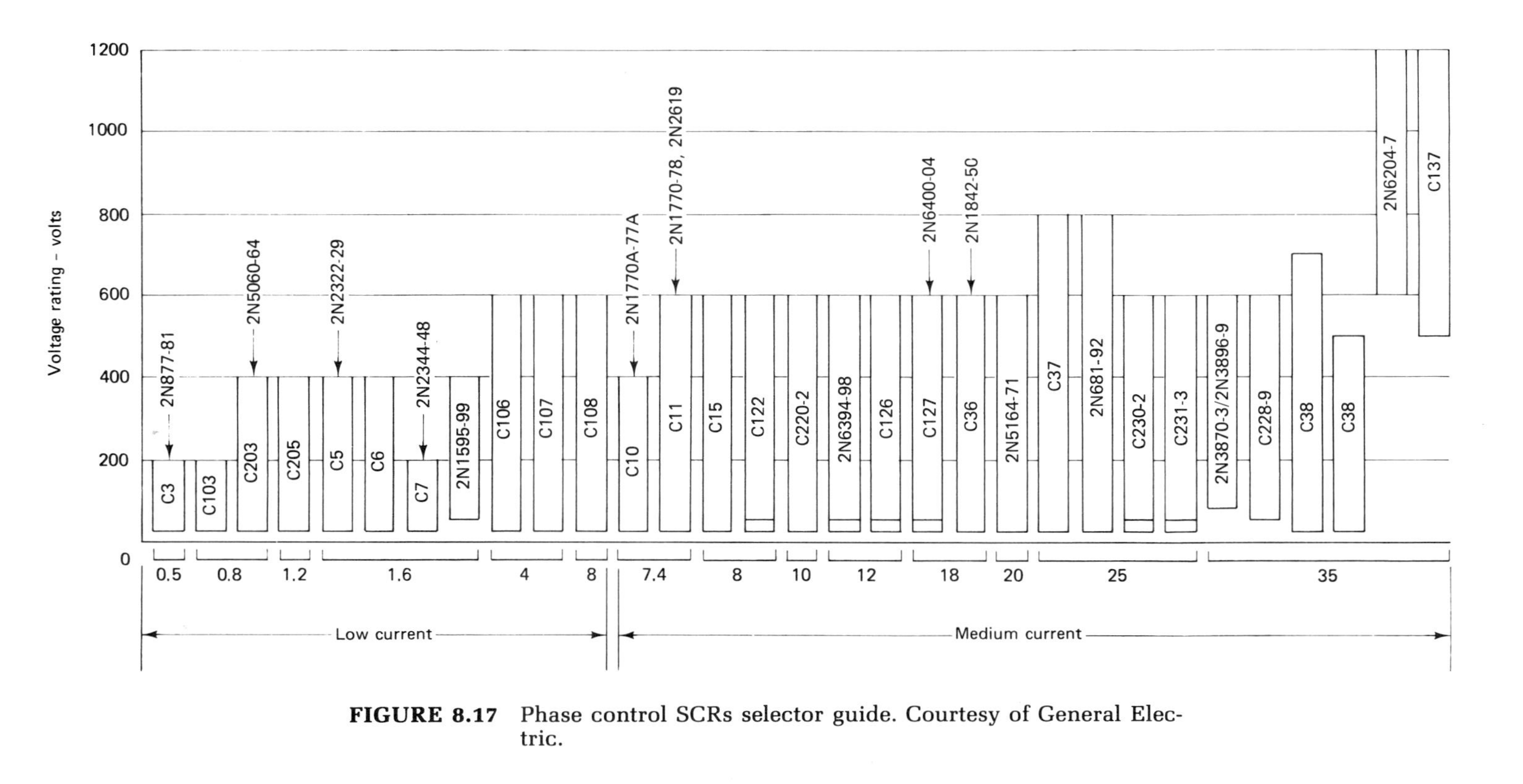

FIGURE 8.17 Phase control SCRs selector guide. Courtesy of General Electric.

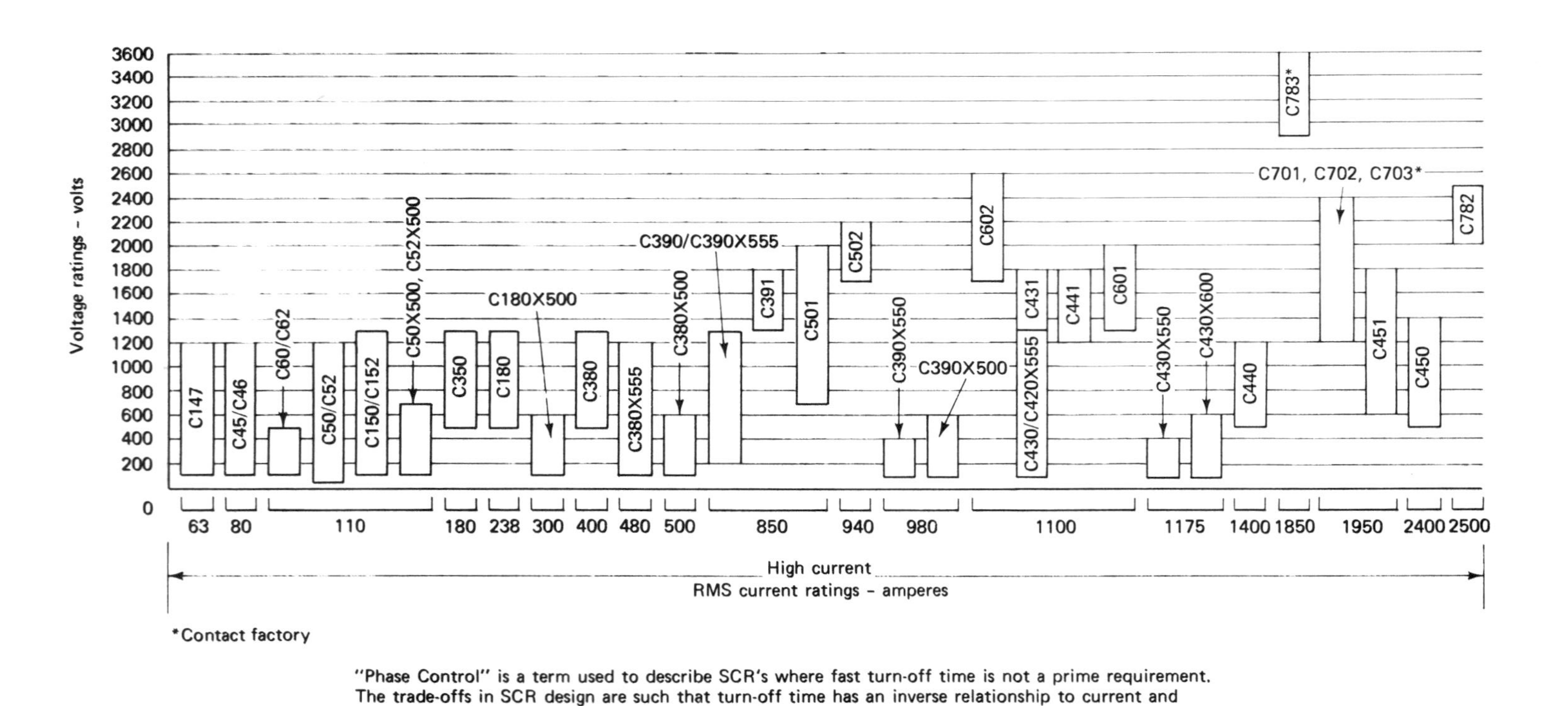

"Phase Control" is a term used to describe SCR's where fast turn-off time is not a prime requirement. The trade-offs in SCR design are such that turn-off time has an inverse relationship to current and voltage capability for any given junction size. Primary applications for a device with relatively slow turn-off are AC phase control–hence the name "Phase Control". This type of device is also used for zero voltage switching and select pulse applications.

FIGURE 8.17 (continued) Phase control SCRs selector guide. Courtesy of General Electric.

8.7.1 Ratings and Characteristics

Because the SCR is primarily used as a switch, its operation is usually nonlinear and intermittent. Because it is often used to control heavy power, it is important to extract the maximum power-handling capability from each device. For both reasons, SCR specifications are thorough and complex. An understanding of these specifications illustrates the considerations involved in SCR circuit design.

Junction Temperature. Low-temperature limits are required to limit stress in the silicon crystal to safe values. This stress is due to the difference in thermal coefficients of expansion of the materials used in SCR construction. Upper operating temperatures are imposed mainly because important parameters of the SCR will go out of range or become unstable. The rated maximum operating junction temperature establishes steady state and periodic (recurrent) limits, but to fit the SCR into systems undergoing transient overloads, higher-than-rated temperature operation must be permitted for short periods.

Power Dissipation. Power is dissipated in the SCR as a result of five operation states: turn-on switching, conduction, turn-off or commutation, blocking, and triggering. Conduction losses in the on state are the main source of junction heating, but for steep current wave forms or high frequencies, turn-on switching losses become significant. Because in AC circuits SCRs are on for only part of a cycle (conduction angles from 0° to 180°, typically), manufacturers supply power dissipation for various conduction angle curves (Figure 8.18). Turn-on switching losses are more difficult to compute, but curves of allowable rms on-state current versus frequency and peak on-state current versus pulse width are supplied. Turn-off losses arise from currents flowing momentarily while depletion regions are reestablished. They are not usually significant. Blocking loss is the straightforward result of the reverse voltage across the SCR and its leakage current. Because of the high reverse voltages that SCRs are subjected to, this requires consideration. Finally, gate triggering losses, usually negligible for pulse triggering, can become significant for high duty cycles or in small packages.

Thermal Resistance. For small transistors, junction temperature can be computed using a thermal resistance approach and average case and junction temperatures; however, to fit the SCR into fused and protected power systems, transient heating must be considered. Also, transient considerations permit the SCR to be used under conditions where average power dissipation limits would be too conservative. Manufacturers supply a transient thermal impedance curve that permits computation of junction temperature under nonrecurrent operation.

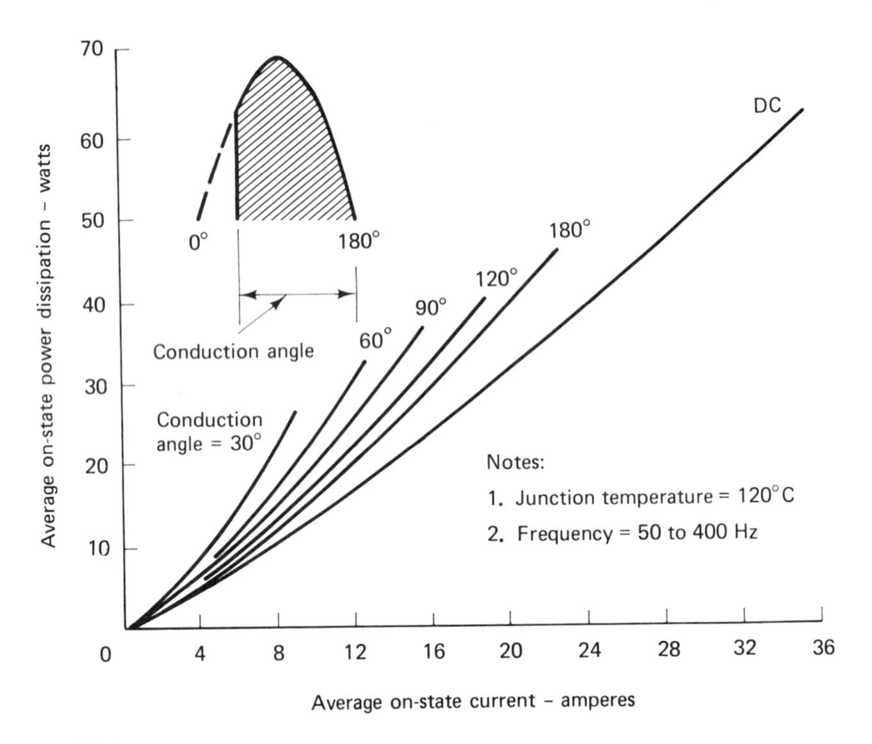

FIGURE 8.18 Maximum average on-state power dissipation for C137 series SCR. Courtesy of General Electric.

Average Current Rating (Recurrent). When controlling AC power, the SCR functions as a rectifier conducting over part of a cycle. The DC power loss is given by the DC anode voltage multiplied by the average current. Curves of maximum allowable case temperature versus average on-state current as a function of conduction angle are available (Figure 8.19).

RMS Current (Recurrent). The wave shapes of Figure 8.19 have an rms value as well as an average value. This rms value is important in high peak current, narrow pulse operation because the heating of resistive elements, such as joints, leads, and interfaces, is a function of this value. Although the average value of the wave form may be within rating, the rms value may be exceeded.

Surge and I^2t Ratings (Nonrecurrent). SCRs operating in power systems are subject to momentary overloads resulting from short circuits and other disturbances. The recurrent ratings can be exceeded in these cases. Figure 8.20 shows the maximum allowable nonrecurrent multicycle surge current at rated load versus 50- and 60-Hz. Figure 8.21 is similar,

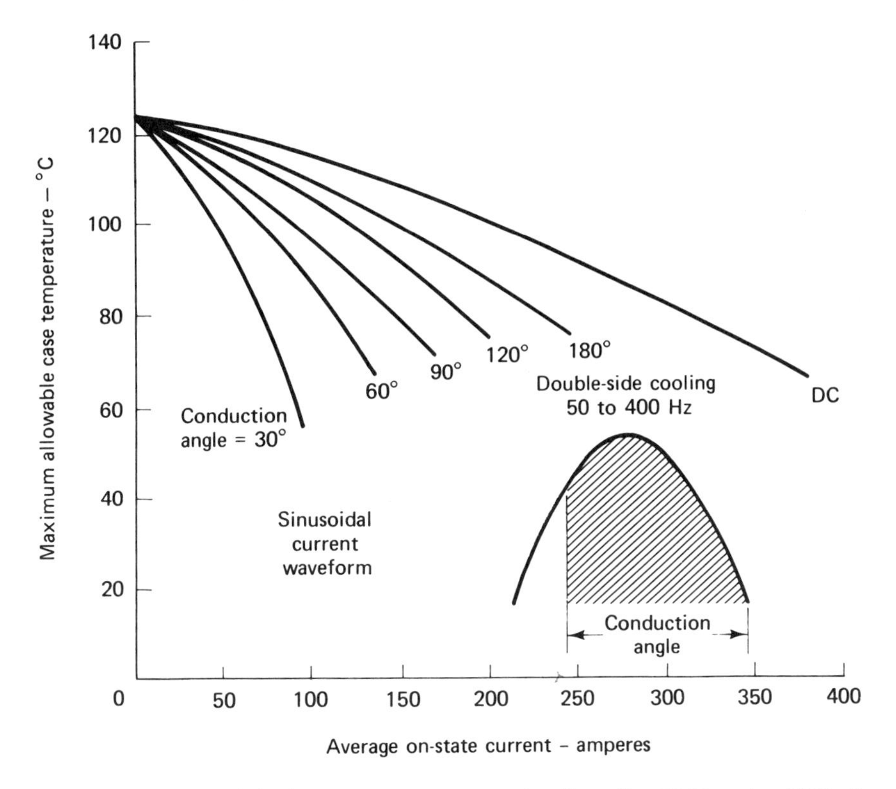

FIGURE 8.19 Maximum average current ratings for C380 series SCR. Courtesy of General Electric.

showing I^2t versus pulse base width. The I^2t rating is derived from the surge current except that I is the rms value. I^2t is a direct measure of heating and is useful in fusing procedures.

The di/dt Rating. The preceding specifications assumed that all junctions turned on instantaneously and completeiy. This assumption is valid provided that the rise of anode current (*di/dt*) is slow compared to the time it takes for the junctions to reach full on with uniform current density. When *di/dt* is rapid compared to junction turn on, current density will not be uniform and local hot spots will occur. This can lead to excessive temperature rise at the hot spot. The *di/dt* rating is based on industry-wide test circuits for recurrent and nonrecurrent turn on. General Electric employs a more stringent test that considers other heating factors, which it labels a "Concurrent *di/dt* Rating (Recurrent)."

Turn-on Voltage ($V_{T\,ON}$). $V_{T\,ON}$ is the voltage across the SCR at the time of peak pulse current. Typical values of the current pulse are 150-A peak

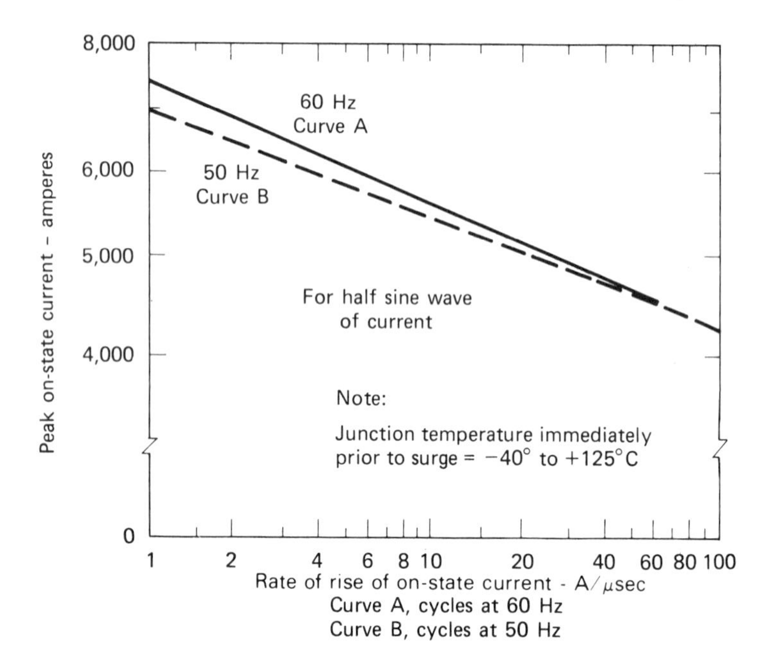

FIGURE 8.20 Maximum allowable multicycle, nonrecurrent, peak surge on-state current for the C398 series SCR. Courtesy of General Electric.

current with a 10-μs pulse width. $V_{T\,ON}$ is a measure of the *di/dt* performance of the SCR, since $V_{T\,ON}$ will fall as more of the junction area turns on. Figure 8.22 shows a commonly used current pulse and typical $V_{T\,ON}$ behavior.

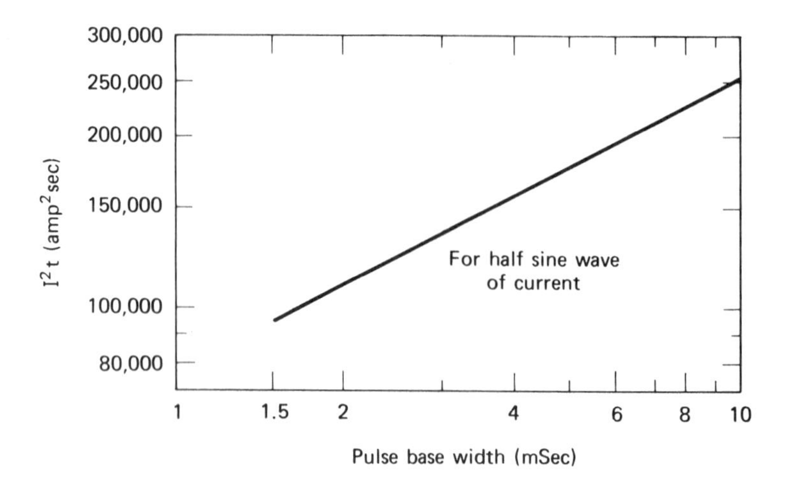

FIGURE 8.21 I^2t rating for the C398 series SCR. Courtesy of General Electric.

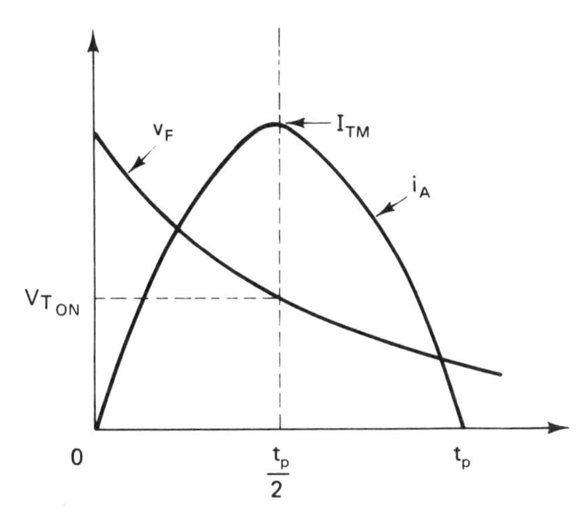

FIGURE 8.22 Definition of turn-on voltage. Courtesy of General Electric.

Reverse Voltage (V_{RRM} and V_{RSM}). In the reverse direction, the SCR acts like a conventional diode. V_{RRM} is the maximum reverse voltage for a recurrent anode voltage with the SCR gate open. V_{RSM} applies to the nonrecurrent case.

Peak Off-State Blocking Voltage (V_{DRM}). In the absence of a gate signal, the SCR will fire if V_{DRM} is exceeded. V_{DRM} is specified at maximum allowable junction temperature (worst case) and with an open gate for the larger SCRs. Smaller SCRs are characterized with a specified gate-to-cathode bias resistor. V_{DRM} is often called V_{FXM}, forward blocking voltage.

Rate of Off-State Voltage Rise (dv/dt). A high rate of rise of the SCRs anode voltage in the off state may cause it to turn on. Since supply voltage transients due to other devices switching off are not unusual, the *dv/dt* withstand specification is important. The specification is either given as a numerical value at fixed conditions or as a time constant (ζ) under specified conditions. It is also supplied as curves of withstand capability versus magnitude of a voltage step (Figure 8.23).

Holding and Latching Current (I_H). Anode current must exceed I_H if the SCR is to remain in the *on* state. Furthermore, if the gate signal does not cause an anode current to remain above I_H, the SCR will not latch on. I_H is given as a minimum current below which the SCR reverts to the *off* state.

Reverse Recovery Characteristic. The SCRs reverse recovery characteristic is specified as either an interval, the familiar diode or transistor reverse recovery time (t_{rr}), or in terms of a quantity of charges (Q_{RR}). The reverse recovery characteristic is a result of the finite interval re-

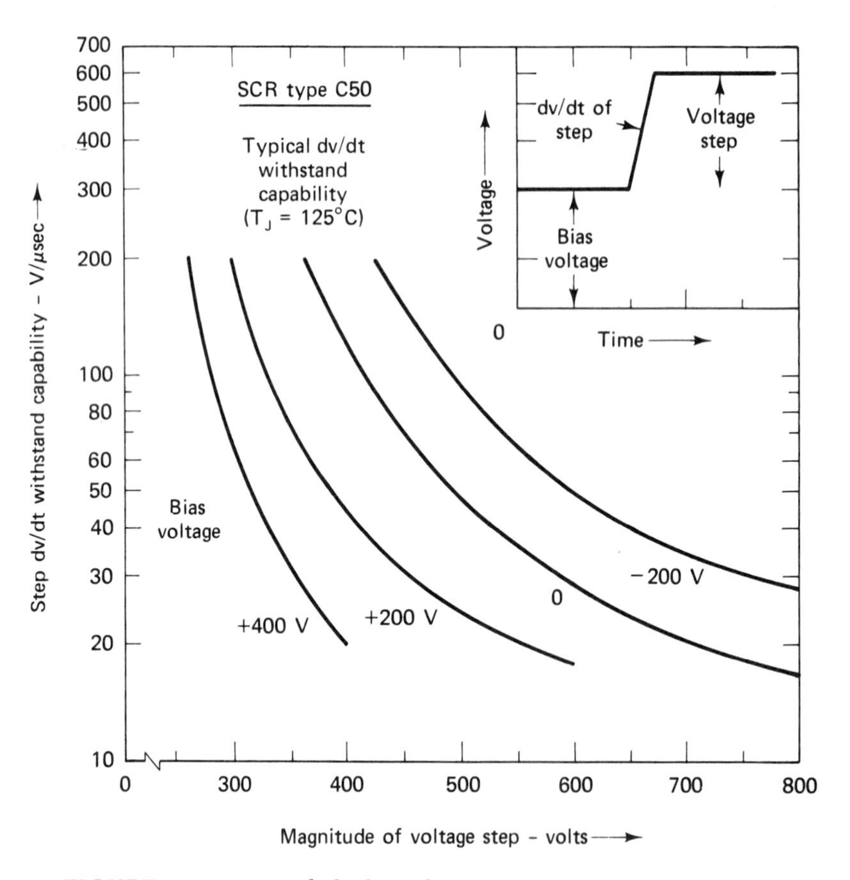

FIGURE 8.23 Typical *dv/dt* withstand capability of C50 SCR. Courtesy of General Electric.

quired to remove carriers in the junction region before the depletion region can be established. Figure 8.24 graphs SCR current versus time for an abrupt anode voltage swing from on to off. The reverse current is initially high and gradually subsides to normal values. T_3 is the time at which $i = i_R/4$. T_4 is defined by the intersection of a line drawn through i_R–T_2 and $L_R/4$–T_3, with the abscissa. The reverse recovery time is defined by

$$t_{rr} = T_4 - T_1 \tag{8.11}$$

Q_{RR} is given by the area contained in the triangle T_1, T_2, and T_4, and is given by

$$Q_{RR} = \frac{i_R t_{rr}}{2} \tag{8.12}$$

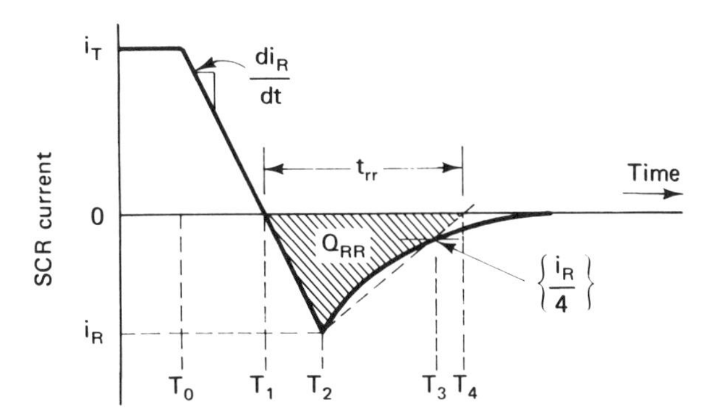

FIGURE 8.24 SCR recovery wave-form definition. Courtesy of General Electric.

8.8 TRIGGERING THE SCR

SCR triggering is a function of gate current and voltage, anode circuit impedance, and supply voltage. Trigger action can be explained by reference to Figure 8.16, which shows an SCR characteristic together with several anode circuit load lines. With zero gate current and supply voltage V_{L1}, the load line shown intersects the SCR characteristic at a stable operating point (1). At gate current I_{G1}, this load line is tangential to the new SCR characteristic at unstable point 2. The SCR now switches abruptly to the stable low-impedance operating point (3) and remains there when I_G returns to 0. If the supply voltage is reduced toward V_{L2}, anode current is reduced. When the load line becomes tangential to the characteristic (point 4), another unstable condition is established and operation shifts back to the stable *off* condition.

Holding current and latch-up (latching action) depend on I_G, V_L, and the slope of the load line. If at V_{L1}, the load resistance were increased to give the 1 through 5 load line, then I_H would be reduced (point 5). If with the original load, I_{G1} were maintained and the supply voltage dropped to V_{L3}, turn off would occur at point 6. A momentary increase in I_G (I_{G2}) would turn the SCR on, but latch-up would not take place since the SCR would switch off upon removal of I_{G2}. The triggering criteria therefore requires meeting at least two conditions, a negative-resistance intercept and a minimum anode current above I_H.

8.8.1 SCR Gate-Cathode Characteristics

Figure 8.25 shows two equivalent gate-cathode circuits for the SCR. In both circuits, R_L (R_{L1}, R_{L2}, etc.) represents the lateral resistance of the *p*-type layer to which the gate terminal is connected and R_S (R_{S1}, R_{S2})

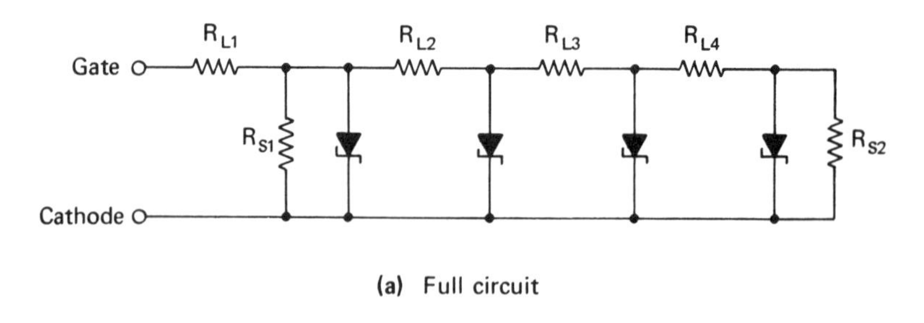

(a) Full circuit

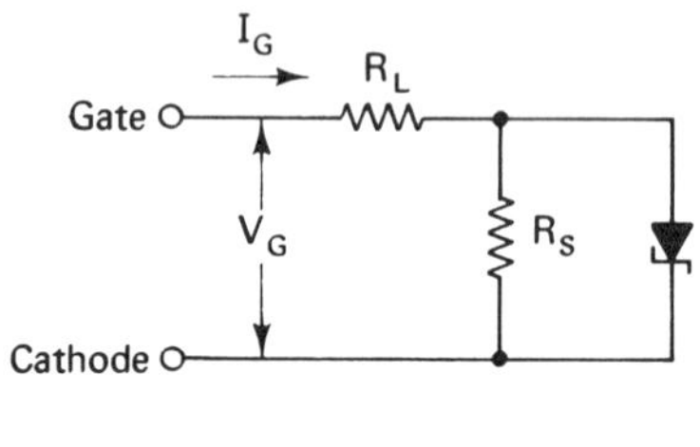

(b) Simplified circuit

FIGURE 8.25 Gate–cathode equivalent circuit for the conventional SCR: (a) full circuit; (b) simplified circuit. Courtesy of General Electric.

represents any resistive bypass to the emitter. R_S and R_L depend on SCR construction; R_S is intentionally made low in some types. The diode(s) are shown as zeners because reverse breakdown of the SCR gate occurs in the 5- to 20-V range. Figure 8.25a takes into account the spatial distribution of the gate-cathode junction; Figure 8.25b does not. Because of R_S, the back resistance of the gate-cathode diode is less than typical, while the forward resistance is greater because of R_L. Figure 8.26 compares the SCR gate-cathode diode to a conventional diode.

The preceding equivalent circuits and characteristics are only valid when the SCR is off (anode current $\cong 0$). At the triggering point the increasing anode current modifies both. Figure 8.27 shows a complete simplified equivalent circuit that takes anode current into account.

Triggering action can be described in terms of Figure 8.27b, which shows the interaction between gate and anode current. For $I_A \cong 0$, the I_G versus V_G characteristic is the same as that of Figure 8.26 (scaled differently). The load line shown is the gate source impedance load line. As anode current increases, the gate-cathode volt drop increases, opposing V_G and slowing I_G. It can be shown that the SCR fires (anode current becomes regenerative) at the point where the gate source impedance load line is tangential to the gate characteristic (point 1). Therefore, I_{GT} measured at the peak of the gate characteristic is specified as the gate current required to trigger the SCR and V_{GT}, the trigger voltage. After

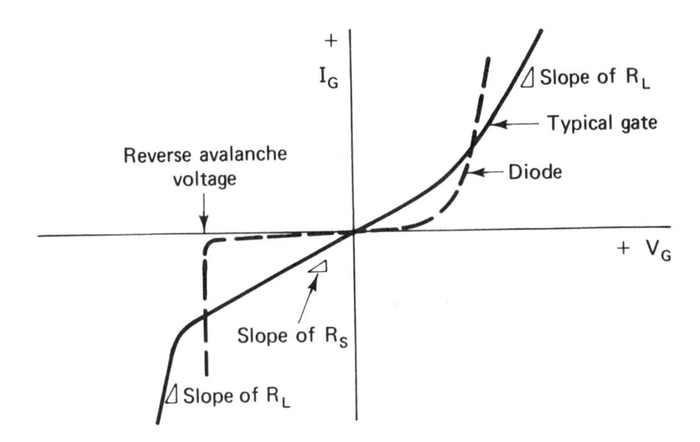

FIGURE 8.26 Gate–cathode characteristic curve ($I_A = 0$). Courtesy of General Electric.

triggering, the gate-cathode volt drop becomes very nearly equal to the SCRs anode cathode drop, the junction becomes a voltage source, the anode current flow across the gate-cathode junction maintains conduction, and the gate loses control. Removal of V_G will therefore have no effect on anode current. The SCR is latched on. Note that the gate source impedance load line plays a critical role in this process.

8.8.2 Determining Trigger Parameters

The trigger circuit must supply a signal that will fire the SCR without exceeding gate power dissipation ratings. Since SCRs vary widely in gate characteristics within types, the trigger parameters cannot, in practice, be computed without an assist from the manufacturer. Figure 8.28 illustrates the variation in gate characteristics for a GE C38 SCR. The C38 characteristic can lie anywhere between the lower and upper curves. Also shown is the locus of all V_G–I_G points that will fire the C38. These are located in the clear area outside the small shaded region near the origin. For a given gate source impedance load line drawn through this region, the specific operating point will depend on the specific gate characteristic.

For DC triggering, the V_G–I_G product must not exceed the average maximum allowable gate dissipation. For narrow pulse triggering, the operating point must lie below the maximum allowable instantaneous gate power curve. For dependable pulse firing, the operating point should lie outside the shaded area and as close to the maximum allowable instantaneous gate power curve as possible. The parameters

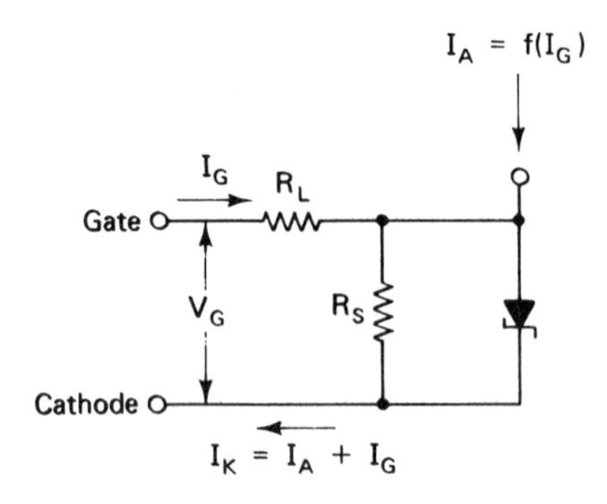

(a) Gate cathode equivalent circuit ($I_A = f(I_G)$)

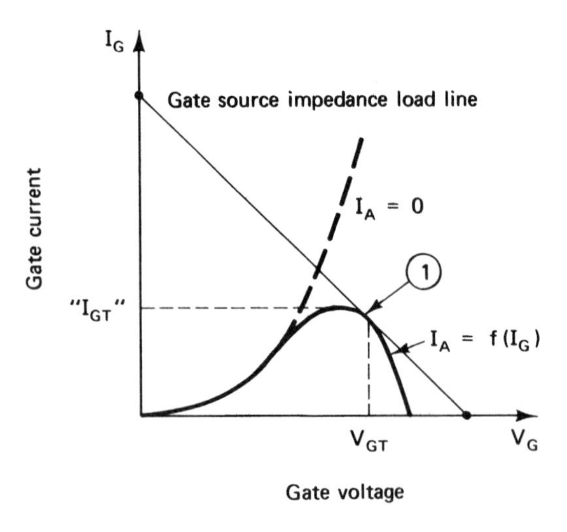

(b) Gate characteristics, anode connected

FIGURE 8.27 Gate characteristics and equivalent circuit of the SCR: (a) gate-cathode equivalent circuit ($I_A = f(I_G)$); (b) gate characteristics, anode connected. Courtesy of General Electric.

of a dependable trigger pulse for the C38 are found by selecting a V_{GO} (open circuit voltage) and an I_{GS} (short-circuit current) based on the capabilities of a particular trigger pulse circuit. Constructing this load line on Figure 8.28 will determine if it is in the preferred firing area. The slope of the load line (V_{GO}/I_{GS}) fixes the output resistance of the trigger circuit. To minimize turn-on anode switching dissipation and jitter, the gate-current rise time should be on the order of several amperes per microsecond.

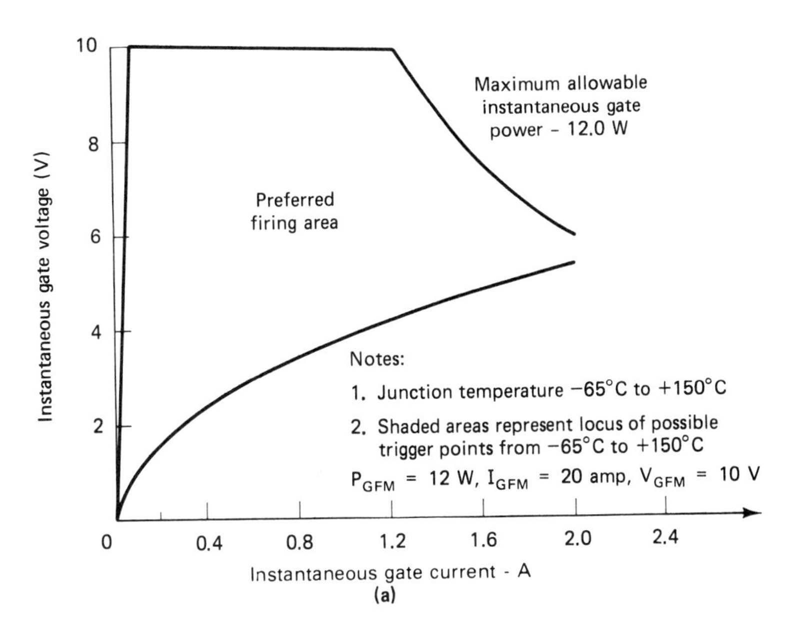

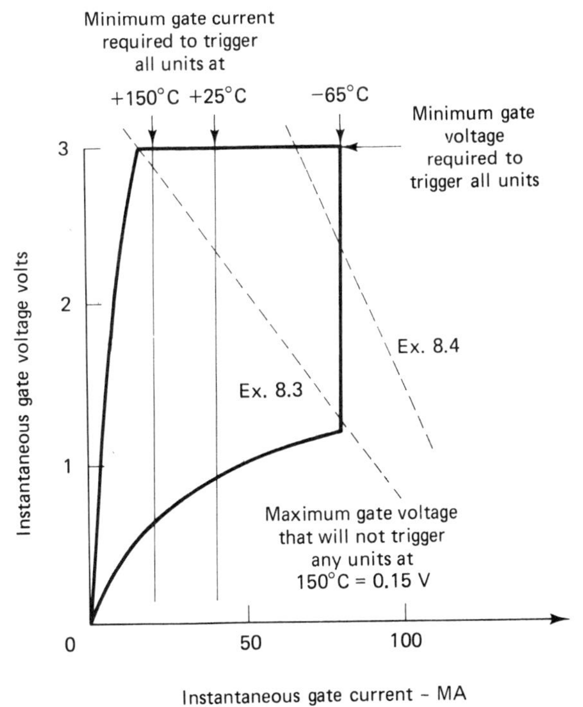

FIGURE 8.28 SCR gate characteristics (GE series C38). Courtesy of General Electric.

Example 8.3 *Testing a Proposed Trigger Circuit*

An SUS relaxation oscillator using a 2N4987 type is proposed as a trigger circuit for a C38 SCR. Is this feasible?

The peak pulse output voltage of 2N4987 measured in the circuit of Figure 8.1a across $R_2 = 20\ \Omega$, is 3.5 V minimum. The output circuit of this circuit is then given by a 3.5-V source in series with 20 Ω. Short-circuit current then is given by

$$I_S = \frac{3.5\ \text{V}}{20\ \Omega} = 0.175\ \text{A}$$

Drawing this load line ($I_{GS} = 0.175$ A, $V_{GO} = 3.5$ V) gives a line that is partially in the shaded area in Figure 8.28; therefore, the SUS trigger will not be dependable.

Example 8.4 *Testing a Unijunction Circuit*

A 2N2647 UJT relaxation oscillator is proposed to trigger the C38. Is this feasible?

The 2N2647 delivers 6 V across 47 Ω. Therefore, $V_{GO} = 6$ V, $I_{GS} = 6$ V/47 Ω = 0.128 A. Drawing this load line on Figure 8.28b, we see that only a small shaded area near −65 °C is passed through. Therefore, the 2N2647 will fire dependably but not below approximately −55 °C.

8.9 GATE TURN-OFF SWITCH (GTO)

The GTO is a four-layer, three-terminal device that physically and conceptually operates like an SCR. However, by controlling the alphas (α's) of the anode and cathode regions, the gate is given a turn-off capability not possessed by the SCR.

GTO operation is explainable in terms of the two-transistor model of the thyristor (Figure 8.29). For thyristor action to take place, the collector current of transistor Q_1 (I_{C1}) must become the base current (I_{B2}) of transistor Q_2. Also, the loop gain of the circuit must be unity to ensure a regenerative action. Note, however, that contrary to SCR theory the device can be turned off if I_{C1} is drawn off through the Q_2 gate, reducing I_{B2} below the level needed to support regeneration. Under these conditions the thyristor could be turned off, as well as on, by gate control. However, because the α of Q_1 equals the α of Q_2, I_{C1} is too large in the SCR to make gate turn off feasible. In the GTO the α of Q_1 is made much less than 1, the α of Q_2 approximately 1, and the conditions for regeneration are maintained while I_{C1} is very much reduced. (In effect, Q_1 becomes a low-gain, low-collector-current amplifier while Q_2's gain is increased.) Gate turn off is now feasible.

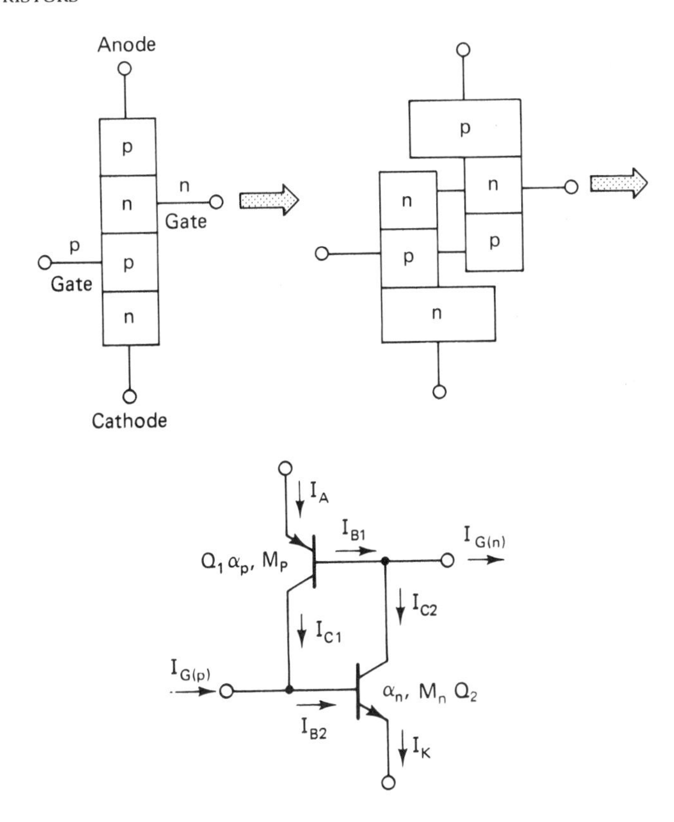

FIGURE 8.29 Two-transistor analog of *P–N–P–N* structures. Courtesy of General Electric.

Gate turn off in the GTO entails some loss of performance. The device operates at lower current density than the SCR and still requires substantial negative I_G to achieve turn off. At present the availability of high-gain silicon power transistors has reduced the demand for the GTO.

8.10 BILATERAL THYRISTORS

To extend the improvement in DC power control to AC power, a group of bilateral thyristors was developed. Conceptually, these devices may be thought of as two unilateral thyristors packaged together with one reversed. Actually, the structural requirements of such packaging often significantly affect the device characteristics.

8.10.1 Silicon Bilateral Switch (SBS)

Essentially, the SBS is composed of two SUS structures arranged in inverse parallel. The symbol, equivalent circuit, and characteristic are displayed in Figure 8.30. When used in the circuit of Figure 8.1 with V_1 replaced by an alternating supply voltage, alternating positive and negative trigger pulses are obtained. These are used mainly to drive a triac, which is the bilateral version of the SCR. Specifications for an SBS are identical to SUSs except for the absence of a reverse voltage rating.

8.10.2 Bilateral Trigger Diode (diac)

Although, conceptually, the diac consists of two four-layer diodes in inverse parallel arrangement, its characteristic curve reflects significant

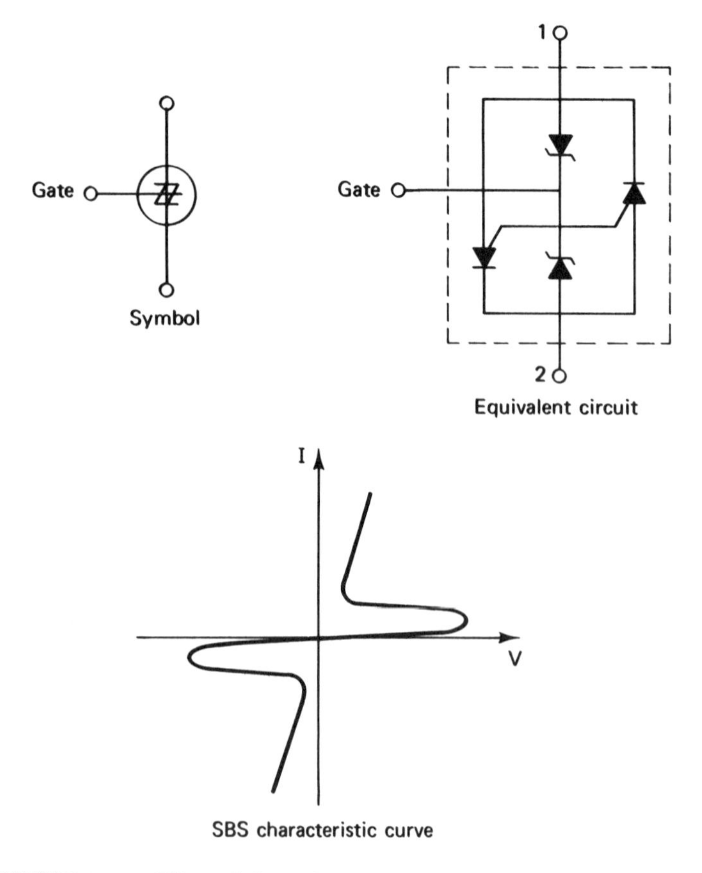

FIGURE 8.30 Silicon bilateral switch (SBS). Courtesy of General Electric.

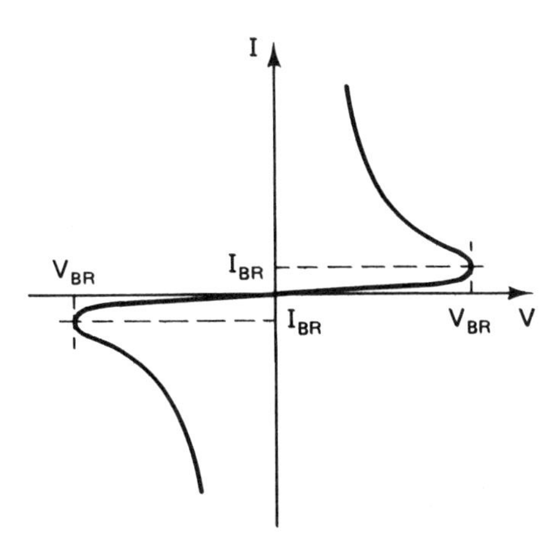

FIGURE 8.31 Diac characteristic curve. Courtesy of General Electric.

differences. Figure 8.31 displays this curve and Figure 8.32 gives the diac symbol. Although the ac curves exhibit the break voltage and negative resistance required for thyristor action, no valley point exists. Instead, the negative resistance region extends over the full operating range of currents above I_{BR}. This means that, unlike the four-layer diode, there is no minimum current (I_H) below which the diac is off. To turn the diac off, its anode voltage must be reduced to zero or reversed. Since the diac is used in circuits like Figure 8.1a with an AC supply, turn off presents no problem.

The diac is typically used to trigger the triac. Therefore, it is designed to ensure proper triggering of all triacs. For example, the GE type ST2 diac has a V_{BR} of 28 to 36 V, an I_{BR} of 200 μA maximum, and an e_p (peak pulse output) of 3 V minimum. This e_p will trigger all correctly designed triac circuits.

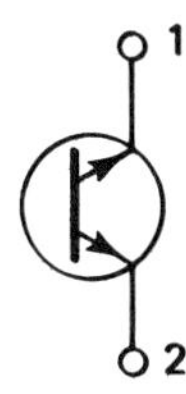

FIGURE 8.32 Symbol of bilateral trigger diode (diac). Courtesy of General Electric.

8.11 TRIAC

The triac (three-electrode ac semiconductor switch) was developed to extend the advantages of the SCR to ac power control. At present, triacs are available (from GE) in current ratings up to 40 A and voltages to 600 V operating at 50, 60, and 400 Hz.

Figure 8.33 shows the triac symbol (reversed SCRs) and the simplified pellet structure. Figure 8.34 shows its characteristic curve. This characteristic is based on MT_1 as a reference. Quadrant I is where MT_2 is positive relative to MT_1, and quadrant III is where MT_2 is negative relative to MT_1. V_{BO} plus and minus is the breakdown voltage. V_{BO} must, in normal operation, be higher than the peak value of the AC voltage appearing across the triac.

The triac may be triggered in four modes. In mode 1 (I+) operation, the gate voltage (relative to MT_1) and current are positive, and MT_2 is also positive (relative to MT_1). This gives first-quadrant operation. In mode 2 (I−), the gate voltage and current are negative and MT_2 is positive. Mode 3 (III+) is similar to mode 1 except that MT_2 is negative relative to MT_1. In mode 4 (III−), gate voltage and current are negative, and MT_2 is also negative. I+ and III− are the most sensitive modes. I− is less sensitive but usable. III+ is least sensitive and not recommended.

The four triggering modes of the triac emphasize the fact that it is more than just two SCRs in inverse-parallel arrangement. In addition to normal SCR action (modes I+ and III−), the package operates as a junction gate thyristor (mode I−) and a remote gate thyristor (mode III+).

I− operation is illustrated by Figure 8.35 in which MT_2 is the anode (MT_1 is the cathode). With the gate negative, junction p_2-n_3 is forward

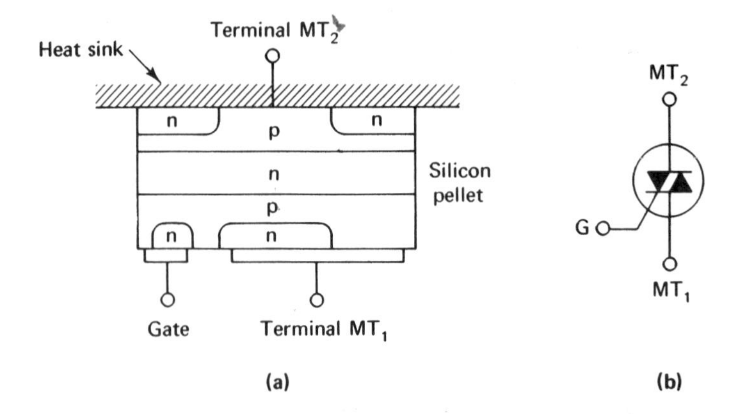

FIGURE 8.33 The triac: (a) simplified pellet structure; (b) circuit symbol. Courtesy of General Electric.

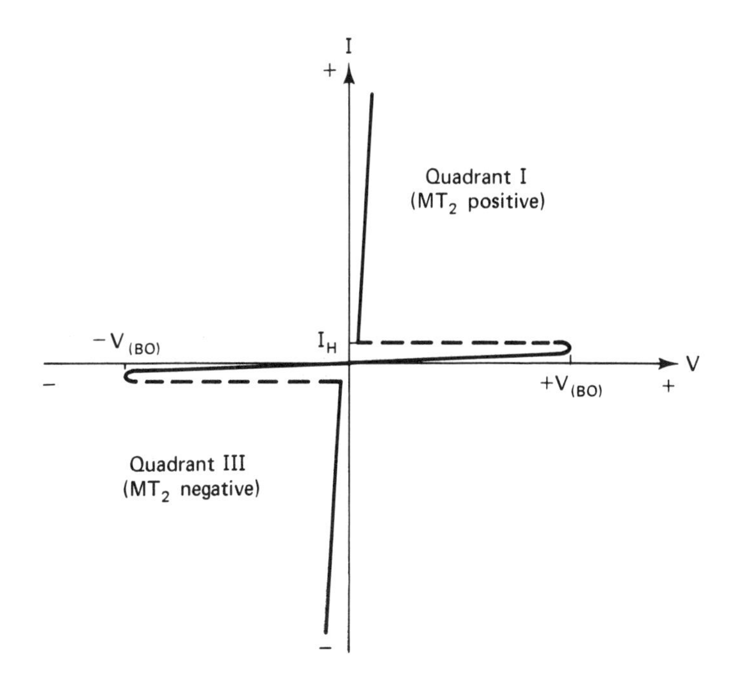

FIGURE 8.34 AC volt-ampere characteristic of the triac. Courtesy of General Electric.

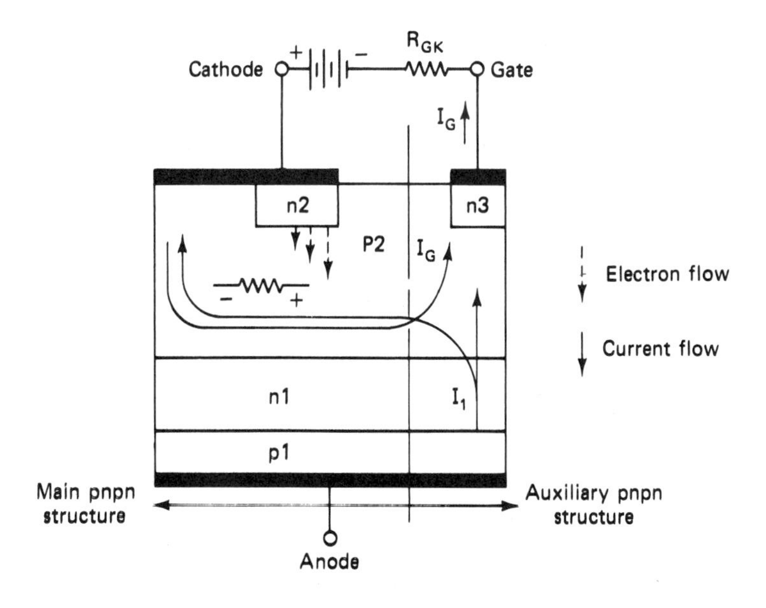

FIGURE 8.35 Junction gate thyristor. Courtesy of General Electric.

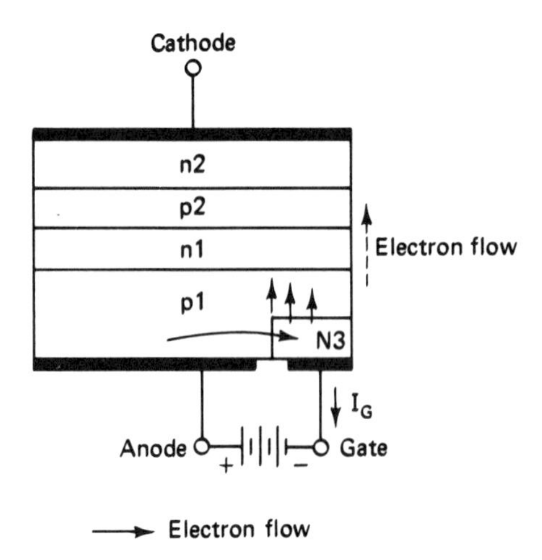

FIGURE 8.36 Remote gate thyristor. Courtesy of General Electric.

biased. An auxiliary structure p_1–n_1–p_2–n_3 turns on in conventional four-layer diode fashion. As a result, the voltage drop across this structure falls, raising the right-hand region of p_2 toward the anode voltage. Current flows laterally through p_2 since its left-hand region is held at cathode potential. The transverse voltage gradient existing across p_2 raises the potential of p_2 at the right-hand edge of $p_2 - n_2$, causing it to become forward biased and turning the main structure p_1–n_1–p_2–n_2 on.

III+ operation is illustrated by Figure 8.36. Here MT_1 is the anode and MT_2 is the cathode. In this case, with the bias shown, p_1–n_3 is forward biased. Current flows from p_1 to n_3. This current causes an increase in current across the p_1–n_1 junction, which starts a regenerative action and turns the structure on.

Turn-off time is more critical in triac operation than it is for the SCR. Typically, the SCR has a half-cycle of AC in which to turn off, while the triac must turn off in the interval in which the load current is passing through zero. This limits the upper frequency of triac operation.

8.12 APPLICATIONS

Basic Triac Static Switch. The circuit of Figure 8.37 permits a small reed switch to control heavy AC power. In the absence of a gate voltage, the triac is off and negligible current flows through the load, R_L. Closing the reed switch delivers a gate voltage to the triac sufficient to switch it to the on state. In its on state, the triac becomes a negligible impedance and current flows through R_L.

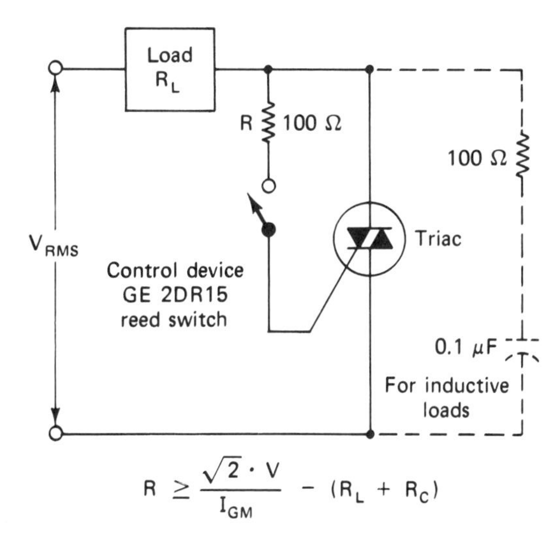

FIGURE 8.37 Static AC switch. Courtesy of General Electric.

Zero Voltage Switching. To avoid shock excitation to equipment and the generation of electrical interference, power supplies must be turned off or on at the instant their AC supply voltages are passing through zero or are at least small. Figure 8.38 shows a half-wave supply that does this.

This action is achieved by controlling the gate current of the SCR by the transistor Q_1. When Q_1 is on and saturated, the gate current of the SCR is diverted and it cannot fire. When Q_1 is off and the value of R_4 is made small enough, the SCR will fire when the AC supply voltage is slightly positive (3 to 5 V).

When S is open, a negative-going supply voltage will charge C through D_1 and R_1 to peak value very quickly. The top of the capacitor will be negative with respect to ground. As the negative AC voltage decreases toward zero, this negative voltage, although it discharges through R_2 and D_2, will keep Q_1 off if its time constant (R_2C) is large enough. With an R_2C time constant large enough to keep Q_1 off until the beginning of the positive cycle, the SCR will fire very close to the start of the cycle. Once the SCR has fired, it will remain on regardless of the state of Q_1. If the switch is opened too late to permit C to charge to peak during the negative half-cycle, nothing will happen. The next negative half-cycle will charge C, and the SCR will fire at the start of the next positive half-cycle.

When S is closed, C will not charge. With a sufficiently low R_3, Q_1 will be driven into saturation and the SCR will be prevented from firing. Since the state of Q_1 has no effect on an *on* SCR, closing S at such a time

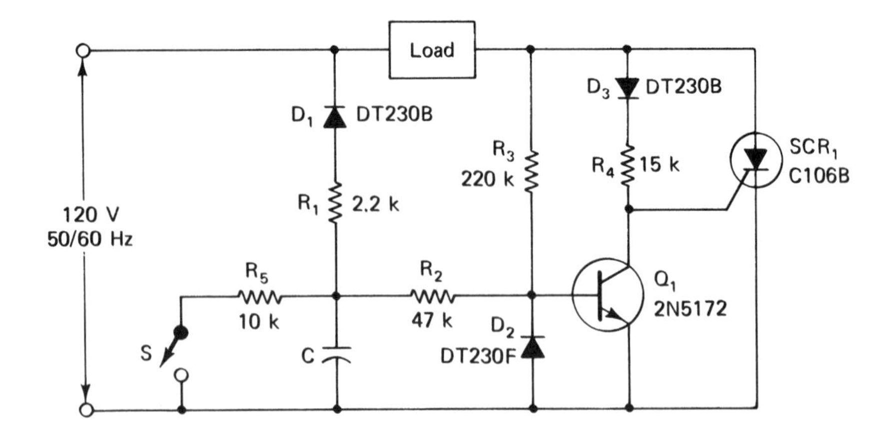

(a) Ideal half wave switching circuit

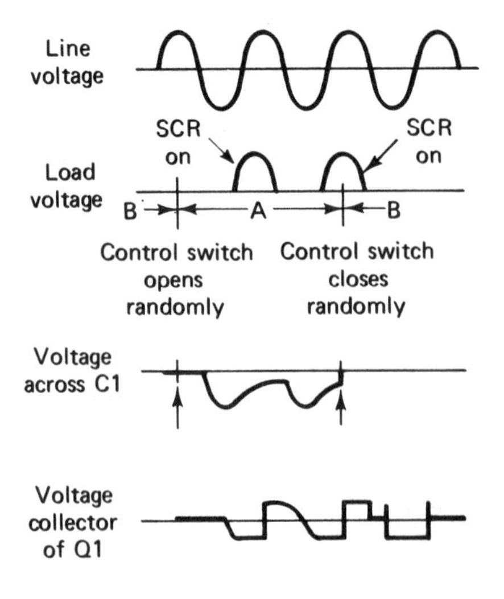

(b) Associated voltage half-wave forms

FIGURE 8.38 Half-wave zero-voltage switching circuit: (a) ideal half-wave switching circuit; (b) associated voltage half-wave forms. Courtesy of General Electric.

will have no immediate effect. The SCR will remain on during the positive half-cycle, turn off at its end, and not go on again until S is opened.

AC Phase Control. The power supplied to a load by an AC power source can be decreased by utilizing only a portion of each cycle. This is

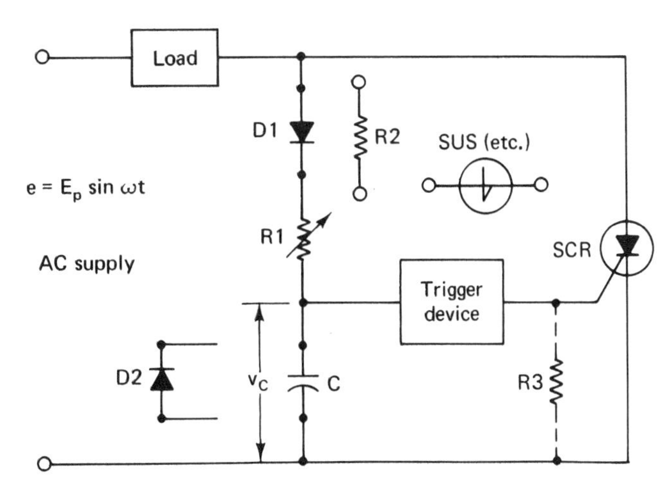

FIGURE 8.39 Basic half-wave phase control circuit. Courtesy of General Electric.

an efficient way to control the average power to lamps, motors, heaters, DC supplies, and the like without using inefficient resistors.

A basic half-wave phase control circuit is shown in Figure 8.39. In this circuit the firing point of the SCR is determined by the voltage across C and the anode voltage of the SCR. For a fixed AC supply, the firing phase angle α_T is determined by the magnitude of R_1, C, and the

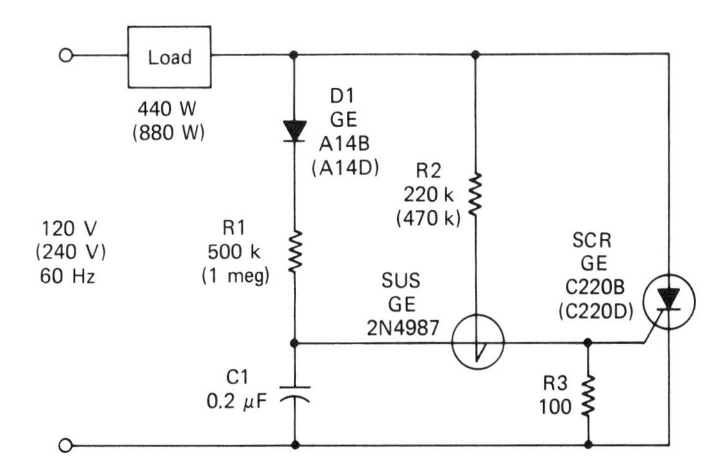

FIGURE 8.40 Half-wave phase control with capacitor reset. Courtesy of General Electric.

characteristics of the trigger device. A small R_1 would cause C to charge rapidly, firing the trigger device and the SCR early in the positive half-cycle. D_1 is used to protect the trigger device during the negative half-cycles when R_1 is small. Conversely, a large R_1 would delay firing until a later part of the positive half-cycle.

For reliable operation, it is necessary that the voltage across C be always returned to the same initial voltage after firing. The circuit of Figure 8.39 does not guarantee this since v_C depends on the minimum holding voltage of the trigger device and the gate-to-cathode source voltage, which appears when the SCR turns on. The circuit of Figure 8.40 resets the capacitor voltage by forcing the SUS to fire at the end of each positive half-cycle, thus discharging the capacitor. The SUS is fired by drawing current out of its gate through R_2 at (or near) the start of each negative half-cycle.

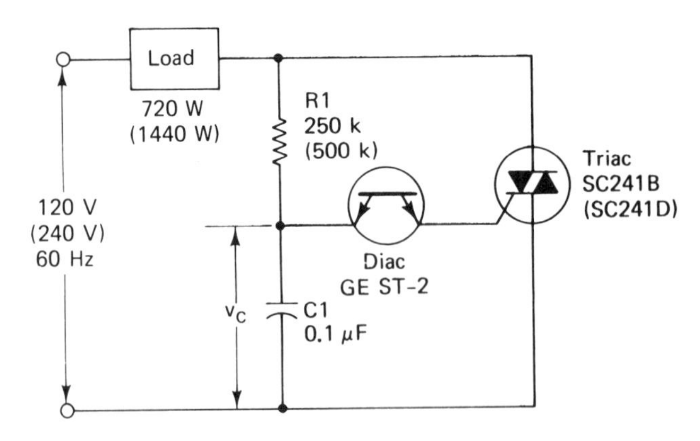

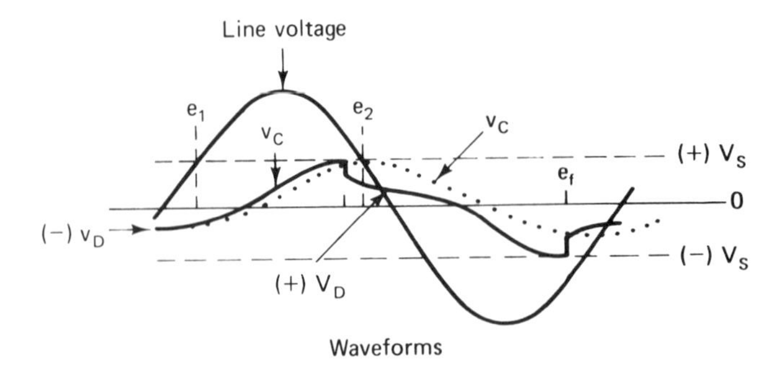

FIGURE 8.41 Basic diac–triac full-wave phase control. Courtesy of General Electric.

Full-Wave Phase Control. Full-wave phase control is achieved by the circuit of Figure 8.41. Supply and capacitor voltages are also shown. The diac (and triac) will fire whenever v_C reaches V_S. When firing occurs, v_C is reduced as shown by the gate current of the triac. Increasing R_1 reduces v_C and reduces load power. Decreasing R_1 increases v_C, increasing the triac on-time and increasing load power.

8.13 OPTOELECTRONIC DEVICES

Electronic devices that respond to or emit light are called *optoelectronic devices*. Excepting light emitters for the moment, they can be categorized as photoemissive, photoconductive, or photovoltaic. Photoemissive devices respond to incident light energy by generating current. The resistance of photoconductive devices is altered by incident light. In photovoltaic devices, incident light generates an output voltage often proportional to intensity.

8.13.1 Measures of Light Intensity

Light is a form of electromagnetic energy occupying a specific region of the electromagnetic spectrum. It is specified in terms of its wavelengths (or spectrum), intensity (irradiance), or, for light-emitting devices, luminous flux or candela. Light-emitting devices produce characteristic spectrums, and light-sensitive devices are most sensitive to specific spectrums. The spectrum of a tungsten lamp, a widely used light emitter, is specified as color temperatures (K) related to the filament temperature of the lamp. Figure 8.42 displays the spectrums of some light-sensitive and light-emitting devices. Wavelengths (λ) are measured in micrometers (μm); 1 μm $= 1 \times 10^{-6}$m. Wavelength may be converted to frequency (f) by

$$f = \frac{c}{\lambda} = \frac{3 \times 10^{8}\ \text{m/s}}{\lambda\ (\text{in meters})} \tag{8.13}$$

where c is the speed of light.

The energy per unit area of light incident on a surface is called *intensity* or *irradiance* (H) and is given in watts per square centimeter. It is also useful to define the irradiance at a given wavelength ($H\lambda$). The relative response of a light-sensitive device as a function of wavelength is called $Y\lambda$ and is supplied by the manufacturer. The effective irradiance (H_E) is the sum of the products $H\lambda Y\lambda$ and is given by

$$H_E = \int H\lambda Y\lambda \, d\lambda \tag{8.14}$$

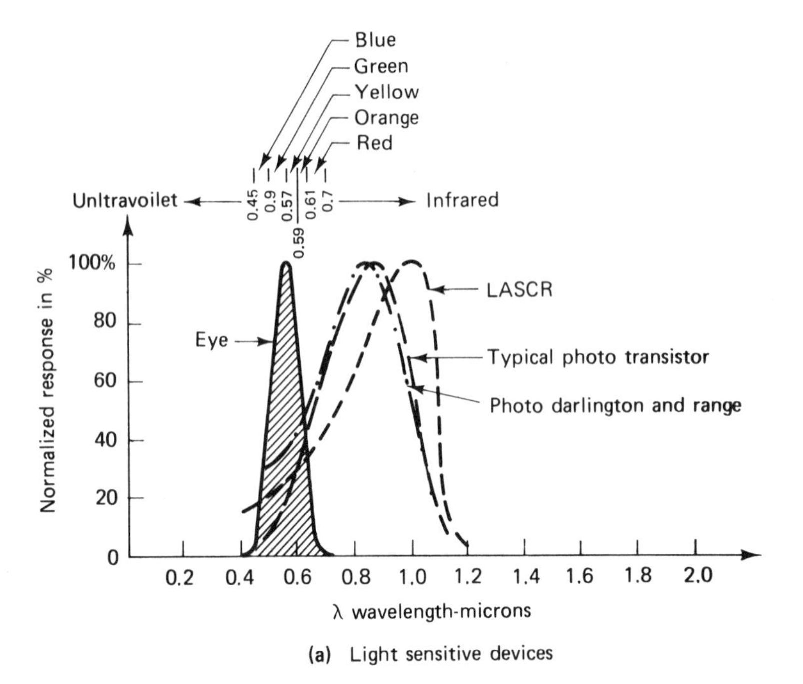

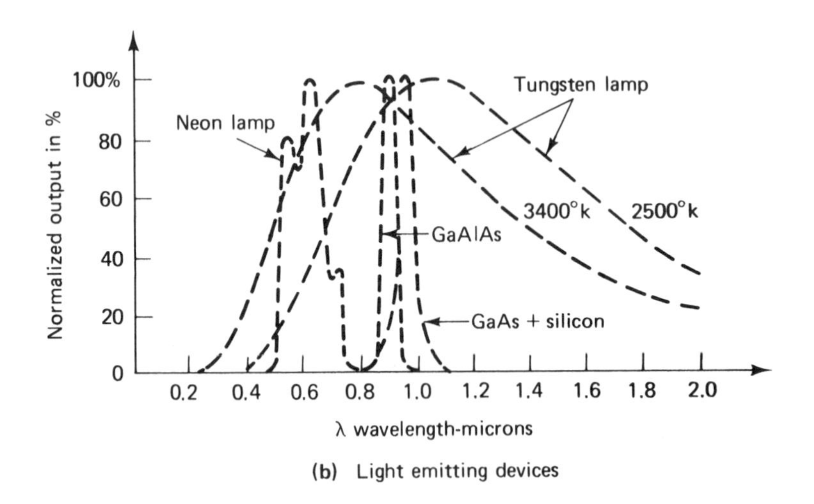

FIGURE 8.42 Spectral distributions of General Electric light-sensitive and light-emitting devices: (a) light-sensitive devices; (b) light-emitting devices. Courtesy of General Electric.

The effective irradiance of the human eye (E) measured in terms of foot candles is given by

$$E = K \int H\lambda Y'\lambda \, d\lambda \tag{8.15}$$

where $Y'\lambda$ is the relative spectral response of the human eye and K, equal to 6.35×10^5, is the conversion factor from milliwatts per square centimeter (mW/cm^2) to foot candles. Manufacturers sometimes relate device sensitivity to E.

Light-sensing devices are usually specified in terms of H_E/H or H_E/E. H_E is the bridge between irradiance and electronic response. Once it is known, the design procedure is transformed into a purely electronic one. The data required for this transformation are supplied by the manufacturer, often in graphical form. Figure 8.43 relates H_E/E to a tungsten lamp source. Example 8.5 illustrates its use in a design problem.

Example 8.5 *Design of a LASCR Trigger Source*

The LASCR (light-activated SCR) is an SCR with a light-sensitive gate. We have available a tungsten lamp with a brightness of 100

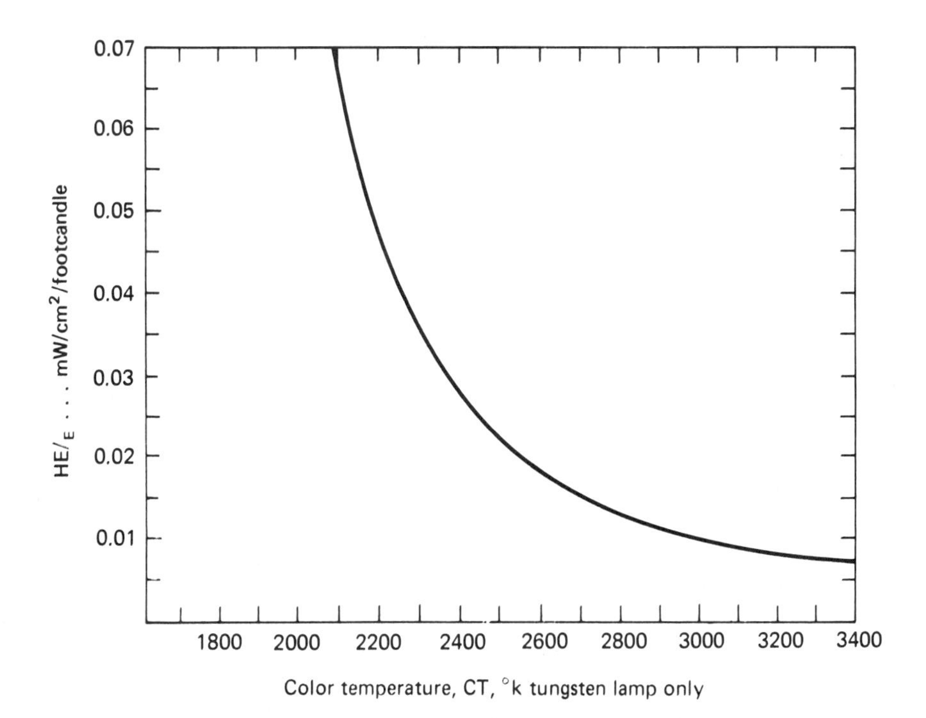

FIGURE 8.43 LASCR effective irradiance per footcandle for a tungsten lamp versus color temperature. Courtesy of General Electric.

candela (candela is a convenient measure of luminous flux or lumens) operating at a color temperature of 2500°K. To trigger a LASCR, H_E must be at least 4.2 mW/cm². What is the maximum distance from which the tungsten lamp can trigger the LASCR?

From Figure 8.43,

$$H_E/E = (0.023\,\text{mW/cm}^2)/\text{foot candle}$$

For $H_E = 4.2\ \text{mW/cm}^2$

$$E = \frac{H_E}{0.023} = \frac{4.2\ \text{mW/cm}^2}{0.023\ \text{mW/(cm}^2\ \text{foot candle)}}$$

$$= 182.61\ \text{foot candles}$$

Since

$$E = \frac{\text{source brightness (candela)}}{(\text{distance ft})^2}$$

$$= \frac{B}{d^2}\ \frac{\text{foot candles ft}^2}{\text{ft}^2}$$

(Note that the units of candela are foot candles ft².)

$$d = \sqrt{\frac{B\ \text{foot candles ft}^2}{E\ \text{foot candles}}} = \sqrt{\frac{100}{182.61}} = 0.74\ \text{ft}$$

8.14 OPTOELECTRONIC DETECTORS

8.14.1 Photodiodes and Phototransistors

Incident light energy will create electron–hole pairs in the depletion region of a reverse-biased diode. As a result, electrons will move through the depletion region, creating a reverse current proportional to H_E. Figure 8.44 illustrates this action and displays the characteristics of the device.

Typical specifications include size of the radiant sensitive area, the wavelength of maximum sensitivity, the spectral sensitivity in amperes per watt of radiant energy, rise time of the photocurrent, junction capacities, the current flowing in the absence of light (dark current), and all other specifications normal for diodes. Physically, photodiodes are typically small cylindrical packages with windows or lenses in one end. Photodiodes are versatile. They function as photoconductive and photo-

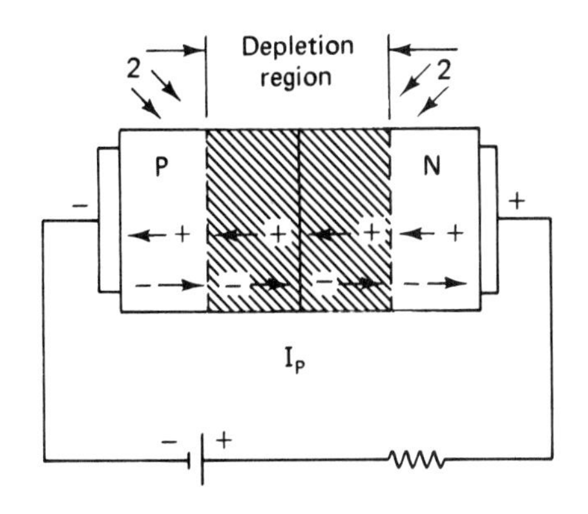

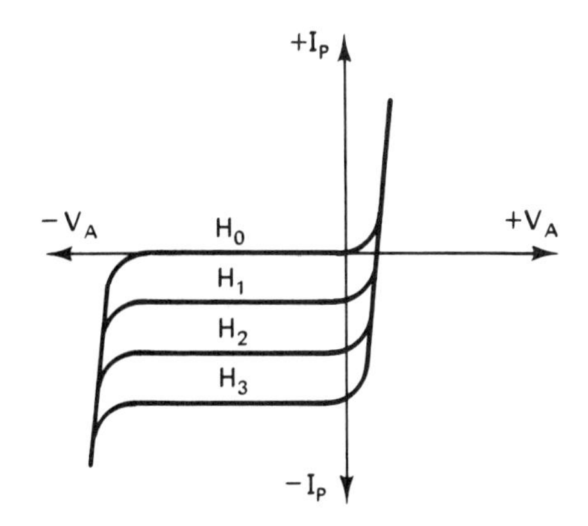

FIGURE 8.44 Photodiode action: (a) hole electron pairs in reverse biased light sensitive *PN* junction; (b) VI characteristic of light sensitive reverse biased *PN* junction. Courtesy of General Electric.

voltaic devices as well. However, their low sensitivity limits their use mainly to applications requiring extremely short rise times.

In the phototransistor, a window permits light to penetrate into the base region. Electron–hole creation generates base current that is amplified in normal transistor action. Physical layout and an equivalent circuit are shown in Figure 8.45. In Figure 8.45b, the photodiode

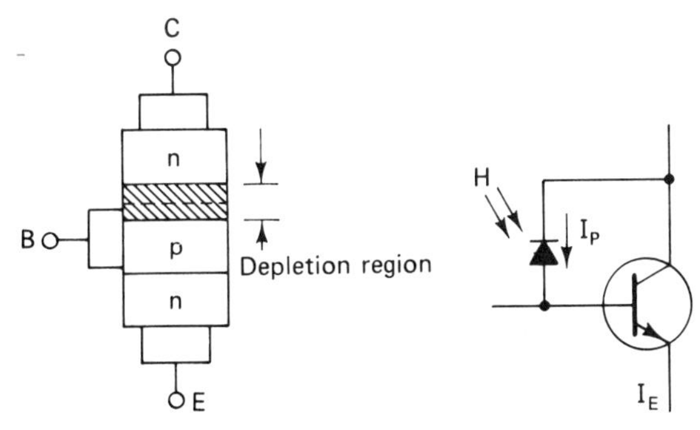

FIGURE 8.45 Phototransistor: (a) simplified physical layout of phototransistor; (b) phototransistor illustrating the effect of photon current generation. Courtesy of General Electric.

transforms light energy to base current. In reality, the collector–base junction of the transistor forms the photodiode.

Figure 8.46 shows a picture of a phototransistor, its circuit symbol, and characteristic. The characteristic curves are similar to the normal case except that the parameter $H\uparrow$ replaces base current. Phototransistor circuit design can follow normal transistor circuit designs by replacing h_{fe} by $I_L/H\uparrow$. This parameter is called the *sensitivity* of the phototransistor and is given by the manufacturer. Example 8.6 illustrates a simplified phototransistor circuit design.

Example 8.6 *Design of a Phototransistor Circuit*

An L14G3 phototransistor is to be used in the circuit of Figure 8.47 to drive a 100-μA/mV meter. The meter is to be calibrated in terms of the intensity (H_E) of the incident light. Set up a calibration scale relating intensity to meter current.

(a) *Calibrate the ordinate of Figure 8.46 in milliamperes:* The ordinate of Figure 8.46 is given as I_L, collector current normalized to (divided by) the I_C at $H\uparrow = 10\ \text{mW/cm}^2$ ($H\uparrow = H_E$).

$$I_C \quad (\text{at } 10\ \text{mW/cm}^2) = SH_E \quad (S = I_L/H_E)$$

Therefore,

$$I_C\,(\text{ordinate}) = I_L I_C \quad (10\ \text{mW/cm}^2) = I_L S H_E$$

From the GE short-form specification, we read

$$S = 1.2\ \frac{\text{mA}}{\text{mW/cm}^2}\ \text{minimum}$$

$$I_C\,(\text{ordinate}) = I_L \times \frac{1.2\ \text{mA}}{\text{mW/cm}^2} \times \frac{10\ \text{mW}}{\text{cm}^2} = 12 I_L\ \text{mA}$$

(b) *Check for linear operation:* For linear operation the meter load line must lie to the right of the knees of the curves.

$$R_M \quad (\text{meter resistance}) = \frac{1\ \text{mV}}{100\ \mu\text{A}} = 10\ \Omega$$

With a supply voltage of 10 V (arbitrary choice), we note that at maximum I_C (24 mA), the load line operates in the linear region.

(c) *Shunt the meter:* Since the maximum meter current is 100 μA, a shunt is needed to reduce the maximum collector current (24 mA) to 100 μA. We leave this to the student.

(d) *Calibration:* The meter calibration can be given analytically:

$$I \quad (\text{meter}) = H_E \times 5 \times 10^{-3}$$

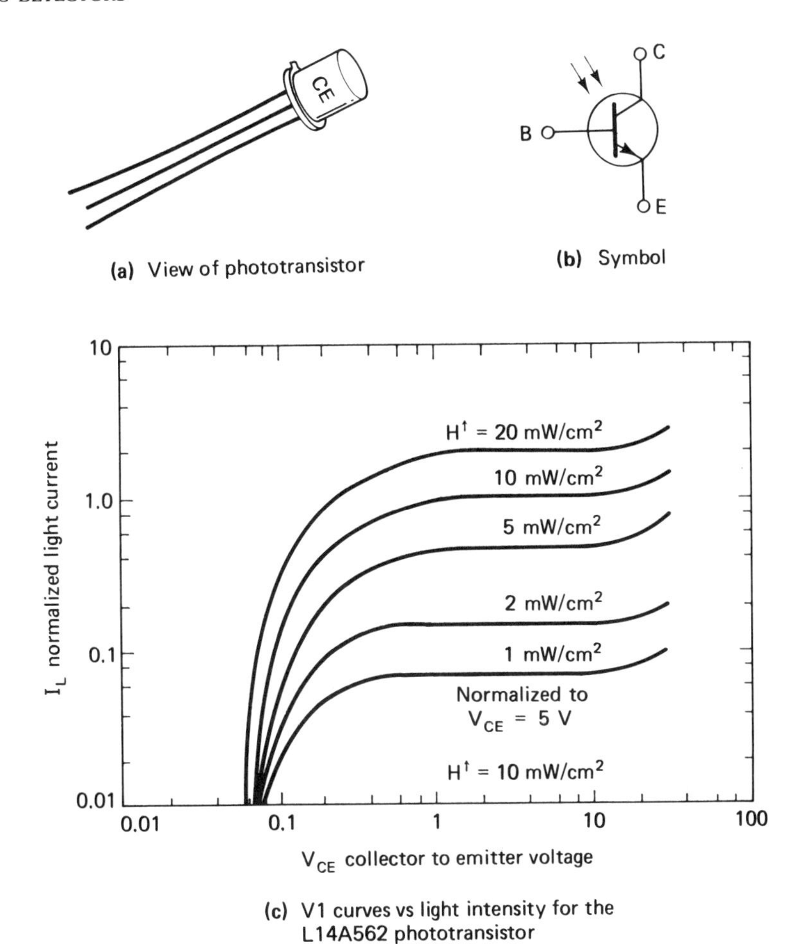

FIGURE 8.46 L14A502 phototransistor: (a) view of phototransistor; (b) symbol; (c) VI curves versus light intensity for the L14A502 phototransistor. Courtesy of General Electric.

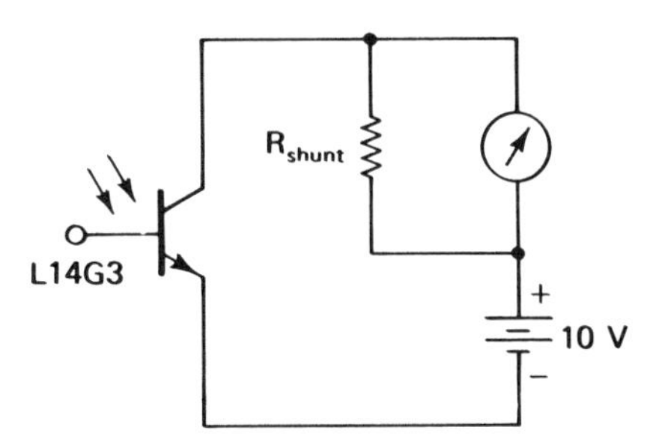

FIGURE 8.47 Phototransistor circuit.

or in tabular form

$$100\,\mu A \rightarrow 20\,mW/cm^2$$

$$50\,\mu A \rightarrow 10\,mW/cm^2$$

etc.

Switching speed is an important characteristic of the photodiode and phototransistor. In one common application, the reading of punched cards, the switching speed is crucial since it limits the reading speed. A brief survey reveals that photodiodes are faster than phototransistors. Photodiodes are available with rise and fall times (measured from 10% to 90% of the output response to a pulse of light) as low as 0.5 ns. Phototransistor response time, on the other hand, typically is above 2 μs. Other factors that limit the speed of phototransistors are delay time (t_d) and storage time (t_s). Figure 8.48 shows the response of a phototransistor to a pulse of light. The delay time is a measure of the interval between the initiation of the light pulse and the 10% point on the response. For the L14G2, t_d is about 2 μs. The storage time is the elapsed time between the termination of the light pulse and the 90% point on the falling slope of the response. Storage is a result of the finite time required to reestablish the depletion region. It is a function of base drive. Also shown in Figure 8.48 are the rise and fall times measured between 10% and 90% of the response.

The photo Darlington amplifier provides an increase in sensitivity at the expense of response times. Figure 8.49 displays a picture of the amplifier, its circuit symbol, and its collector characteristic. (Note that this figure is plotted with linear axes, as opposed to Figure 8.46 with log–log axes.) As is normal for Darlingtons, the configuration provides a higher input impedance (unimportant in this application) and greater gain. However, rise and fall times are much longer. For example, for the

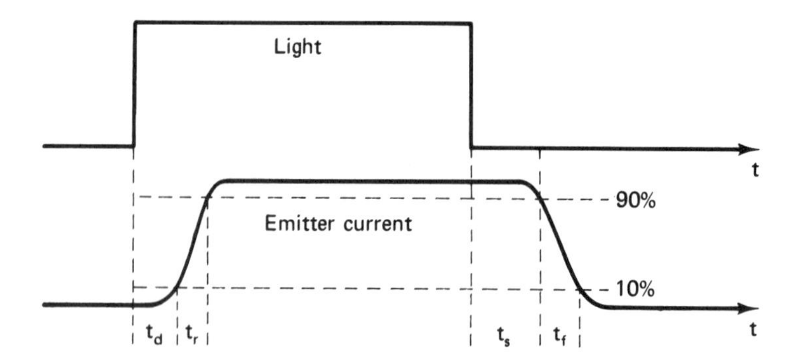

FIGURE 8.48 Time relationships between input and output of a phototransistor.

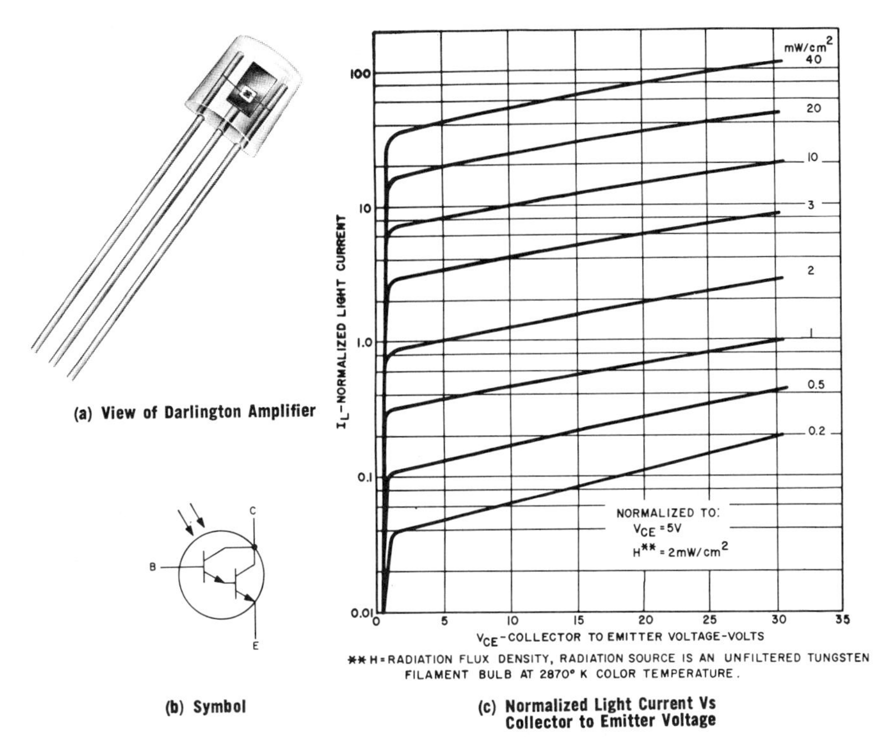

(a) View of Darlington Amplifier

(b) Symbol

(c) Normalized Light Current Vs Collector to Emitter Voltage

FIGURE 8.49 Light-sensitive Darlington amplifier: (a) view of Darlington amplifier; (b) symbol; (c) normalized light current versus collector-to-emitter voltage. Courtesy of General Electric.

2N5777–2N5780 family, rise time is typically 75 μs. Figure 8.50 gives the short-form specifications for GE phototransistors and photo Darlingtons. Note the differences in sensitivity and switching times.

8.14.2 Light-Activated SCR (LASCR)

The creation of electron–hole pairs in the p-region above the cathode (p_2) by incident light replaces gate current in the LASCR. Otherwise, LASCR operation is similar to that of the SCR. However, to obtain usable sensitivities, the LASCR must be constructed from a relatively thin silicon pellet of small dimensions. Therefore, high-current devices are not available for light triggering nor are they considered practical at this time.

The high sensitivity of the LASCR to electron–hole creation by light or any other means makes it respond more sensitively to temperature

PHOTO TRANSISTORS

GE TYPE	SENSITIVITY (ma/mw/cm²) MIN.	SENSITIVITY (ma/mw/cm²) MAX.	BV_{CEO} (V)	BV_{CBO} (V)	I_D (nA) MAX.	SWITCHING TYP. t_r (μSEC.)	SWITCHING TYP. t_f (μSEC.)	TYP. $V_{CE(SAT)}$	PKG
BPW36	.6	–	45	45	100	5	5	.4	55
BPW37	.3	–	45	45	100	5	5	.4	55
L14G1	.6	–	45	45	100	5	5	.4	55
L14G2	.3	–	45	45	100	5	5	.4	55
L14G3	1.2	–	45	45	100	5	5	.4	55
L14H1	.05	–	60	60	100	5	5	.4	263
L14H2	.2	–	30	30	100	5	5	.4	263
L14H3	.2	–	60	60	100	5	5	.4	263
L14H4	.05	–	30	30	100	5	5	.4	263

PHOTO DARLINGTONS

GE TYPE	SENSITIVITY (ma/mw/cm²) MIN.	SENSITIVITY (ma/mw/cm²) MAX.	BV_{CEO} (V)	BV_{CBO} (V)	I_D (nA) MAX.	SWITCHING TYP. t_r (μSEC.)	SWITCHING TYP. t_f (μSEC.)	TYP. $V_{CE(SAT)}$	PKG
2N5777	.25	–	25	25	100	75	50	.8	263
2N5778	.25	–	40	40	100	75	50	.8	263
2N5779	1.0	–	25	25	100	75	50	.8	263
2N5780	1.0	–	40	40	100	75	50	.8	263
BPW38	15.0	–	25	25	100	75	50	.8	55
L14F1	15.0	–	25	25	100	75	50	.8	55
L14F2	5.0	–	25	25	100	75	50	.8	55

FIGURE 8.50 Short-form specifications for phototransistors and photo Darlingtons. Courtesy of General Electric.

change, applied voltage, and rate of change of applied voltage. It also exhibits a longer turn-off time than the SCR.

Example 8.5 illustrated the design of the tungsten lamp irradiated LASCR for which the trigger irradiance (H_{ET}) is known. The designer's problem is to determine an H_{ET} reliable over the range of operating temperatures and anode voltage surges to which the LASCR will be subjected. Manufacturer-supplied data assist this determination.

Figure 8.51 displays the pertinent characteristics of an SCR (L8–L9 series). These are useful in determining a reliable H_{ET} that will trigger the LASCR for all specified conditions of its environment.

Figure 8.51a gives the effective irradiance (H_E) required to trigger the L8–L9 family over a temperature range of $-80°$ to $100°C$. Note, however, that these data apply for an anode voltage of 6 V. Corrections are required for higher anode voltages. An H_E between 1 and 6 mW/cm^2 will reliably trigger the family at an anode voltage of 6 V from $-60°$ to $100°C$. Anything lower may work at higher temperatures but will fail at lower ones. H_E's exceeding 6 mW/cm^2 may be satisfactory at lower temperatures but will overload the device at higher temperatures. An H_E of less than 0.1 mW/cm^2 is needed to prevent firing.

Figure 8.51b relates the angular orientation of the LASCR to a tungsten source. When the source is at $0°$ relative to a line perpendicular to the LASCR window, sensitivity is at maximum. Primary response is limited to $\pm 25°$ but with sensitivity reductions of 50% at the extremes. Angular deviations of $\pm 15°$ will give sensitivities within 90% of maximum.

Figure 8.51c displays a typical spectrum. This spectrum is displaced toward the infrared region when compared with the human eye

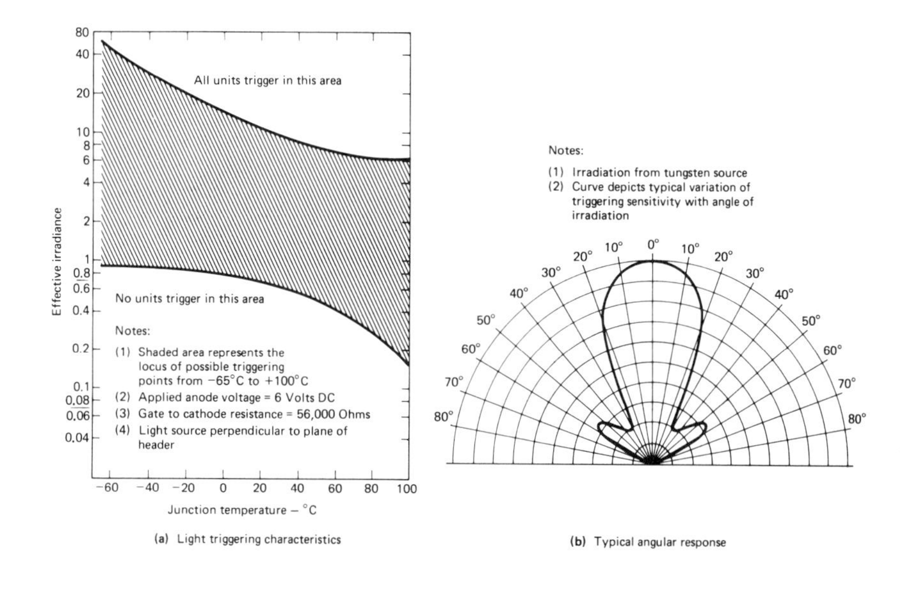

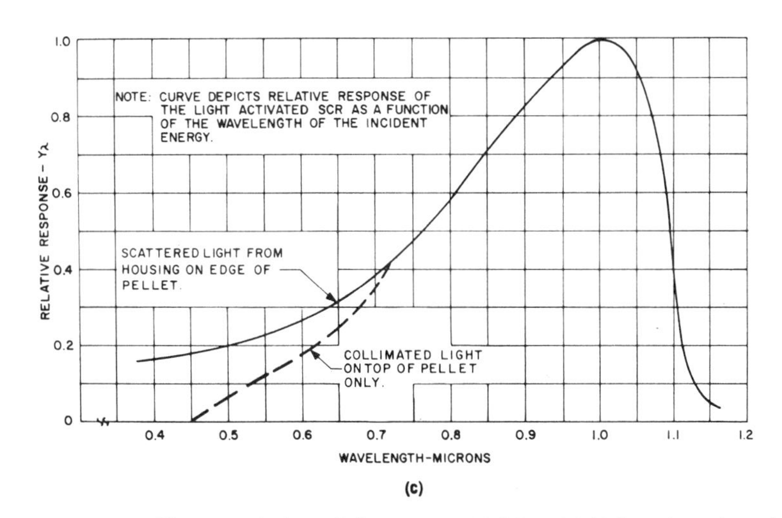

FIGURE 8.51 Characteristics of the L8–L9 LASCR; (a) light triggering characteristics; (b) typical angular response; (c) typical spectral response;

and to a lesser degree compared to the photodiode and phototransistor. However, it is a good match with a tungsten lamp with a 2500°K color temperature (see Figure 8.41).

Figure 8.51d relates the H_{ET} for an anode voltage of 6 V to higher

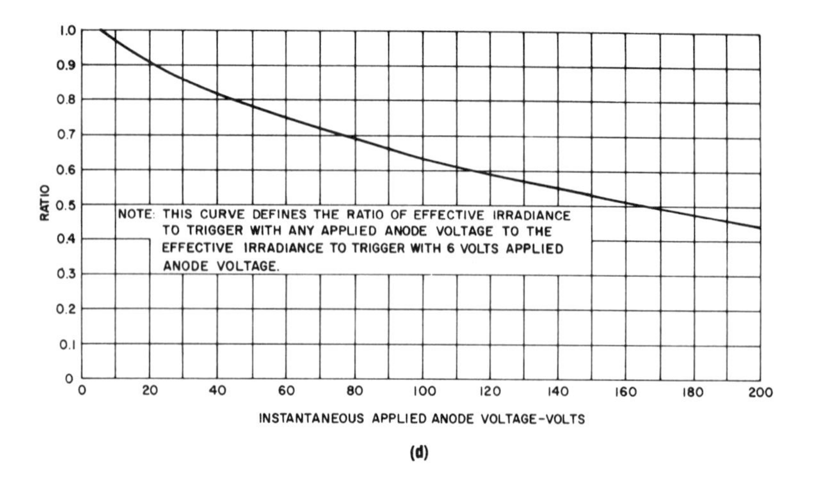

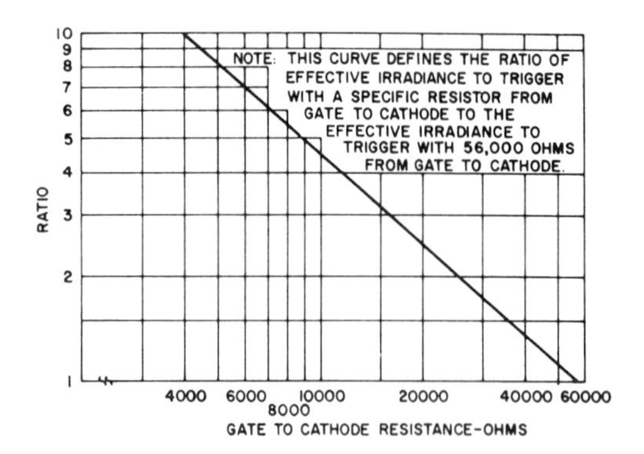

FIGURE 8.51 (continued) (d) typical variation of light sensitivity with anode voltage; (e) typical variation of light sensitivity with gate-to-cathode resistance. Courtesy of General Electric.

anode voltages. If at 6 V an H_{ET} of 1 mW/cm^2 is needed to reliably trigger the LASCR, then the H_{ET} at 80 V is given by

$$\frac{H_{ET} \quad (80\text{ V})}{H_{ET} \quad (6\text{ V})} = 0.68 \quad \text{(from curve)}$$

$$H_{ET} \quad (80\text{ V}) = H_{ET} \quad (6\text{ V}) \times 0.68 = 1 \times 0.68 = 0.68\,\text{mW/cm}^2$$

This increase in sensitivity with anode voltage can be exploited for phase control by light. When a sinusoidal voltage is applied to the LASCR anode, triggering will advance from the peak of the sinusoid (90°) toward the start (0°).

Figure 8.51e illustrates how the sensitivity of the LASCR may be controlled by a gate-to-cathode resistance. The resistor reduces the irradiation current flowing through the lower *n–p–n* sections of the LASCR, reducing gain. This useful characteristic provides a simple way to adjust the LASCR firing point.

The curve relates H_{ET} with a given resistor to H_{ET} with a resistor large enough to have very small effect on sensitivity (56 kΩ). For example, if an H_{ET} of 1 mW/cm² is sufficient to fire the LASCR with a 56-kΩ gate–cathode resistor, then the H_{ET} required to fire with an 8-kΩ resistor is given by

$$\frac{H_{ET}\quad (8\ \text{k}\Omega)}{H_{ET}\quad (56\ \text{k}\Omega)} = 5.5$$

$$H_{ET}\quad (8\ \text{k}\Omega) = 5.5 \times H_{ET}\quad (56\ \text{k}\Omega) = 5.5 \times 1 = 5.5\ \text{mW/cm}^2$$

The specifications for L8–L9 series LASCR, except for those described previously, are similar to the SC type SCR. The device can handle up to 1.6-A rms anode current with a blocking voltage up to 200 V peak.

Example 8.7 *Determining* H_{ET}

Find H_{ET}'s for a LASCR operating at a blocking (anode) voltage of 100 V with a gate potentiometer adjustable from 10 to 60 kΩ. The H_{ET} at 6 V, open gate is given as 2 mW/cm².

(a) *Correction for anode voltage:* From Figure 8.51d,

$$\frac{H_{ET}\quad (100\ \text{V})}{H_{ET}\quad (6\ \text{V})} = 0.64$$

$$H_{ET}\quad (100\ \text{V}) = 0.64 \times H_{ET}\quad (6\ \text{V}) = 0.64 \times 2 = 1.28\ \text{mW/cm}^2$$

(b) *Correction for gate resistance:* From Figure 8.51e,

$$\frac{H_{ET}\quad (10\ \text{k}\Omega)}{H_{ET}\quad (56\ \text{k}\Omega)} = 4.6$$

$$H_{ET}\quad (10\ \text{k}\Omega) = 4.6 \times H_{ET}\ (56\ \text{k}\Omega) = 4.6 \times 1.28 = 5.89\ \text{mW/cm}^2$$

Therefore, H_{ET}'s can vary from 5.89 to 1.28 mW/cm² by adjustment of the pot.

8.15 LIGHT-EMITTING DEVICES

Light-emitting devices include the tungsten incandescent lamp, the visible light-emitting diode (LED), the infrared light-emitting diode (IRED), and the laser. Also included in this section are the liquid crystal display, which is a display device rather than a light emitter, and the optocoupler.

8.15.1 Tungsten Lamp

Light is emitted by the tungsten lamp as a result of electrically heating the tungsten filament to incandescence. Lamps may be evacuated, but efficiency is increased when the filament operates in a gas-filled environment. The spectrum of the tungsten lamp is broader than other sources and is a function of the filament temperature. The spectrum is specified in terms of a color temperature, the color temperature being almost identical with filament temperature. Increasing color temperature shifts the spectrum toward shorter wavelengths.

Tungsten lamps have the advantage of having a spectrum encompassing the LASCRs. Therefore, they are effective in irradiating LASCRs. However, the light output responds extremely slowly to changes in electrical current. Therefore, their use in electronic applications is usually limited to providing steady or slowly changing illumination.

However, because the resistance of the tungsten filament increases rapidly with temperature, they are often used as current-sensitive resistances, particularly for stabilizing oscillators. Figure 8.52 shows the relationship between color temperature and resistance.

8.15.2 Light-Emitting Diodes (LED and IRED)

The light-emitting diode (LED) and the infrared light-emitting diode (IRED) are *p–n* junction diodes constructed so that light radiation generated in their junctions is emitted through a window or lens. Junction electroluminescence is a result of applying a low forward-biased voltage across a suitably doped *p–n* junction. Forward-bias current in the junction injects holes into the *n*-type material and electrons into the *p*-type material (minority carriers). Recombination of these carriers generates both light and heat. The higher the energy level produced by the recombination process, the shorter the wavelength of the light emitted. Control of doping levels and the materials used permits the production of visible light with wavelengths longer than 500 nm (10^{-9} m) and infrared light above 800 nm.

LEDs and IREDs have important advantages over conventional light sources. They have extremely fast response times (a few nanoseconds to a few microseconds). They are mechanically rugged and have a long life. They are low-impedance devices similar to conventional forward-biased diodes and are therefore compatible with solid-state circuitry. Their light output is monochromatic and lies within the sensitivity spectrum of silicon detectors. They are available in various shapes and colors (red, yellow, green, and infrared), and they are easily configurated in packages displaying numbers and/or numerals.

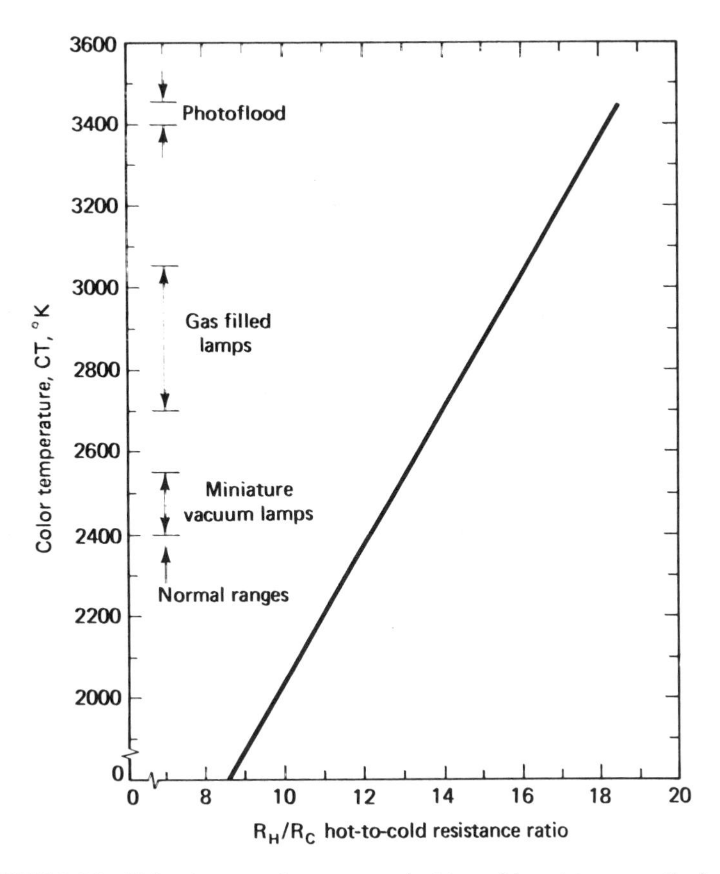

FIGURE 8.52 Color temperature versus hot-to-cold resistance ratio for tungsten. Courtesy of General Electric.

General Electric's IREDs are made either from gallium arsenide or gallium aluminum arsenide, which emit infrared at 940 and 880 nm, respectively. Figure 8.53 compares the IRED spectrums to the spectral response of silicon. Note that both IREDs, but particularly gallium aluminum arsenide (GaAlAs), are close to the peak sensitivity of silicon. The GaAlAs IRED is typically about five times more sensitive than the other, but it is also much more difficult to process.

Electrical characteristics of LEDs and IREDs are similar to conventional silicon diodes except for having higher forward voltages and lower reverse voltages. Circuits are also designed in a similar fashion except that a transformation from input electrical power to output light is required. Example 8.8 illustrates this process in the design of an optocoupler.

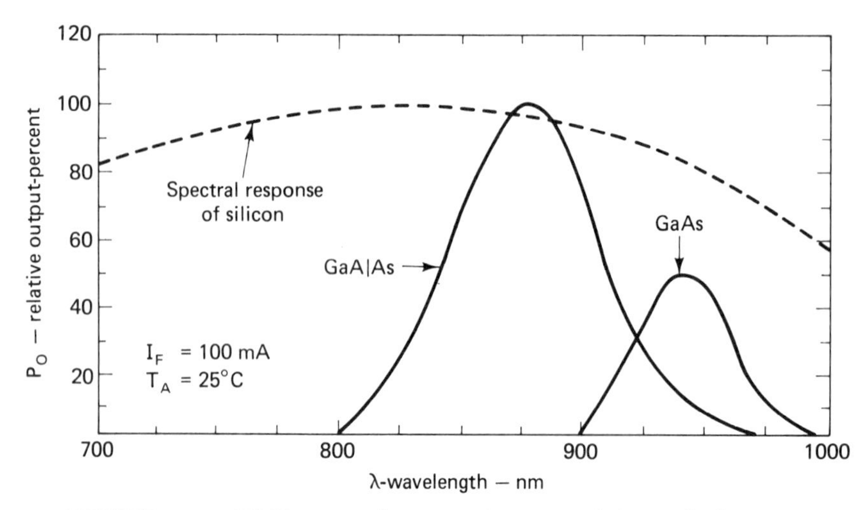

FIGURE 8.53 IRED spectral output. Courtesy of General Electric.

Example 8.8 *Design of an Optocoupler*

Semiconductor optocouplers usually consist of an IRED and a silicon light-sensitive semiconductor coupled by a transparent high-breakdown voltage path and enclosed in an opaque housing. They are primarily used for isolation in high-power, high-voltage circuits. Their action is straightforward; the input signal drives the IRED, which then produces light and drives the detector. The major design consideration is to choose an IRED that will reliably drive the light-sensitive device (in this problem an L14G phototransistor). Figure 8.54 illustrates the structure of an optocoupler.

Determine the forward current (I_F) of a GaAs IRED that is to drive an L14G phototransistor in an optocoupler. Collector current of the L14G must equal 12 mA. The IRED is to be positioned on the L14G axis, 0.8 in. away, and isolated by a perfectly transparent material.

(a) *Find the output power required of the IRED:* Figure 8.55 relates the spacing of a GE IRED from an L14G phototransistor (IRED positioned on lens axis of the phototransistor) to the parameter *percent* (p). The output power P_o is related to H_E and percent by

$$P_o = \frac{H_E}{30 \times p}$$

From Example 8.6 we find that, for a phototransistor collector current of 12 mA, H_E = 10 mW/cm². From Figure 8.55 we read, for a spacing of 0.8 in., p = 10% or 0.1. Therefore,

$$P_o \text{ (output power)} = \frac{H_E}{30 \times p} = \frac{10\ \text{mW/cm}^2}{30 \times 0.1} = 3.33\ \text{mW}$$

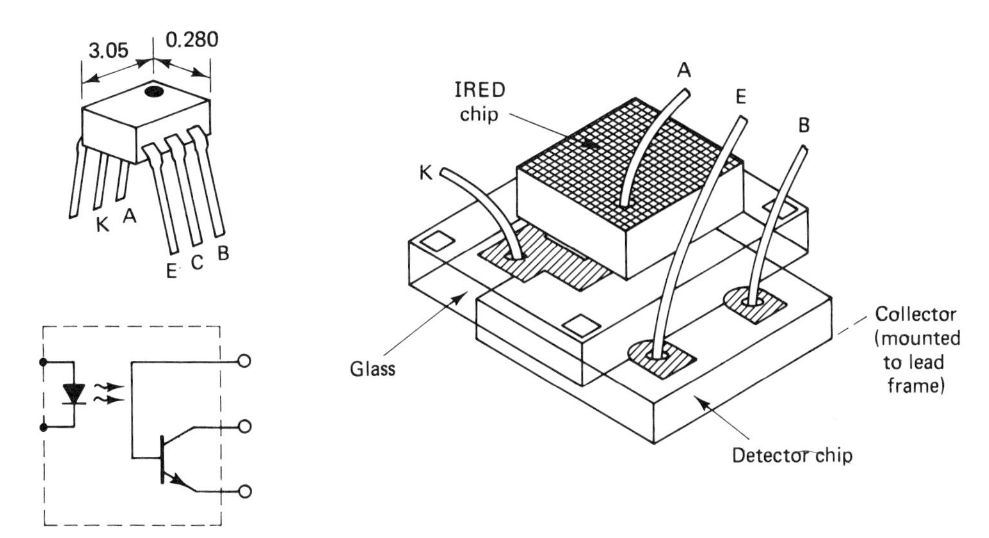

FIGURE 8.54 Dual-in-line package (DIP) optocoupler illustrating glass dielectric. Courtesy of General Electric.

(b) *Find IRED input current (I_F):* From Figure 8.56, using the GaAs curve for 3.33 mW, we read

$$I_F = 62\,\text{mA}$$

(c) *Choose an IRED:* From a set of IRED specifications (GE) (Figure 8.57) we choose the LED55B with P_o min = 3.5 mW and I_F max = 100 mA. (Note the high I_F's required.)

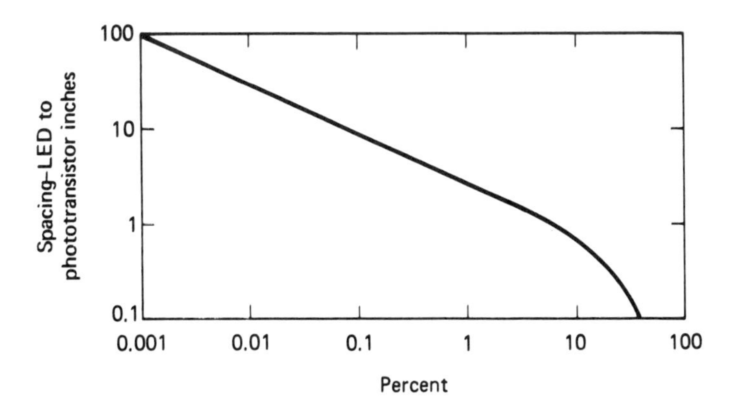

FIGURE 8.55 IRED-to-phototransistor coupling chart. Courtesy of General Electric.

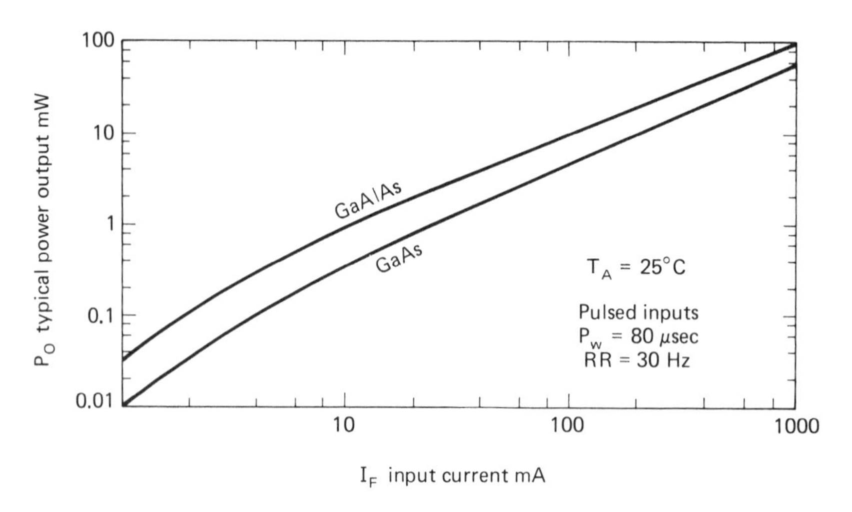

FIGURE 8.56 IRED power output. Courtesy of General Electric.

8.15.3 Liquid Crystal Display (LCD)

Unlike ordinary liquids, in which molecules are randomly oriented, the molecules of liquid crystals are normally fixed into definite orientations. However, when current is caused to flow through the liquid crystal, this orientation is destroyed or at least altered. The molecules of the liquid crystals used in LCDs (cholesteryl nonanote and *p*-azoxyanisole) tend to be rodlike. In the presence of a voltage field they tend to orient their short axis perpendicular to the field. This orientation can also be affected by the presence of other organic compounds. Figure 8.58 shows the molecular orientation of the liquid crystal for the undisturbed and disturbed conditions. These characteristics of liquid crystals are exploited in two modes of operation, dynamic scattering and field effect.

INFRARED EMITTERS

GE TYPE	MIN. P_O@ I_F=100mA	MAX. V_F@ I_F=100mA	PEAK EMISSION WAVELENGTH TYP.n.METERS	RISE TIME TYP.μ.SEC.	FALL TIME TYP.μ.SEC.	MAX. P_D mW	MAX. I_F CONT. mA	PKG
IN6264	6.0mW	1.7V	940	1.0	1.0	1300	100	54A
IN6265	6.0mW	1.7V	940	1.0	1.0	1300	100	54
IN6266	25mW/sr	1.7V	940	1.0	1.0	1300	100	54A
CQX14	5.4mW	1.7V	940	1.0	1.0	1300	100	54A
CQX15	5.4mW	1.7V	940	1.0	1.0	1300	100	54
CQX16	1.5mW	1.7V	940	1.0	1.0	1300	100	54A
CQX17	1.5mW	1.7V	940	1.0	1.0	1300	100	54
F5D1	12mW	1.7V	880	1.5	1.5	1300	100	54A
F5D2	9mW	1.7V	880	1.5	1.5	1300	100	54A
F5E1	12mW	1.7V	880	1.5	1.5	1300	100	54
F5E2	9mW	1.7V	880	1.5	1.5	1300	100	54
LED55C	5.4mW	1.7V	940	1.0	1.0	1300	100	54A
LED55B	3.5mW	1.7V	940	1.0	1.0	1300	100	54A
LED56	1.5mW	1.7V	940	1.0	1.0	1300	100	54A
LED55CF	5.4mW	1.7V	940	1.0	1.0	1300	100	54
LED55BF	3.5mW	1.7V	940	1.0	1.0	1300	100	54
LED56F	1.5mW	1.7V	940	1.0	1.0	1300	100	54

54A

54

FIGURE 8.57 IRED data (short form). Courtesy of General Electric.

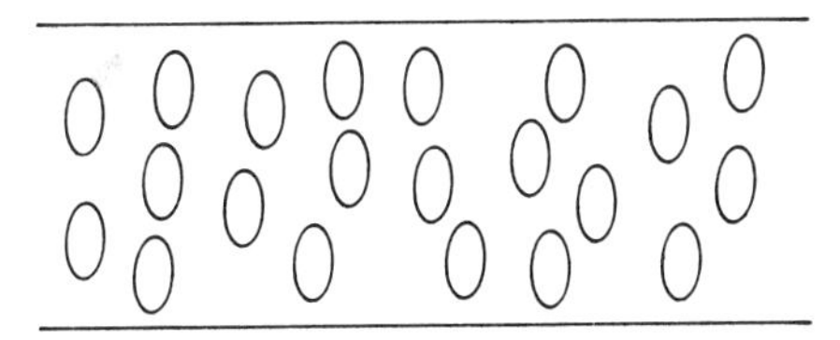

(a) Undisturbed liquid crystal (no current)

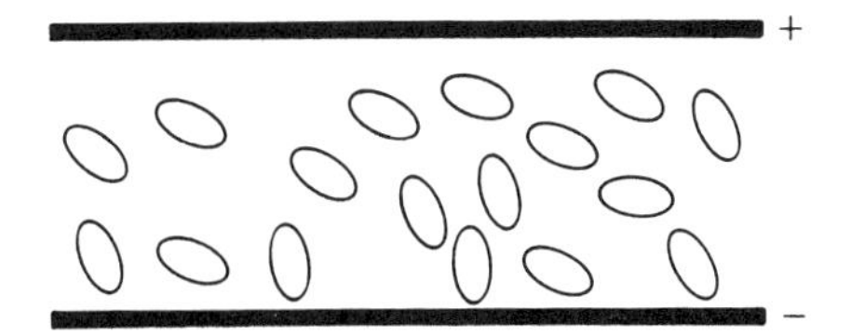

(b) Disturbed (current and voltage present)

FIGURE 8.58 Effect of current and voltage on liquid crystal molecules: (a) undisturbed liquid crystal (no current); (b) disturbed liquid crystal (current and voltage present). Courtesy of General Electric.

Dynamic Scattering LCDs. Figure 8.59 illustrates the construction of an LCD using dynamic scattering. This is also called a *nematic liquid crystal display.* As shown, the liquid crystal is confined between two transparent sheets of glass. Transparent conductive coatings con-

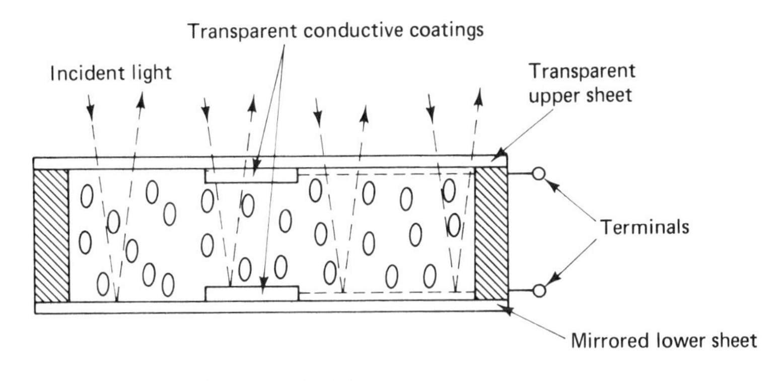

(a) No applied bias

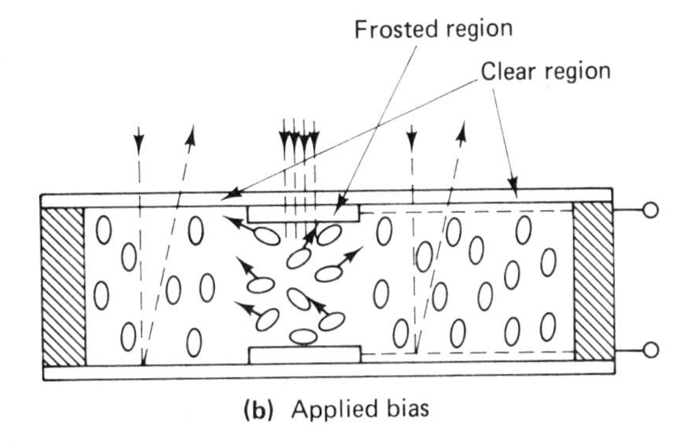

(b) Applied bias

FIGURE 8.59 Nematic liquid crystal: (a) no applied bias; (b) applied bias.

figurated in any desired way are plated onto the inner surfaces of the glass sheets with leads drawn out to accessible terminals. (Usually these coatings form a set of segments that can be selectively energized to form letters or numerals.)

In the absence of an applied voltage, the liquid crystal remains in its undisturbed state and incident light passes through it and is reflected from its background (often a mirrored surface plated onto the lower sheet). However, when a voltage is applied to a segment, the molecular shifting that results produces increased reflection and scattering of light in the region below the segments, giving it a frosted appearance. Energizing appropriate segments will cause the selected symbol to stand out from its background. Symbols can be enhanced with edge lighting.

Field-Effect LCD. Figure 8.60 illustrates the construction of a field-effect (twisted nematic) LCD. The conducting surface on the top is covered with an organic film (etched or coated) that orients the molecules of the liquid crystal at right angles to their undisturbed state with their long axis parallel to the long axis of the cell wall. The lower conducting surface is chemically treated to orient the molecules at right angles to their undisturbed state, but with their long axis parallel to the short axis of the cell wall. Also, the enclosing transparent sheets of glass are polarized.

In the transmissive field-effect LCD (Figure 8.60a), both transparent sheets are vertically polarized. In the absence of an applied bias, vertically polarized light entering the liquid on the upper side is transformed into horizontally polarized light by the twisted configuration of the liquid crystal's molecules. This light is blocked by the vertically polarized sheet. The viewer below sees a uniformly dark display. On application of a bias (above 2 to 8 V), the twisted configuration is destroyed in the region between the conducting plates and the vertically polarized light is unaffected by the liquid crystal. It therefore passes through the bottom sheet of glass unhindered. The viewer now sees the selected symbols as a bright pattern against a dark background.

In the reflexive field-effect LCD, the viewer and the incident light are on the same side (below in Figure 8.60b). The top sheet of glass is mirrored, and its inner surface is covered with a horizontally polarized film. The bottom sheet is transparent and vertically polarized. In the absence of bias, vertically polarized light entering the liquid crystal region from below is transformed to horizontal polarization on passing through the liquid. Being horizontally polarized it passes through the film on the top, is reflected back through the liquid, transformed back to vertical polarization, and passed unhindered through the bottom wall. When bias is applied, the molecular pattern under the conducting plates is destroyed. Vertically polarized light retains its vertical polarization on

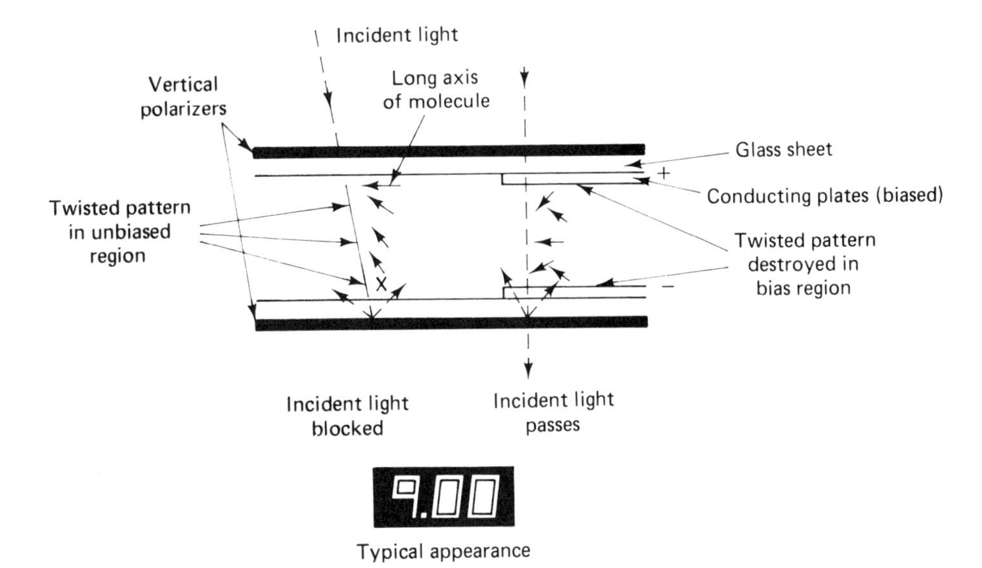

(a) Transmittive field-effect LCD

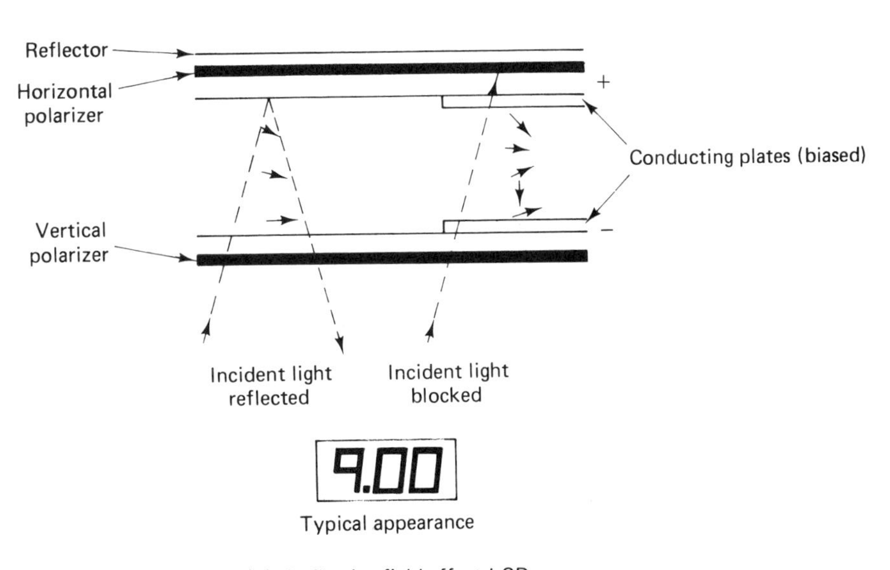

(b) Reflective field-effect LCD

FIGURE 8.60 Field-effect LCDs: (a) transmittive field-effect LCD; (b) reflective field-effect LCD.

passing through the liquid crystal between the plates. This light is blocked by the top side film and not reflected. The viewer sees dark symbols against a brighter background.

LCD Biasing and General Characteristics. LCDs require AC biasing. A continuous (direct) current would plate the cell electrodes and degrade or damage the device. Typically, a square-wave supply is used, about 30 V, 60 Hz for a scattering LCD, and about 8 V for a field-effect LCD. A back plane encompassing all the conducting segments on one side of the LCD is constructed, and the square wave is placed across these segments and the back plane. Segments on the other side (opposing segments) that are required to be activated will have a 180° out-of-phase square wave placed across them and the back plane. Therefore, twice the peak value of the square wave will appear across opposed activated segments. Activated segments on the other side will have an in-phase square wave placed across them and the back plane, giving a net voltage of zero for opposed segments.

The major advantage of the LCD over the LED is its low current requirement. LCDs operate at about 25 μA for dynamic scattering cells and 300 μA for field-effect cells (typical for a small seven-segment display). Compare this to the 20 mA typical for a LED. The power requirement for an LCD is about 140 μW per seven-segment numeral, as opposed to 400 mW per numeral of the LED. Because of this low power requirement, 8-in.-square LCD cells are feasible. They are at present replacing CRT displays in small portable computers.

The LCD's major disadvantage is its slow decay time (150 ms or more). This disqualifies its use in anything but slow-speed applications.

8.15.4 Laser Diode

Laser is the acronym for "light amplification by stimulated emission of radiation." Lasers produce virtually single wavelength (coherent) light, as opposed to the multiwavelength bands of light (incoherent) produced by other light-emitting devices. Because only the light radiation propagating along the major axis of the laser diode is amplified, it emits a narrow, high-energy density beam without significant divergence.

Figure 8.61 shows the laser diode to be a specially constructed LED manufactured from gallium arsenide or gallium arsenide plus other materials. The junction length (L) is precisely related to the operating wavelength. The nonemitting end of the junction is highly reflective; the emitting end is only partially reflective. As forward current (I_F) is increased from zero, the diode at first acts like a LED. The charge carriers entering the depletion region excite the atoms they strike, and photons of random energy levels are generated. At low current levels, the diode

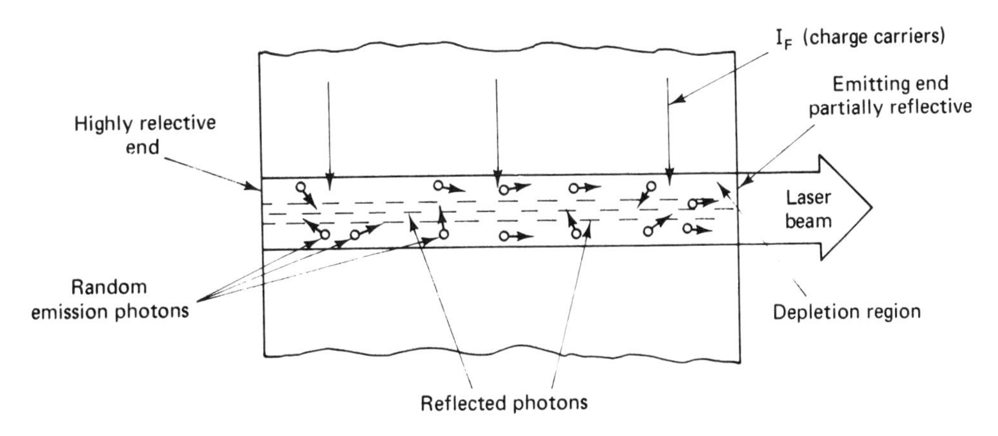

FIGURE 8.61 Laser diode.

emits incoherent light like a LED. At higher current levels, photons will strike the reflective ends of the junction and be reflected back. Those photons that are moving in a path perpendicular to the ends will be reflected back and forth along their original paths many times. Passing back and forth through the depletion region lengthwise, they will generate other photons in increasing amounts with precisely equal energy. The reflection process involves resonance and is related to the wavelength of the light and the length of the diode. Only one wavelength will propagate and be increased in energy. This process builds up to extremely high levels, emitting a monochromatic (coherent) beam from its emitting end.

Problems

1. In Figure 8.1, let $V_S = 6$ V, $V_F = 1.5$ V, and $i_P = 200$ mA. Compute R_2 and e_p (the pulse voltage across R_2).
2. If, in Figure 8.1, $I_H = 1.5$ mA, find V_H.
3. For $V_1 = 10$ V and $I_H = 1.5$ mA, compute R_1 minimum (Figure 8.1).
4. For $V_1 = 10$ V, $I_S = 0.5$ mA, compute R_1 maximum (Figure 8.1).
5. A UJT operates with $V_{BB} = 10$ V. Assuming $\eta = 0.07$ and $V_D = 0.5$, find V_p.
6. For the UJT of problem 5, plot V_p vs V_{BB}. Let V_{BB} vary from 5 to 30 V.
7. Using the circuit of Figure 8.4 and a 2N2647 UJT, determine C_1, R_{B2}, and V_1 for the following specifications: $R_{B1} = 27\ \Omega$, SCR is a C40, $f = 1$ kHz.

8. Replace R_{B_1} in problem 7 with a Sprague 31Z204 pulse transformer and repeat for the same specifications. If V_1 is too high, choose another UJT and repeat.

9. Manufacturer's data specify that a gate charge accumulation of 0.25 $\mu C/\mu s$ is required to trigger an SCR. Plot charge versus time for this case.

10. The analytical expression for an exponentially decaying current pulse is given by

$$i_t = I_p e^{-t/RC}$$

where RC is the time constant (T) and I_p is the peak value. The total charge conveyed as a function of time for this wave shape is given by

$$Q(t) = \int_0^{T_1} I_p e^{-t/RC}\, dt = I_p RC(1 - e^{-T_1/RC})$$

Plot $Q(t)$ versus t for (a) $I_p = 50$ mA and $RC = 5$ μs; (b) $I_p = 50$ mA and $RC = 10$ μs; (c) $I_p = 70$ mA and $RC = 5$ μs; (d) $I_p = 100$ mA, $RC = 5$ μs.

11. An SCR requires a 0.40 $\mu C/\mu s$ rate of gate charging to fire. For an exponentially decaying pulse with $RC = 5$ μs, determine the I_p required to fire the SCR. How many microseconds of pulse will be required?

12. Design a relaxation oscillator using the PUT circuit of Figure 8.6 to trigger a C20 SCR from a 12-V supply at an operating frequency of 10 pulses/min. Find R_T, C_T, R_1, and R_2. Let $R_S = 33$ Ω. Use a 2N6027 PUT and check I_H.

13. For a Schockley diode, $V_F' = 15$ V, $V_F = 1$ V, and $R_S = 10$ Ω. Compute I_A (on) when $V_S = V_F'$. Can you compute I_A for $V_S = 2V_F'$ without using the characteristic?

14. Compute the on resistance of the 2N4987 SUS at rated conditions.

15. An open-gate 2N4987 SUS is operated in the circuit of Figure 8.1. If the switching voltage (V_S) is 8 V, what must R_2 be to limit I_F(max) to 175 mA?

16. Using the equivalent circuit of the SCS (Figure 8.11), explain why the device will fire when the anode voltage exceeds the anode gate voltage by approximately 0.5 to 0.7 V and the cathode gate is open. Also explain the turn-off action of the anode gate.

17. A relaxation oscillator can deliver 6 V across 120 Ω. Will this reliably trigger a C38 SCR?

18. A relaxation oscillator delivers 0.8 A across 7.5 Ω. Will this reliably trigger a C38 SCR?

19. Can a relaxation oscillator with a maximum output pulse voltage of 3 V be used to reliably trigger a C38 SCR? Explain.

20. Relate the wave forms of Figure 8.38b to the operation of the half-wave zero-voltage switching circuit.

21. For the circuit of Figure 8.40, compute the average voltage across the load when $R_1 = 500$ kΩ and 1 MΩ. Assume the forward voltage drops of D_1 and the SCR to be $\cong 0$. Also assume that the voltage across C_1 is 0 at the start of each positive cycle.

22. Convert as requested (a) 1 foot candle to x mW/c^2, (b) 20 mW/c^2 to x foot candles, (c) 100 μm to frequency, (d) 3×10^{12} Hz to micrometers.

23. A tungsten lamp operating at 200 candela, 2200°K, is to be used to trigger a LASCR (4.2 mW/cm^2 is the triggering irradiance). Find the maximum distance from which the tungsten lamp can affect triggering.

24. A tungsten lamp is positioned 1 ft away from a LASCR on its axis. The color temperature of the lamp is 2200°K. What must its brightness in candela be to trigger the LASCR (4.2 mW/cm^2 triggering irradiance)?

25. Determine H_{ET} for a LASCR operating at an anode voltage varying from 50 to 100 V with a gate resistor of 22 kΩ.

26. In problem 25, what must be the range of the gate potentiometer (lowest value 22 kΩ) to keep H_{ET} constant?

27. The resistance of a tungsten lamp increases 12 times from the instant the circuit turns on to the time it reaches a stable operating temperature. What is the color temperature of the lamp?

28. A 3-Ω cold resistance tungsten lamp operates at 2400°K color temperature. What is its hot resistance?

29. An L14G phototransistor is to be driven by a LED 55B IRED in an optocoupler. The phototransistor current is to be held to 10 mA and the IRED current to 80 mA. What should the spacing be between the L14G and the LED 55B?

30. An L14G phototransistor is at 2000 V above ground. It is to be driven by a LED 55B at ground potential. The L14G current is to be held at 12 mA and the LED 55B at 110 mA. What must the insulation rating in volts/m be between the two?

References

Bell, David. *Electronic Devices and Circuits*, 2nd ed., Reston Publishing Co., Reston, Virginia, 1980.

General Electric Semiconductors: Short Form Catalog, General Electric, Auburn, N.Y., 1980.

Graham, O. R., and others. *SCR Manual*, 6th ed., General Electric, Auburn, N.Y., 1979.

Nashelsky, Louis, and Robert Boylestad. *Devices: Discrete and Integrated*, Prentice-Hall, Inc., Englewood Cliffs, New Jersey, 1981.

9

GATES AND DIGITAL CIRCUITS

9.1 INTRODUCTION

The introduction of digital (discrete) circuitry some years ago started a continuing and profound revolution. Digital circuit concepts are fundamentally different from the analog circuits we have studied so far. Familiar devices (diodes, transistors, FETs, op-amps, etc.) are used in basically different ways. As a result, digital circuits extend our technical capabilities remarkably, permitting the realization of circuits functioning in ways heretofore considered impossible or at least impractical.

The most fundamental difference between analog and digital circuits lies in the ways information is coded into the input and output signals. For the analog case, information resides in the instantaneous magnitudes of the signal or, put another way, in the shape of the signal. The analog circuit amplifies or reshapes the signal in some desired way. For example, the encoded information in an audio signal is contained in its instantaneous voltage magnitudes. An analog audio amplifier increases the magnitudes of this input signal in a linear and virtually instantaneous operation, producing an output signal of the same shape. A differentiator, integrator, or clipping circuit also reacts to the instantaneous magnitudes of the input, changing its shape in some predictable way.

On the other hand, information in a digital signal resides not in its shape, but in its levels at discrete intervals of time. A specific period of time must elapse before this information can be retrieved. Usually, the signal levels represent sequences of 1's and 0's decoding into a stream of numbers. The digital circuit acts mathematically on this stream, adding, multiplying, and inverting. Its output is also an encoded stream of numbers.

Memory as an essential element also differentiates the digital circuit from the analog circuit. Memory in the form of capacitors requiring finite discharge times also exists in analog circuits. But there it is usually a secondary element that disturbs the normal instantaneous operation of the circuit. In digital circuits the capability of retaining previous inputs and intermediate results is essential. In fact, at the highest level, all digital circuits can be analyzed into three elements or components, an adder, which adds numbers, a multiplier, which multiplies a number by a constant, and a delay (memory).

Adders, multipliers, delay, and other digital circuits are high-level electronic devices composed primarily of gates. Gates themselves are hardware devices composed of diodes, transistors, FETs, and other ordinary circuit components. In this chapter we study gates at two levels, as abstract ideal devices and as hardware. The ideal device level prepares us for the study of combinational digital circuits leading up to the microcomputer. The hardware level leads to a design capability and an understanding of the limitations imposed by reality. Also, in preparation for succeeding chapters, we study signal coding.

9.2 IDEAL GATES

The ideal gate is an abstract circuit device having one or more input terminals and one output. The input impedances are infinitely large; the output impedance is zero. Input and output signals are either 1's or 0's, where 1's and 0's represent logical states, not voltages. Gates are completely defined by truth tables that list the output states for all combinations of input states.

Figure 9.1 shows the circuit symbols and truth tables for the AND, OR (inclusive), exclusive OR, NOT, NAND, and NOR gates. Note that the AND gate outputs a 1 only when both inputs are 1's. The OR gate outputs a 1 when either or both inputs are 1's. The exclusive OR outputs a 1 when either but not both inputs are 1's. The NOT gate inverts the input ($1 \rightarrow 0$ and $0 \rightarrow 1$), and the NOR and NAND gates can be derived from the OR and AND gates by adding an inverter (NOT gate) to the output.

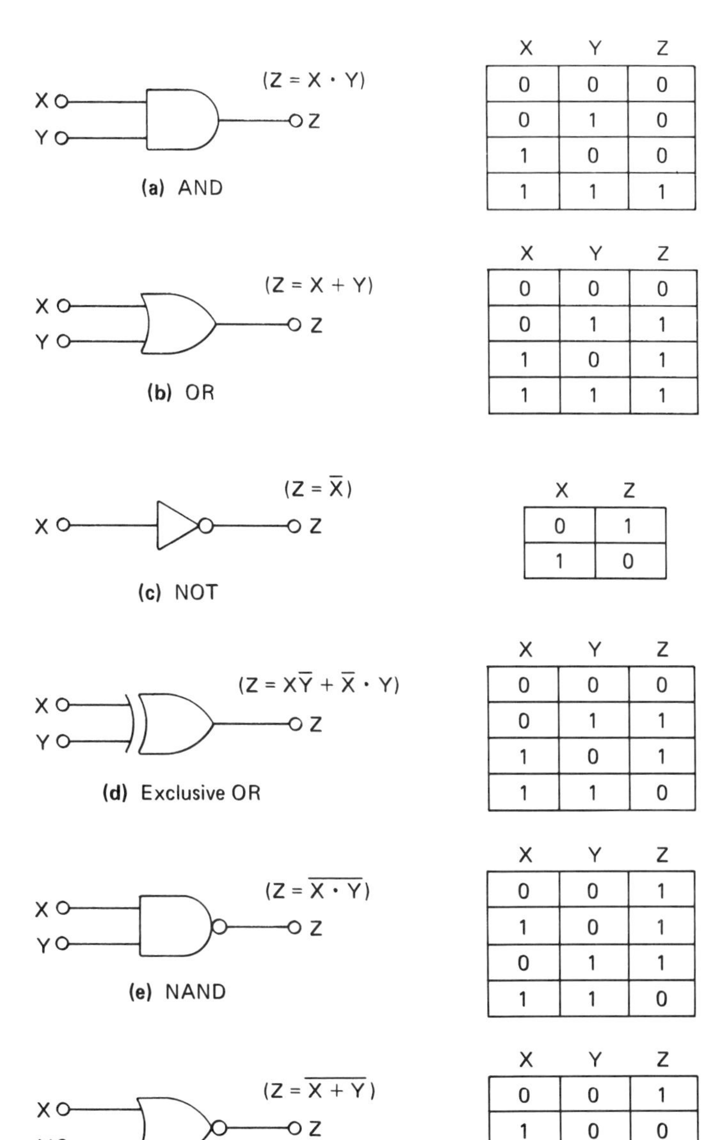

X	Y	Z
0	0	0
0	1	0
1	0	0
1	1	1

X	Y	Z
0	0	0
0	1	1
1	0	1
1	1	1

X	Z
0	1
1	0

X	Y	Z
0	0	0
0	1	1
1	0	1
1	1	0

X	Y	Z
0	0	1
1	0	1
0	1	1
1	1	0

X	Y	Z
0	0	1
1	0	0
0	1	0
1	1	0

FIGURE 9.1 Ideal gates.

9.2.1 Truth Table Analysis of Gate Circuits

Not all the gates of Figure 9.1 are needed to form an independent system. All the functions shown can be obtained from circuits composed solely of NOR gates. (This is also true for circuits composed solely of NAND gates.) Although circuits composed of one type of gate may not appear to be economical, in this era of ICs in which transistors are very inexpensive, there are advantages in having a standardized building block with universal application.

Figure 9.2 illustrates how combinational NOR gate circuits provide all the gate functions. Tying the two inputs together converts the NOR gate to a NOT gate. Feeding the output of a NOR gate into an inverter (NOT gate) gives an OR gate. It is less obvious that the configuration of Figure 9.2c constitutes an AND gate, but this can be established by the procedure illustrated by Table 9.1, which uses combined truth tables.

Table 9.1 shows all possible inputs and outputs for all NOR gates.

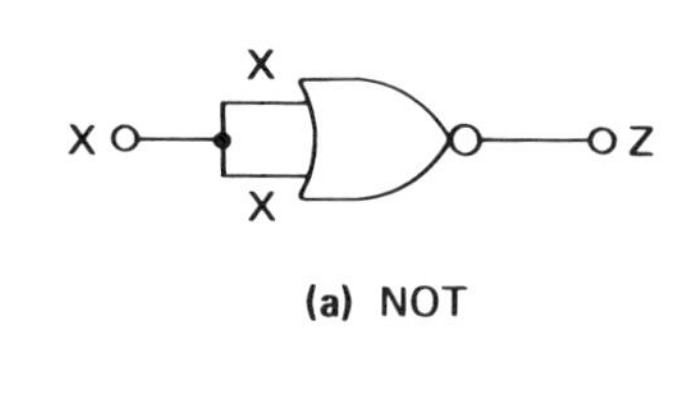

(a) NOT

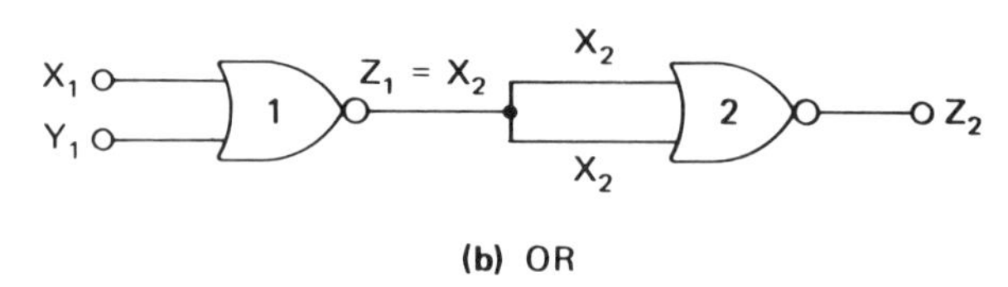

(b) OR

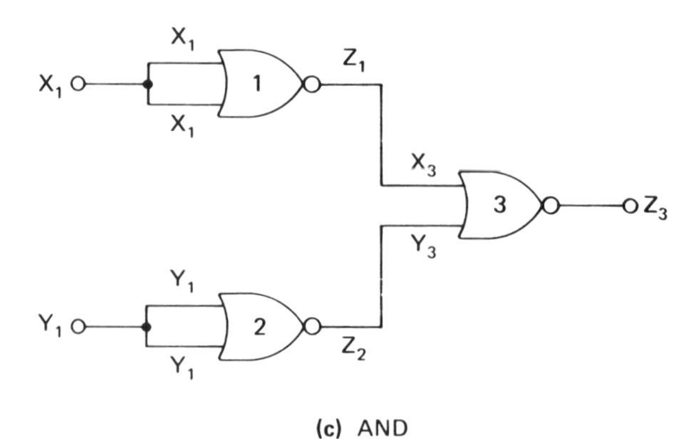

(c) AND

FIGURE 9.2 Implementing logic functions with NOR gates.

TABLE 9.1 *Truth Table for Figure 9.2c*

1	2	3	4	5	6	7
X_1	Y_1	Z_1	Z_2	$X_3 = Z_1$	$Y_3 = Z_2$	Z_3
0	0	1	1	1	1	0
0	1	1	0	1	0	0
1	0	0	1	0	1	0
1	1	0	0	0	0	1

TABLE 9.2 *Overall Truth Table for Figure 9.2c*

1	2	7
X_1	Y_1	Z_3
0	0	0
0	1	0
1	0	0
1	1	1

Also, the table relates the output of preceding gates to their successors (cascaded gates). Z_1 is affected only by X_1, and Z_2 only by Y_1, since they are both inverters. This is shown by columns 1 through 4. Column 5 reflects the fact that the output of 1 (Z_1) is the X_3 input of 3. Column 6 does the same for 2 (Z_2). Columns 5, 6, and 7 constitute the truth table for the NOR gate (3), given X_3 and Y_3. Combining the input columns X_1 and Y_1 with the output column Z_3 gives the overall input/output (I/O) truth table shown in Table 9.2. This is seen to be the AND gate truth table.

Example 9.1 illustrates how this procedure can be used to find the I/O truth table for any combinational gate circuit. In this example, inverters and AND gates are used in addition to OR gates, but in practice the circuit could be entirely composed of NOR gates.

Example 9.1 *Analysis of a Gate Circuit*

Find the I/O truth table for the gate circuit of Figure 9.3 and identify the truth table if possible.

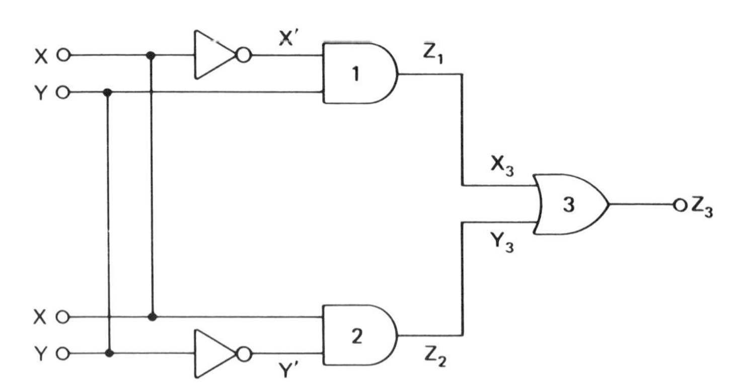

FIGURE 9.3 Circuit for Example 9.1 (exclusive OR gate).

(a) *Set up the combined truth table.*

1	2	3	4	5	6	7	8	9
X	Y	X'	Y'	Z_1	Z_2	$X_3 = Z_1$	$Y_3 = Z_2$	Z_3
1	0	0	1	0	1	0	1	1
0	1	1	0	1	0	1	0	1
0	0	1	1	0	0	0	0	0
1	1	0	0	0	0	0	0	0

(b) *Set up an I/O table (columns 1, 2 and 9).*

1	2	9
X	Y	Z_3
1	0	1
0	1	1
0	0	0
1	1	0

The I/O truth table represents the exclusive OR gate.

9.2.2 Boole-Schroeder Analysis of Gate Circuits

Gate circuits may be analyzed more efficiently using Boole–Schroeder analysis (Boolean algebra). In Boolean algebra, the variables are restricted to values of either 1 or 0. All the conventional laws of simple algebra remain unchanged except for the addition of 1's. This becomes

$$1 + 1 = 1 \tag{9.1}$$

Division is not defined and two additional terms are defined:

$$\bar{0} = 1 \tag{9.2}$$

and

$$\bar{1} = 0 \tag{9.3}$$

where the overbar ($\bar{\ }$) means "the inversion of." Table 9.3 summarizes some simple Boolean operations.

The theorems of Boolean algebra are easily proved by substituting the range of values (0 or 1) for the variables. Some theorems and their proofs are given next.

Commutative law:

$$X + Y = Y + X \tag{9.4}$$

Proof:

$0 + 0 = 0 + 0\ (X = 0, Y = 0)$,
$0 + 1 = 1 + 0\ (X = 0, Y = 1)$
$1 + 0 = 0 + 1\ (X = 1, Y = 0)$
$1 + 1 = 1 + 1 = 1\ (X = 1, Y = 1)$

Associative laws:

$$(X + Y) + Z = X + (Y + Z) = X + Y + Z \tag{9.5}$$

Proof:

$(0 + 0) + 0 = 0 + (0 + 0) = 0 + 0 + 0 = 0\ (X, Y, Z = 0)$
$(1 + 0) + 0 = 1 + (0 + 0) = 1 + 0 + 0 = 1\ (X = 1, Y, Z = 0)$
etc.

$$(X \cdot Y) \cdot Z = X \cdot (Y \cdot Z) = X \cdot Y \cdot Z \tag{9.6}$$

Proof:

$(0 \cdot 0) \cdot 0 = 0 \cdot (0 \cdot 0) = 0 \cdot 0 \cdot 0 = 0\ (X, Y, Z = 0)$
$(1 \cdot 0) \cdot 0 = 1 \cdot (0 \cdot 0) = 1 \cdot 0 \cdot 0 = 0\ (X = 1, Y, Z = 0)$
etc.

TABLE 9.3 *Sample of Boolean Operations*

$0 + 0 = 0$	$1 \times 1 = 1$
$1 + 1 = 1$	$0 \times 0 = 0$
$0 + 1 = 1 + 0 = 1$	$1 \times 0 = 0 \times 1 = 0$
$\bar{0} = 1$	$\bar{1} = 0$

Distributive laws:

$$X \cdot (Y + Z) = X \cdot Y + X \cdot Z \tag{9.7}$$

Proof:

$0 \cdot (0 + 0) = 0 \cdot 0 + 0 \cdot 0 = 0\,(X, Y, Z = 0)$
$1 \cdot (1 + 1) = 1 \cdot 1 + 1 \cdot 1 = 1\,(X, Y, Z = 1)$
etc.

$$X + Y \cdot Z = (X + Y) \cdot (X + Z) \tag{9.8}$$

Note this is not valid for ordinary algebra.

Proof:

$0 + 0 \cdot 0 = (0 + 0) \cdot (0 + 0) = 0\,(X, Y, Z = 0)$
$1 + 1 \cdot 1 = (1 + 1) \cdot (1 + 1) = 1\,(X, Y, Z = 1)$
(note $1 + 1 = 1$)
etc.

Two more theorems known as DeMorgan's laws are important.

$$\overline{(X + Y)} = \overline{X} \cdot \overline{Y} \tag{9.9}$$

$$\overline{(X \cdot Y)} = \overline{X} + \overline{Y} \tag{9.10}$$

We leave these proofs to the student (problems 5 and 6).

Boolean algebra permits us to express the truth tables of Figure 9.1 analytically. The AND gate is given by the Boolean expression

$$Z = X \cdot Y \tag{9.11}$$

For either $X = 0$ or $Y = 0$ or both, the expression generates $Z = 0$. Only for $X, Y = 1$ does Eq. (9.11) give $Z = 1$. These results duplicate the AND gate truth table. The Boolean expression for the OR gate is given by

$$Z = X + Y \tag{9.12}$$

For either X or Y or both equal to 1, Eq. (9.12) gives $Z = 1 \cdot Z = 0$ only when both X and $Y = 0$. This is the OR gate truth table. NOR and NAND gates are given, respectively, by

$$Z = \overline{X + Y} \quad \text{(NOR gate)} \tag{9.13}$$

and

$$Z = \overline{X \cdot Y} \quad \text{(NAND gate)} \tag{9.14}$$

The inverter truth table is given by

$$Z = \overline{X} \tag{9.15}$$

and the exclusive OR by

$$Z = X \cdot \overline{Y} + \overline{X} \cdot Y \qquad (9.16)$$

These expressions are included (in parentheses) in Figure 9.1.

To illustrate the analysis of gate circuits by Boolean algebra, Example 9.1 is repeated analytically (Example 9.2).

Example 9.2 *Boolean: Analysis of a Gate Circuit*

Find the I/O truth table for the gate circuit of Figure 9.3 using Boolean algebra. Identify the truth table if possible. Set up the Boolean expression:

Working backward from Z_3,

$$Z_3 = X_3 + Y_3 = Z_1 + Z_2$$

But

$$Z_1 = X' \cdot Y \quad \text{and} \quad Z_2 = X \cdot Y'$$

Therefore,

$$Z_3 = X' \cdot Y + X \cdot Y'$$

or since $X' = \overline{X}$ and $Y' = \overline{Y}$

$$Z_3 = \overline{X} \cdot Y + X \cdot \overline{Y}$$

This is an OR gate. Its truth table is given in Figure 9.1b.

9.2.3 Design of Gate Circuits

The design of gate circuits may be accomplished in either of two ways. In the first case, a desired truth table is established and expressed as a Boolean equation. This equation is then reduced to its simplest form and translated into a gate circuit. Often this simplification is complicated and dependent on experience. The second procedure uses a representation called a Karnaugh map, which leads to a simpler and more direct reduction. We give examples of both procedures.

Example 9.3 *Gate Circuit Design Using Boolean Analysis*

For this procedure, it is useful to have a table of Boolean relations. This is given in Table 9.4. All relations can be proved using truth tables.

Design a gate circuit with the following truth table.

	X	Y	Z
1	1	0	0
2	0	1	1
3	1	1	1
4	0	0	1

(a) We express this table as a sum of products of X and Y operating on one row at a time. When X or Y have the value 1, they appear in the product term as X or Y. When they equal 0, they appear as $\overline{X}$ or $\overline{Y}$. Only the non-zero Z terms are summed.

Row	*Relationship*	*Term*
1	$1 \cdot 0 = 0$	$X \cdot \overline{Y}$
2	$0 \cdot 1 = 1$	$\overline{X} \cdot Y$
3	$1 \cdot 1 = 1$	$X \cdot Y$
4	$0 \cdot 0 = 1$	$\overline{X} \cdot \overline{Y}$

Therefore, summing rows 2, 3, and 4 gives

$$Z = \overline{X} \cdot Y + X \cdot Y + \overline{X} \cdot \overline{Y}$$

(The student should verify this expression using a truth table.)

(b) *Simplification:*

$$\begin{aligned} Z &= Y \cdot (X + \overline{X}) + \overline{X} \cdot \overline{Y} \qquad \text{(Theorem 3a)} \\ &= Y + \overline{X} \cdot \overline{Y} \qquad (\text{since } X + \overline{X} = 1, \text{Theorem 6a}) \\ &= Y + \overline{X} \qquad (X + \overline{X} \cdot Y = X + Y, \text{Theorem 9a}) \end{aligned}$$

(c) *Transformation into a gate circuit:* The circuit is seen to be an OR gate with a direct input (Y) and an inverted input ($\overline{X}$). This circuit is shown in Figure 9.4.

Gate Circuit Design Using the Karnaugh Map. The Karnaugh map provides a visual guide for the simplification of logical equations. Figure 9.5

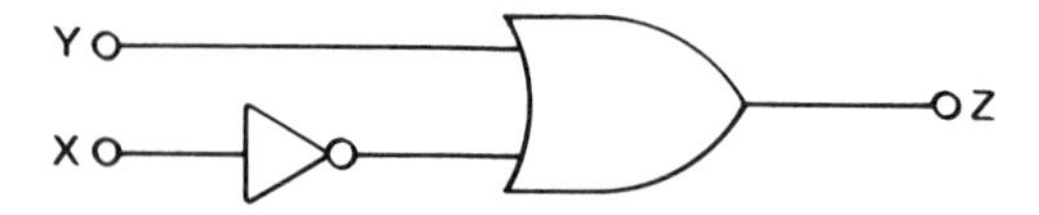

FIGURE 9.4 Gate circuit of Example 9.3.

TABLE 9.4 *Theorems of Boolean Algebra*

No.	*Theorem*	*No.*	*Theorem*
1a	$X + Y = Y + X$	7a	$X + X \cdot Y = X$
1b	$X \cdot Y = Y \cdot X$	7b	$X \cdot (X + Y) = X$
2a	$(X + Y) + Z = X + (Y + Z)$	8a	$0 + X = X$
2b	$(X \cdot Y) \cdot Z = X \cdot (Y \cdot Z)$	8b	$1 \cdot X = X$
3a	$X \cdot (Y + Z) = X \cdot Y + X \cdot Z$	8c	$1 + X = 1$
3b	$X + (Y \cdot Z) = (X + Y) \cdot (X + Z)$	8d	$0 \cdot X = 0$
4a	$\overline{X} = \overline{X}$	9a	$X + \overline{X} \cdot Y = X + Y$
4b	$\overline{\overline{X}} = X$	9b	$X \cdot (\overline{X} + Y) = X \cdot Y$
5a	$X + X = X$	10a	$\overline{X} + \overline{Y} = \overline{X \cdot Y}$
5b	$X \cdot X = X$	10b	$\overline{X} \cdot \overline{Y} = \overline{X + Y}$
6a	$\overline{X} + X = 1$		
6b	$\overline{X} \cdot X = 0$		

shows a Karnaugh map for a two-variable truth table. The two possible values of X are listed alongside their respective rows, and the two values of Y, above their respective columns. The map contains four cells, one for each row of the corresponding truth table. In each cell a product term is placed. The term is determined by assuming X and Y both equal to 1 and inverting as needed. For example, the term in the upper-left cell is the product of $X = 0$ and $Y = 0$. Since X and $Y = 1$, $0 \cdot 0$ is given by $\overline{X} \cdot \overline{Y}$. Similarly, the upper-right cell is the product of $X = 0$, $Y = 1$, or $\overline{X} \cdot Y$, and so on for the rest of the cells. These terms are invariant and constitute the name of each cell.

The unsimplified sum-of-products expression is obtained by placing a 1 in each cell for which a 1 appears in the corresponding row of the truth table and ORing (+) the cell names. For example, using the truth table of Figure 9.5b, the Boolean expression is given by

$$Z = \overline{X} \cdot Y + X \cdot \overline{Y} + X \cdot Y \tag{9.17}$$

The simplified (minimal) expression is obtained by pairing adjacent cells either horizontally or vertically. (Cells may overlap.) Each horizontal pair is, in our layout, a simple X, each vertical pair a single Y.

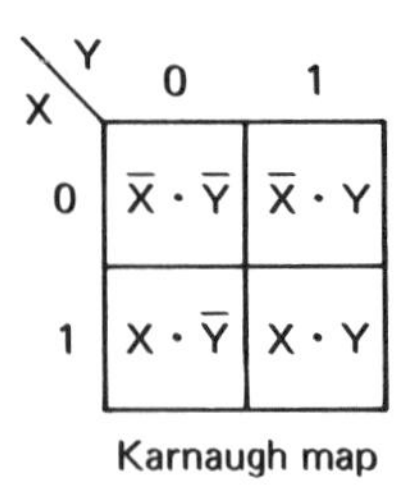

X	Y	Cell term
0	0	$\overline{X} \cdot \overline{Y}$
0	1	$\overline{X} \cdot Y$
1	0	$X \cdot \overline{Y}$
1	1	$X \cdot Y$

Truth table

(a) Cell terms

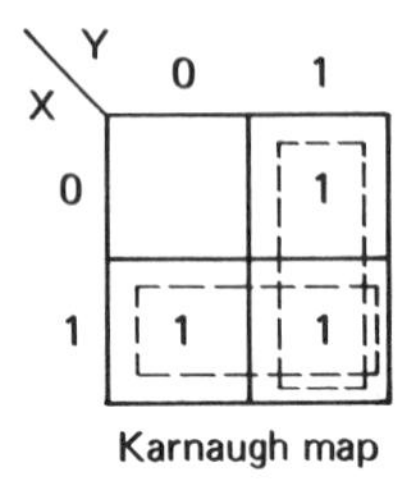

X	Y	Z
0	0	0
0	1	1
1	0	1
1	1	1

Truth table

Raw expression

$$Z = \overline{X} \cdot Y + X \cdot \overline{Y} + X \cdot Y$$

Minimal expression (from pairs)

$$Z = X + Y \text{ (OR gate)}$$

(b)

FIGURE 9.5 Karnaugh mapping.

The expression is the sum (ORing) of the pairs, in this case, an OR gate ($Z = X + Y$).

Example 9.4 *Gate Design Using Karnaugh Mapping*

Using the truth table of Example 9.3, obtain the minimal logical expression.

(a) Draw the Karnaugh map and its corresponding truth table (Figure 9.6a, b). Encircle adjacent pairs (Figure 9.6a).

(b) The horizontal pair on the $X = 0$ row gives the component $\overline{X}$. The vertical pair on the $Y = 1$ column gives the component (Y). Therefore, $Z = \overline{X} + Y$.

In addition to pairs, Karnaugh mapping may yield single cells. A single cell will produce a two-variable term. Example 9.5 illustrates such a case.

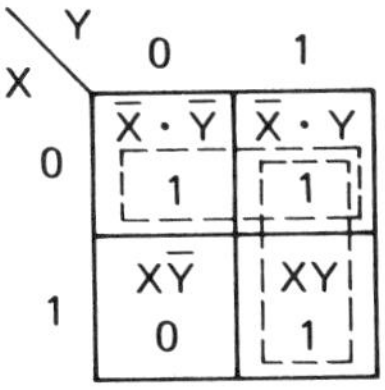

X \ Y	0	1
0	$\bar{X}\cdot\bar{Y}$ 1	$\bar{X}\cdot Y$ 1
1	$X\bar{Y}$ 0	XY 1

(a) Karnaugh map

X	Y	Z
0	0	1
0	1	1
1	0	0
1	1	1

(b) Truth table

$$Z = \bar{X} + Y$$

(c) Minimal expression

FIGURE 9.6 Example 9.4.

Example 9.5 *Design of a Gate Circuit (Karnaugh Mapping)*

Design a gate circuit with the following truth table

X	Y	Z
0	0	0
1	0	1
0	1	1
1	1	0

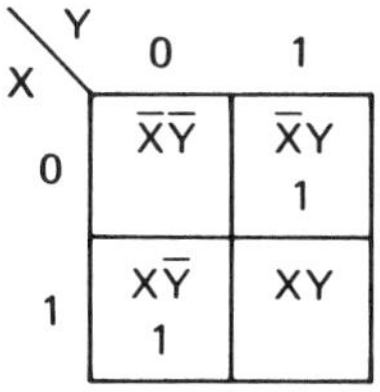

X \ Y	0	1
0	$\bar{X}\bar{Y}$	$\bar{X}Y$ 1
1	$X\bar{Y}$ 1	XY

(a) Karnaugh map

X	Y	Z
0	0	0
1	0	1
0	1	1
1	1	0

(b) Truth table

$Z = X\bar{Y} + \bar{X}Y$ (exclusive OR gate)

(c) Minimal expression

FIGURE 9.7 Example 9.5.

(a) Draw the truth table and its corresponding Karnaugh map (Figure 9.7).

(b) Note there are no adjacent pairs. Instead there are two irreducible single cells. Therefore,

$$Z = X\overline{Y} + \overline{X}Y$$

an exclusive OR gate.

9.3 GATE HARDWARE

The physical (hardware) gate used to construct the gates discussed in the preceding section are almost always etched into integrated-circuit chips. Over the years, three families (types) have emerged: transistor–transistor logic (TTL), emitter-coupled logic (ECL), and complementary symmetry MOS logic (CMOS or COS/MOS). TTL logic is the most popular at present because it satisfies the requirements for low power drain and is fast enough for most applications. The CMOS family dissipates the least power but is significantly slower than the rest. The ECL family is the fastest of the three, but also dissipates the most power. Usually, a logic system employs only gates from a single family, but occasionally families may be mixed.

9.3.1 Common Specifications

Specifications for all gates, regardless of family type, include maximum and minimum output levels, which will be recognized reliably as either a logic 1 or 0, susceptibility to noise pulses and the number of like family gates that may be driven. These common specifications, called logic levels, noise margin, and fan-out, respectively, are defined next.

Maximum and Minimum Voltage Levels

1. V_{IH} minimum gate input voltage, which will be reliably recognized as a logic 1 or HIGH.
2. V_{IL} maximum gate input voltage, which will be reliably recognized as a logic 0 or LOW.
3. V_{OH} minimum voltage at gate output when output is supposed to be a logic 1.
4. V_{OL} maximum voltage at gate output when output is supposed to be a logic 0.

The manufacturer in effect guarantees that there will be a region between V_{OH} and V_{OL} in which the gate will never operate, provided the input voltage (V_i) is above V_{IH} for an intended logic 1 and below V_{IL} for a logic 0.

Noise Margin. A practical gate must offer protection against impulse noise, which can invert logics and introduce serious errors. The noise margin of a gate is expressed in terms of the effects of a preceding gate. For example, for a gate to recognize the low output of a preceding gate (V_{OL}) as a low, this output must be below V_{IL} of the second (following) gate. The degree of protection, or *noise margin* (NM_L), depends on how much lower V_{OL} is, and is given by

$$NM_L = V_{IL} - V_{OL} \tag{9.18}$$

Obviously, if V_{IL} is very close to V_{OL}, a small positive noise pulse on V_{IL} will cause its output to be recognized as a logic 1 rather than a logic 0. Similarly, the high output of the driving gate should be far enough above V_{IH} to prevent errors caused by negative noise pulses. The high noise margin (NM_H) is given by

$$NM_H = V_{OH} - V_{IH} \tag{9.19}$$

The higher the noise margin, the less susceptible is the gate to noise.

Fan-Out. Typically, a gate drives other gates. The number of gates of the same family that can be reliably driven is called the *fan-out* of the driver gate. Fan-out is determined by the tolerable values of V_{IL}, I_{IL}, V_{IH}, I_{IH}, V_{OL}, I_{OL}, V_{OH}, and I_{OH}. When the driver gate is high, it must supply current to the driven gates. The greater the number of driven gates, the higher the output current (I_{OH}) from the driver. This current produces an internal voltage drop in the driver, decreasing V_{OH} and may reduce the noise margin below the manufacturer's specifications. A lower limit on V_{OH} therefore establishes an upper limit on fan-out. When the driver gate is low, the current out of the driven gates passes through the driver gate to ground. This current raises V_{OL}, which again decreases the noise margin. The maximum current the driver gate can pass (sink) establishes another limit on fan-out.

9.3.2 Transistor-Transistor Logic (TTL)

Figure 9.8 shows a basic TTL gate. The configuration is simple, requiring only two transistors (one possessing two or more emitter leads) and two resistors. This simple configuration occupies very little area on a chip (real estate) and permits the etching of many gates into a small chip. Also

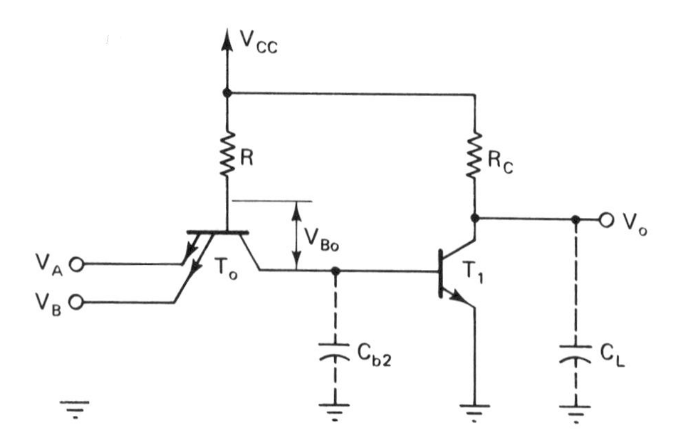

FIGURE 9.8 Basic TTL gate.

shown by the dotted lines are the unavoidable stray capacities, which appear even though they are not etched and are unwanted.

The gate depicted is a NAND gate and performs the following. When either V_A or V_B is LOW (0.2 V), the base–emitter junction of T_O is forward biased. Current flows from V_{CC} through R into V_A and/or V_B. The base-to-ground voltage (V_{BO}) is given by

$$V_{BO} = V_{BE} + V' \tag{9.20}$$

where V' is either V_A or V_B, both assumed to be 0.2 V. Therefore, assuming V_{BE} to be about 0.7 V,

$$V_{BO} = 0.7\,\text{V} + 0.2\,\text{V} = 0.9\,\text{V} \tag{9.21}$$

In this state, T_O is driven to saturation, making its collector-to-emitter voltage (V_{CE}) about 0.2 V. (This is the typical V_{CE} for a saturated transistor.) Since V_{BE} of T_1 equals V_{CE} of T_O plus V', then

$$V_{BE}(T_1) = V_{CE}(T_O) + V' = 0.2\,\text{V} + 0.2\,\text{V} = 0.4\,\text{V} \tag{9.22}$$

and T_1 is cut off. A cutoff T_1 outputs a HIGH. When both inputs (V_A and V_B) are HIGH (equal to V_{CC} and typically 5 V), $V_{BE}(T_O)$ is reversed biased and no emitter current flows. However, because the base-to-ground voltage of T_O [$V_{BO}(T_O)$] attempts to rise to V_{CC} while the maximum base-to-ground voltage of T_1 is about 0.7 V, the base–collector circuit of T_1 becomes forward biased. A relatively high current flows into the base of T_1, saturating T_1 and outputting a LOW.

One advantage of the TTL gate over gates previously used arises from its ability to quickly remove the base charge of T_1 when T_1 is switching from LOW to HIGH. Recall that this is called *turn-off time* and is a significant limitation on switching speeds. When all inputs drop from HIGH to LOW, $V_{BE}(T_O)$ becomes strongly forward biased. With

$V_C(T_O)$ equal to $V_{BO}(T_1)$ equal to 0.7 V, transistor T_O is in the active on state. In this state I_L, the collector current of T_O, is large and rapidly discharges the base of T_1.

The remaining limitation on switching speed is the capacitive load on T_1. Before V_O can rise to V_{CC} (T_1 off), C_L must be charged through R_C. With typical values, $V_{CC} = 5$ V, $R_C = 1$ kΩ, and $C_L = 5$ pF, rise times are limited to about 5.2 ns, an unacceptable delay. Decreasing R_C will increase rise times but at the expense of increased power dissipation when V_O is low. A better solution is to replace R_C by an active resistance that is large when the output is LOW and becomes small when C_L is being charged. This action is called *active pull-up*. Figure 9.9 shows a two-input NAND gate incorporating this circuit. This circuit is developed from Figure 9.8 by interposing a phase splitter (T_3) between T_O and T_1 and replacing R_C of Figure 9.8 by T_2.

In this circuit T_2 acts as an active resistor switched on or off by the phase splitter T_3. When both inputs are high, I_{CO} flows into the base of T_3, driving it into saturation. In turn, the voltage across R_2 drives T_1 into saturation, giving a low V_O. During this period, V_{B2} (base-to-ground voltage of T_2) is low because T_3 is saturated. Therefore, T_2 is off or almost off. T_2 then looks like a high resistance and draws little current from V_{CC}. When one or both inputs switch to LOW, I_{CO} diverts from T_3 to the two emitters of T_O. T_3 turns off, T_1 turns off, and since V_{B2} rises almost to V_{CC}, T_2 turns on. C_L is now charged through T_2 (effectively a low-impedance

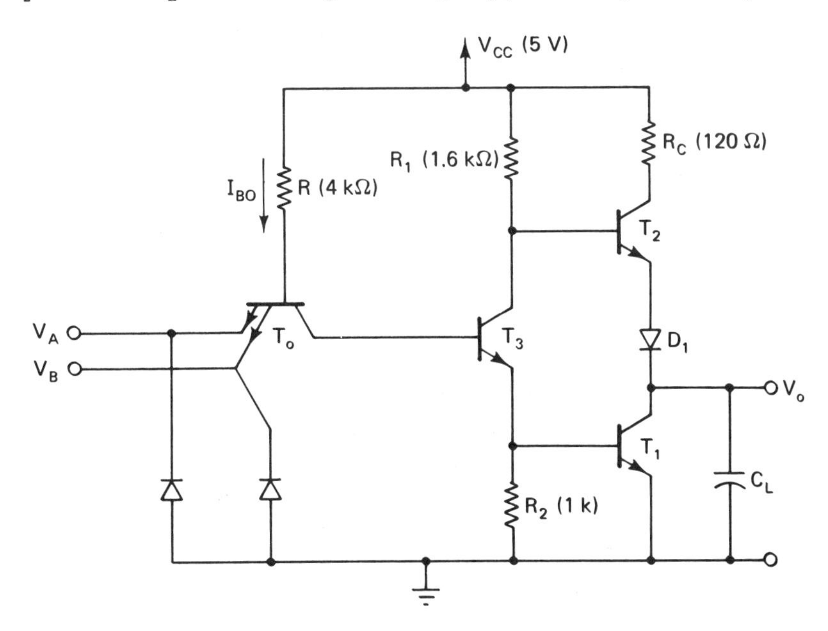

FIGURE 9.9 NAND gate with active pull-up.

source) and very rapidly reaches maximum value (about 3.6 V). V_O is now HIGH. Decreases in rise times and propagation delays by a factor of 30 over the circuit of Figure 9.8 can be realized.

Manufacturers Specification. Figure 9.10 shows a manufacturer's specification sheet for two TTL NAND gates, SN5400 and SN7400. These gates are available in two packages, a flat package and a dual-in-line package. Each package contains four dual input gates. Both types operate from a supply voltage of 5 V and each gate has a fan-out of 10, provided they are driving other 5400/7400 gates.

The manufacturer supplies the following electrical characteristics: input voltages V_{IH} and V_{IL}. V_{IH} is the minimum input voltage required at both input terminals to give a logical 0. V_{IL} is maximum input voltage for a logical 1. The range of variation of output voltages is given, as is the currents required at the input terminals. Output current for the short-circuit condition (output to ground) is given. (These gates have a short-circuit current limiter). Also given are the logical 0 and 1 supply currents, this characteristic being important to the design of multiple-gate circuits.

Propagation delay, an unavoidable characteristic, is specified under switching characteristics. Because each transistor in the gate circuit has a finite switching time, the output of the gate lags the input. Figure 9.11 shows how these delays are measured using simplified trapezoidal output wave forms. The high- to low-level delay (t_{pHL}) is measured from the 50% point of the input to the 50% point of the output. The low- to high-level delay (t_{pLH}) is measured similarly. These propagation delays set an upper limit on switching speed.

9.3.3 Emitter-coupled Logic (ECL)

The ECL gate is basically a difference amplifier with one input returned to a reference voltage. Propagation delays are minimized because neither transistor is driven into saturation. There are some disadvantages. The voltage range (V_{OH} to V_{OL}) is only 0.8 V instead of the 3-V range of the TTL. Also, the ECL gate dissipates more power. Figure 9.12 shows a simplified, one-input ECL gate.

The circuit of Figure 9.12 is a difference amplifier designed so that when T_1 is *on* T_2 is *off*, and vice versa. In addition the values of V_{CC}, R_{C2}, and R_{C1} are chosen so that the *on* transistor is not saturated. Typically, V_I swings from -0.4 V to $+0.4$ V, -0.4 V turning T_1 off and T_2 on and $+0.4$ V turning T_1 on and T_2 off. Two outputs are available. V_{O2} gives an output in phase with the input (in-phase output), and V_{O1} gives an out-of-phase output.

A commercial ECL gate provides multiple inputs by adding tran-

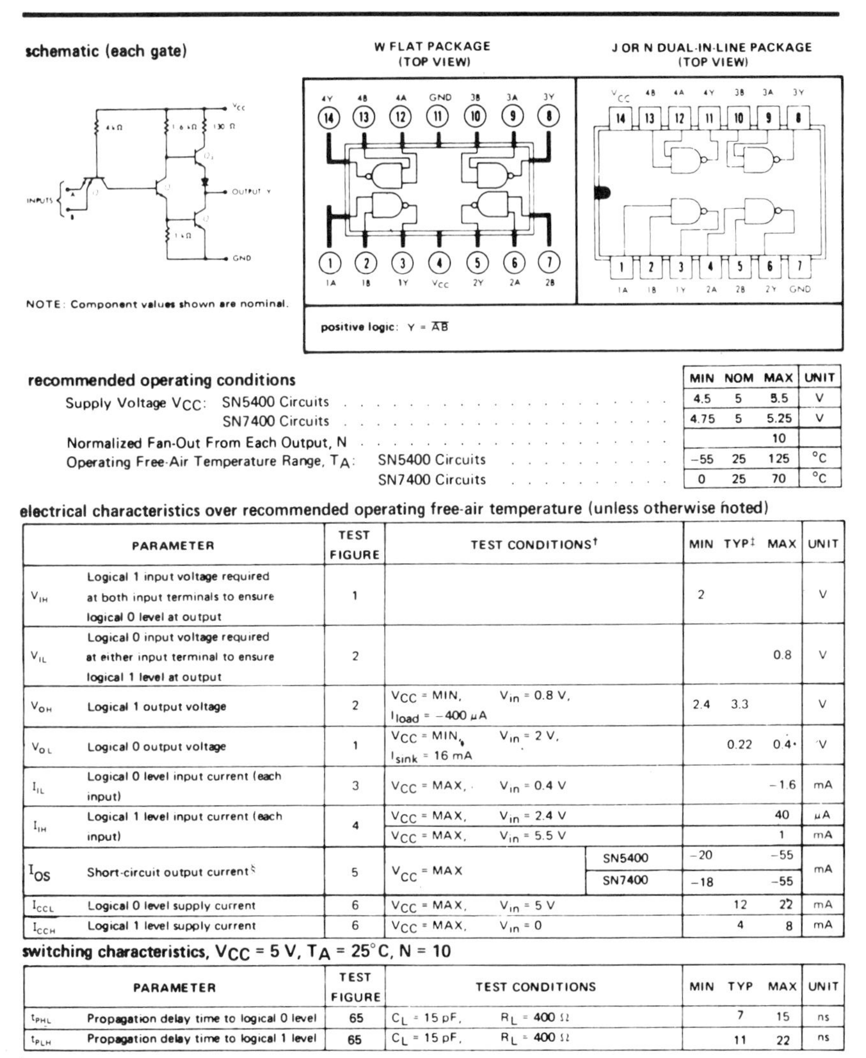

CIRCUIT TYPES SN5400, SN7400
QUADRUPLE 2-INPUT POSITIVE NAND GATES

recommended operating conditions

	MIN	NOM	MAX	UNIT
Supply Voltage V_{CC}: SN5400 Circuits	4.5	5	5.5	V
SN7400 Circuits	4.75	5	5.25	V
Normalized Fan-Out From Each Output, N			10	
Operating Free-Air Temperature Range, T_A: SN5400 Circuits	−55	25	125	°C
SN7400 Circuits	0	25	70	°C

electrical characteristics over recommended operating free-air temperature (unless otherwise noted)

	PARAMETER	TEST FIGURE	TEST CONDITIONS†		MIN	TYP‡	MAX	UNIT
V_{IH}	Logical 1 input voltage required at both input terminals to ensure logical 0 level at output	1			2			V
V_{IL}	Logical 0 input voltage required at either input terminal to ensure logical 1 level at output	2					0.8	V
V_{OH}	Logical 1 output voltage	2	V_{CC} = MIN, V_{in} = 0.8 V, I_{load} = −400 μA		2.4	3.3		V
V_{OL}	Logical 0 output voltage	1	V_{CC} = MIN, V_{in} = 2 V, I_{sink} = 16 mA			0.22	0.4	V
I_{IL}	Logical 0 level input current (each input)	3	V_{CC} = MAX, V_{in} = 0.4 V				−1.6	mA
I_{IH}	Logical 1 level input current (each input)	4	V_{CC} = MAX, V_{in} = 2.4 V				40	μA
			V_{CC} = MAX, V_{in} = 5.5 V				1	mA
I_{OS}	Short-circuit output current§	5	V_{CC} = MAX	SN5400	−20		−55	mA
				SN7400	−18		−55	
I_{CCL}	Logical 0 level supply current	6	V_{CC} = MAX, V_{in} = 5 V			12	22	mA
I_{CCH}	Logical 1 level supply current	6	V_{CC} = MAX, V_{in} = 0			4	8	mA

switching characteristics, V_{CC} = 5 V, T_A = 25°C, N = 10

	PARAMETER	TEST FIGURE	TEST CONDITIONS	MIN	TYP	MAX	UNIT
t_{PHL}	Propagation delay time to logical 0 level	65	C_L = 15 pF, R_L = 400 Ω		7	15	ns
t_{PLH}	Propagation delay time to logical 1 level	65	C_L = 15 pF, R_L = 400 Ω		11	22	ns

† For conditions shown as MIN or MAX, use the appropriate value specified under recommended operating conditions for the applicable device type.
‡ All typical values are at V_{CC} = 5 V, T_A = 25°C.
§ Not more than one output should be shorted at a time.

FIGURE 9.10 5400/7400 data sheet. Courtesy of Texas Instruments.

sistors in parallel with T_1 and including emitter-follower output stages for level shifting and for low output impedances. The two outputs provide either OR or NOR logics. The reference voltage is usually set at -1.3 V and V_{EE} at -5.2 V. This provides a HIGH output of approxi-

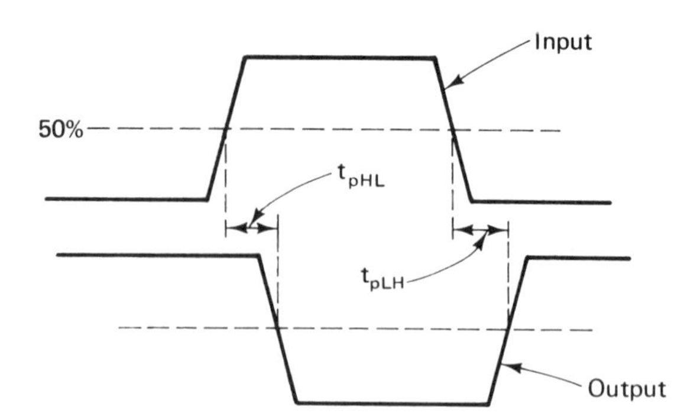

FIGURE 9.11 Propagation delays in the TTL gate.

mately −0.8 to 0.9 V and a LOW of about −1.7 V. This swing is symmetrical about 1.3 V. Figure 9.13 shows a typical ECL OR/NOR gate.

Manufacturer's specification sheets for the ECL gate are similar to those for the TTL. Propagation delays range from 1 to 7 ns for the ECL as compared to 7 to 15 ns for the TTL. ECL gates are available as dual-input, single-output quads (four in a case) in 14-pin ceramic packages.

9.3.4 Complementary MOSFET Logic (CMOS)

MOSFET logics have several advantages and one disadvantage relative to transistor logics. On the negative side, they are considerably slower (at present) than either TTL or ECL. On the positive side, because of their

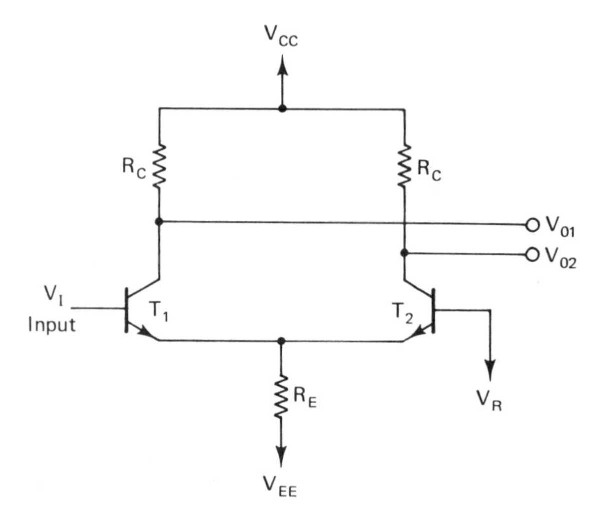

FIGURE 9.12 Simplified ECL gate (one input).

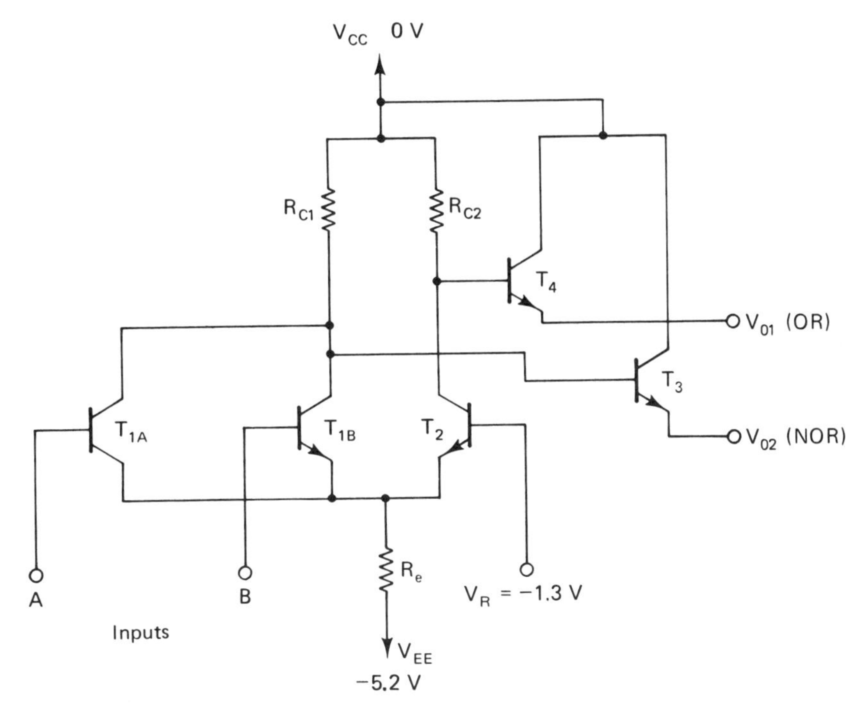

FIGURE 9.13 Typical commercial OR/NOR gate.

simpler geometry and smaller physical size, they can be packed more densely into a chip than the transistor logics. Thousands of MOSFETs can be contained in a fraction of a square inch. Because MOSFETs are voltage-driven devices requiring virtually no input power, fan-out is much higher than for the transistor logics. Also, dc power dissipation and consequently heat generation are extremely low. This is an important consideration when MOSFETs are densely packed.

A basic CMOS NAND gate is shown in Figure 9.14. It consists of two *p*-channel FETs in parallel connected to two *n*-channel FETs in series. The *p*-channel FET requires a negative gate-to-drain voltage exceeding its threshold voltage to turn it on. The *n*-channel FET requires a positive gate-to-drain voltage to turn it on. When either or both inputs are LOW, one or both *p*-channel FETs are turned on and one or both *n*-channel FETs, off. A low impedance exists between V_{SS} and V_O and a high impedance between V_O and ground. V_O is therefore HIGH. For V_O to drop to LOW, both inputs must be HIGH. In this state, both *p*-channel FETs are off and both *n*-channel FETs are on. A high impedance exists between V_{SS} and V_O and a low impedance from V_O to ground. V_O is LOW. Interpreting HIGH as logic 1 and LOW as logic 0 gives the NAND gate.

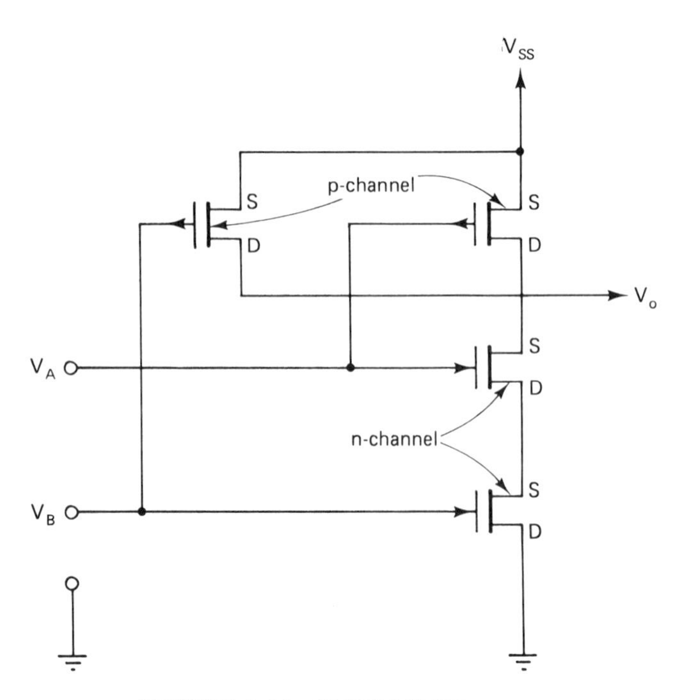

FIGURE 9.14 CMOS NAND gate.

The configuration shown in Figure 9.15 gives a CMOS NOR gate. In this case the two *p*-channel FETs are in series and the two *n*-channel FETs are in parallel. A LOW input (both A and B) turns both *p*-channel FETs on and both *n*-channel FETs off. This provides a low impedance from V_{SS} to V_O and a high impedance from V_O to ground. V_O is HIGH. If only one input is LOW and the other HIGH, a high impedance exists from V_{SS} to V_O and a low impedance from V_O to ground. V_O is LOW. For HIGH equal to logic 1 and LOW equal to logic 0, the truth table of the NOR gate is generated.

Figures 9.14 and 9.15 reveal another advantage of the CMOS gate over transistors relative to chip density. The CMOS gates require no resistors, while the transistor gates do. Etched resistors occupy more space than etched FETs and also dissipate power. The CMOS gates achieve high efficiency by using FETs as resistors, which are switched from high to low as needed.

Manufacturer's Specifications. Manufacturer's data for four types of COS/MOS (CMOS) NOR gates are shown in Figure 9.16. These are either dual or triple input gates available in packages containing two, three, or four gates. These specifications are similar to those provided for the TTL and ECL gates. Notice that, whereas the input current specification for

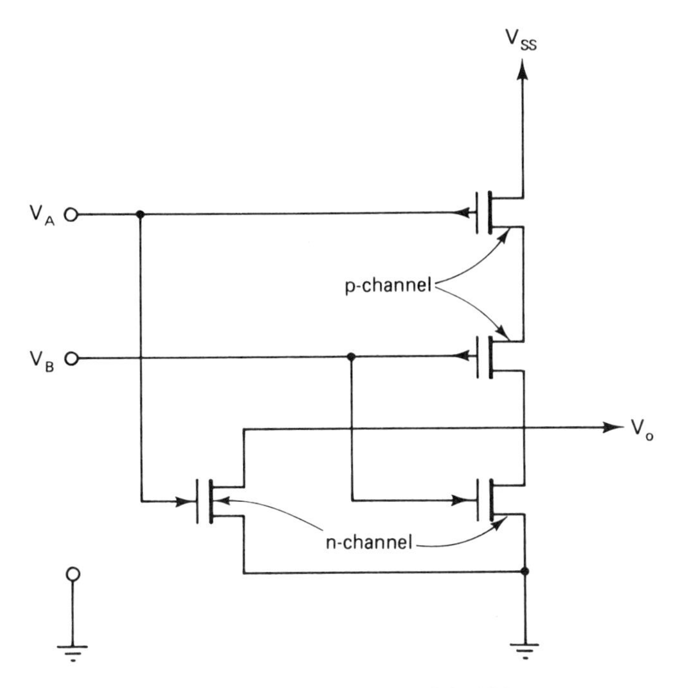

FIGURE 9.15 CMOS NOR gate.

logic 1 (high input) is 1 mA for the SN5400, SN7400 TTL gates, it is typically $\pm 10^{-5}$ μA for the COS/MOS gates, with a maximum of ± 0.1 μA. However, also notice that propagation delays are higher for the COS/MOS gate than the TTL gates.

9.4 BINARY NUMBERS

The numbering system we are all most accustomed to is based on a decimal code of ten digits (0 to 9). Each digit represents one of ten possible states, and the position of a digit in a number represents the power of 10 it is multiplied by. For example, the number 192 is a decimal code representation of

$$1 \times 10^2 + 9 \times 10^1 + 2 \times 10^0 \tag{9.23}$$

which equals 100 + 90 + 2 or 192. It is important to understand that without a knowledge of the rules of expansion the decimal code would be meaningless.

Digital circuits (at present) recognize only two states; therefore, they require numbers made up of two digits, 0 and 1. The position (place)

CD4000B, CD4001B, CD4002B, CD4025B Types

COS/MOS NOR Gates

High-Voltage Types (20-Volt Rating)
Dual 3 Input
plus Inverter – CD4000B
Quad 2 Input – CD4001B
Dual 4 Input – CD4002B
Triple 3 Input – CD4025B

RCA-CD4000B, CD4001B, CD4002B, and CD4025B NOR gates provide the system designer with direct implementation of the NOR function and supplement the existing family of COS/MOS gates. All inputs and outputs are buffered.

The CD4000B, CD4001B, CD4002B, and CD4025B types are supplied in 14-lead hermetic dual-in-line ceramic packages (D and F suffixes), 14-lead dual-in-line plastic packages (E suffix), 14-lead ceramic flat packages (K suffix), and in chip form (H suffix).

Features:

- **Propagation delay time = 60 ns (typ.) at C_L = 50 pF, V_{DD} = 10 V**
- **Buffered inputs and outputs**
- **Standardized symmetrical output characteristics**
- **100% tested for maximum quiescent current at 20 V**
- **5-V, 10-V, and 15-V parametric ratings**
- **Maximum input current of 1 μA at 18 V over full package-temperature range; 100 nA at 18 V and 25°C**
- **Noise margin (over full package temperature range):**
  1 V at V_{DD} = 5 V
  2 V at V_{DD} = 10 V
  2.5 V at V_{DD} = 15 V
- **Meets all requirements of JEDEC Tentative Standard No.13A, "Standard Specifications for Description of "B" Series CMOS Devices"**

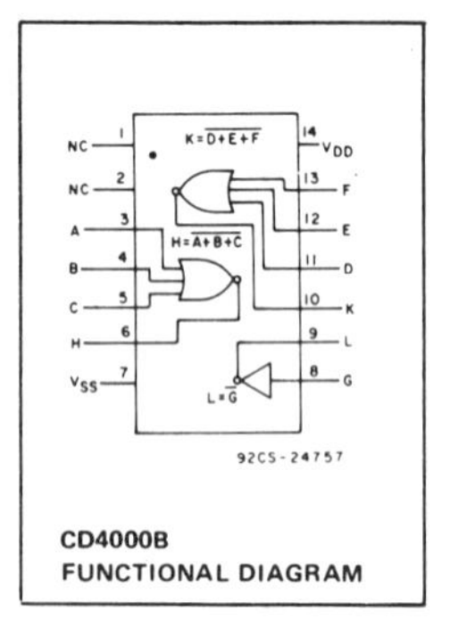

CD4000B
FUNCTIONAL DIAGRAM

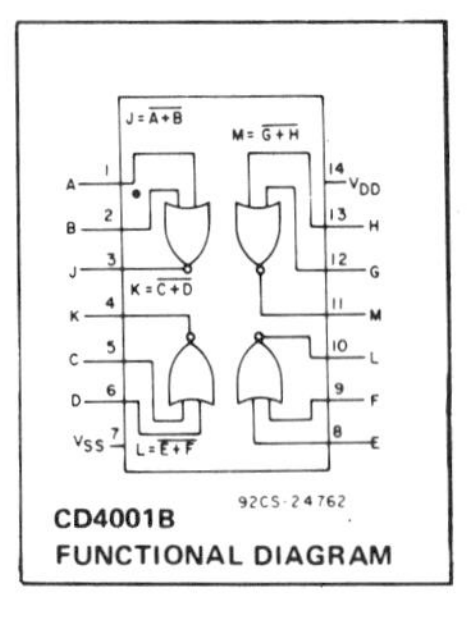

CD4001B
FUNCTIONAL DIAGRAM

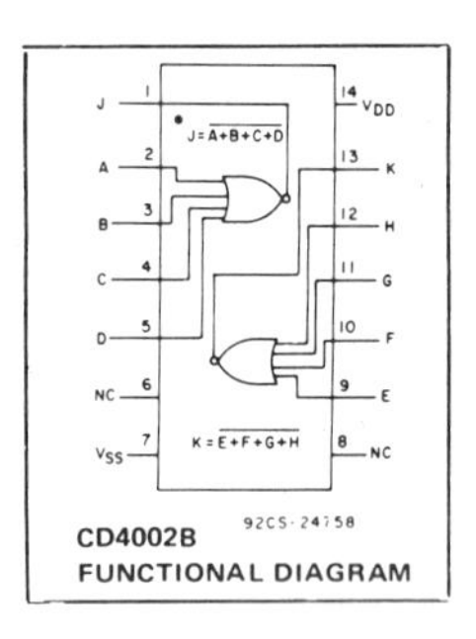

CD4002B
FUNCTIONAL DIAGRAM

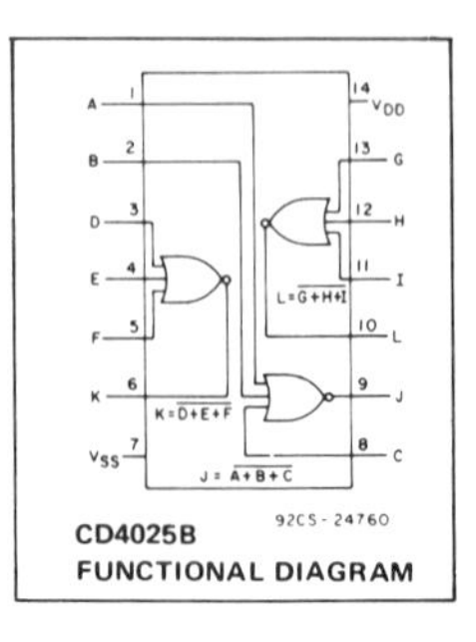

CD4025B
FUNCTIONAL DIAGRAM

STATIC ELECTRICAL CHARACTERISTICS

CHARACTERISTIC	CONDITIONS			LIMITS AT INDICATED TEMPERATURES (°C) Values at –55, +25, +125 Apply to D,F,H Packages Values at –40, +25, +85 Apply to E Package							UNITS
	V_O (V)	V_{IN} (V)	V_{DD} (V)	–55	–40	+85	+125	+25 Min.	+25 Typ.	+25 Max.	
Quiescent Device Current, I_{DD} Max.	–	0,5	5	0.25	0.25	7.5	7.5	–	0.01	0.25	μA
	–	0,10	10	0.5	0.5	15	15	–	0.01	0.5	
	–	0,15	15	1	1	30	30	–	0.01	1	
	–	0,20	20	5	5	150	150	–	0.02	5	
Output Low (Sink) Current I_{OL} Min.	0.4	0,5	5	0.64	0.61	0.42	0.36	0.51	1	–	mA
	0.5	0,10	10	1.6	1.5	1.1	0.9	1.3	2.6	–	
	1.5	0,15	15	4.2	4	2.8	2.4	3.4	6.8	–	
Output High (Source) Current, I_{OH} Min.	4.6	0,5	5	–0.64	–0.61	–0.42	–0.36	–0.51	–1	–	
	2.5	0,5	5	–2	–1.8	–1.3	–1.15	–1.6	–3.2	–	
	9.5	0,10	10	–1.6	–1.5	–1.1	–0.9	–1.3	–2.6	–	
	13.5	0,15	15	–4.2	–4	–2.8	–2.4	–3.4	–6.8	–	
Output Voltage: Low-Level, V_{OL} Max.	–	0,5	5	0.05				–	0	0.05	V
	–	0,10	10	0.05				–	0	0.05	
	–	0,15	15	0.05				–	0	0.05	
Output Voltage: High-Level, V_{OH} Min.	–	0,5	5	4.95				4.95	5	–	
	–	0,10	10	9.95				9.95	10	–	
	–	0,15	15	14.95				14.95	15	–	
Input Low Voltage, V_{IL} Max.	0.5,4.5	–	5	1.5				—	—	1.5	V
	1,9	–	10	3				—	—	3	
	1.5,13.5	–	15	4				—	—	4	
Input High Voltage, V_{IH} Min.	0.5	–	5	3.5				3.5	—	—	
	1	–	10	7				7	—	—	
	1.5	–	15	11				11	—	—	
Input Current I_{IN} Max.		0,18	18	±0.1	±0.1	±1	±1	–	$\pm 10^{-5}$	±0.1	μA

FIGURE 9.16 CMOS/MOS NOR gates. Courtesy of RCA.

CD4000B, CD4001B, CD4002B, CD4025B Types

RECOMMENDED OPERATING CONDITIONS

For maximum reliability, nominal operating conditions should be selected so that operation is always within the following ranges:

CHARACTERISTIC	LIMITS		UNITS
	MIN.	MAX.	
Supply-Voltage Range (For T_A = Full Package Temperature Range)	3	18	V

MAXIMUM RATINGS, *Absolute-Maximum Values:*

DC SUPPLY-VOLTAGE RANGE, (V_{DD})
(Voltages referenced to V_{SS} Terminal) . . . . . −0.5 to +20 V
INPUT VOLTAGE RANGE, ALL INPUTS . . . . . −0.5 to V_{DD} +0.5 V
DC INPUT CURRENT, ANY ONE INPUT . . . . . ±10 mA
POWER DISSIPATION PER PACKAGE (P_D):
For T_A = −40 to +60°C (PACKAGE TYPE E) . . . . . 500 mW
For T_A = +60 to +85°C (PACKAGE TYPE E) . . . . . Derate Linearly at 12 mW/°C to 200 mW
For T_A = −55 to +100°C (PACKAGE TYPES D, F) . . . . . 500 mW
For T_A = +100 to +125°C (PACKAGE TYPES D, F) . . . . . Derate Linearly at 12 mW/°C to 200 mW
DEVICE DISSIPATION PER OUTPUT TRANSISTOR
FOR T_A = FULL PACKAGE-TEMPERATURE RANGE (All Package Types) . . . . . 100 mW
OPERATING-TEMPERATURE RANGE (T_A):
PACKAGE TYPES D, F, H . . . . . −55 to +125°C
PACKAGE TYPE E . . . . . −40 to +85°C
STORAGE TEMPERATURE RANGE (T_{stg}) . . . . . −65 to +150°C
LEAD TEMPERATURE (DURING SOLDERING):
At distance 1/16 ± 1/32 inch (1.59 ± 0.79 mm) from case for 10 s max. . . . . +265°C

DYNAMIC ELECTRICAL CHARACTERISTICS
At T_A = 25°C; Input t_r, t_f = 20 ns, C_L = 50 pF, R_L = 200kΩ

CHARACTERISTIC	TEST CONDITIONS	V_{DD} VOLTS	ALL TYPES LIMITS TYP.	ALL TYPES LIMITS MAX.	UNITS
Propagation Delay Time, t_{PHL}, t_{PLH}		5	125	250	ns
		10	60	120	
		15	45	90	
Transition Time, t_{THL}, t_{TLH}		5	100	200	ns
		10	50	100	
		15	40	80	
Input Capacitance, C_{IN}	Any Input		5	7.5	pF

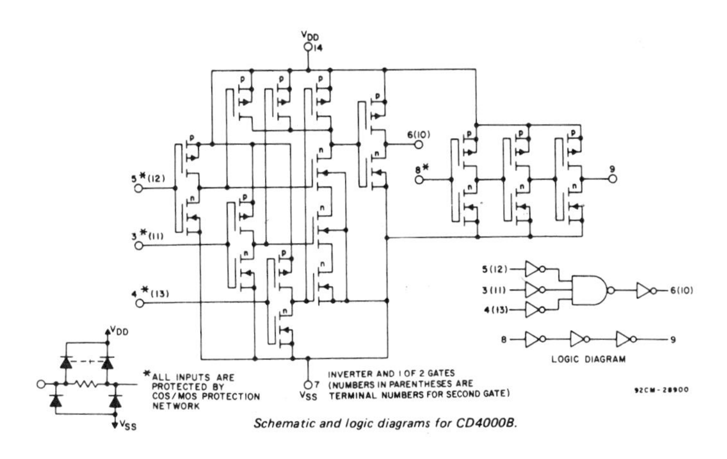

Schematic and logic diagrams for CD4000B.

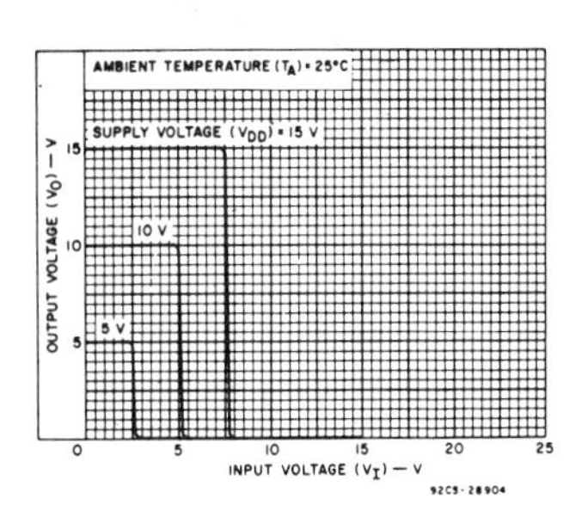

Typical voltage transfer characteristics.

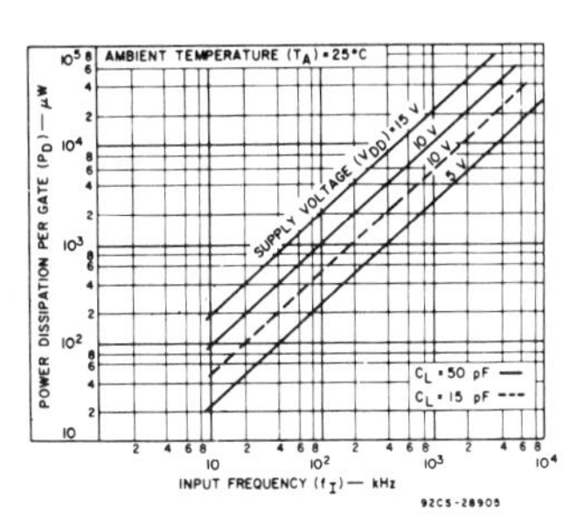

Typical power dissipation vs. frequency.

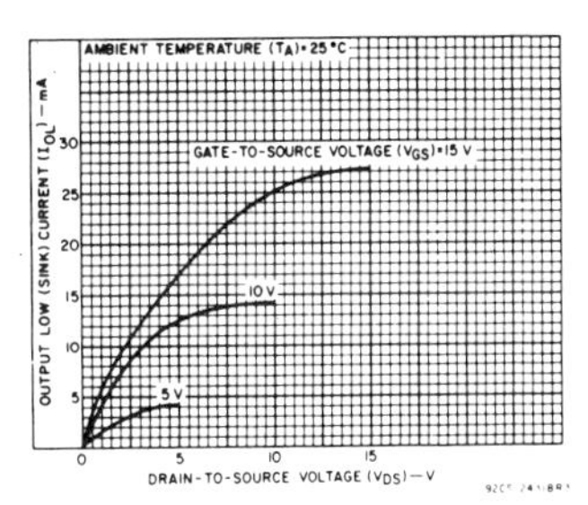

Typical output low (sink) current characteristics.

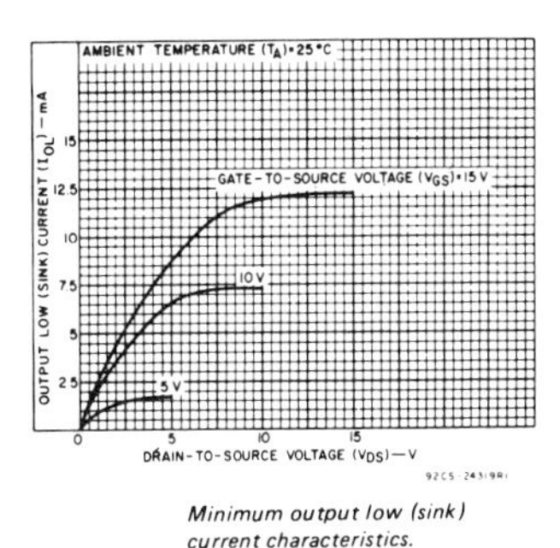

Minimum output low (sink) current characteristics.

FIGURE 9.16 (continued) CMOS/MOS NOR gates. Courtesy of RCA.

of the digit in this binary system plays the same role as it does in the decimal system, except that place represents the power of 2 a digit is multiplied by. For example, the binary number 1101 represents the number

$$1 \times 2^3 + 1 \times 2^2 + 0 \times 2^1 + 1 \times 2^0 \tag{9.24}$$

which in decimal code is equal to $8 + 4 + 0 + 1$ or 13.

Binary numbers can always be converted to decimal code by expanding the number in powers of 2 as in Eq. (9.24). For example, compare counting in binary numbers to counting in decimal code (Table 9.5).

Because we are more conversant with decimal code than binary code, it is normal to stress conversion into decimal code. However, the binary code is a complete numbering system in itself and does not require conversion into decimal code. It should be understood as such, and conversion procedures should be de-emphasized as soon as possible.

9.4.1 Binary Operations

Addition. Binary addition follows four basic rules.

$$0 + 0 = 0 \tag{9.25}$$

$$1 + 0 = 1 \tag{9.26}$$

$$0 + 1 = 1 \tag{9.27}$$

$$1 + 1 = 10 \tag{9.28}$$

These rules, together with the carry procedure (a result exceeding 1 digit is added to the next left column), are sufficient to add any two binary numbers. For example,

TABLE 9.5 *Binary–Decimal Conversions*

Binary No.	*Expansion in Powers of 2*	*Decimal Code*
0	0×2^0	0
1	1×2^0	1
10	$1 \times 2^1 + 0 \times 2^0$	2
11	$1 \times 2^1 + 1 \times 2^0$	3
100	$1 \times 2^2 + 0 \times 2^1 + 0 \times 2^0$	4
101	$1 \times 2^2 + 0 \times 2^1 + 1 \times 2^0$	5
110	$1 \times 2^2 + 1 \times 2^1 + 0 \times 2^0$	6
111	$1 \times 2^2 + 1 \times 2^1 + 1 \times 2^0$	7
1000	$1 \times 2^3 + 0 \times 2^2 + 0 \times 2^1 + 0 \times 2^0$	8
1001	$1 \times 2^3 + 0 \times 2^2 + 0 \times 2^1 + 1 \times 2^0$	9
1010	$1 \times 2^3 + 0 \times 2^2 + 1 \times 2^1 + 0 \times 2^0$	10

$$\begin{array}{r} 101 \\ +\ 110 \\ \hline 1011 \end{array} \quad \text{and} \quad \begin{array}{r} 10111 \\ +\ 00111 \\ \hline 11110 \end{array}$$

Subtraction. Binary subtraction also observes four rules.

$$0 - 0 = 0 \tag{9.29}$$

$$1 - 0 = 1 \tag{9.30}$$

$$0 - 1 = -1 \tag{9.31}$$

$$1 - 1 = 0 \tag{9.32}$$

These rules, together with the borrowing procedure (borrowing a digit from the next left column when needed), suffice for all subtractions. For example,

$$\begin{array}{r} 1010 \\ -\ 0110 \\ \hline 0100 \end{array} \quad \text{and} \quad \begin{array}{r} 1110 \\ -\ 1001 \\ \hline 0101 \end{array}$$

Subtracting by Complements. Digital systems seldom use the direct subtraction procedures just illustrated. Using complements converts subtraction to adding and requires less circuitry. Two types of complements are used, the 1's complement and the 2's complement.

The 1's complement of a number is obtained by changing each 1 to a 0 and each 0 to a 1. Subtraction is performed by obtaining the 1's complement of the subtrahend (the number to be subtracted) and adding it to the minuend (the number to be subtracted from). Before the operation is performed, both numbers (subtrahend and minuend) are given the same number of digits (bits). If the operation results in an additional bit (a 1), this bit is added to the result. This additional bit is called a *carry*, and the addition of this bit to the resultant is called an *end-around carry*. If there is a carry, the answer is positive and in binary form. Otherwise, the answer is negative and in 1's complement form. For example, to subtract 1001 from 0101 (0101 − 1001), find the 1's complement of 1001 and add.

$$\begin{array}{r} \text{Binary form} \\ 0101 \\ -\ 1001 \\ \hline -\ 0100 \end{array} \qquad \begin{array}{r} \text{1's complement form} \\ 0101 \\ +\ 0110 \\ \hline 1011 \end{array}$$

Since there is no carry, the result is negative and in 1's complement form. The binary form result is found by reversing the 1's complement (actually by complementing the result again). The 1's complement of 1011 is 0100, and adding a minus sign gives the final result, −0100.

For a second example, subtract 0110 from 1100.

Binary form		1's complement form
1100		1100
− 0110		1001
0110	carry→ [1]	0101

Since there is a carry, it is added to the result.

```
  0101
+ 0001
------
  0110
```

This procedure can be simplified by using the 2's complement of the subtrahend. This procedure is similar to the foregoing except that an end-around carry is not performed. If a carry (1) results, the answer is positive and in binary form. If not, then the answer is − and is in 2's complement form. The 2's complement is found by adding a 1 to the 1's complement. For example, the subtraction 1010 − 0110 is performed as follows:

Binary form	1's complement form	2's complement form
1010	1010	1010
− 0110	+ 1001	+ 1010
0100		10100

Since the carry is 1, the answer is positive and in binary form, 0100.

Binary Multiplication. Binary multiplication observes the following rules:

$$0 \times 0 = 0 \tag{9.33}$$

$$0 \times 1 = 0 \tag{9.34}$$

$$1 \times 0 = 0 \tag{9.35}$$

$$1 \times 1 = 1 \tag{9.36}$$

For example,

```
   1001
 ×  101
 ------
   1001
  0000
 1001
 ------
 101101
```

Digital circuits rarely perform binary multiplication in this way. Instead, a procedure that reduces multiplication to successive additions is used. This is possible since multiplying by 1 reproduces the original number and multiplying by 0 gives 0. For example, to multiply 11011 by 101, we perform the following steps:

```
    11011
 ×    101
 --------
    11011     Since multiplying by 1 reproduces the number, we
              repeat the number
   00000      Multiplying by 0 gives 0
  11011       Multiplying by 1, 3 places to the left repeats the
 --------     number again, shifted 3 places to the left.
 10000111     Adding the three results together
```

Binary Division. Binary division may also be performed like decimal division. However, again taking advantage of the simplicity of binary multiplication, division is reduced to successive shifting and subtraction. The procedure becomes somewhat involved, however, and will be skipped over. The student can pursue this procedure in any digital text.

9.5 BINARY-DERIVED SYSTEMS

Number systems that have a base (radix) that is a power of 2 are called binary-derived systems. Although digital systems primarily carry out arithmetic in the binary system, the binary-derived systems are often used to communicate with system users. Conversion back and forth from the binary-derived systems is very simple. For example, to convert a binary number into an octal number (base 8), the binary number is divided into groups of 3 digits each, starting from the right-most bit. Zeros are added if necessary to complete the left-most group. Each group is converted to decimal code and the task is completed. For example, to convert the binary number 101101011100 to octal code, divide as follows:

	101	101	011	100
Evaluating each group gives	5	5	3	4

The resulting octal number is 5534_8, where the subscript 8 signifies an octal number. Conversion from octal to binary just inverts this procedure. The octal number is separated into digits and each digit is converted into binary code.

Hexadecimal numbers behave in a similar way. To convert from binary to hexadecimal code, the binary number is divided in groups of four each, and each group is converted into decimal code. However, to avoid ambiguity the numbers 10, 11, 12, 13, 14, and 15 are symbolized by A, B, C, D, E, and F, respectively.

9.6 BINARY CODES

In addition to the binary and binary-derived codes, digital systems use codes that are a compromise between binary and decimal codes. These are called *binary-coded decimal codes* (BCD). Some of these are described in the following.

8421 Code. Although arithmetic operations in digital systems are performed most efficiently in pure binary form, conversion to decimal code for communication to users is costly in hardware terms. The 8421 BCD code can be used for system computation effectively and also simplifies conversion into decimal code. Therefore, it is often used.

In this code, bits are divided into groups of four. The right-most group represents the right-most digit of the decimal number (digit $\times 10^0$). The next group represents the digit $\times 10^1$ decimal number, and so on. For example, the BCD number

	0011	0110	1001
represents the decimal number	3	6	9

Notice that, although a 4-bit number can range to 15_{10} (1111), only numbers to 9 are used. Therefore, 8421 code groups never exceed 1001.

Excess-3 Code. Excess-3 code is a BCD code that can make use of ordinary binary addition and yet retain the interface advantages of BCD. It therefore supports more efficient arithmetic than the 8421 code.

The excess-3 code is formed by adding 3 to each decimal digit before it is converted to binary form. For example, to convert 15 to its excess-3 form, proceed as follows:

1	5	number 15
+ 3	+ 3	add 3 to each digit
4	8	result
0100	1000	transform each digit to binary form

Gray Code. The Gray code is a BCD code that can detect transmission errors. Although it is not suitable for computation, it is widely used for transmitting data to analog-to-digital converters and other peripherals. Like the other BCD codes, groups of 4 bits represent decimal digits. However, the bit groupings are arbitrarily chosen so that successive groups differ from the preceding group by a single bit. This arrangement is illustrated in Table 9.6.

TABLE 9.6 *Gray Code*

Decimal	*Gray Code*	*Binary*
0	0000	0000
1	0001	0001
2	0011	0010
3	0010	0011
4	0110	0100
5	0111	0101
6	0101	0110
7	0100	0111
8	1100	1000
9	1101	1001
10	1111	1010
11	1110	1011
12	1010	1100
13	1011	1101
14	1001	1111

9.7 DIGITAL CIRCUITS

This section introduces the basic digital circuits that are used to build digital systems, including the digital computer. These circuits provide memory, delay, counting, coding and encoding, and arithmetic. In our discussion, the gates with which these circuits are constructed will be considered ideal and activated by ideal voltage sources.

9.7.1 Flip-Flops

The flip-flop (bistable multivibrator) is the first circuit studied so far that possesses memory. All flip-flops have two output terminals. At any time, one output is at logic 1 (HIGH), the other at logic 0 (LOW). These are usually labeled Q and $\overline{Q}$ (not Q). An input signal shifting from HIGH to LOW (or LOW to HIGH) will cause Q and $\overline{Q}$ to reverse states. In the absence of any appropriate input, these new states will be maintained. Flip-flops see important applications in memories, counters, registers, adders, control circuits, and the like.

Cross-Coupled Inverters. Figure 9.17 shows the simplest flip-flop, two cross-coupled inverters. When Q is raised to a logic 1, the output of the lower inverter $\overline{Q}$ is driven to logic 0. Since a logic 0 at the input of the upper inverter produces the initial state, $Q = 1$, this state is stable. Forcing a logic 0 at Q or a logic 1 at $\overline{Q}$ will reverse states. This method of control

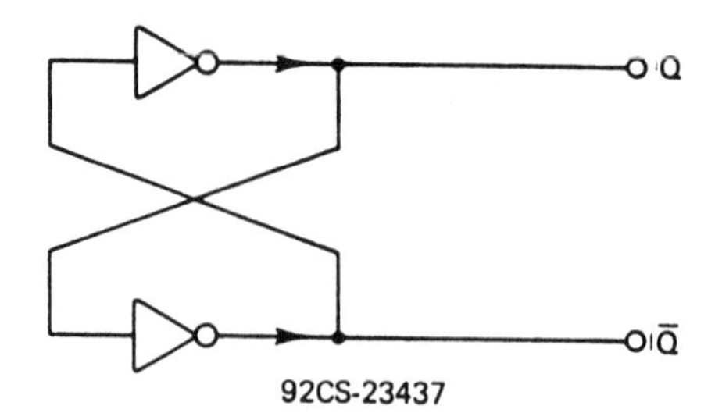

FIGURE 9.17 Two cross-coupled inverters as a flip-flop device. Courtesy of RCA Solid-State Division.

is called *jamming*. Reset and set of a flip-flop (setting Q to 0 and 1) by jamming requires a control signal source with very low impedance.

A basic CMOS cross-coupled inverter flip-flop is shown in Figure 9.18. Setting Q at logic 1 causes *F*1 to turn off and *F*2 to turn on. This places $\overline{Q}$ at logic 0. A LOW at the gates of *F*3 and *F*4 places *F*3 in the on state and *F*4 in the off state, reproducing the initial state. Because of its compact structure, this device is widely used in memories and registers.

R–S Flip-Flop. An R–S (set–reset or reset–set) flip-flop is shown in Figure 9.19. It consists essentially of two cross-connected NOR gates and operates as follows. In its initial state, Q is at logic 1 and $\overline{Q}$ at logic 0. A logic 1 at R (reset) will cause Q to go to logic 0. This output delivered to the set gate input will shift $\overline{Q}$ to 1. This is a stable condition, which will

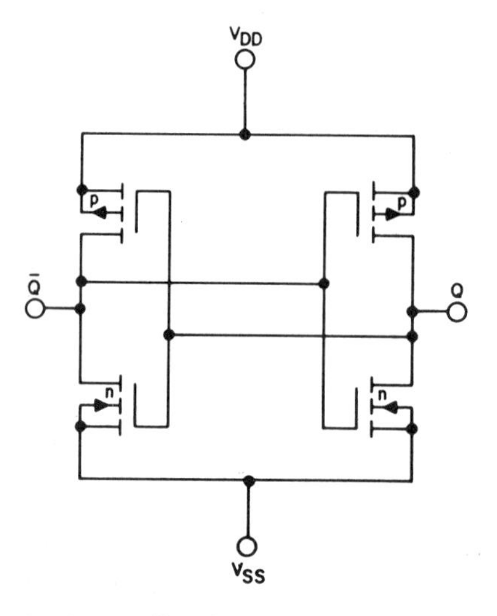

FIGURE 9.18 A basic CMOS flip device. Courtesy of RCA Solid-State Division.

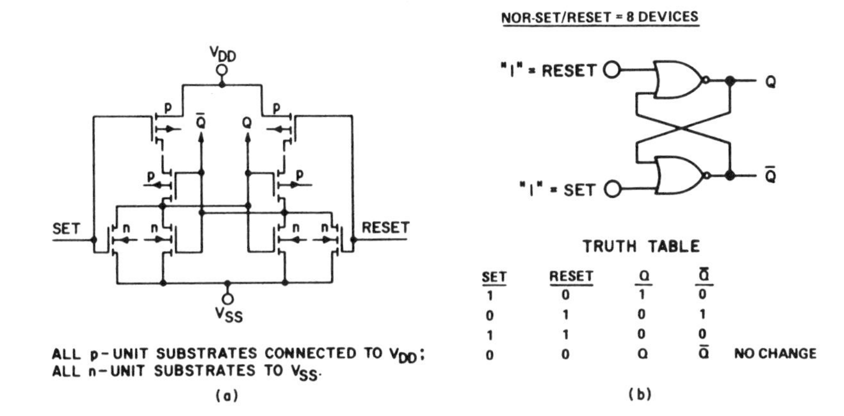

SET	RESET	Q	$\bar{Q}$	
1	0	1	0	
0	1	0	1	
1	1	0	0	
0	0	Q	$\bar{Q}$	NO CHANGE

FIGURE 9.19 Set–reset flip-flop: (a) schematic diagram; (b) logic diagram and truth table. Courtesy of RCA Solid-State Division.

be retained when the R input disappears. The R input is used to establish this latter state and is called the *reset*. Now a logic 1 at S (set) will send $\bar{Q}$ to logic 0 and Q to 1. Again this stable state is retained when the input vanishes. (If, instead of a logic 1, a logic 0 had been delivered to S, $\bar{Q}$ would not have changed and neither would Q.)

The truth table for the R–S flip-flop is also shown in Figure 9.19. Note that with Q at logic 1 initially a reset 1 is required to shift the output. With Q at logic 0, a set 1 will then shift the output. The use of NOR gates provides specific terminals for the set and reset functions. An RCA CMOS set/reset flip-flop is also shown.

Toggled R–S Flip-Flop (RST Flip-Flop). The RST flip-flop is a R–S flip-flop with a third input terminal (T). Terminal T (toggle) will always cause a change of state. The R and S inputs act as they do for the R–S flip-flop. Since the Q output initially at logic 1 will go to logic 0 at the next T input and then return to logic 1 at the following T logic 1 input, the RST flip-flop may be used to divide by 2. Figure 9.20 shows the schematic symbol.

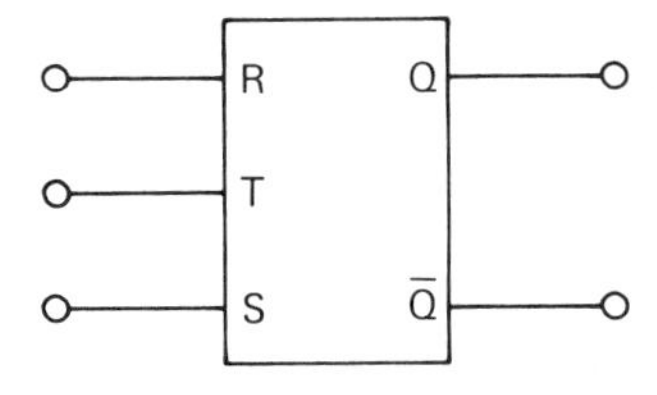

FIGURE 9.20 RST flip-flop.

Clocked R–S Flip-Flop. It is frequently necessary to synchronize an R–S flip-flop to a free-running oscillator (clock). Figure 9.21 illustrates the circuit logic. A logic 1 at either S or R will not produce a logic 0 at the output of the two NAND gates until a logic 1 is input at C. When this occurs, the flip-flop will behave as a normal R–S flip-flop. (Notice because of the NAND gates S and R are interchanged.) Synchronization is used to control digital circuits and prevent mistimed operation errors.

J–K Flip-Flop. The *J–K* flip-flop is the most versatile and is consequently the most popular in digital systems. This device is clock controlled but responds only to the negative-going edge of the clock pulses. It has three or four input terminals, *J*, *K*, *C*, and sometimes an auxiliary *R*. Four events dependent on *J* and *K* can occur.

1. $J = 1, K = 0$, flip-flop sets.
2. $J = 0, K = 1$, flip-flop resets.
3. $J = 1, K = 1$, flip-flop toggles.
4. $J = 0, K = 0$, flip-flop remains in previous state.

The *R* input if present may be used to reset the flip-flop.

The *J–K* flip-flop is configured as a master–slave unit. Figure 9.22 illustrates the circuit logic and truth table. With *J* and *K* equal to 0, both master AND gates are disabled. Input clock pulses will have no effect on the state of the flip-flop ($Q_{n+1} = Q_n$). With $J = 0$, $K = 1$, and $Q_{out} = 0$ (reset state), the upper input AND gate is disabled because $J = 0$. The lower AND gate is also disabled because Q_{out} is fed back to its input. Therefore, the next downward pulse will not change Q_{out}. For the same initial conditions except that $Q_{out} = 1$ (set state), the upper input AND gate is still disabled, but the lower one is not. Therefore, a clock pulse will reset the flip-flop ($Q_{out} = 0$). With $J = 1$ and $K = 0$, the upper AND

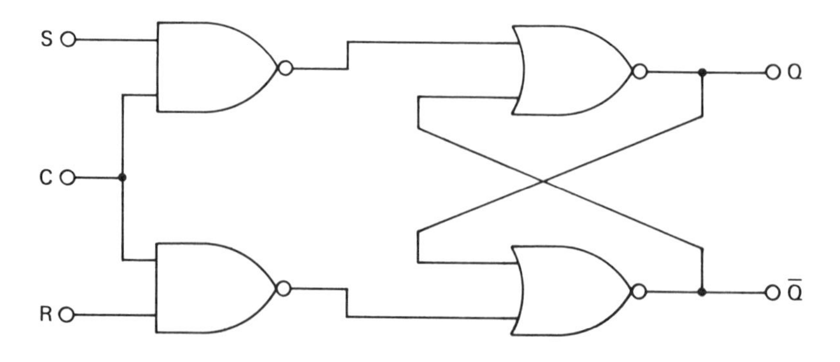

FIGURE 9.21 Clocked R–S flip-flop.

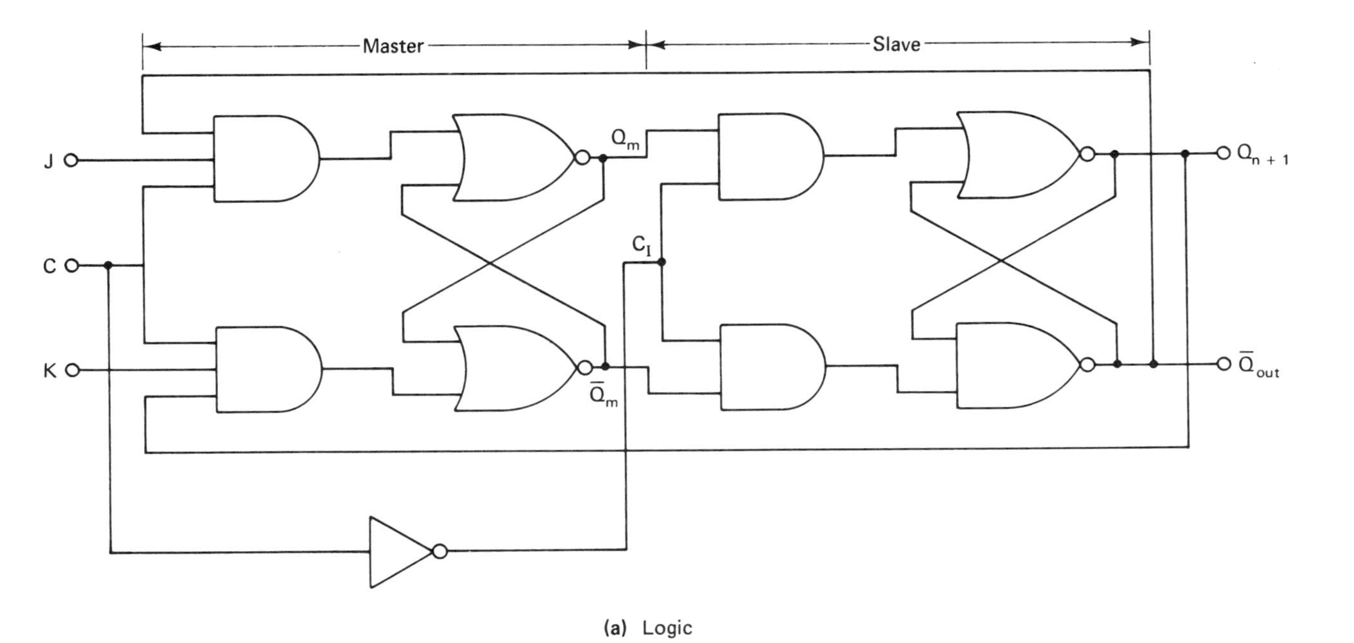

(a) Logic

J	K	Q_{n+1}	Cond.
0	0	Q_n	No change
0	1	0	Reset
1	0	1	Set
1	1	$\bar{Q}_n$	Toggle

(b) Truth table

FIGURE 9.22 *J–K* flip-flop.

gate is enabled. A clock pulse will set the flip-flop ($Q_{out} = 1$) if it is not already set. When J and K are both 1, the AND gate response depends on Q_{out} and $\overline{Q}_{out}$. If $Q_{out} = 0$, $\overline{Q}_{out} = 1$, the upper AND gate is enabled. If $Q_{out} = 1$, $\overline{Q}_{out} = 0$ the lower AND gate is enabled. In either case, a control pulse changes the output state (toggles). A control pulse at logic 1 triggers the master unit, but because the slave control pulse is inverted, the device triggers on the down-going edge of the control pulse. Figure 9.22 gives the logic circuit and truth table.

D-Type Flip-Flop[a]. The D-type (delay-type) flip-flop stores a previous input and outputs it one clock pulse later. For example, if at clock pulse n a logic 1 was input, a logic 1 would be output at clock pulse $n + 1$. A block diagram of a D-type flip-flop is shown in Figure 9.23a; the schematic diagram is shown in Figure 9.23b.

The block diagram shows a master flip-flop formed from two inverters and two transmission gates (shown as switches) that feed a slave

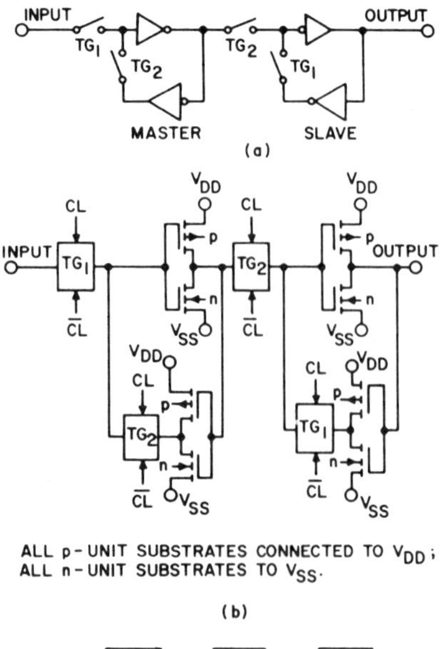

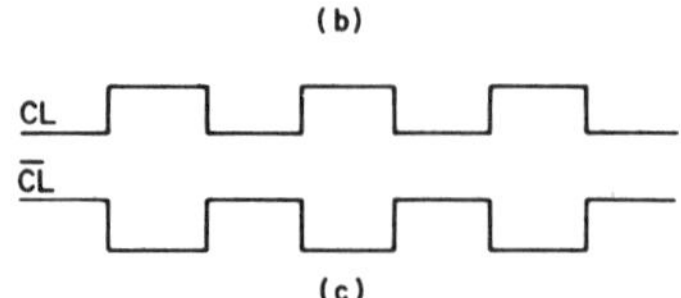

FIGURE 9.23 (a) Block, (b) schematic, and (c) clock-pulse diagrams for a D-type flip-flop. Courtesy of RCA Solid-State Division.

[a] The material in this section is taken from "Understanding CMOS" © 1974, RCA Corp.

flip-flop having a similar configuration. When the input signal is at a low level, the TG_1 transmission gates are closed and the TG_2 gates open. This configuration allows the master flip-flop to sample incoming data and the slave to hold the data from the previous input and feed it to the output. When the clock is high, the TG_1 transmission gates open and the TG_2 transmission gates close so that the master holds the data entered and feeds it to the slave. The D flip-flop is static and holds its state indefinitely if no clock pulses are applied (i.e., it stores the state of the input prior to the last clocked input pulse).

A clock pulse is the pulse applied to the logical elements of a sequential digital system to initiate logical operations. Both the clock, *CL*, and inverted clock, $\overline{CL}$, as shown in Figure 9.23c, are required; clock inversion is accomplished by an inverter internal to each D flip-flop.

Figure 9.24 shows the logic diagram and truth table for a D-type flip-flop.

9.7.2 Counters

All counters count the number of pulses entering the device. This operation is basic to digital systems where counters control decoders. There are many types of counters, but here we will consider only the ripple, synchronous (parallel), ring, and Johnson counters.

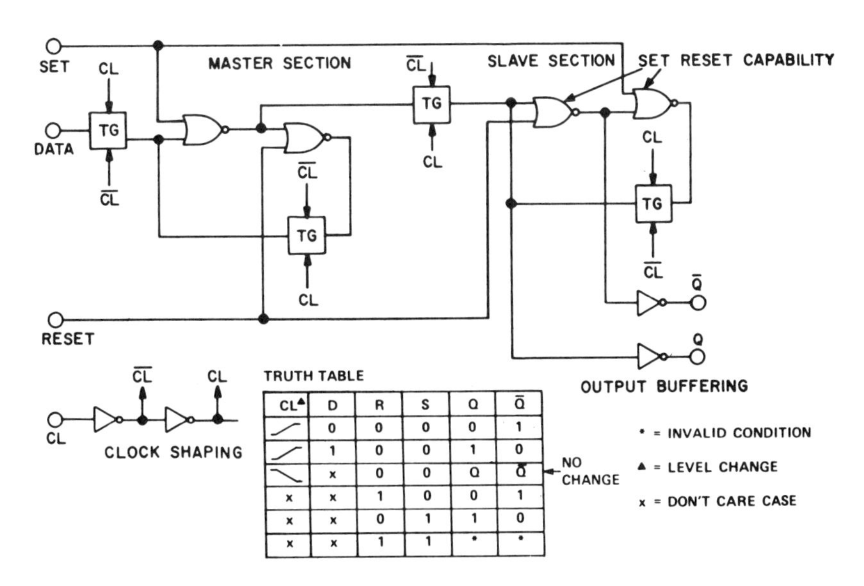

CL▲	D	R	S	Q	$\overline{Q}$
/	0	0	0	0	1
/	1	0	0	1	0
\	x	0	0	Q	$\overline{Q}$
x	x	1	0	0	1
x	x	0	1	1	0
x	x	1	1	•	•

FIGURE 9.24 Logic diagram and "truth" table for a D-type flip-flop. Courtesy of RCA Solid-State Division.

Ripple Counter. Figure 9.25 shows a 4-bit ripple counter using RST flip-flops. Each flip-flop is triggered by the negative-going pulse of the preceding one, except the first, which is triggered by the clock pulse. Initially, a pulse delivered to R will set all Q's to zero. The first negative-going clock pulse will set Q_1 to 1. This change of state, however, will not affect Q_2 since the flip-flop responds only to the negative-going edge of pulses. The next clock pulse will switch Q_1 to 0. This change will cause Q_2 to change state to 1. Because the flip-flops change state only on the negative-going edge of a pulse, each succeeding flip-flop will change state when the preceding one goes from 1 to 0 (at Q).

This action generates a binary count provided the number is considered to be ordered as Q_4, Q_3, Q_2, Q_1. A single clock pulse would set Q_1 to 1 and would not affect Q_2 or those following. This gives 0001. The next clock pulse would send Q_1 to 0, causing Q_2 to become 1 (Q_3 would be unaffected). This gives 0010. A third clock pulse would send Q_1 to 1, but this state change would not affect Q_2, giving 0011. A fourth clock pulse would send Q_1 to 0, which in turn would send Q_2 to 0, which would send Q_3 to 1. This gives 0100. This procedure continues until all Q's are 1's, giving 1111 or decimal 15. The next pulse sends all Q's to zero and counting resumes from zero. Table 9.7 summarizes the counting mechanism.

Parallel Counter. The parallel counter is designed to correct a serious disadvantage inherent to the ripple counter. Because each flip-flop is triggered by the preceding flip-flop, an inherent time delay, proportional

TABLE 9.7 *Four-Bit Counting*

Q_4	Q_3	Q_2	Q_1	*Clock Pulses*
0	0	0	0	0
0	0	0	1	1
0	0	1	0	2
0	0	1	1	3
0	1	0	0	4
0	1	0	1	5
0	1	1	0	6
0	1	1	1	7
1	0	0	0	8
1	0	0	1	9
1	0	1	0	10
1	0	1	1	11
1	1	0	0	12
1	1	0	1	13
1	1	1	0	14
1	1	1	1	15
0	0	0	0	16 = 0

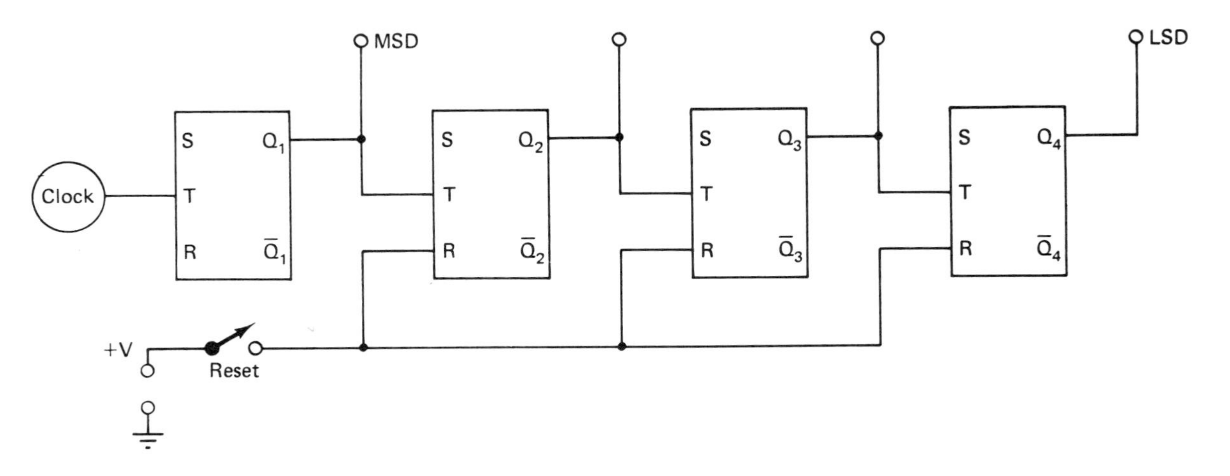

FIGURE 9.25 Ripple counter.

to the number of flip-flops, is encountered. The time required to complete a count is called *settling time,* and this can be excessive for the ripple counter. In the parallel counter, all flip-flops switch together under control of a clock pulse, thus reducing settling time.

Figure 9.26 shows a 4-bit synchronous parallel counter. For this counter to work, the following events must occur:

1. FF_0 must toggle with each clock pulse. To do this, J_0 and K_0 must be fixed at logic 1, as shown.
2. FF_1 must change state whenever Q_0 was previously at logic 1; a clock pulse arrives and J_1 and K_1 are at logic 1 (next count after 0001). This is guaranteed by tying J_1 and K_1 to Q_0 and the clock pulse to C. After a first pulse has switched Q_0 to logic 1, FF_1 is ready to toggle the instant a second clock pulse arrives. It will toggle before Q_0 switches to logic 0.
3. FF_2 must change state only when $Q_0 = Q_1 = 1$ (next count after 0011). This is guaranteed by connecting Q_0 and Q_1 to the AND gates feeding FF_2. With J_2 and K_2 high for FF_2, it will toggle at the next clock pulse.
4. FF_3 must change state only when $Q_0 = Q_1 = Q_2 = 1$ (next count after 0111). This requires 2 three-input AND gates feeding J_3 and K_3, as shown.

Ring Counter. The ring or shift register counter circulates a word (pattern of bits) repetitively in synchronism with clock pulses. It has numerous applications, multiplexing being one of them. It is constructed by returning Q_{out} to J_1 and $\overline{Q}_{out}$ to K_1, as shown in Figure 9.27. Notice that each J–K flip-flop contains set and reset terminals. These terminals are used to set up an initial string of bits (words). On each clock pulse, this word will shift to the right by one step, the Q_n bit moving to the Q_{n+1} bit and the output of the last flip-flop looping back to the first. For example, suppose Q_1 is set to logic 1 and the other Q's to zero. Since J_1 is 0 and K_1 is 1, FF_1 will reset on the control pulse. Q_1 becomes 0; $\overline{Q_1}$ becomes 1. J_2 and K_2 were, respectively, 1 and 0 before the control pulse. This being the set position, Q_2 is driven to 1 by the control pulse. J_3, J_4 and K_3, K_4 were, respectively, 0 and 1 prior to the control pulse. This is the reset position, and no change occurs upon application of the control pulse. Therefore, the logic 1 has moved from Q_1 to Q_2, and Q_1, Q_3, and Q_4 are zero. Similar reasoning shows the same progression to the right (shift right) for any word.

If the $\overline{Q_4}$ output is led back to J_1 and Q_4 to K_1, the circuit becomes a divide by 8 counter in which only one stage changes state at each count. This is called a Johnson, Mobius, or twisted-ring counter and cycles

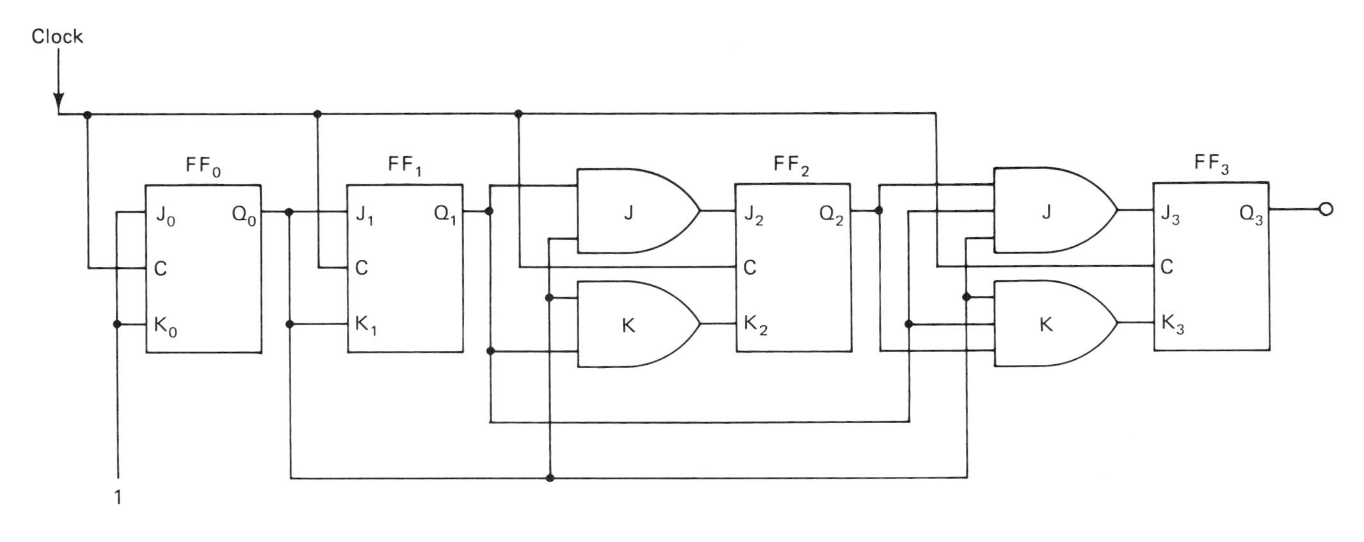

FIGURE 9.26 Parallel binary counter.

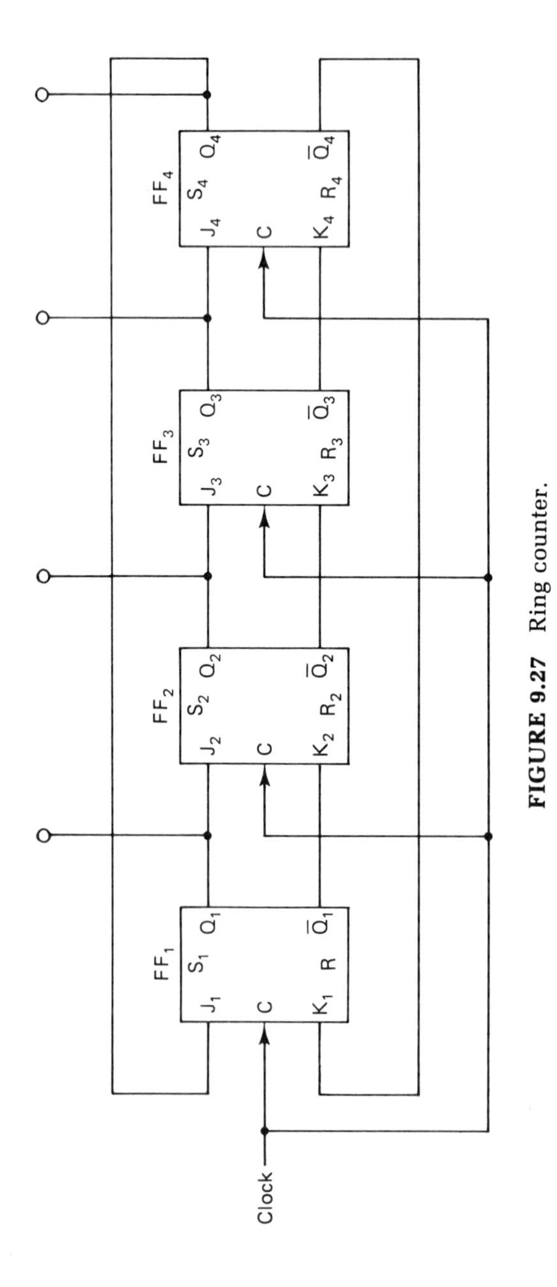

FIGURE 9.27 Ring counter.

through a modulus equal to twice the number of stages. For example, a three-stage counter will count with a modulus of 6 (0, 1, 2, 3, 4, 5), a four-stage with a modulus of 8, and so on. This enhances circuit performance.

9.7.3 Decoders

In digital systems, it is necessary to open and close switches (gates) in response to binary words. This is the function of the decoder. Decoders are used to convert the output of binary counters to decimal code, to control input and output lines, to select cells in memory, and so on. There are several decoder configurations, among them the parallel decoder, the tree-type decoder, and the balanced multiplicative decoder. Here we discuss the most straightforward type, the parallel decoder.

Figure 9.28 shows a parallel decoder that decodes the output of three flip-flops. The decoder is connected to the Q and $\overline{Q}$ terminals, as

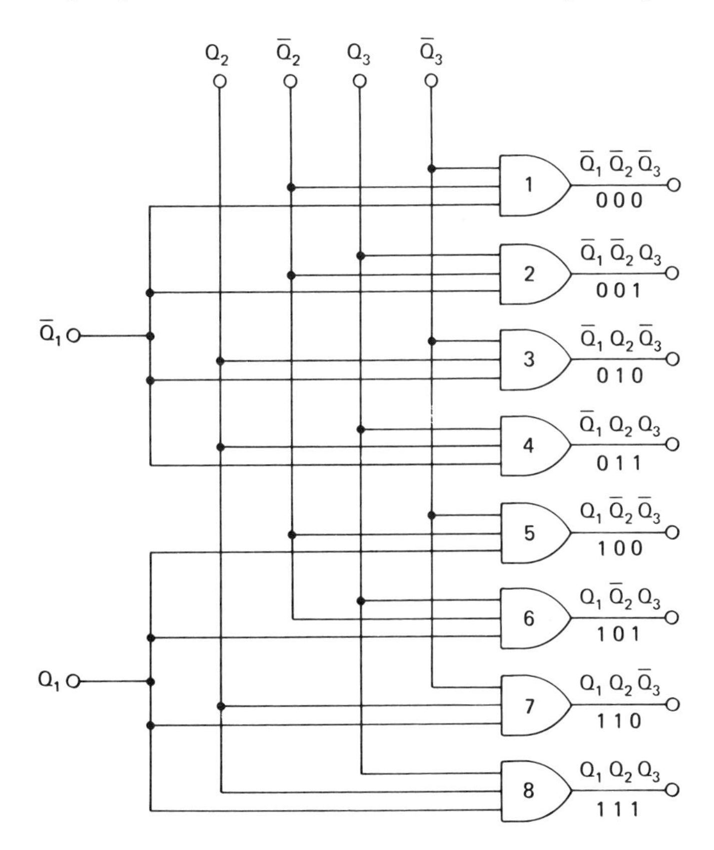

FIGURE 9.28 Parallel decoder.

shown, either directly or indirectly. The operating principle is straightforward. Two binary states from three sources give 2^3 (8) combinations. The outputs of the three flip-flops are connected to the AND gates in such a way that each combination turns on a specific gate. For example, the output $\overline{Q_1} = 1$, $\overline{Q_2} = 1$, and $\overline{Q_3} = 1$, which represents the number 000, turns on AND gate 1; the output $\overline{Q_1} = 1$, $\overline{Q_2} = 1$, and $Q_3 = 1$ (001) turns on AND gate 2.

9.7.4 Shift Registers

Shift registers are used to temporarily store binary words and to perform complementation, multiplication, and division. They operate either in serial or parallel modes or combinations of both. Basically, a register is a series of flip-flops, usually one flip-flop per bit.

Serial Shift Register. Figure 9.29 shows a 5-bit serial register with serial and parallel output. For serial operation, data are entered 1 bit at a time in response to a clock pulse, each bit moving toward the right as another is entered. When the register is full, clock pulses are stopped. When the clock resumes, data leave the register in the same way. For parallel output, the five lines Q_1 through Q_5 are gated on by a clock pulse. Parallel output is five times faster than serial output but requires more circuitry.

Operation of the serial shift register is similar to the ring counter. Initially, the register is cleared by a logic 1 on the clear line. This resets each flip-flop, setting all Q's to logic 0. Suppose the word 10101 is to be entered. Prior to clock pulse 1, the input data set $J_1 = 1$, $K_1 = 0$. This is the set mode so that clock pulse 1 sets Q_1 to logic 1. All other J's and K's are at 0 so that no state change occurs. Prior to clock pulse 2, $J_1 = 0$, $K_1 = 1$ and $J_2 = 1$, $K_2 = 0$. Therefore, clock pulse 2 will send Q_1 to logic 0 (reset) and Q_2 to logic 1 (set). Other flip-flops remain unaffected. Prior to clock pulse 3, $J_1 = 1$, $K_1 = 0$, $J_2 = 0$, $K_2 = 1$, and $J_3 = 1$, $K_3 = 0$. Therefore, clock pulse 3 sends Q_1 to logic 1 (set), Q_2 to logic 0 (reset), and Q_3 to logic 1 (set). Succeeding clock pulses move this 101 toward the right while setting Q_1 and Q_2 at 10. The register is now filled with 10101 and clock pulses are stopped. Resuming clock pulses will continue the shifting to the right and output the number serially.

Figure 9.30 shows a commercial two-stage shift register using CMOS inverter flip-flops and transmission gates.[b] Each transmission gate is driven by two out-of-phase clock signals arranged, as shown in Figure 9.30b, so that when alternate transmission gates are turned on the others are turned off. When the first transmission gate in each stage is turned on, it couples the signal from the previous stage to the inverter

[b] The material in this section is taken from "Understanding CMOS" © 1974, RCA Corp.

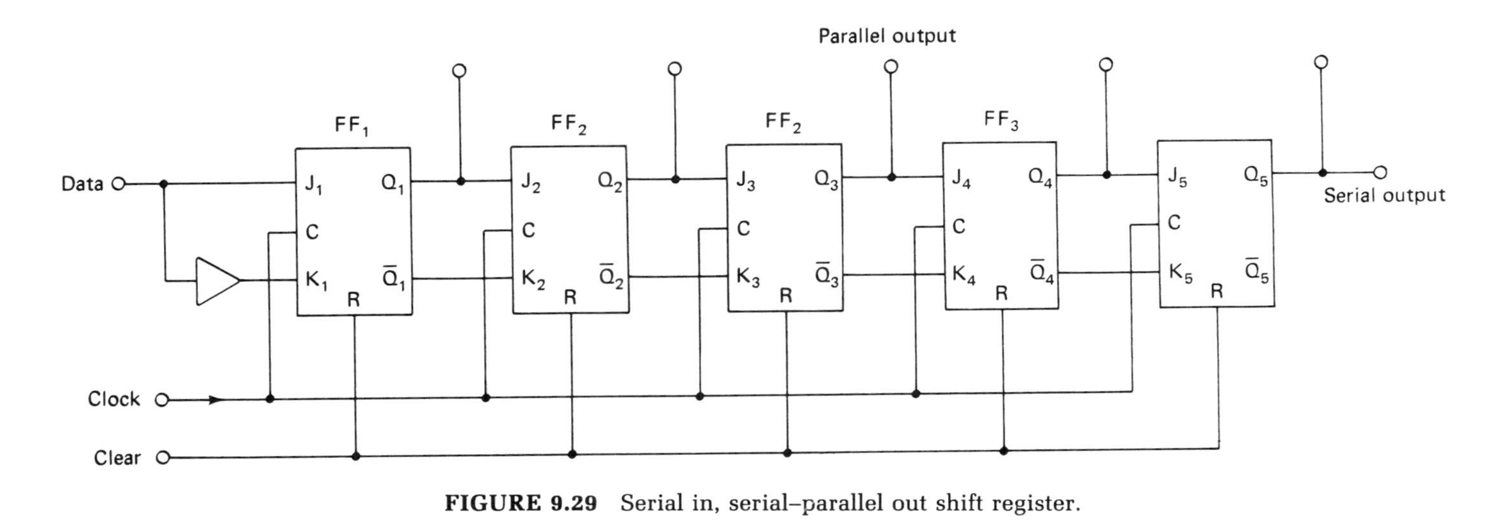

FIGURE 9.29 Serial in, serial–parallel out shift register.

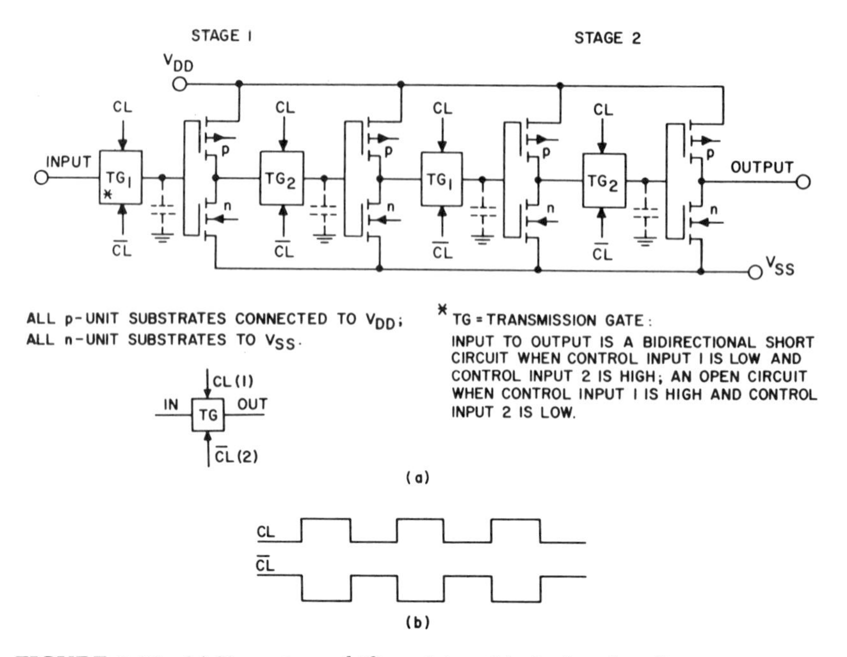

FIGURE 9.30 (a) Two-stage shift register; (b) clock-pulse diagram. Courtesy of RCA Solid-State Division.

and causes the signal to be stored on the input capacitance of the inverter. The shift register utilizes the input of the inverter for temporary storage.

When the transmission gate is turned off on the next half-cycle of the clock, the signal is stored on this input capacitance, and the signal remains at the output of the inverter where it is available to the next transmission gate, which is now turned on. Again, this signal is applied to the input of the next inverter, where it is stored on the input capacitance of the inverter, making the signal available at the output of the stage. Thus a signal progresses to the right by one half-stage on each half-cycle of the clock, or by one stage per clock cycle.

Because the shift register is dependent on stored charge, which is subject to slow decay, there is a minimum frequency at which it will operate; reliable operation can be expected at frequencies as low as 5 kHz.

Parallel Shift Register. Figure 9.31 shows a parallel in/parallel out shift register. Data are present at terminals *A*, *B*, and *C* of the AND gates. When the shift line is enabled (goes to logic 1) and the clear line is disabled (goes to logic 0), all *J*'s are at the data values and all *K*'s are at logic 0. If the datum is a logic 1, the next clock pulse sets the *Q* of the corre-

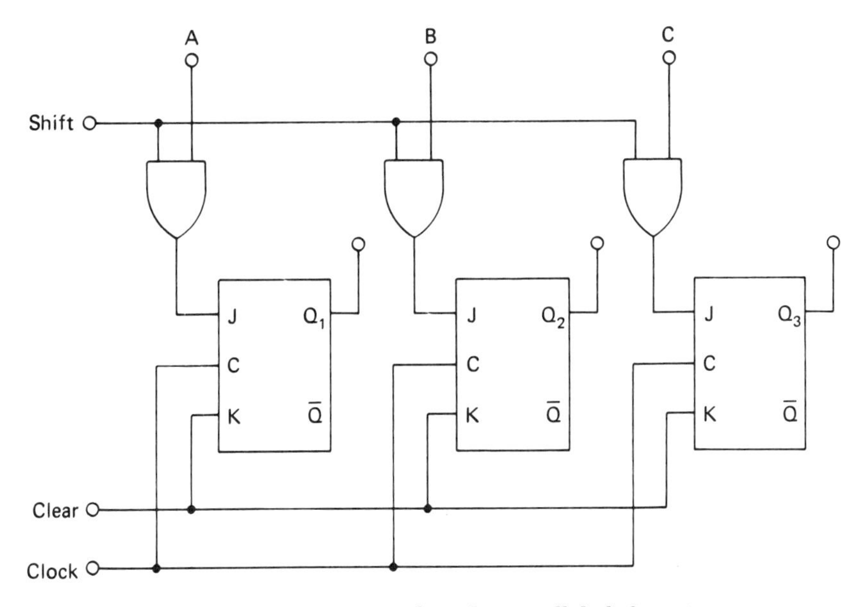

FIGURE 9.31 Three-bit parallel shift register.

sponding flip-flop to logic 1, or else it sets the corresponding Q to logic 0. A single clock pulse thus places the word in the register. When the shift line is disabled, all *J*'s and *K*'s are zero, so continuing clock pulses will not affect the contents of the register. Output is taken in parallel by gating on lines Q_1, Q_2, and Q_3.

Register Arithmetic. Complementing (1's complement) can be performed in the *J–K* registers by enabling the *J–K* inputs of each flip-flop for one clock pulse. This toggles each flip-flop, changing all 1's to 0's, and vice versa. If the register is of sufficient size to accommodate shifting to the left or right a given number of times without losing bits, multiplication and division by 2 can be performed. Shifting a word to the left is equivalent to multiplying by 2 for each left shift. Shifting a word to the right divides by 2 for each right shift. For example, 01010 becomes 10100 for a left shift and 00101 for a right shift. The 2's complement is obtained by adding 1 to the 1's complement.

9.7.5 Arithmetic Circuits

Fundamental to digital systems are circuits that test for equality, add, accumulate, and indirectly multiply and divide. These circuits are discussed in this section.

Equality Checking. Figure 9.32 shows a circuit that can determine if two 4-bit numbers are equal. Basically, corresponding bits are compared in four exclusive NOR gates. Exclusive NOR gates will output a logic 1 only when both inputs are logic 1's or logic 0's. In both cases the inputs must be the same. The AND gate will only output a logic 1 when all inputs are at logic 1, signifying equality.

Half-Adder. The half-adder (HA) adds two binary digits and outputs the result as a single digit sum plus a carry. For example, the half-adder result of 1 + 1 gives a sum (S) equal to 0 and a carry (C) equal to 1. The half-adder result of 1 + 0 is S = 1, C = 0. The schematic symbol, logic circuit, and truth table for a half-adder are shown in Figure 9.33.

Full Adder. The full adder adds 2 bits like the half-adder, but it also includes a carry in (C_I) from a previous stage. It outputs a sum and a carry. Figure 9.34 shows the schematic symbol, logic circuit, and truth table.

Parallel Adder. Figure 9.35 shows a 3-bit parallel adder that can add two 3-bit numbers. The output of the adder is a 4-bit number since it includes a carry. The least significant digits (LSD) are entered into the half-adder since they include no carry. The next 2 bits are entered into a full adder. The carry of the half-adder is also input into this full adder. The most significant digits (MSD) plus the carry from the preceding stage are entered into the left-most full adder. The S outputs of each stage are the result.

Figure 9.35 illustrates the case where $B = 110$ and $A = 111$. Adding the 2 LSDs gives 1 and no carry. Adding A_2 and B_2 with 0 carry in gives 0 plus a C_o (carry out) = 1. The sum of the MSDs plus a C_I (carry in)

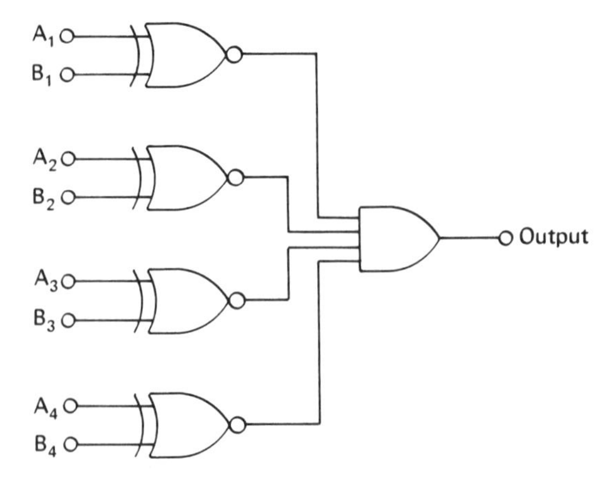

FIGURE 9.32 Equality checker.

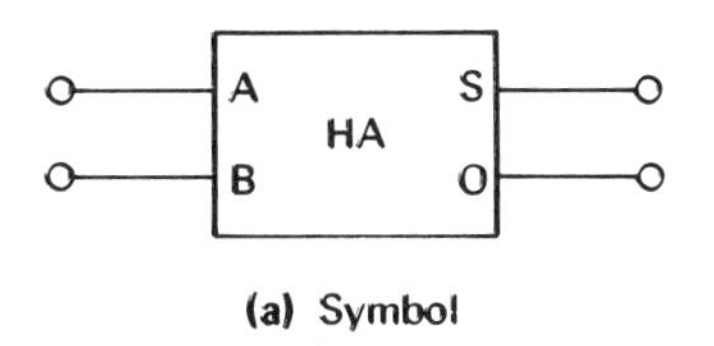

(a) Symbol

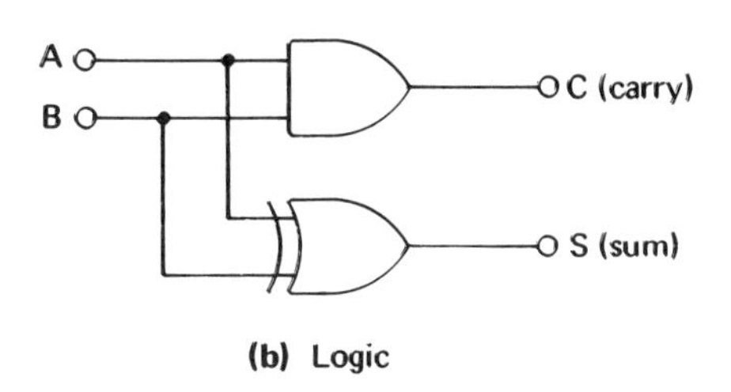

(b) Logic

A	B	C	S
0	0	0	0
0	1	0	1
1	0	0	1
1	1	1	0

(c) Truth table

FIGURE 9.33 Half-adder.

gives 1 plus $C_o = 1$ $(1 + 1 + 1 = 11)$. The final sum, S_4, S_3, S_2, S_1, is 1101.

Serial Adder. Figure 9.36 shows a serial adder. This device adds binary numbers bit by bit working from LSDs to MSDs. The numbers to be added are typically stored in two input registers and the output stored in an output register. The circuit contains two half-adders, an OR gate, and a D flip-flop used as a delay. The serial adder, although slower than the parallel adder, can add numbers of any length, limited only by the size of the registers.

Figure 9.36 also illustrates the addition of numbers, 111 and 100. Initially, the numbers are read into registers *A* and *B* and the output register is cleared. A clock pulse shifts the bits in registers *A* and *B* into the half-adder. The first half-adder produces a sum (S) and a carry (C). S is transmitted to the second half-adder on the first clock pulse, but C is delayed one pulse. Therefore, on the first clock pulse only the sum of the two LSDs is placed in the output register. On the second clock pulse, the first half-adder generates the sum of the second 2 bits. These, plus the

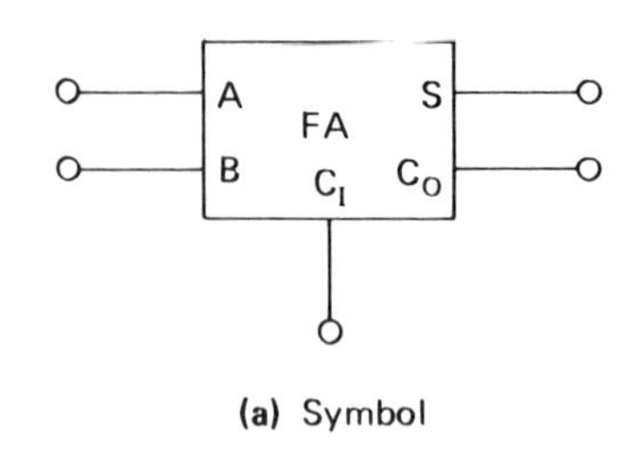

(a) Symbol

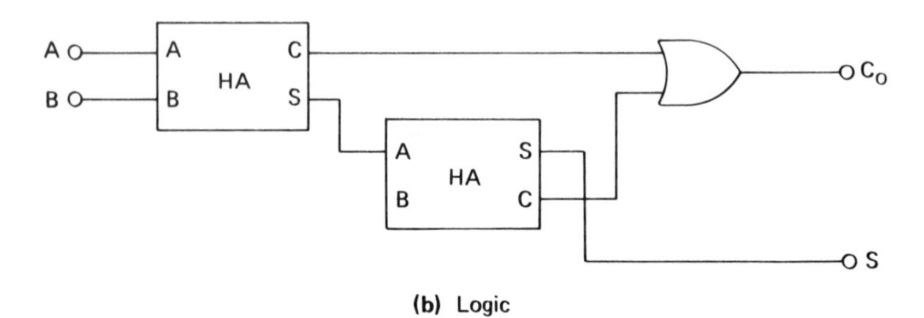

(b) Logic

A	B	C	S	C_O
0	0	0	0	0
1	0	0	1	0
0	1	0	1	0
1	1	0	0	1
0	0	1	1	0
1	0	1	0	1
0	1	1	0	1
1	1	1	1	1

(c) Truth table

FIGURE 9.34 Full-adder.

carry from the first addition, are entered into the second half-adder and output into the output register. This procedure continues for one clock pulse more than the number of bits in the input register.

Accumulator. Often the output of adders (parallel or serial) is stored in one of the input registers. This register is called an *accumulator*. This procedure is particularly simple for the serial adder, since after the first clock pulse the left-most cell of both input registers is empty and available for the LSD of the output, and so on. The accumulator, how-

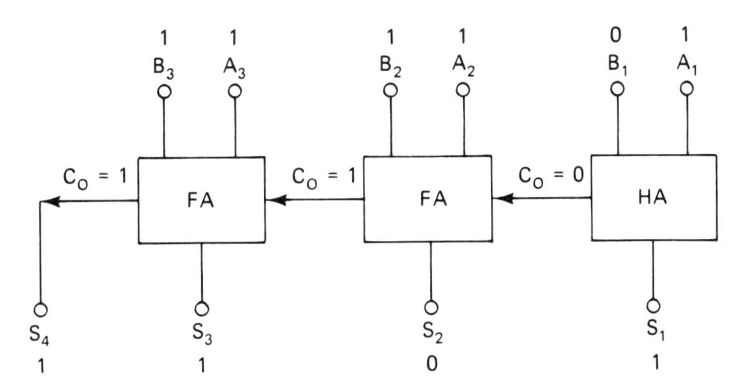

FIGURE 9.35 Parallel adder.

ever, must be at least 1 bit larger than the numbers added. Accumulators can carry out successive additions.

Subtractor. Digital subtraction is implemented by complementing one of the numbers (either 1's or 2's complements), adding and performing the required end-around carry. The 1's complement of a number may either be obtained from the $\overline{Q}$ terminals of the register or by means described previously. The number may also be inverted at the input to the subtractor. Figure 9.37 shows a 4-bit subtractor that receives non-complemented numbers, 1's complements one of the numbers, and performs an end-around carry.

Commercial Arithmetic Devices.[c] Commercial integrated units that can perform all arithmetic operations are available. This section

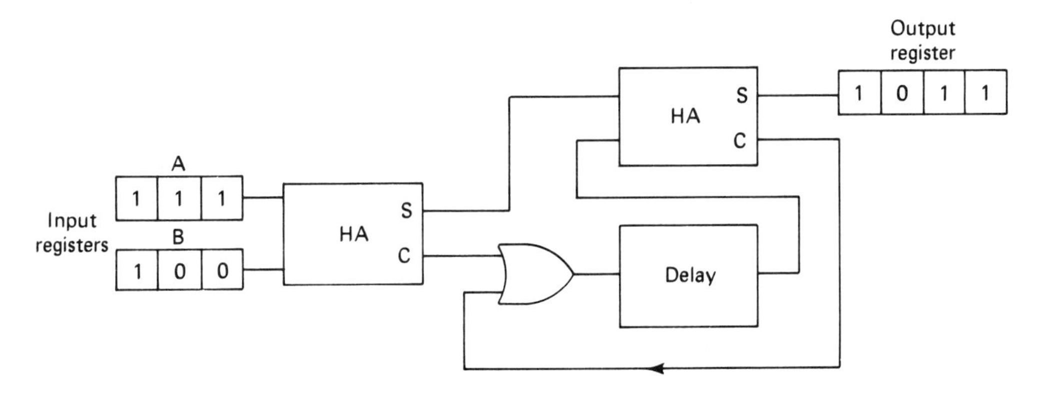

FIGURE 9.36 Serial adder.

[c] Material in this section is taken from "Understanding CMOS" © 1974 RCA Corp.

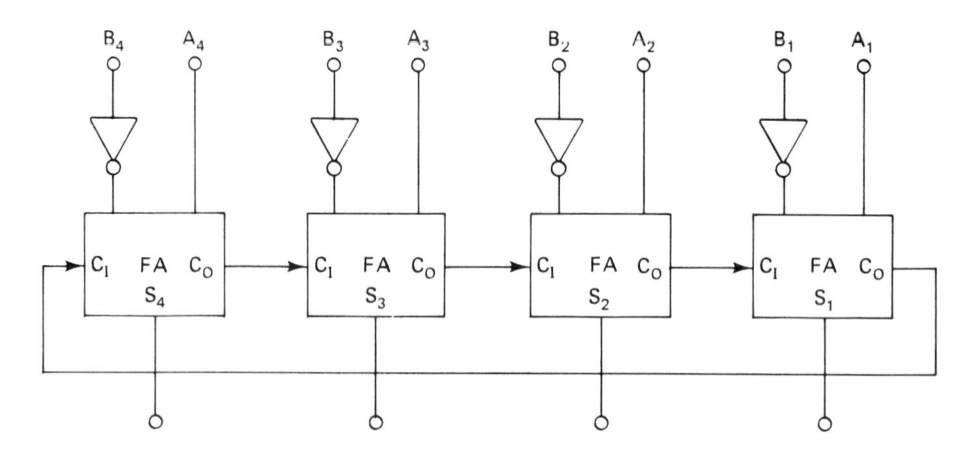

FIGURE 9.37 Parallel binary subtractor.

describes a COS/MOS arithmetic unit capable of adding, subtracting, multiplying, and dividing, as well as performing the logical functions OR, AND, and exclusive-OR on two 4-bit words. Three 4-bit registers permit either of two words to perform the desired operation with the third. The unit consists of COS/MOS devices including registers, AND–OR select gates, a full adder, and NOR and NAND logic gates.

A block diagram of the 4-bit arithmetic unit is shown in Figure 9.38. The required package count and the function performed by each package are shown in Table 9.8. A brief description of each of the COS/MOS devices used to configure the arithmetic unit is given next.

The RCA CD4008A or CD4008B consists of four full-adder stages with fast look-ahead carry provision from stage to stage. The logic diagram for this device is shown in Figure 9.39. Circuitry is included to provide a fast parallel carry-out bit to permit high-speed operation in arithmetic sections that use several CD4008As or CD4008Bs.

The RCA CD4008A or CD4008B inputs include the four sets of bits to be added, A_1 to A_4 and B_1 to B_4, in addition to the carry-in bit from a previous section. RCA CD4008A outputs include the four sum bits, S_1 to S_4, in addition to the high-speed parallel carry-out, which may be used with a succeeding RCA CD4008A or CD4008B section.

The RCA CD4019A or CD4019B, as shown in the logic diagram in Figure 9.40, consists of two 2-input AND gates driving a single 2-input OR gate. Selection is accomplished by control bits K_a and K_b. In addition to simple selection of either channel A or channel B information, the control bits can be applied in combination to accomplish a third selection of data.

The RCA CD4013A or CD4013B consists of two identical indepen-

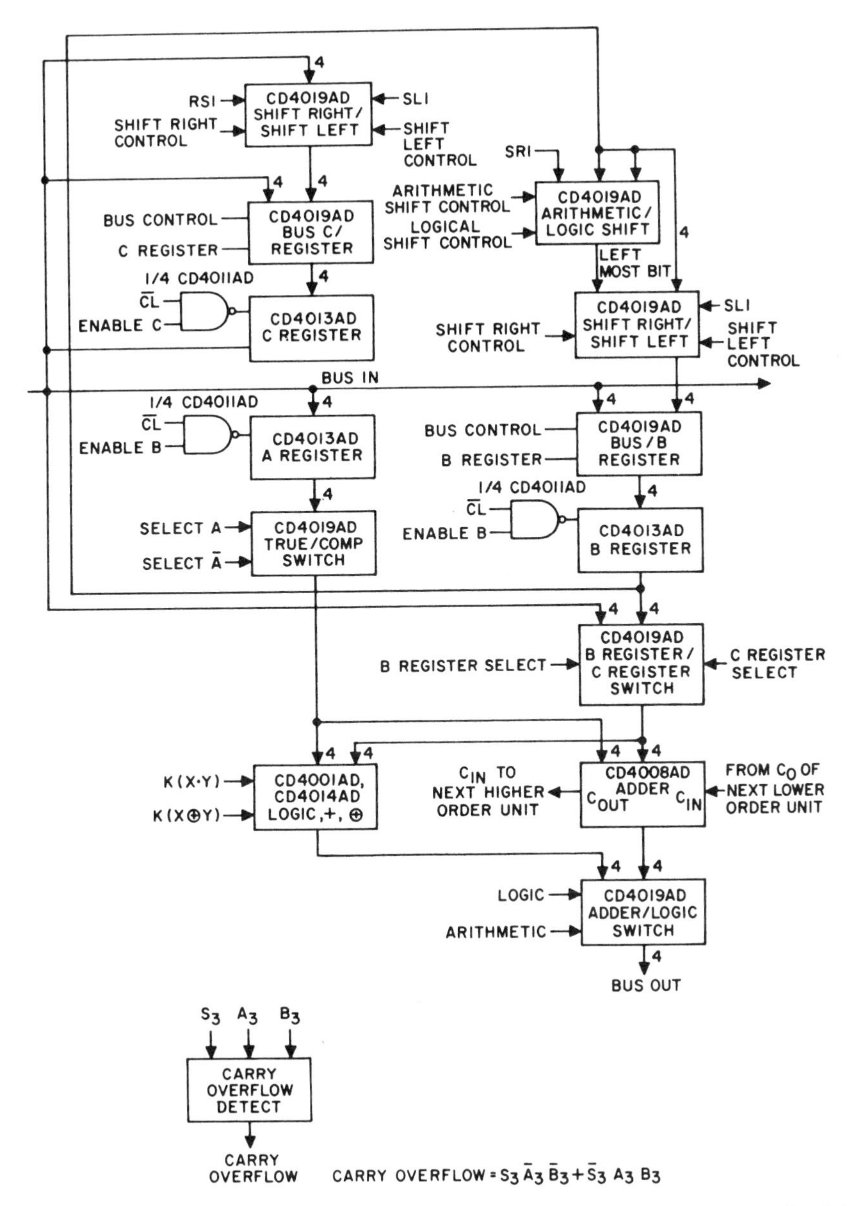

FIGURE 9.38 Block diagram of the 4-bit arithmetic unit. Courtesy of RCA Solid-State Division.

dent data-type flip-flops on a single monolithic silicon chip; the logic diagram for one flip-flop is shown in Figure 9.41. Each flip-flop has independent data, reset, set, and clock inputs and complementary buffered outputs. The RCA CD4013A or CD4013B can be used in shift-

TABLE 9.8 *Function and Package Count of Devices in Arithmetic Unit*

	Packages				
Function	*CD4008A or B*	*CD4019A or B*	*CD4013A or B*	*CD4011A or B*	*CD4001A or B*
A Register (includes A, $\bar{A}$ select capability and buffered outputs)		1	2		
B Register (includes shift capability and buffered outputs for B and $\bar{B}$)		3	2		
C Register (includes shift capability and buffered outputs)		2	2		
Select B or C		1			
Add	1				
Perform logic		1			5
Select logic or addition		1			
Overflow detector					1
Clock inhibit				1	
Total	1	9	6	1	6

Courtesy of RCA Solid-State Division

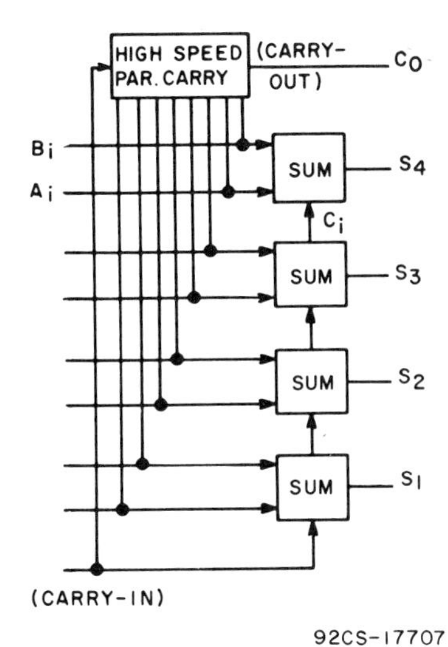

FIGURE 9.39 RCA CD4008A or CD4008B 4-bit full adder. Courtesy of RCA Solid-State Division.

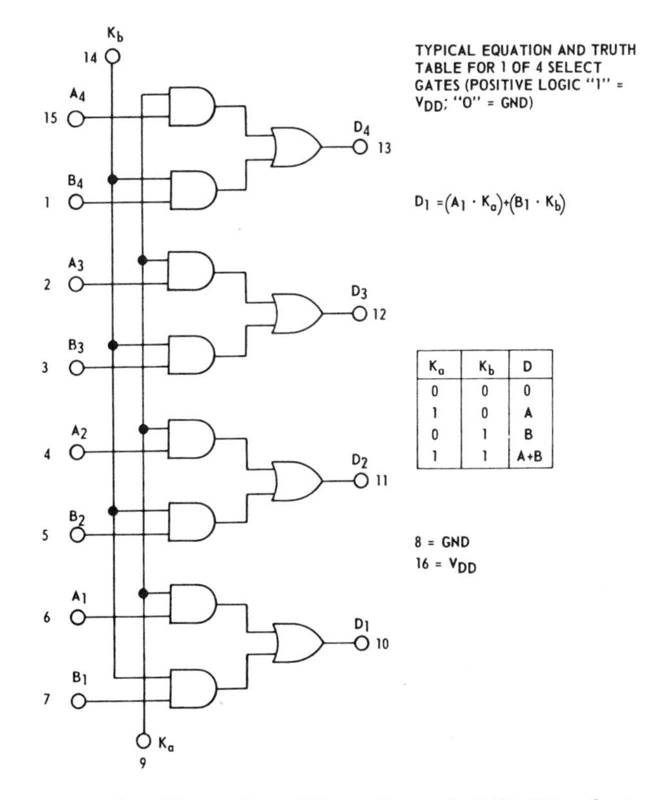

K_a	K_b	D
0	0	0
1	0	A
0	1	B
1	1	A+B

FIGURE 9.40 RCA CD4019A or CD4019B quad AND–OR select gate. Courtesy of RCA Solid-State Division.

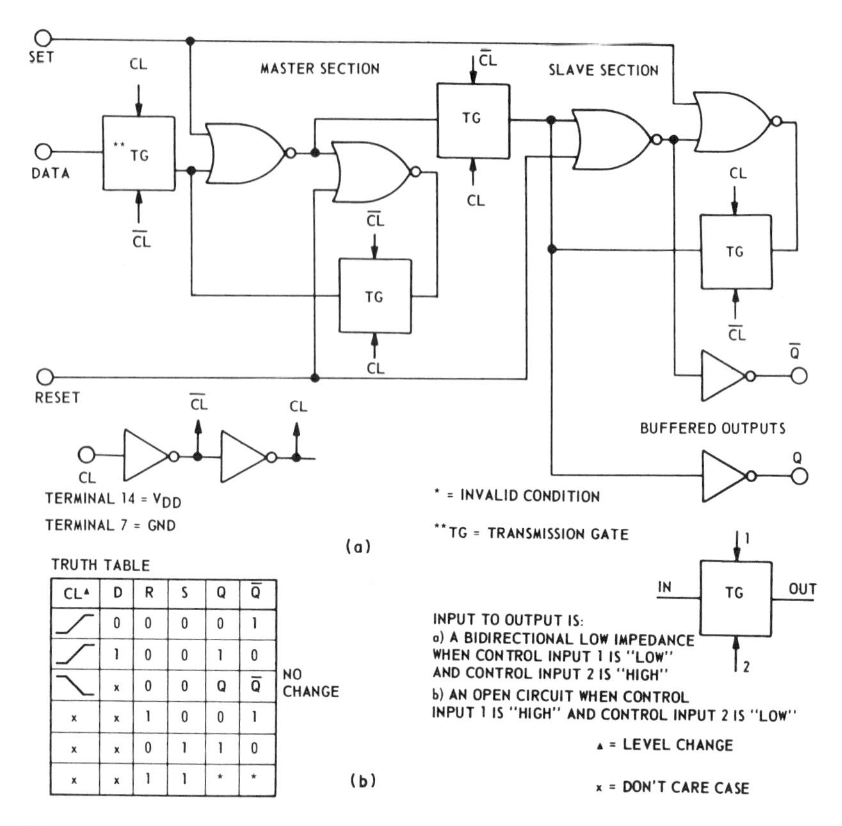

TRUTH TABLE

CL▲	D	R	S	Q	$\bar{Q}$	
↗	0	0	0	0	1	
↗	1	0	0	1	0	
↘	x	0	0	Q	$\bar{Q}$	NO CHANGE
x	x	1	0	0	1	
x	x	0	1	1	0	
x	x	1	1	*	*	

FIGURE 9.41 (a) One of two identical flip-flops composing the CD4013A dual D-type flip-flop; (b) circuit truth table. Courtesy of RCA Solid-State Division.

register applications and for counter and toggle applications by connecting the Q output to the data input. The logic level present at the D input is transferred to the Q output during the positive-going transition of the clock pulse. Resetting or setting is accomplished by a high level on the reset or set line, respectively.

Problems

1. Construct the truth tables for the NOT, OR, and AND gates of Figure 9.2.
2. Using a truth table, show that Figure 9.2b is an OR gate.
3. Using truth table analysis, determine the overall truth tables for the logic circuits of Figure 9.42.

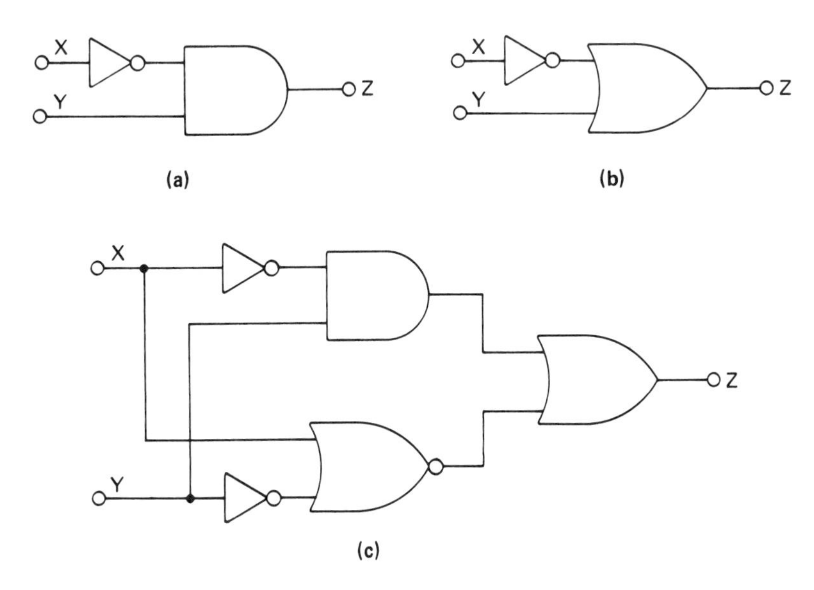

FIGURE 9.42 Problem 3.

4. Prove $\overline{(X + Y)} = \overline{X} \cdot \overline{Y}$ by substituting values for X and Y.

5. Prove $\overline{(X \cdot Y)} = \overline{X} + \overline{Y}$ by substitution.

6. Determine the truth tables for Eqs. 9.11 through 9.16 and show that they represent AND, OR, NOR, NAND, and exclusive OR gates.

7. Verify that the expression $Z = X \cdot Y + \overline{X} \cdot Y + \overline{X} \cdot \overline{Y}$ represents the truth table of Example 9.3.

8. Simplify the following Boolean equations as far as possible.
   (a) $\overline{X} \cdot Y \cdot Z + \overline{X} \cdot \overline{Y}$
   (b) $\overline{X} \cdot \overline{Z} + \overline{Y} \cdot \overline{Z}$
   (c) $(A + \overline{B})(\overline{A} \cdot \overline{B} \cdot \overline{C})$
   (d) $X \cdot (\overline{Y} + \overline{Z}) + X \cdot Y$
   (e) $X \cdot Y + X \cdot Y \cdot Z + X \cdot Y \cdot \overline{Z} + \overline{X} \cdot Y \cdot Z$

9. Using the Karnaugh map, simplify the following expressions:
   (a) $X \cdot (X + Y) + Y \cdot (X + Y)$
   (b) $X \cdot (\overline{X} + Y) + Y \cdot (X + Y)$
   (c) $X \cdot (\overline{X} + Y + X) + Y \cdot X$

10. The following data are supplied by a manufacturer for a TTL gate: $V_{IL} = 0.8\ V_{max}$, $V_{IH} = 2\ V_{min}$, $V_{OH} = 2.4\ V_{min}$, and $V_{OL} = 0.4\ V_{max}$. Compute the noise margins, NM_L and NM_H.

11. Write the following decimal numbers in place-independent form: (a) 173, (b) 4001, (c) 26793.

12. Write the following binary numbers in place-independent form: (a) 1101, (b) 100101, (c) 0101.

13. Convert the following binary numbers to decimal code: (a) 1110, (b) 10101, (c) 100001, (d) 11.01.

14. Convert the following decimal codes to binary numbers: (a) 123, (b) 78, (c) 44, (d) 2.24.

15. Perform the following binary operations:
(a) $1010 + 1110$
(b) $10.10 + 10.10$
(c) $1111 - 1010$
(d) 1101×111

16. Perform the following subtractions using both 1's and 2's complements for each case. Check the results by direct subtraction.
(a) $1111 - 1010$
(b) $1001 - 1100$
(c) $1010 - 0011$

17. Perform the following multiplication by shifting.
(a) 1101×101
(b) 11011×11
(c) 11011×10

18. Perform the following conversions.
(a) 110111011_2 to base 8
(b) 11010110_2 to base 16
(c) 472_8 to binary
(d) 121_{16} to binary
(e) 10010110 (8421 code) to decimal code
(f) 124_{10} to excess-3 code

19. Draw the logic diagram for an RST flip-flop and give its truth table.

20. What kind of counting takes place if the output is taken from the $\overline{Q}$ terminals for Figure 9.25?

21. Draw a 3-bit ripple counter and prove that it counts clock pulses.

22. Draw a 3-bit parallel counter and prove that it accurately counts clock pulses.

23. Place the number 1011 in the ring counter of Figure 9.27. Describe how this number circulates to the right when the counter is driven by clock pulses.

24. Describe the response of a parallel decoder (Figure 9.28) to the following numbers: (a) 110, (b) 011, (c) 101.

25. Describe how the number 11011 is stored in the serial shift register of Figure 9.29.

26. Design a digital circuit that tests a number for zero.

27. Show how 100 and 101 are added in a parallel adder.

28. Show how 101 and 101 are added in a serial adder.

29. Show how the operation 1001 − 0101 is performed by the subtractor of Figure 9.37.

References

Bartee, Thomas C. *Digital Computer Fundamentals,* 4th ed., McGraw-Hill Book Company, New York, 1977.

COS/MOS Integrated Circuits, SSD-250B, RCA Solid-State Division, Somerville, N.J., 1980.

Libes, Sol. *Digital Logic Circuits,* Hayden Book Company, Rochelle Park, New Jersey, 1975.

QMOS High-Speed CMOS Logic, RCA Solid-State Division, Somerville, N.J., 1982.

Schilling, Donald L. *Electronic Circuits Discrete and Integrated,* 2nd ed., McGraw-Hill Book Company, New York, 1979.

10

THEORY AND APPLICATIONS CONVERTERS AND AUXILIARY EQUIPMENT PERIPHERALS

10.1 INTRODUCTION

Data acquisition and conversion systems interface the analog world signal to a digital world. The devices that perform the interfacing are the analog-to-digital converter (A/D or ADC) and the digital-to-analog converter (D/A or DAC). Together they are known as data converters. These devices are used in the following:

1. Data telemetry systems
2. Automated electric manufacturing
3. Signal processing
4. Data logging systems
5. Medical instrumentation
6. Data communication systems

10.2 DIGITAL CONVERTER CODES[a]

The inputs to D/A converters and the outputs from A/D converters are digital numbers in various binary codes; they are so called because they

[a] The material in Sections 10.2 to 10.14 is used courtesy of the ILC Data Device Corporation.

consist of binary bits each of which has two possible states, designated on and off, true and false, or 1 and 0. Each state is represented by a discrete voltage level with respect to ground. The off, false, or 0 state is represented by the logic 0 level; and the on, true, or 1 state is represented by the logic 1 level.

D/A converters accept inputs of digital numbers in one or more of the following binary codes: straight or natural binary (BIN), offset binary (OBN), inverted OBN, one's complement (1SC), two's complement (2SC), and binary-coded decimal (BCD). The input circuits are generally DTL/TTL compatible, positive true, which means that they will interface with DTL and/or TTL logic (logic 0 = 0 to +0.8 V and logic 1 = +2.0 to +5.5 V). (DTL refers to diode-transistor logic.)

A/D converters generate digital-number outputs in one or more of the following binary codes: BIN, inverted BIN, OBN, 2SC, and BCD. The outputs are generally DTL/TTL compatible, positive true, which means that each logic 0 output bit is at a level between 0 and +0.4 V, and each logic 1 bit is at a level between +2.4 and +5.5 V.

Straight BIN and BCD are designated unipolar codes, because they represent analog quantities having only one polarity. However, since the two polarities (+) and (−) can be represented by 1 bit (e.g., logic 0 = minus and logic 1 = plus), BIN and BCD can be made bipolar by adding a sign bit, usually in front of the digital number. They are then called sign-magnitude BIN or BIN plus sign and sign-magnitude BCD or BCD plus sign. OBN, inverted OBN, 1SC, and 2SC are designated bipolar codes because they can represent both positive and negative analog quantities. All bipolar codes, including BIN + sign and BCD + sign, are 1 bit longer than equivalent unipolar codes having the same resolution, but have twice the range.

Although binary digital codes are capable of representing both integers and fractions, the codes employed in D/A and A/D converters, whether unipolar or bipolar, are fractional codes consisting of binary bits having the weighted values $A_n/2^n$ (in straight BIN) or $A_n/2^{n-1}$ (in many other binary codes). The presence or absence of any bit in a digital number depends on the code relationship between the digital number and the particular analog quantity it represents. If the code calls for the presence of the nth bit, then $A_n = 1$, and the nth bit has the value $1/2^n$ (or $1/2^{n-1}$); if the code calls for the absence of the nth bit, then $A_n = 0$, and the value of the nth bit is 0. An N-bit fractional number in straight binary, for example, is the sum of N terms of the form $A_n/2^n$, as follows:

$$\frac{A_1}{2^1} + \frac{A_2}{2^2} + \frac{A_3}{2^3} + \cdots + \frac{A_n}{2^n} + \cdots + \frac{A_N}{2^N} \tag{10.1}$$

$$= \frac{A_1}{2} + \frac{A_2}{4} + \frac{A_3}{8} + \cdots + \frac{A_n}{2^n} + \cdots + \frac{A_N}{2^N}$$

The first term, $A_1/2$, is the most-significant bit or MSB; the term $A_n/2^n$ is the nth bit; and the last term, $A_N/2^N$, is the least-significant bit or LSB. Since the LSB, $A_N/2^N$, is the smallest increment by which the digital number can change, it is equal to the resolution. Since each bit is half the weight of the previous bit, the resolution doubles for each additional bit. Table 10.1 lists the weights and resolutions of 1 to 20 binary bits.

The value of a digital number is the sum of the weights of all terms having coefficients $A = 1$. The maximum value of a digital number is the sum of all the terms when all coefficients $A_n = 1$ and is less than unity by the weight of the LSB, as follows:

$$\sum_{n=1}^{N} \frac{A_n}{2^n} = \left(1 - \frac{1}{2^N}\right) = (1 - \text{LSB}) \tag{10.2}$$

Thus, the larger the number of bits N, the closer the value of a fractional digital number can approach unity. It is also noteworthy that the sum of all the terms to the right of any term can never exceed the value of that term. Since the weight of each bit in any binary code is related only to its position, it is not necessary to state the denominators, but only the value of the coefficient, 0 or 1. Thus, in straight BIN, the 10-bit digital

TABLE 10.1 *Binary Bit Weights and Resolutions*

Bit No. n	*Fraction* $1/2^n$	*Decimal* $(0.5)^n$	*Percent Resolution*	*PPM Resolution*
1	1/2	0.5	50.0	500000
2	1/4	0.25	25.0	250000
3	1/8	0.125	12.5	125000
4	1/16	0.0625	6.25	62500
5	1/32	0.03125	3.125	31250
6	1/64	0.015625	1.5625	15625
7	1/128	0.007812	0.7812	7812
8	1/256	0.003906	0.3906	3906
9	1/512	0.001953	0.1953	1953
10	1/1024	0.0009766	0.0977	977
11	1/2048	0.00048828	0.0488	488
12	1/4096	0.00024414	0.0244	244
13	1/8192	0.00012207	0.0122	122
14	1/16384	0.000061035	0.0061	61
15	1/32768	0.0000305176	0.0031	30.5
16	1/65536	0.0000152588	0.0015	15.3
17	1/131072	0.00000762939	0.00076	7.6
18	1/262144	0.000003814697	0.00038	3.8
19	1/524288	0.000001907349	0.00019	1.9
20	1/1048576	0.0000009536743	0.000095	0.95

number 1001100110 has the value 0.599609, calculated from Table 10.1 as follows:

Bit	*Weight*	*Code*	*Value*
1	0.5	1	0.5
2	0.25	0	0.0
3	0.125	0	0.0
4	0.0625	1	0.0625
5	0.03125	1	0.03125
6	0.015625	0	0.0
7	0.007812	0	0.0
8	0.003906	1	0.003906
9	0.001953	1	0.001953
10	0.0009766	0	0.0
			0.599609

Tables 10.2 and 10.3 give examples of 12-bit unipolar codes representing analog voltage ranges of 0 to +5 V FS (full-scale) and 0 to +10 V FS. Table 10.4 gives examples of 12-bit bipolar codes representing analog voltage ranges of ±5 V FS. Note that 1SC and BIN + sign

TABLE 10.2 *Twelve-Bit Unipolar Binary Codes*

Scale	*+10 V FS*	*+5 V FS*	*Straight Binary*	*Complementary Binary*
+FS − 1 LSB	+9.9976	+4.9988	1111 1111 1111	0000 0000 0000
+7/8 FS	+8.7500	+4.3750	1110 0000 0000	0001 1111 1111
+3/4 FS	+7.5000	+3.7500	1100 0000 0000	0011 1111 1111
+5/8 FS	+6.2500	+3.1250	1010 0000 0000	0101 1111 1111
+1/2 FS	+5.0000	+2.5000	1000 0000 0000	0111 1111 1111
+3/8 FS	+3.7500	+1.8750	0110 0000 0000	1001 1111 1111
+1/4 FS	+2.5000	+1.2500	0100 0000 0000	1011 1111 1111
+1/8 FS	+1.2500	+0.6250	0010 0000 0000	1101 1111 1111
0 + 1 LSB	+0.0024	+0.0012	0000 0000 0001	1111 1111 1110
0	0.0000	0.0000	0000 0000 0000	1111 1111 1111

TABLE 10.3 *Twelve-Bit Unipolar BCD Codes*

Scale	*+10 V FS*	*+5 V FS*	*Binary-Coded Decimal*	*Complementary BCD*
+FS − 1 LSB	+9.99	+4.95	1001 1001 1001	0110 0110 0110
+7/8 FS	+8.75	+4.37	1110 0000 0000	0001 1111 1111
+3/4 FS	+7.50	+3.75	1100 0000 0000	0011 1111 1111
+5/8 FS	+6.25	+3.12	1010 0000 0000	0101 1111 1111
+1/2 FS	+5.00	+2.50	1000 0000 0000	0111 1111 1111
+3/8 FS	+3.75	+1.87	0110 0000 0000	1001 1111 1111
+1/4 FS	+2.50	+1.25	0100 0000 0000	1011 1111 1111
+1/8 FS	+1.25	+0.62	0010 0000 0000	1101 1111 1111
0 + 1 LSB	+0.01	+0.00	0000 0000 0001	1111 1111 1110
0	+0.00	0.00	0000 0000 0000	1111 1111 1111

TABLE 10.4 *Twelve-Bit Bipolar Codes*

Scale	*5 V FS*	*Offset Binary*	*Two's Complement*	*One's Complement*	*Magnitude + Sign*
+FS −1 LSB	+4.9976	1111 1111 1111	0111 1111 1111	0111 1111 1111	1111 1111 1111
+3/4 FS	+3.7500	1110 0000 0000	0110 0000 0000	0110 0000 0000	1110 0000 0000
+1/2 FS	+2.5000	1100 0000 0000	0100 0000 0000	0100 0000 0000	1100 0000 0000
+1/4 FS	+1.2500	1010 0000 0000	0010 0000 0000	0010 0000 0000	1010 0000 0000
0	0.0000	1000 0000 0000	0000 0000 0000	0000 0000 0000* 1111 1111 1111*	1000 0000 0000* 0000 0000 0000*
−1/4 FS	−1.2500	0110 0000 0000	1110 0000 0000	1101 1111 1111	0010 0000 0000
−1/2 FS	−2.5000	0100 0000 0000	1100 0000 0000	1011 1111 1111	0100 0000 0000
−3/4 FS	−3.7500	0010 0000 0000	1010 0000 0000	1001 1111 1111	0110 0000 0000
−FS+1 LSB	−4.9976	0000 0000 0001	1000 0000 0001	1000 0000 0000	0111 1111 1111
−FS	−5.0000	0000 0000 0000	1000 0000 0000	—	—

* ISC and Magnitude + Sign each have two words for zero, the upper of which is called 0+ and the lower of which is called 0−.

each have two words for zero, the upper of which is called 0+ and the lower of which is called 0−. Because 1SC and BIN + sign have two words for zero, their range is one LSB smaller than that of OBN and 2SC.

10.2.1 Translation between Codes

When digital input data are not in a code acceptable to the input circuits of a particular D/A converter, or when digital output data from an A/D converter are not in the code required by subsequent equipment, it becomes necessary to translate the data to the required code. Some D/A converters can be externally connected so as to accept either of two codes. The following list tells how to translate from any one to any other of four bipolar codes: BIN magnitude + sign, OBN, 1SC, and 2SC.

1. To convert from BIN magnitude + sign to:
   1SC: If MSB = 1, complement other bits.
   2SC: If MSB = 1, complement other bits, add 00 . . . 01.
   OBN: Complement MSB. If new MSB = 1, complement other bits, add 00 . . . 01.
2. To convert from 1SC to:
   2SC: If MSB = 1, add 00 . . . 01.
   OBN: Complement MSB. If new MSB = 0, add 00 . . . 01.
   BIN magnitude + sign: If MSB = 1, complement other bits.
3. To convert from 2SC to:
   1SC: If MSB = 1, add 11 . . . 11.
   OBN: Complement MSB.
   BIN magnitude + sign: If MSB = 1, complement other bits, add 00 . . . 01.
4. To convert from OBN to:
   1SC: Complement MSB. If new MSB = 1, add 11 . . . 11.
   2SC: Complement MSB.
   BIN magnitude + sign: Complement MSB. If new MSB = 1, complement other bits, add 00 . . . 01.

10.3 DIGITAL-TO-ANALOG CONVERTERS

D/A converters or DACs generate analog output voltages that are digitally controlled, binary weighted, discrete fractions of some reference voltage V_{REF}. Unless otherwise designated, D/A converters usually have a regulated internal reference voltage source, so V_{REF} is fixed, and the analog output voltage has a single value for each digital input code.

When designated ratio, multiplying, external-reference or universal-reference D/A converters, the reference voltage V_{REF} is externally supplied as a separate input. The analog output voltage is then the product of the variable analog voltage V_{REF} and the discrete fraction represented by the digital input code.

Contemporary D/A converters employ networks of precision resistors, controlled by bipolar transistors or FET switches turned on and off in accordance with the digital input code. Figure 10.1 shows a D/A converter employing a network of summing resistors, having the relative weightings R, $2R$, $4R$, $8R$, . . ., $2^N R$, connected to the (−) summing point of an operational amplifier. The input end of resistor R is connected to the output of the amplifier, and the input ends of resistors $2R$, $4R$, $8R$ . . . $2^N R$ are all connected to a source of reference voltage, V_{REF}. The high amplifier gain and the large negative feedback operate to maintain the summing point of virtual ground; so if switches S1, S2, S3, . . ., S(N) are closed, the currents

$$I_1 = \frac{V_{REF}}{2R} \quad (=\text{MSB}) \qquad I_3 = \frac{V_{REF}}{8R}$$

$$\vdots$$

$$I_2 = \frac{V_{REF}}{4R} \qquad I_N = \frac{V_{REF}}{2^N R} \quad (=\text{LSB})$$

flow through the resistors $2R$, $4R$, $8R$, . . ., $2^N R$ to the summing point. The amplifier has high input resistance, so the current into the (−) input is

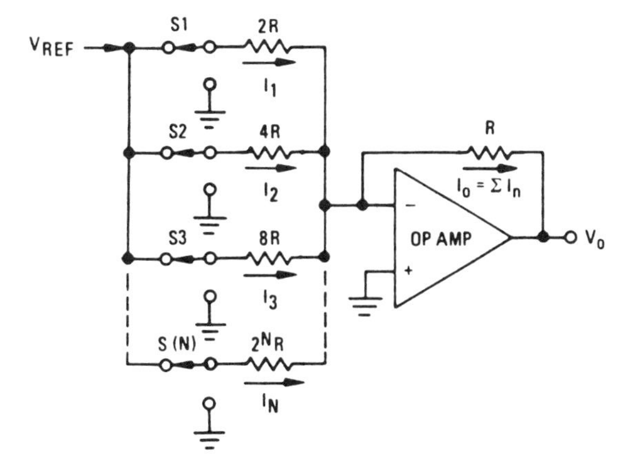

FIGURE 10.1 D/A converter employing $2^N R$ summing network. Courtesy of ILC Data Device Corp.

negligible, and the sum of the input currents is equal to the feedback current I_O out of the summing point. The output voltage is therefore

$$V_O = I_O R = -V_{REF}\left(\frac{A_1}{2} + \frac{A_2}{4} + \frac{A_3}{8} + \cdots + \frac{A_N}{2^N}\right)$$

$$= -V_{REF}\left[\sum_{n=1}^{N} \frac{A_n}{2^n}\right] \tag{10.3}$$

where $A_n - 1$ if switch n is closed, and $A_n = 0$ if switch n is open. This configuration therefore generates analog output voltages equal to the binary-weighted values of the switch positions.

D/A conversion by means of $2^N R$ summing networks is limited to 12 or 13 bits because more bits would require impractically high resistances. Resistance R_{12}, for example, must be 4096 times R, and R_{20} must be 1,048,576 times R. Also, since the network resistance varies greatly with the combinations of switch settings, the amplifier design cannot be optimized to achieve fastest settling time.

The preceding limitations are overcome by means of the R–$2R$ ladder network shown in Figure 10.2, for which it can be shown that switches S1, S2, S3, . . . , S(N) cause output increments $-V_{REF}/2$, $-V_{REF}/4$, $-V_{REF}/8, \ldots, -V_{REF}/2^N$. This network is economically realizable because the resistor values are either R or $2R$.

Also, the network resistance seen from the summing points is con-

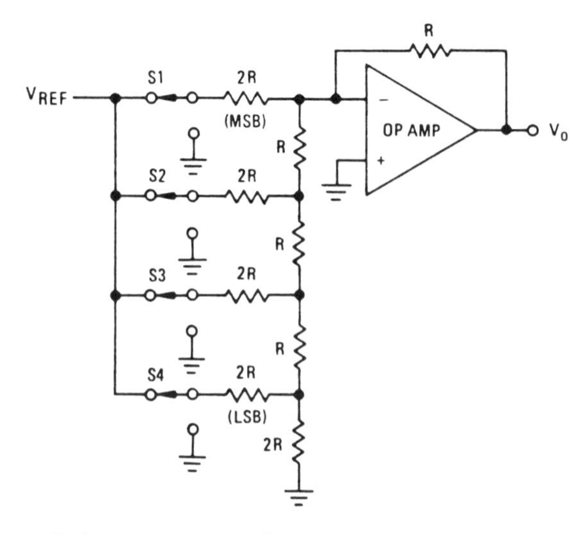

FIGURE 10.2 D/A converter employing R–$2R$ ladder network. Courtesy of ILC Data Device Corp.

stant at the low value of R, so the operational amplifier may be designed for fastest settling time. Most of the D/A converters listed in this section therefore employ R–2R ladder networks, with low-resistance FET switches that turn on and off in nanoseconds.

10.4 APPLICATION OF D/A CONVERTERS

An important application of D/A converters is shown in Figure 10.3, where multiplexed digital data are fed to several D/A converters, and it is desired that each converter respond only to its channel. The i-channel multiplexed data are fed to a register that is strobed when the ith channel data appear and holds these data until the next strobe. The output of the register is fed to the D/A converter, which generates the ith channel analog output. This output is held constant until updated by the next strobe. This system is equivalent to a long-term sample–hold circuit with digital input and analog output.

Figure 10.4 shows a bipolar D/A converter employed in a digital servo for automatic frequency control of a voltage-tuned receiver. The bipolar discriminator output (e.g., negative when the receiver is tuned below the signal, positive when the receiver is tuned above the signal) is fed to a control logic circuit that sends clock pulses and either an up-count or a down-count command to an up–down counter, the digital output of which drives the D/A converter. The analog output voltage is proportional to the frequency error, and its polarity is such as to reduce that error.

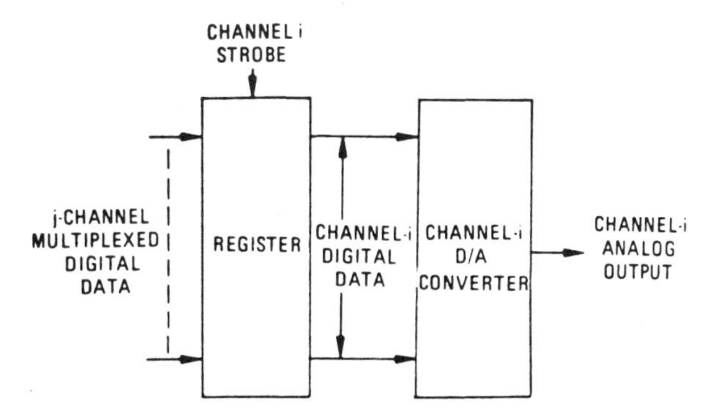

FIGURE 10.3 Digital data distribution systems. Courtesy of ILC Data Device Corp.

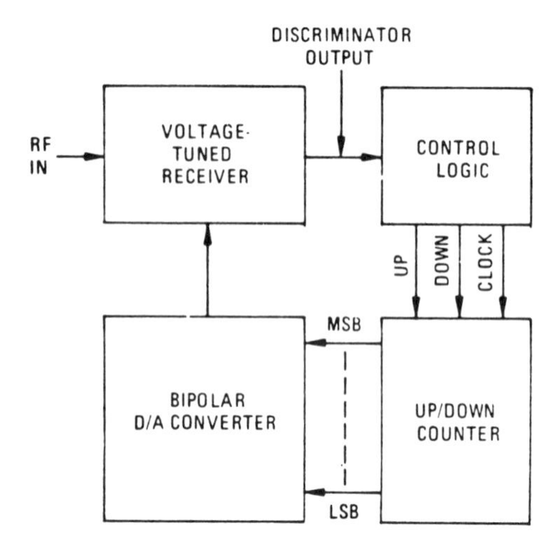

FIGURE 10.4 Digital AFC employing D/A converter. Courtesy of ILC Data Device Corp.

10.5 MULTIPLYING D/A CONVERTERS

In the D/A converters described so far, the output voltage is the product of the reference voltage V_{REF} and the binary weighted fraction represented by the switch settings. If the circuits are designed to operate with wide-ranging values of V_{REF}, as in Figure 10.5, they are designated multiplying, ratio, external-reference, or universal-reference D/A converters. If both the analog input V_{REF} and the digital input are unipolar, the output is in one quadrant; if either input is unipolar and the other is bipolar, the output is in two quadrants; if both digital and analog inputs are bipolar, the output is in four quadrants.

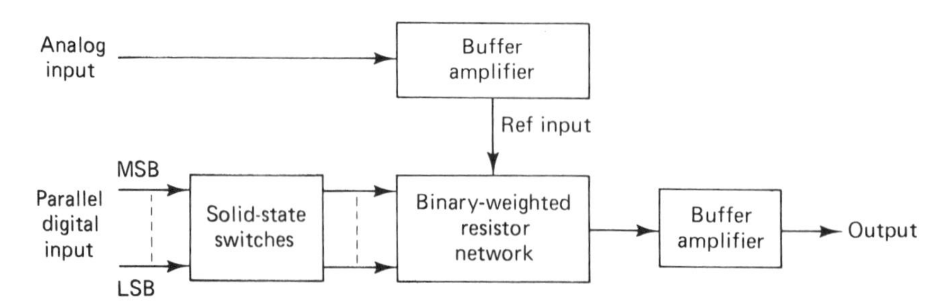

FIGURE 10.5 Simplified block diagram of multiplying D/A converter. Courtesy of ILC Data Device Corp.

10.6 APPLICATIONS OF MULTIPLYING D/A CONVERTERS

Multiplying Analog and Digital Inputs. The basic block diagram of Figure 10.5 shows that a multiplying D/A converter can be used to generate an analog output voltage XV_{REF}, where X is a digital input and V_{REF} is an analog input.

Digitally Programmable Attenuator. The D/A converter in the preceding application can be considered a digitally programmable attenuator, the output of which (XV_{REF}) is a digitally controlled fraction (X) of the analog input voltage (V_{REF}). If the converter accepts bipolar values of V_{REF}, it will accept AC and will thus function as a digitally programmable AC attenuator.

Multiplying Two Digital Inputs. Figure 10.6 shows how an analog output voltage XY can be obtained from the digital inputs X and Y by use of an internal-reference D/A converter and an external reference D/A converter.

Polar-to-Rectangular Conversion. Figure 10.7 shows how one internal-reference D/A converter and two multiplying D/A converters can be used to obtain the analog rectangular coordinates $Y = R \sin \theta$ and $X = R \cos \theta$ from the polar coordinates R and θ. Digital R is fed to the D/A converter, and digital sin θ and digital cos θ are fed to the multiplying D/A converters.

PPI Sweep Generator. Figure 10.8 shows how a rotating radial sweep, as used in a PPI radar, can be generated from an analog linear sweep and digital sin θ and digital cos θ inputs.

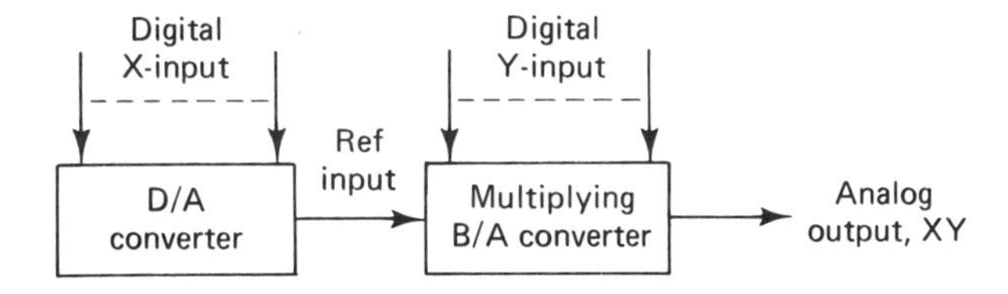

FIGURE 10.6 Multiplying two digital inputs. Courtesy of ILC Data Device Corp.

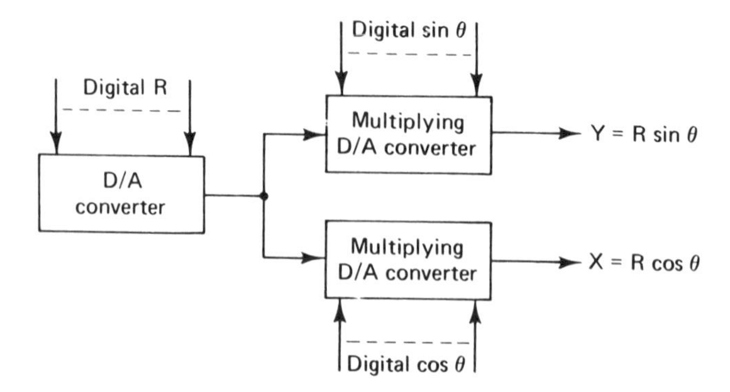

FIGURE 10.7 Polar-to-rectangular conversion. Courtesy of ILC Data Device Corp.

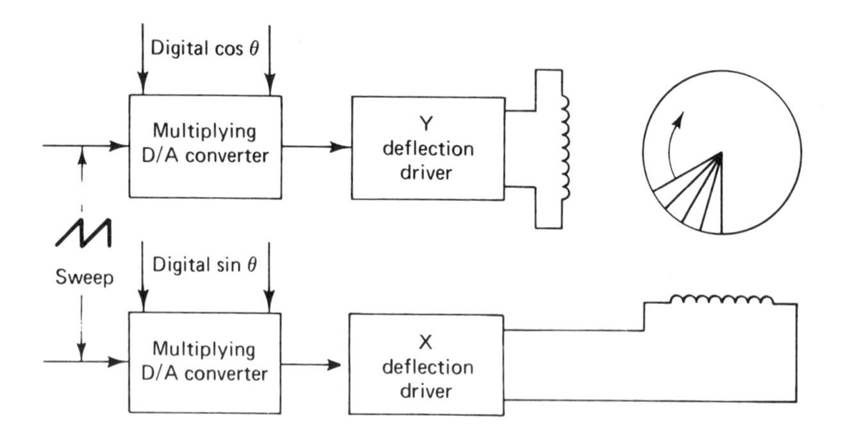

FIGURE 10.8 PPI sweep generator. Courtesy of ILC Data Device Corp.

10.7 ANALOG-TO-DIGITAL CONVERTERS

Converter Terminology and Parameters. Although A/D and D/A converters perform inverse operations, their input–output or transfer characteristics, shown in Figures 10.9 and 10.10, are not exactly inverse. In the D/A converter, each digital input produces a single value of the analog output, and all the input–output points fall on a straight line; but in the A/D converter, any analog value within $\pm\frac{1}{2}$ LSB of the ideal value will give the same digital output, so the digital output is a step approximation to the ideal straight-line relationship. A plot of quantizing error

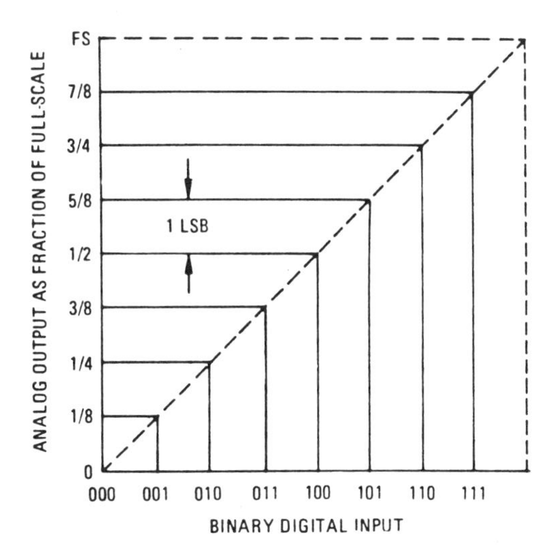

FIGURE 10.9 Transfer characteristic of a 3-bit D/A converter. Courtesy of ILC Data Device Corp.

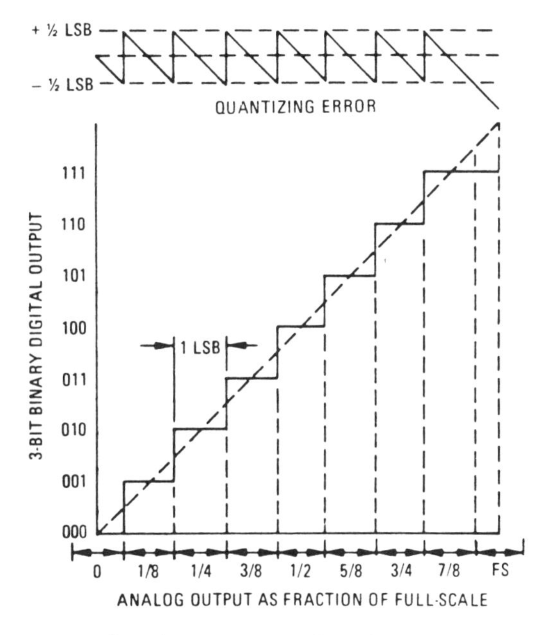

FIGURE 10.10 Transfer characteristic of a 3-bit A/D converter. Courtesy of ILC Data Device Corp.

versus analog input is shown at the top of Figure 10.10. This plot shows that the quantizing error varies cyclically in a sawtooth fashion from $+\frac{1}{2}$ LSB through 0 to $-\frac{1}{2}$ LSB. The rms value of that sawtooth is

$$\text{rms error} + \frac{\frac{1}{2}\,\text{LSB}}{2\sqrt{3}} = \frac{\text{LSB}}{4\sqrt{3}} \tag{10.4}$$

The maximum quantizing error of $\pm\frac{1}{2}$ LSB (called the quantizing uncertainty) is inherent in the discrete conversion process and can only be reduced by reducing the size of the LSB. Since LSB $= 1/2^N$, the size of the LSB is reduced by increasing N, the number of bits. The theoretical resolution of an N-bit converter is 2^N, which means that, ideally, a D/A converter will be able to generate 2^N discrete levels of analog output voltage, and an A/D converter will be able to respond with separate digital code numbers to 2^N discrete levels of analog input voltage. If internal noise and/or drift in a converter exceed quantizing step levels, missing codes will occur, and the actual resolution will be less than the theoretical resolution.

Monotonicity in a converter means that the output continuously increases with the input. A monotonic A/D converter will generate successively increasing codes as the analog input increases, and a monotonic D/A converter will generate a continuously increasing analog output voltage with increasing values of the digital input code.

The linearity of a converter is defined as the maximum deviation of the transfer function from a best straight line through that transfer function. The relative accuracy of a converter is the maximum deviation of the analog input or output voltage from a straight line referred to the source of reference voltage, V_{REF}. Absolute accuracy is the maximum deviation of the analog input or output voltage from a straight line referred to the NBS absolute volt.

The major types of A/D converters employ (in order of increasing speed) counter or ramp, dual-slope integration, successive approximation, serial–parallel, and parallel techniques. The successive-approximation technique has the highest performance–cost ratio in the present state of the art and permits a wider range of trade-offs between speed and resolution than any other type. Each A/D converter described in this section is a successive-approximation type, offering a high resolution–speed product at low cost. Video types are parallel and two-step (serial–parallel).

Figure 10.11 is a simplified block diagram of a successive-approximation A/D converter. The analog input voltage, V_{in}, to be digitized and the output voltage, V_{da}, of a D/A converter are fed to a comparator. The comparator output is fed to a control logic circuit that drives a register, which in turn controls the D/A converter and thus determines the voltage V_{da}. The closed loop thus formed operates by suc-

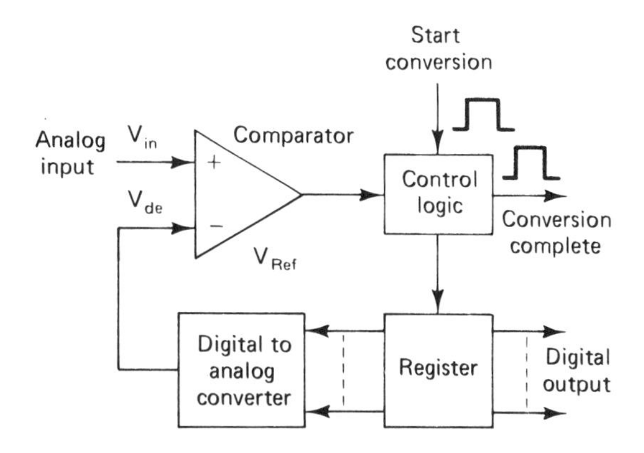

FIGURE 10.11 Simplified block diagram of a successive approximation A/D converter. Courtesy of ILC Data Device Corp.

cessive approximations to reduce the difference between the D/A output voltage V_{da} and the analog input voltage V_{in} to less than the value of the LSB. In this condition, a number represented by the digital output of the register is equal (within the value of the LSB) to the analog input to the comparator.

The system operates as follows: a start conversion command clears the register and converter. Bit 1 (the MSB) is set to 1, and its weight ($\frac{1}{2}$ FS) is compared to the analog input V_{in}. If V_{in} is greater than the MSB, the MSB remains at 1, and bit 2 ($\frac{1}{4}$ FS) is set to 1. If the combined weights of bits 1 and 2 ($\frac{1}{2}$ FS + $\frac{1}{4}$ FS = $\frac{3}{4}$ FS) now exceed V_{in}, bit 2 is set to 0, and bit 3 ($\frac{1}{8}$ FS) is set to 1. Any succeeding bit that does not cause the combined bit weight to exceed V_{in} remains set at 1, while any succeeding bit that causes the combined bit weight to exceed V_{in} is set to 0. When the LSB has been tried, the control logic transmits a conversion complete signal. The output of the register at that time is a digital number representing the quantized analog voltage V_{in} as a fraction of the D/A reference voltage V_{REF}. There is one value of V_{REF} for which the digital output represents V_{in} exactly in volts, so the A/D converter is a digital voltmeter. However, if V_{REF} is a separate input variable, the A/D converter becomes a ratiometric converter, the digital output of which represents the fraction V_{in}/V_{REF}.

Since the A/D conversion process takes time, the digital output will be in error if the analog input voltage changes during the conversion. A sample–hold circuit (S/H) is therefore often employed at the analog input to sample V_{in} before conversion and to hold the A/D input at the value of V_{in} that existed at the instant of the start conversion command, until conversion is complete. The conversion complete signal may then be used to

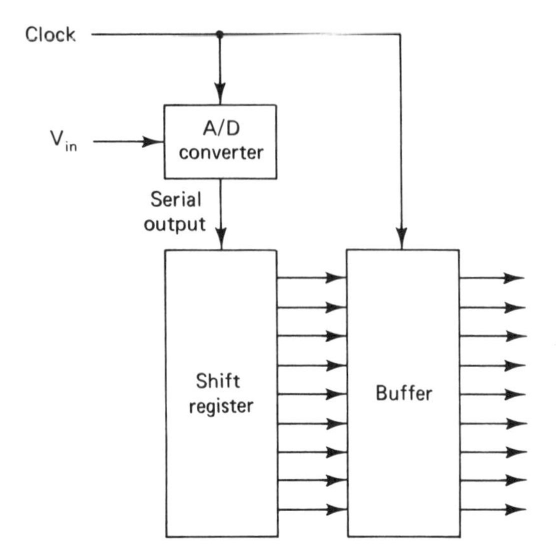

FIGURE 10.12 Conversion of A/D serial output to parallel output. Courtesy of ILC Data Device Corp.

return the S/H to its sampling mode. If the converter is a ratiometric type with variable V_{REF}, it will be necessary to apply a S/H to the input, too.

The digital output of an A/D converter may appear simultaneously on *N* lines in parallel or serially on one line. Where it is necessary to convert such serial data to equivalent parallel data, the arrangement of Figure 10.12 may be used. Note that serial data may be transmitted in either of two ways. In return-to-zero (RZ) transmission, the levels return to ground between successive bits; in non-return-to-zero (NRZ), the levels change only when the leading edge of a clock pulse is present.

10.8 APPLICATIONS OF A/D CONVERTERS

The fundamental application of A/D converters is the conversion of analog signals to digital form (i.e., digitizing), and this application is included in all the following applications of A/D converters.

Figure 10.13 shows how an A/D converter with an internal reference, V_{REF}, and a multiplying D/A converter can be used to multiply two analog voltages with excellent precision. If both are 12-bit devices, the overall accuracy will be on the order of 0.1%. Input V_1 is fed to the A/D converter, the digital output of which corresponds to V_1/V_{REF}.

This digital number and input V_2 are fed to the multiplying D/A converter, which generates the analog output voltage V_1V_2/V_{REF}. Figure

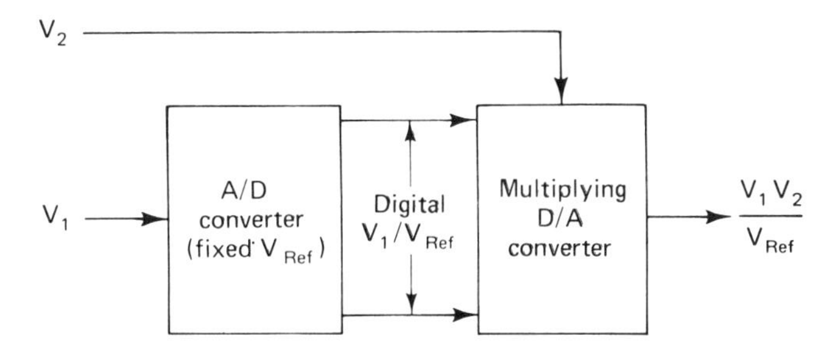

FIGURE 10.13 Precision multiplication of analog voltages. Courtesy of ILC Data Device Corp.

10.14 shows a similar arrangement, employing a ratiometric A/D converter dividing an analog input V_1 by a second input V_2.

The DDC 5200 series A/D converters shown in Figure 10.15 are 12-bit, successive-approximation devices in hermetically sealed 24-pin double DIP hybrid packages. All specifications are met with an externally applied 240-kHz clock, providing a conversion time of 50-μs. Functional laser trimming of a thin-film resistor network results in extremely accurate and highly stable adjustment-free converters. Linearity error is guaranteed to be better than $\frac{1}{2}$ LSB, with no missing codes over the specified temperature range.

These converters are available in two unipolar and two bipolar input ranges and will operate with full accuracy at temperatures of 0° to +70°C or −55° to +125°C. The DDC 5200 series converters are form-fit-function replacements for the MN5200 series A/D converters.

Because of their high reliability, hermetically sealed package, low power consumption, and wide temperature and dynamic range, the DDC 5200 series will meet the most demanding military and industrial requirements. Typical applications include data-acquisition systems, automatic test equipment, and electronic countermeasures.

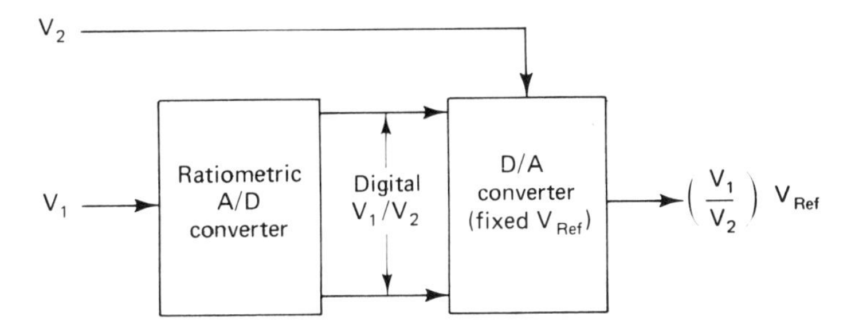

FIGURE 10.14 Precision division of analog voltages. Courtesy of ILC Data Device Corp.

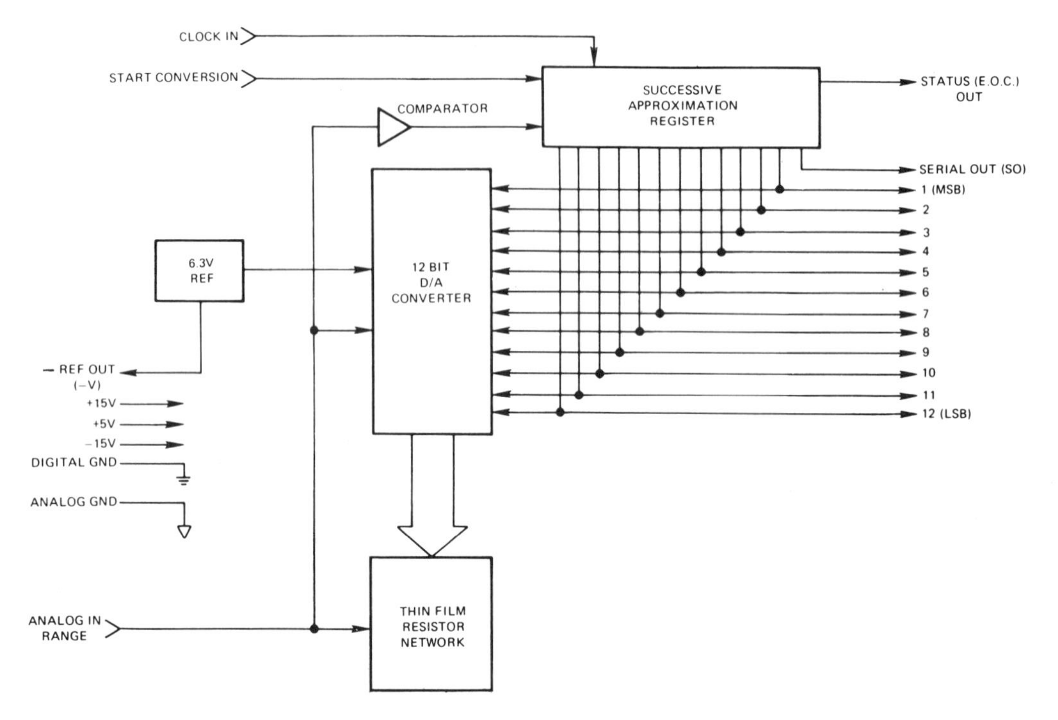

FIGURE 10.15 Block diagram of the DDC 5200 series, 12-bit, 50-μs hybrid A/D converter. Courtesy of ILC Data Device Corp.

Table 10.5 provides the reader with the specifications of the DDC 5200 series A/D converter. Parameters of the specifications include the following:

1. Analog inputs
2. Resolution
3. Digital inputs
4. Reference output
5. Power supply characteristics
6. Physical characteristics

The DDC 5200 series A/D converters reset internal logic on a low-to-high clock transition, while the start conversion (pin 1) is low. The start line must be low 25 ns (min) prior to a low-to-high clock transition (see Figure 10.16). The converter will remain reset until the start line is allowed to go high. Conversion commences upon the next low-to-high clock transition. Data are available 50 μs after the clock transition (commencing the conversion) occurs. A status (EOC) output signal (logic 1) is presented on pin 22 at a maximum of 120 ns after the start line is driven

TABLE 10.5. *Specifications for the DDC 5200 Series*

Parameter	*Unit*	*Value*			
ANALOG INPUTS		DDC 5200	DDC 5201	DDC 5202	DDC 5206
Input Ranges	V	0 to −10	−5 to +5	−10 to +10	0 to +10
Input Impedance	kΩ	6.7	6.7	13.4	13.4
Maximum without damage	V	±25	±25	±25	±25
RESOLUTION	bits		12		
ACCURACY AND DYNAMICS					
Linearity Error	LSB		TYP	MAX	
+25°C			±1/4	±1/2	
0°C to +70°C (−3)			±1/4	±1/2	
−55°C to +125°C (−1)				±1/2	
Differential Linearity Error	LSB		1/2		
No Missing Codes		Guaranteed over temperature			
Full Scale Absolute Accuracy Error					
+25°C	%FSR		±0.025	±0.05	
0°C to +70°C (−3)	%FSR		±0.2	±0.4	
−55°C to +125°C (−1)	%FSR			±0.4	
Zero Error					
+25°C	%FSR		±0.01	±0.025	
0° to +70°C (−3)	%FSR		±0.025	±0.05	
−55°C to +125°C (−1)	%FSR			±0.05	
Gain Error	%		±0.025		
Gain Drift	ppm/°C		±10		
Conversion Time	μs			50	
DIGITAL INPUTS		MIN	TYP	MAX	
Clock Input (External)					
High Pulse	ns	125			
Low Pulse	ns	175			
High Loading	μA		2	20	
Low Loading	mA		−0.25	−0.4	
Frequency	kHz			240	
Start Conversion Input					
High Loading	μA		4	40	
Low Loading	mA		−0.25	−0.4	
Setup Time	ns	25			
Logic Levels					
Logic "1"	V	2.0			
Logic "0"	V			0.7	
DIGITAL OUTPUTS					
Logic Coding					
Unipolar Ranges		Complementary Straight Energy			
Bipolar Ranges		Complementary Offset Binary			
Logic Levels		MIN	TYP	MAX	
Logic "1"	V	2.4	3.6		
Logic "0"	V		0.15	0.3	

TABLE 10.5 (*Continued*)

Parameter	*Unit*	*Value*		
Output Drive Capability				
Logic "1"	TTL Loads	8		
Logic "0"	TTL Loads	2		
REFERENCE OUTPUT		MIN	TYP	MAX
Internal Reference				
Voltage	V		−6.3	
Accuracy	%		±2	
Drift Tempco	ppm/°C		±5	
Max External Current (without buffer)	μA			100
POWER SUPPLIES				
Power Supply Range				
+15 V Supply	%			±3
+5 V Supply	%			±5
Power Supply Rejection				
+15 V Supply	%FSR/%ps		±0.005	±0.02
−15 V Supply	%FSR/%ps		±0.01	±0.05
Current				
+15 V Supply	mA		+13	+18
−15 V Supply	mA		−20	−27
+5 V Supply	mA		+35	+48
Power Consumption	mW		670	915
PHYSICAL CHARACTERISTICS				
Size	in	1.3 × 0.8 × 0.2 (33.9 × 21.1 × 0.5mm)		
Weight	oz	0.2		

Courtesy of ILC Data Service Corp.
SPECIFICATIONS
Typical values at +25°C ambient temperature and nominal power supply voltages

low. Twelve bits are tied in succession, MSB through LSB, and serial data are immediately available on pin 3 during the conversion. Parallel data are available 30 ns (max) after the status (EOC) line returns to logic 0. Output data are held in the latch until another start conversion signal is input on pin 1. The drive capability of output data is two TTL loads. Continuous conversions may be accomplished by connecting the status (EOC) output (pin 22) to the start conversion (pin 22).

10.9 MULTIPLEXERS

When it is required to perform A/D conversion on multiple analog inputs, a separate A/D converter may be employed for each, or an analog MUX may be used to sample the analog inputs and feed the sampled

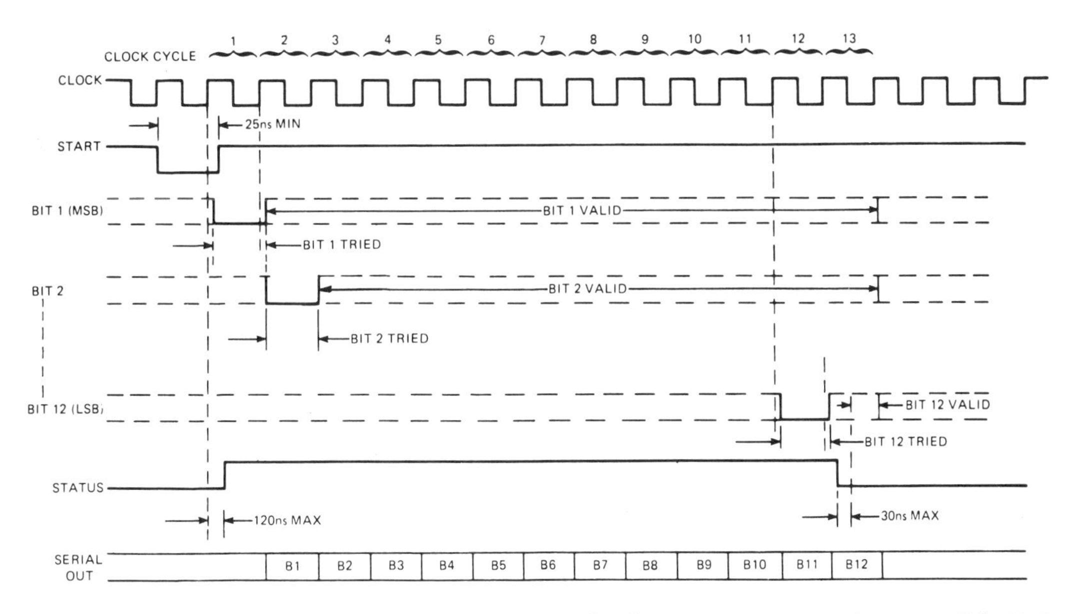

FIGURE 10.16 Converter timing for the DDC 5200 series. Courtesy of the ILC Data Device Corp.

analog data sequentially to a single A/D converter, which develops a corresponding sequential digital output. Conversely, an analog MUX may be used to distribute the sequential analog output of a D/A converter to multiple analog output channels. Figure 10.17 shows both processes. Although not shown, means must be provided for relating each segment of the sequential data to the corresponding analog channel so that data from each analog input channel are always distributed to the same-numbered analog output channel.

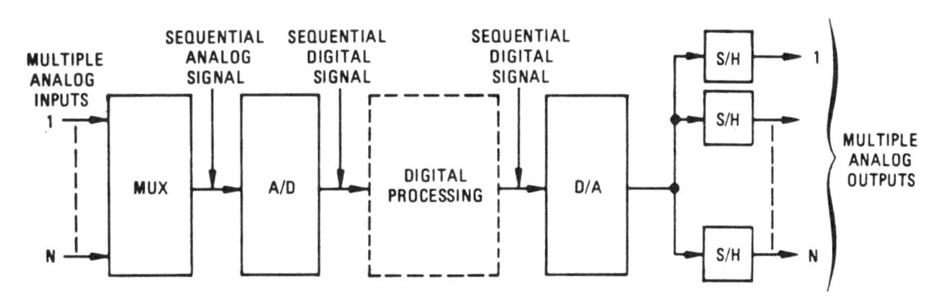

FIGURE 10.17 Use of MUXs for multiplexing and distribution. Courtesy of ILC Data Device Corp.

10.9.1 Multiplexer Terminology and Parameters

The transfer accuracy of any channel of a MUX is a measure of the difference between the input voltage, e_{in}, and the output voltage, e_o:

$$\text{Transfer accuracy} = 100\frac{e_{in} - e_o}{e_{in}} \% \tag{10.5}$$

Throughput rate is the highest rate at which the MUX will switch from channel to channel while maintaining a specified transfer accuracy. Throughput rate is the inverse of the sum of switching and settling times, as follows:

$$\text{Throughput rate} = \frac{1}{\text{switching time} + \text{settling time}}$$

Crosstalk attenuation is a measure of the MUX's ability to prevent signals at the *off* inputs from appearing at output. When a test voltage e_{test} of specified amplitude and frequency is applied to all the *off* channel inputs and the output voltage e_o is measured, the crosstalk attenuation in decibels is given by:

$$\text{Crosstalk attenuation} = \left(\frac{20 \log_{10} e_{test}}{e_o}\right) \text{dB} \tag{10.6}$$

10.10 SAMPLE-HOLD AND TRACK-HOLD AMPLIFIER CONCEPTS

Sample–hold (S/H) and track–hold (T/H) amplifiers are used when the voltage level of a signal must be held or stored for a short period of time. A signal that varies rapidly may have to be held constant to allow time for an analog-to-digital or synchro-to-digital conversion to take place, or it may have to be stored while multiplexed signal channels are converted sequentially.

The main components of a S/H or T/H amplifier are shown in the simplified diagram of Figure 10.18. When the switch is closed, the output signal follows the analog input, and when the switch is opened, the output voltage is ideally fixed at the level of the input because it is stored on the holding capacitor. The difference between a S/H and T/H amplifier is that a S/H can only sample the input signal briefly, whereas a T/H can track the input continuously. In a S/H the gate is AC coupled to the switch, and the switch can remain closed for only a brief time. In a T/H the gate is DC coupled to the switch, and the switch remains closed as long as the gate signal stays in the track mode.

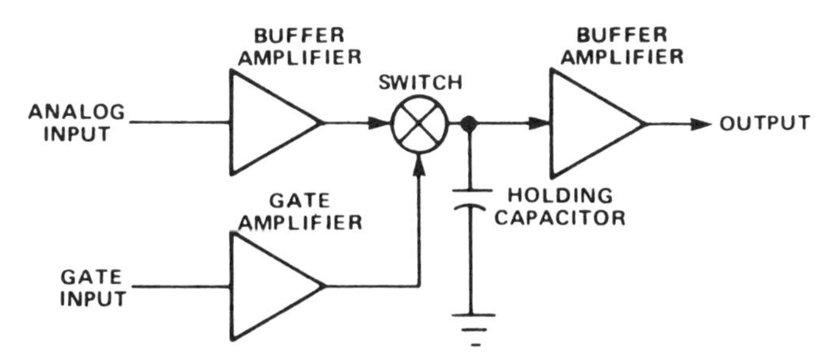

FIGURE 10.18 Simplified diagram of track–hold or sample–hold amplifier. Courtesy of ILC Data Device Corp.

The relationships between the output, input, and gate signals of the S/H or T/H amplifier are complex. In these devices there are delays between the gate signal changes and the opening or closing of the switch. Time is required for the output to settle, the output can have offsets, and input signals can feed through to the output even when the switch is open. These effects are illustrated in Figure 10.19 and are discussed in the following sections.

The definitions that follow describe the key terms used to characterize S/H and T/H amplifiers. The terminology is depicted graphically in Figure 10.19.

A logic gate signal initiates the track or sample period. After this signal has been applied, the holding capacitor requires an interval of time called the *acquisition time* to charge to the point where it can track the input signal. The acquisition time depends on the magnitude of the voltage change required. It includes the *acquisition turn-on-delay,* which

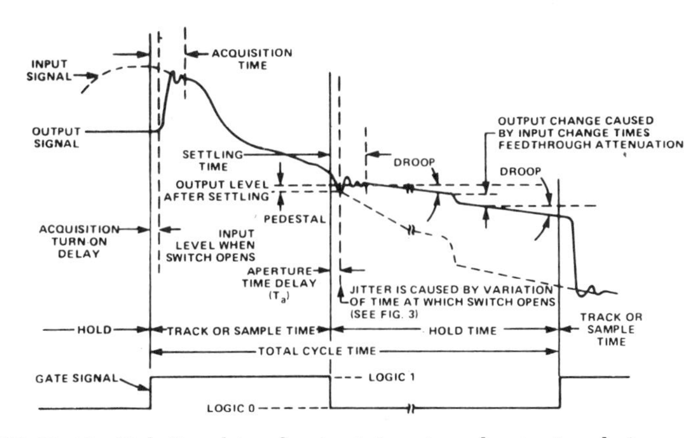

FIGURE 10.19 Relationship of output, input, and gate signals (assuming gate logic 0 = hold). Courtesy of ILC Data Device Corp.

is the time interval between application of the logic signal and the switch closure, plus in some cases output buffer activation time. The minimum track or sample period must be at least as long as the acquisition time.

When a signal is applied to the gate to initiate the hold period, there is a short delay, the *aperture-time-delay* (T_a), before the switch opens to isolate the holding capacitor from the analog input. The delay itself can be compensated for, but the *aperture time delay uncertainty* (*jitter*) associated with the delay can lead to errors, as shown in Figure 10.20.

The time needed for the S/H amplifier output to settle out to the holding capacitor voltage after the gate changes from the sample mode to the hold mode is the *settling time,* which includes the aperture time delay. Settling time is defined as the time required for an amplifier to settle to its final value $\pm$ a specified error band. If the output voltage is used at a shorter time interval after switching to the hold mode, errors will be introduced.

After the switch is opened and the output has settled, the output voltage generally increases or decreases linearly with time. This *droop* is caused mainly by the bias current of the output buffer amplifier and by leakage through the switch. The bias current and leakage can be positive or negative. Droop can also be caused by leakage from the holding capacitor. The droop rate will determine how long a hold time can be tolerated to maintain the required accuracy.

When the switch opens after the sample period, any unbalanced charge in the switch is transferred to the holding capacitor, creating a small positive or negative step error, the *pedestal,* in the output voltage during the hold period. The SH-8518 has the provision for fine trimming the pedestal error.

There will be a dc offset or difference in dc voltage level between the input and output voltages of the S/H or T/H because of offset errors

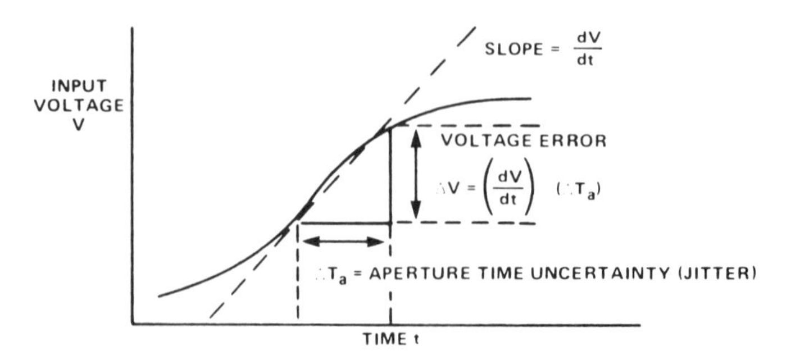

FIGURE 10.20 Aperture time uncertainty error (jitter). Courtesy of ILC Data Device Corp.

in the input and output amplifiers. Such a DC offset is present during both the track and the hold modes.

While the switch is open and the signal is being held, a step change in the input signal will still feed through the open switch capacitance to the output. The small output signal change that results is equal to the input voltage divided by the *feedthrough attenuation* of the open switch circuit.

10.11 SAMPLING RATE LIMITATIONS

The sampling rate is the rate at which an entire sample-and-hold cycle is completed.

$$\text{Sampling rate} = \frac{1}{\text{total cycle time}}$$

$$= \frac{1}{\text{track (or sample) time} + \text{hold time}} \quad (10.7)$$

The maximum sampling rate is therefore determined by the minimum sample time plus minimum hold time. The minimum sample time is the acquisition time, while the minimum hold time will include the settling time plus whatever additional time is required for information processing.

A S/H amplifier has a minimum sampling rate as well as a maximum one. The minimum rate is determined by the maximum sample time plus the maximum hold time. The sample time has a maximum value (which should be specified) because the gate is ac coupled.

The hold time may also have a maximum value because of droop. The output error will exceed the specified accuracy if the hold time is greater than the maximum error divided by the droop rate.

Droop may eventually cause the output voltage to exceed the limits of the specified output voltage range. The time to acquire a new signal may then exceed the specified acquisition time because of saturation effects.

10.12 ERROR CAUSED BY APERTURE TIME UNCERTAINTY

The uncertainty in the moment of time at which the signal is acquired and held (the aperture time jitter) causes an error whose magnitude depends on the rate at which the input is changing (Figure 10.20). The following discussion relates the maximum percent of error in the output V caused by this effect with the aperture time uncertainty ΔT_a and the analog signal frequency f.

The error ΔV in the voltage is

$$\Delta V = \left(\frac{dV}{dt}\right) \Delta T_a \tag{10.8}$$

For a sinusoidal signal, $V - V_o \sin(2\pi ft)$,

$$\Delta V = 2\pi f V_o \cos(2\pi ft)\, \Delta T_a \tag{10.9}$$

Since the maximum value of the cosine function is 1, the maximum percent of error is

$$\text{maximum \% error} = 100\left(\frac{\Delta V}{V_o}\right) = 200\pi(f)(\Delta T_a) = 628(f)(\Delta T_a)$$

As an example, if the aperture time uncertainty is 500 ps, then at a frequency of 0.5 MHz

$$\text{maximum \% error} = (628)(0.5 \times 10^6)(500 \times 10^{-12}) = 0.16\%$$

This percent of error is the percentage of the maximum input signal amplitude, not the percentage of allowable full scale. Note also that, if the input rate dV/dt is constrained only by slew rate limiting, then the maximum percent of error should be calculated directly as

$$\text{maximum \% error} = \frac{100}{V_o}\left(\frac{dV}{dt}\right)_{\text{max slew}} \Delta T_a \tag{10.10}$$

The error caused by aperture time uncertainty may be obtained from the nomograph in Figure 10.21 if the input signal is sinusoidal and the output is not slew rate limited. Any sloping straight line drawn on the nomograph from top to bottom relates the corresponding aperture uncertainty, percent of error, and input frequency. As an example, suppose the maximum allowed error caused by aperture time is specified to be 0.1%, and the aperture time is 500 ps. A straight line drawn through these values on the nomograph intersects the frequency scale at 300 kHz. This would be the highest input frequency consistent with a maximum error of 0.1%.

The nomograph can also be used to show that voltage acquisition errors in A/D converters can be greatly reduced if a T/H amplifier precedes the converter. Assume an A/D converter with a conversion time of 2.2 μs and an input with a maximum frequency of 20 kHz. In this situation the 2.2 μs can be treated as an aperture time uncertainty. Connecting this with the 20-kHz frequency on the nomograph gives a maximum error of 28%. This error can easily be reduced by several orders of

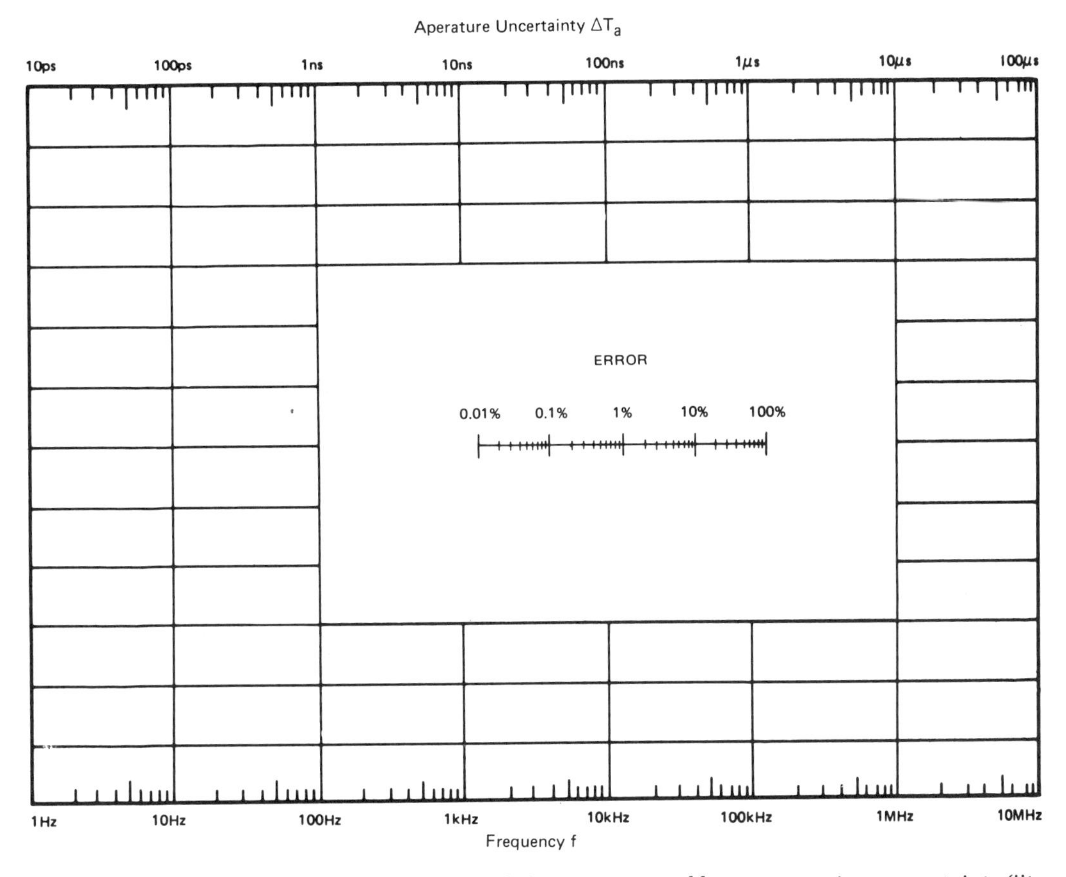

FIGURE 10.21 Nomograph for error caused by aperture time uncertainty (jitter). Courtesy of ILC Data Device Corp.

magnitude by a T/H or S/H whose aperture time uncertainty is much less than 2.2 μs.

10.13 MATCHING A S/H OR T/H TO AN A/D CONVERTER

When a S/H or T/H amplifier is used at the input of an A/D converter, the performance of the system will depend on the combined specifications of both units, as well as on their compatibility. Since the signal should be sampled at a frequency at least twice that of the maximum signal frequency, the maximum signal frequency is equal to one-half of the maximum sampling rate of the system. As discussed in Section 10.11, the sampling rate of the system depends on the total cycle time. The cycle time is at least as long as the acquisition time + settling time + A/D conversion time, so

$$(f_{SIG})_{MAX} = \frac{1/2}{t_{acquisition} + t_{settling} + t_{A/D\ conversion}} \tag{10.11}$$

The feedthrough attenuation of the S/H or T/H at f_{SIG} should represent less than $\frac{1}{4}$ LSB of the A/D converter. For instance, suppose the converter has 8-bit resolution (1 LSB = 0.4%) and an 18-MHz clock rate. Then f_{SIG} is 9 MHz max, and at this frequency the S/H should have a maximum feedthrough attenuation of 0.1% (60 dB).

The acquisition time of the S/H or T/H is specified both for the magnitude of the input voltage change and for the percentage to which the final value will be accurate. The acquisition time should be specified both for a sufficiently large input voltage and for settling to within at least $\frac{1}{2}$ LSB of the A/D converter.

Aperture time uncertainty (jitter) creates a sample-to-sample ambiguity in the output voltage when the S/H is processing high-frequency signals. As shown in the previous section, the jitter time ΔT_a is related to the maximum percent of error at a signal frequency f_{SIG} by

$$\text{maximum \% error} = 6.28(f_{SIG})(\Delta T_a) \tag{10.12}$$

The maximum percent of error should be less than $\frac{1}{2}$ LSB of the A/D converter, so the maximum uncertainty in the aperture time of the S/H or T/H should be

$$(\Delta T_a)_{max} = \frac{1/2 \text{ LSB (expressed as a \%)}}{628(f_{SIG})} \tag{10.13}$$

10.14 DEGLITCHING DIGITAL-TO-ANALOG CONVERTERS[b]

Digital-to-analog converters (DACs) produce output noise called *glitch* whenever major carry code changes are made. The worst case occurs at half-scale when input code transition is between 011-1 and 100-0. Data skew and turn on or off of DAC bit switches at this transition often cause a momentary erroneous code of 000-0 or 111-1. The output response to

[b] The material in Section 10.14 is from David C. Pinkowitz, Product Manager, Data Converters of ILC Data Corp., "A Designer's Guide to Deglitching DAC's," *Instrument and Control Systems*, Vol. 55, No. 4, April 1982, and is used courtesy of the I&CS Magazine and ILC Data Device Corp.

this code is glitch. Figure 10.22 shows that the DAC output slews toward the erroneous code for the length of the skew time. Since glitch amplitude decreases binarily for each lower-order code transition, glitch becomes harder to filter effectively.

For many applications, DAC output glitch deteriorates the overall performance of equipment. Vector stroke CRT displays and digital signal generators are typical applications requiring low glitch. In CRT displays, a DAC glitch causes intensity and position nonuniformities. And in digital wave-form generators, a DAC glitch causes harmonic distortion.

For less critical applications, a DAC designed for low glitch may be suitable. When driven from a latch to minimize data skew, a low-glitch DAC offers lower glitch than a general-purpose DAC. In more sensitive applications, where low-glitch DACs are not good enough, a deglitcher will be needed.

A deglitcher is a special T/H amplifier that masks DAC output glitch. You decrease glitch energy by placing the deglitcher in *hold* mode at the same time output glitch will occur. After DAC output glitch has settled, the deglitcher is returned to *track* mode. In track mode, the deglitcher is used as an output amplifier until the next DAC code change.

The output glitch can be obtained from the combined DAC and deglitcher. This is caused by the switch charge transfer during the track-to-hold mode change. Because the glitch is the same for every code change and occurs at the DAC update rate, it can be filtered out. Therefore, DAC output glitch is reduced by an order of magnitude and is the same for all code transitions.

Many deglitcher circuit approaches are available and can be grouped as open-loop or closed-loop methods. Closed-loop methods offer greater accuracy at slightly higher cost and complexity.

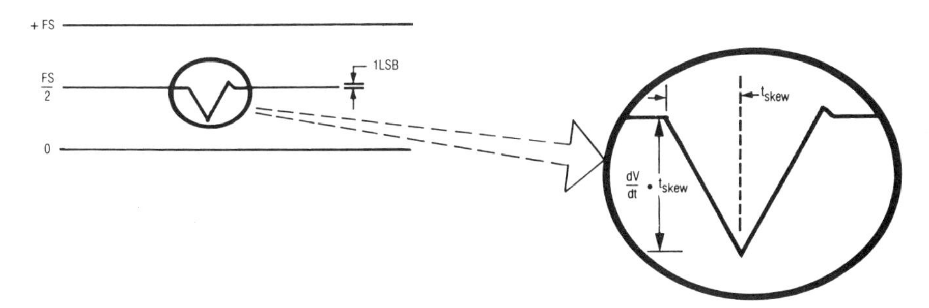

FIGURE 10.22 The largest glitch in a DAC is usually caused by the digital data skew at the half-scale major carry transition (011–1 and 100–0). Courtesy of the ILC Data Device Corp.

Figure 10.23 shows the most common open-loop deglitcher. It is similar to a T/H having an electronic switch (FET or diode bridge), a hold capacitor, and a follower op-amp. Voltage input charges the capacitor when the switch is closed and freezes the output when the switch is open. Although this configuration performs a deglitching function, it is unsuitable for most applications. Since the circuit is open loop, its errors add directly to DAC errors. This results in deteriorated performance. Errors include switch nonlinearities, op-amp offset, and common-mode errors. Besides the error problem, the DAC needs a redundant op-amp because the circuit requires a voltage input.

10.14.1 Deglitched DAC Example

An example of a compatible DAC and deglitcher combination is the DDAC. Although the DDAC has its own particular specifications, its timing diagram (Figure 10.24) will apply to any well-chosen combination of DAC, deglitcher, and latch with similar specifications.

As shown in Figure 10.24, in response to a strobe pulse, the deglitcher timing circuits put the DAC in a hold mode, as well as latch new input data into the registers. The hold interval lasts 50 ns to mask the DAC glitch. After 50 ns the DAC glitch has settled, and the deglitcher is put in track mode to acquire the new DAC output voltage. The DDAC can be updated at a 10-MHz rate for LSB DAC changes, since they can be acquired in less than 50 ns.

10.14.2 Applications Hints for Low Glitch

The following fabrication methods will help you in your deglitching efforts:

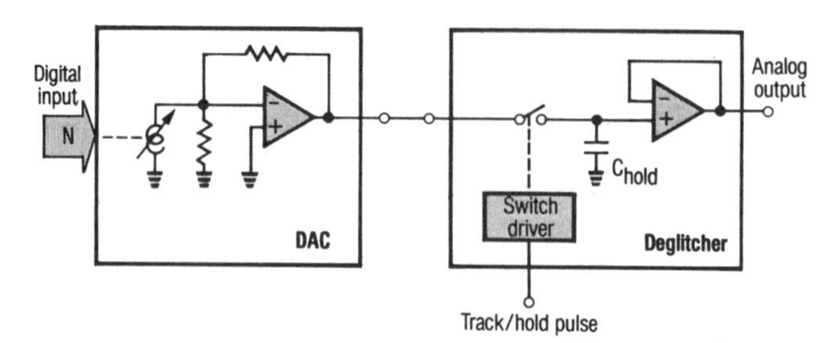

FIGURE 10.23 In this open-loop deglitcher, switch nonlinearity, offset, and common-mode errors are added to DAC output. Because the DAC must have a voltage output, an extra op-amp is also required. Courtesy of the ILC Data Device Corp.

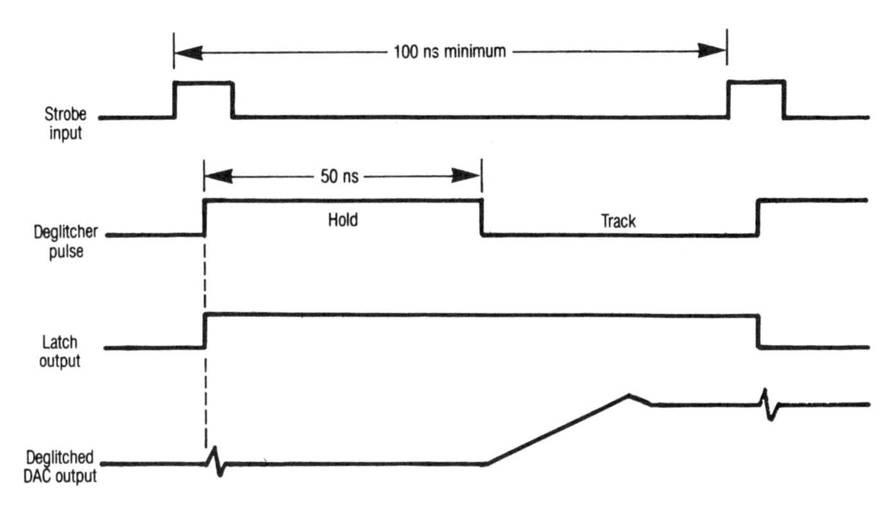

FIGURE 10.24 This deglitched DAC timing diagram shows that the deglitcher is in the hold mode when the DAC glitch occurs. This minimizes the output glitch. Courtesy of the ILC Data Device Corp.

1. *Use latches.* DAC input latches reduce unequal digital delays. Put the latch close to the DAC, with equal-length printed circuit runs for each bit. LS logic latches have slower transition times and almost equal rise and fall times, which minimize capacity coupling and data skew. LS logic also creates less ground current noise.
2. *Use a ground plane.* The PC (printed circuit) board should have a large area ground plane to maintain a low impedance ground at all frequencies of interest. This results in less ground noise and eliminates ground loops. (If a ground plane is not used, wide printed circuit runs are needed as well as care that no ground loops exist.)
3. *Keep signal runs short.* PC signal runs must be kept as short as possible to reduce inductive effects and line reflections. When layout constraints prevent this, controlled impedance methods such as microstrip or coaxial cable can be used.
4. *Keep analog and digital lines apart.* Be sure that your PC layout has digital input signals separated from analog output signals. This includes internal digital signals such as latch outputs and strobes, as well as the DAC output that drives the deglitcher. Crosstalk and extraneous signal coupling must be eliminated for minimum noise performance.
5. *Decouple all power supplies.* To reduce noise on power supply lines, decoupling capacitors can be put on each supply. Tantalum

capacitors of greater than 1 μF are used for low-frequency noise filtering. For high frequencies, ceramic capacitors from 0.01 to 0.1 μF are used. Decoupling capacitors should be placed as close as possible to the deglitcher and DAC circuits.

6. *Remote sense the deglitcher load.* Connect deglitcher capacitor and resistor feedback elements at the load. This puts the op-amp output line inside the feedback loop and decreases errors due to load currents and line drops.

Problems

1. Given a sample-and-hold amplifier, define the following terms:
   (a) Acquisition time
   (b) Application turn-on delay
   (c) Aperture-time-delay uncertainty (jitter)
   (d) Droop
   (e) DC offset
   (f) Sampling rate
2. Discuss the meaning of a glitch circuit for a DAC.
3. Draw two types of analog-to-digital converters.
4. Draw two types of digital-to-analog converters.
5. Given two transducers and amplifiers, draw a circuit to show how the signal from the output of the transducer can be fed through a complete A/D and D/A conversion system to a readout device. Where would you place a microcomputer to record the output signal?

References

Allocca, John A. and Allen Stuart, *Electronic Instrumentation*, Reston Publishing Co., Reston, Virginia, 1983.

Product Catalogue, ILC Data Device Corp., Bohemia, N.Y.

11 SEMICONDUCTOR MEMORIES

11.1 INTRODUCTION

Semiconductor memories are modern electronic devices used in microcomputers and microprocessors. Microcomputer registers can also be thought of as computer memories. Memory devices are also used in computer peripheral equipment, control systems, measurement equipment, communication and telecommunication equipment, office and business machines, and hundreds of other applications. Metal-oxide semiconductor large-scale integrated circuits have been used with semiconductor memories in over 3000 different memory devices now available from manufacturers.

Memory devices are classified as volatile and nonvolatile. Volatile memories retain their data only as long as power is applied. In many applications, this limitation presents a design problem. The random-access memory (RAM) used in microprocessors is typically a volatile memory in which there is a constant rewriting of stored data.

In other applications, nonvolatile devices are desired because the nonvolatile memory retains its data whether or not power is applied. Examples include tape and disk storage.

11.2 READ-WRITE MEMORY[a]

First we examine read–write memory (RAM), which permits access to stored memory (reading) and transformation of the stored data (writing).

Before the advent of solid-state read–write memory, active data (data being processed) were stored and retrieved from a nonvolatile core memory (a magnetic-storage technology). These memories were large and consumed large amounts of energy. Solid-state RAMs solved the size and power consumption problems, but added the problem of volatility. Because RAMs lose their memory when you turn off their power, you must leave systems on all the time, add battery backup, or store important data on a nonvolatile medium before or in case the power goes down.

Despite their volatility, RAMs have become very popular, and an industry was born that primarily fed the computer's insatiable appetite for higher bit capacities and faster access speeds.

11.3 RAM TYPES

Two basic RAM types have evolved since 1970. Dynamic RAMs (DRAMs) are noted for high capacity, moderate speeds, and low power consumption. Their memory cells are basically charge-storage capacitors with driver transistors. The presence or absence of charge in a capacitor is interpreted by the RAM's sense line as a logical 1 or 0. Because of the charge's natural tendency to distribute itself into a lower energy-state configuration, however, dynamic RAMs require periodic charge refreshing to maintain data storage.

Traditionally, this requirement has meant that system designers had to implement added circuitry to handle the dynamic RAM subsystem refresh. And if, at certain times, refresh procedures made the RAM unavailable for writing or reading, the memory's control circuitry had to arbitrate access. However, there are now available two alternatives that largely offset this disadvantage. For relatively small memories in microprocessor environments, the *integrated RAM*, or iRAM, provides all the complex refresh circuitry on chip, thus greatly simplifying the system design. For larger storage requirements, LSI dynamic memory controllers reduce the refresh requirement to a minimal design by offering a monolithic controller solution.

---

[a] The material in Sections 11.2 to 11.8 is taken from *Memory Component Handbook*, Chapter 1, "Memory Overview," by Joe Althether, © 1983 by the Intel Corp., and is reprinted courtesy of Intel Corp.

Where engineers are less concerned with space and cost than with speed and reduced complexity, the second RAM type, *static RAMs* (SRAMs), generally prove best. Unlike their dynamic counterparts, static RAMs store ones and zeros using traditional flip-flop logic-gate configurations. They are faster and require no refresh. A user simply addresses the static RAM and, after a very brief delay, obtains the bit stored in that location. Static devices are also simpler to design with than dynamic RAMs, but the static cell's complexity puts these nonvolatile chips far behind dynamics in bit capacity per square mil of silicon.

11.4 THE iRAM

There is a way, however, to gain the static RAM's designing simplicity but with the dynamic RAM's higher capacity and other advantages. An integrated RAM or iRAM integrates a dynamic RAM and its control and refresh circuitry on one substrate, creating a chip that has dynamic RAM density characteristics, but looks like a static RAM to users. You simply address it and collect your data without worrying about refresh and arbitration.

Before iRAM's introduction, users who built memory blocks smaller than 8K bytes typically used static RAMs because the device's higher price was offset by the support-circuit simplicity. On the other hand, users building blocks larger than 64K bytes usually opted for dynamic RAMs because density and power considerations began to take precedence over circuit complexity issues.

11.5 READ-ONLY MEMORY

Another memory class, read-only memory (ROM), is similar to RAM in that a computer addresses it and then retrieves data stored in that address. However, ROM includes no mechanism for altering the data stored at that address, hence, the term read-only.

ROM is basically used for storing information that is not subject to change, at least not frequently. Unlike RAM, when the system power goes down, ROM retains its contents.

ROM devices became very popular with the advent of microprocessors. Most early microprocessor applications were dedicated systems; the system's program was fixed and stored in ROM. Manipulated data could vary and were therefore stored in RAM. This application split caused ROM to be commonly called program storage, and RAM, data storage.

The first ROMs contained cell arrays in which the sequence of ones and zeros was established by a metallization interconnect mask step during fabrication. Thus, users had to supply a ROM vendor with an interconnect program so the vendor could complete the mask and build the ROMs. Setup charges were quite high, in fact even prohibitive, unless users planned for large volumes of the same ROM.

To offset this high setup charge, manufacturers developed a user-programmable ROM (or PROM). The first such devices used fusible links that could be melted or "burned" with a special programmer system.

Once burned, a PROM was just like a ROM. If the burn program was faulty, the chip had to be discarded. But PROMs furnished a more cost-effective way to develop program memory or firmware for low-volume purposes than did ROMs.

As one alternative to fusible-link programming, Intel pioneered an erasable MOS-technology PROM (termed an EPROM) that used charge-storage programming. It came in a standard ceramic DIP package but had a window that permitted die exposure to light. When the chip was exposed to ultraviolet light, high-energy photons could collide with the EPROM's electrons and scatter them at random, thus erasing the memory.

The EPROM was obviously not intended for use in read–write applications, but it proved very useful in research and development for prototypes, where the need to alter the program several times is quite common. Indeed, the EPROM market consisted almost exclusively of development laboratories. As the fabrication process became mature, however, and volumes increased, EPROM's lower price made it attractive even for medium-volume production-system applications.

Another ROM technology advance occurred in 1980 with the introduction of Intel's 2816, a 16K ROM that is user programmable and electrically erasable. Thus, instead of removing it from its host system and placing it under ultraviolet light to erase its program, the 2816 can be reprogrammed in its socket. Moreover, single bits or entire bytes can be erased in one operation instead of erasing the entire chip.

Such E^2PROMs (for electrically erasable programmable ROM) are opening up new applications. In point-of-sale terminals, for example, each terminal connects to a central computer, but each can also handle moderate amounts of local processing. An E^2PROM can store discount information to be automatically figured in during a sales transaction. Should the discount change, the central computer can update each terminal via telephone lines by reprogramming that portion of the E^2PROM (Figure 11.1).

In digital instrumentation, an instrument could become self-calibrating using an E^2PROM. Should the instrument's calibration drift outside specification limits, the system could employ a built-in

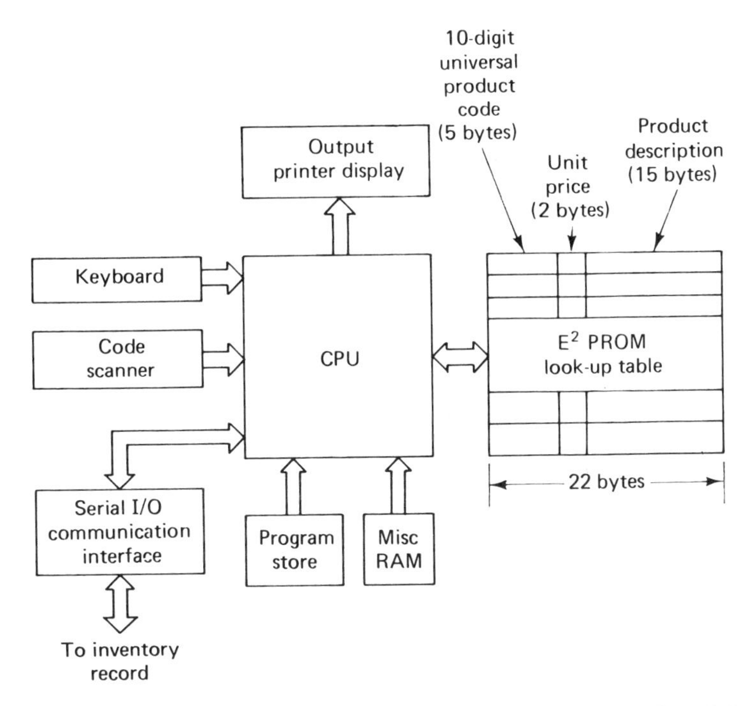

FIGURE 11.1 Typical E^2PROM application. Courtesy of Intel Corp.

diagnostic to reprogram a parametric setting in an E^2PROM and bring the calibration back within limits.

E^2PROMs contain a floating-gate tunnel-oxide (Flotox) cell structure. Based on electron tunneling through a thin (less than 200 angstroms) layer of silicon dioxide, these cells permit writing and erasing with 21-V pulses. During a read operation, the chips use conventional + 5-V power.

11.6 BUBBLE MEMORY

A very different device type, bubble memory was once considered the technology that would make RAM components obsolete. This view failed to consider the inherent features and benefits of each technology. There is no question that RAMs have staked out a read–write applications area that is vast. Nevertheless, their volatility presents severe problems in more than a few applications. Remote systems, for example, might be unable to accept a memory that is subject to being wiped out should a power failure occur.

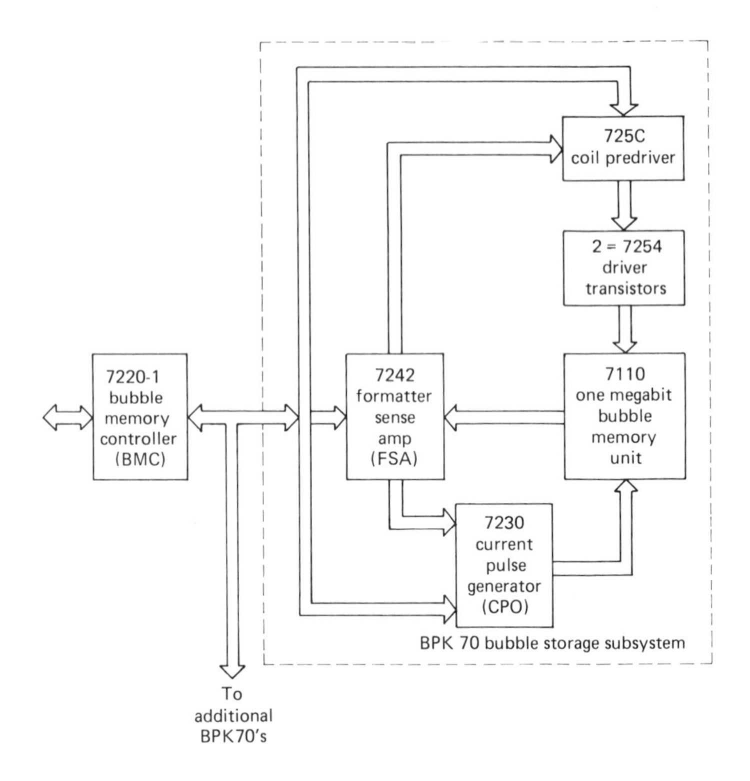

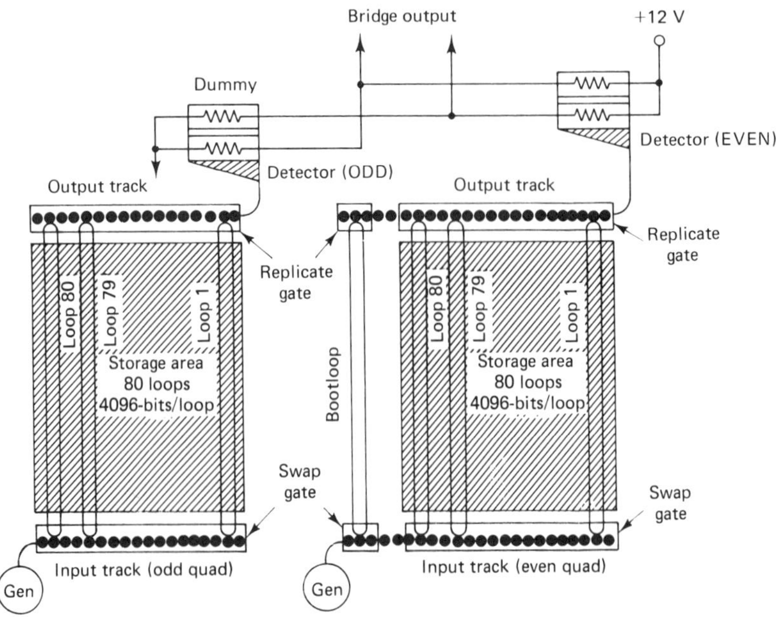

FIGURE 11.2 Intel Model 7110 bubble memory. Courtesy of Intel Corp.

Bubble memories use a magnetic storage technique, roughly similar to the core memory concept but on a much smaller size and power-consumption scale. They are nonvolatile and physically rugged. Thus, their first clear application target has been in severe environment and remote system sites. Portable terminals represent another application area in which bubbles provide unique benefits.

Intel's latest bubble memory design provides 1,048,576 bits of data storage via a defect-tolerant technique that makes use of 1,310,720 total bits (Figure 11.2). Internally, the product consists of 256 storage loops of 4086 bits each. Coupled with available control devices, this single chip can implement a 128K byte memory subsystem.

11.7 SELECTION OF THE MEMORY DEVICE

Other factors besides the memory characteristics must enter into the selection of a memory product. These include:

1. Performance
2. Cost
3. Memory Size
4. Power consumption
5. Memory architecture and organization

11.7.1 Performance

Generally, the term performance relates to how fast the device can operate in a given system environment. This parameter is usually rated in terms of the access time. Fast SRAMs can provide access times as short as 20 ns, while the fastest DRAM cannot go much below the 100-ns mark. A bipolar PROM has an access time of 35 ns. RAM and PROM access is usually controlled by a signal most often referred to as *chip select* ($\overline{CS}$). $\overline{CS}$ often appears in device specifications. In discussing access times, it is important to remember that in SRAMs and PROMs the access time equals the cycle time of the system, whereas in DRAMs the access time is always less than the cycle time.

11.7.2 Cost

There are many ramifications to consider when evaluating cost. Cost can be spread over factors such as design-in time, cost per device, cost per bit, size of memory, and power consumption.

Cost of design time is directly proportional to design complexity. For example, SRAMs generally require less design-in time than DRAMs because there is no refresh circuitry to consider. Conversely, the DRAM provides the lowest cost per bit because of its higher packing density.

11.7.3 Memory Size

Memory size is generally specified by number of bytes (a byte is a group of 8 bits). The memory size of a system is usually segmented depending on the general equipment category. Computer mainframes and most of today's minicomputers use blocks of RAM substantially beyond 64K bytes, usually in the hundreds of thousands of bytes. For this size of memory, the DRAM (dynamic RAM) has a significantly lower cost per bit. The additional costs of providing the refresh and timing circuitry are spread over many bits. The microprocessor user generally requires memory sizes ranging from 2K bytes up to 64 bytes.

11.7.4 Power Consumption

Power consumption is an important consideration because the total power required for a system directly affects overall cost. Higher power consumption requires bigger power supplies, more cooling, and reduced device density per board, all affecting cost and reliability. All things considered, the usual goal is to minimize power. Many memories now provide automatic power-down. With today's emphasis on saving energy and reducing cost, the memories that provide these features will gain an increasingly larger share of the market.

In some applications, extremely low power consumption is required, such as battery operation. For these applications, the use of devices made by the CMOS technology has a distinct advantage over the NMOS products. CMOS devices offer power savings of several magnitudes over NMOS. Nonvolatile devices such as E^2PROMs are usually independent of power problems in these applications.

Power consumption also depends on the organization of the device in the system. Organization usually refers to the width of the memory word. At the time of their inception, memory devices were organized as nK × 1 bits. Today, they are available in various configurations, such as 4K × 1, 16K × 1, 1K × 4, and 2K × 8. As device width increases, fewer devices are required to configure a given memory word, although the total number of bits remains constant. The wider organization can provide significant savings in power consumption, because a fewer number of devices are required to be powered up for access to a given memory word. In addition, the board layout design is simpler due to

TABLE 11.1 *Segmentation of Memory Devices*

Operating From	*Read Speed* *Fast*	*Read Speed* *Slow*	*Write Speed* *Fast*	*Write Speed* *Slow*	*Down Load*	*Size* *Small*	*Size* *Large*	*Removable (Archive)*	*System Level*
Mass		Bubbles Disk		Bubbles Disk	N/A		Bubbles Disk	Bubbles Disk	Add on RAM Bubbles
Boot	EPROM		N/A	N/A	N/A	All[2]			N/A
Monitor	EPROM		N/A	N/A	N/A	All			N/A
Buffer	Bytewide	Bubbles	Bytewide	Bubbles	N/A	All	Bubbles × 1		Add in RAM Bubbles
Diagnostics	E^2/EPROM/ RAM		Bytewide	E^2PROM	Disk[1] Bubbles	All	All[3]	Bubbles Disk	N/A
Operating System			N/A	N/A	Disk[1] Bubbles	All	× 1	Bubbles Disk	Add in/ Add on RAM
APP/PGM/ Data store	E^2/EPROM/ RAM		Bytewide × 1	E^2PROM	Disk[1] Bubbles	All	× 1	Bubbles Disk	Add in/ Add on RAM

Courtesy of Intel Corp.
[1] Down loaded from add on/add in bubbles.
[2] E^2/EPROM bytewides.
[3] × 1 DRAM bubbles disk.

fewer traces and better layout advantages. The wider width is of particular advantage in microprocessors and bit-slice processors because most microprocessors are organized in 8- or 16-bit architectures. A memory chip configured in the nK × 8 organization can confer a definite advantage, especially in universal site applications. All nonvolatile memories other than bubble memories are organized nK × 8 for this very reason.

Table 11.1 is a summary of the various Intel memory devices.

11.8 SYNCHRONOUS AND ASYNCHRONOUS MEMORIES

The synchronous memory possesses an internal address register that latches the current device address, but the asynchronous device lacks this capability. The logic of this definition is easy to follow: Register transfer or sequential logic is considered synchronous because it is clocked by a common periodic signal, the system clock. Memories with internal address registers are also internally sequential logic arrays clocked by a signal, common throughout the memory system, and are, therefore, synchronous.

Asynchronous memories require the device address be held valid on the bus throughout the memory cycle. Static RAMs fall into this category. In contrast, synchronous memories require the address to be valid only for a very short period of time just before, during, and just after the arrival of the address register clock. DRAMs and clocked static RAMs fall into this category.

With the introduction of the 2186 and 2187 iRAMs, the preceding definition no longer fits, because both devices have on-chip address latches. Yet with respect to the system, one device operates synchronously and the other asynchronously.

Therefore, in considering memory devices or systems that operate within a specified cycle time, Intel defines a synchronous memory as one that responds in a predictable and sequential fashion, always providing data within the same time frame from the clock input. This allows a system designer to take advantage of the predictable access time and maximize system performance by reducing or eliminating wait states.

Intel defines an asynchronous memory as one that (within the framework of the memory cycle specifications) does not output data in a predictable and repeatable time frame with respect to system timing. This is generally true of DRAM systems, where a refresh cycle, which occurs randomly skewed to the balance of the system timing, may be in progress at the time of a memory cycle request by the CPU. In this case, provision must be made to resynchronize the system to the memory, usually with a READY signal. The 2186 iRAMs fit into this category, while the 2187 iRAMs are considered synchronous devices.

11.9 INTEL MEMORY CHIP EXAMINED[b]

There are three major MOS technology families, PMOS, NMOS, and CMOS (Figure 11.3). They refer to the channel type of the MOS transistors made with the technology. PMOS technologies implement *p*-channel transistors by diffusing *p*-type dopants (usually boron) into an *n*-type silicon substrate to form the source and drain. *P*-channel is so named because the channel is comprised of positively charged carriers. NMOS technologies are similar, but use *n*-type dopants (normally phosphorus or arsenic) to make *n*-channel transistors in *p*-type silicon substrates. *N*-channel is so named because the channel is composed of

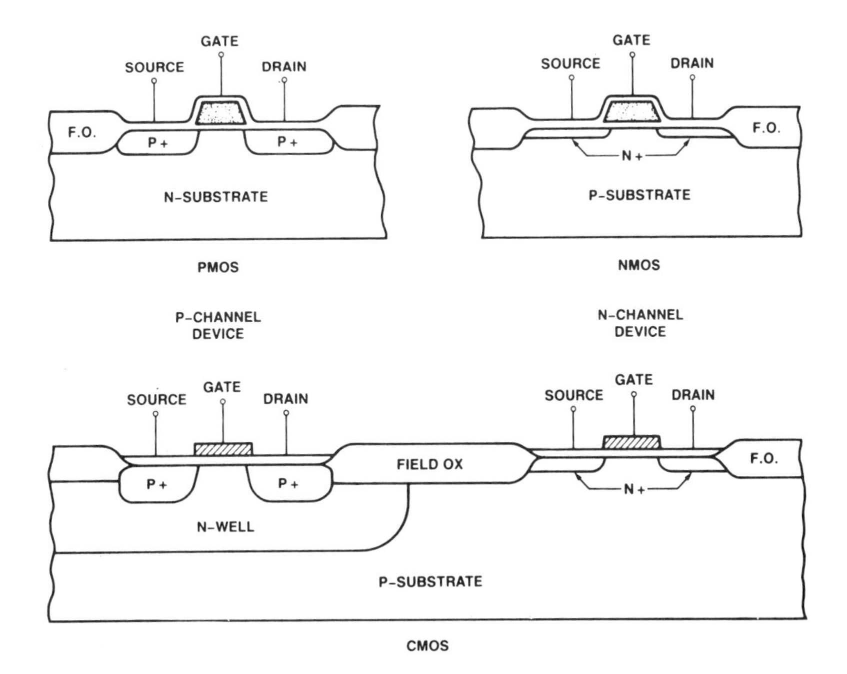

FIGURE 11.3 MOS process cross-sections. Courtesy of Intel Corp.

[b] The material in Section 11.9 is taken from *Memory Component Handbook*, Chapter 2, "Intel Memory Technologies," by Larry Brigham, Jr., © 1983 by the Intel Corp. and is reprinted courtesy of Intel Corp.

negatively charged carriers. CMOS or complementary MOS technologies combine both *p*-channel and *n*-channel devices on the same silicon. Either *p*- or *n*-type silicon substrates can be used; however, deep areas of the opposite doping type (called wells) must be defined to allow fabrication of the complementary transistor type.

Most of the early semiconductor memory devices, like Intel's pioneering 1103 dynamic RAM and 1702 EPROM, were made with PMOS technologies. As higher speeds and greater densities were needed, most new devices were implemented with NMOS. This was due to the inherently higher speed of *n*-channel charge carriers (electrons) in silicon along with improved process margins. The majority of MOS memory devices in production today are fabricated with NMOS technologies. CMOS technology has begun to see widespread commercial use in memory devices. It allows for very low power devices, and these have been used for battery-operated or battery backup applications. Historically, CMOS has been slower than any NMOS device. Recently, however, CMOS technology has been improved to produce higher-speed devices. Up to now, the extra cost processing required to make both transistor types has kept CMOS memories limited to those areas where the technology's special characteristics would justify the extra cost. In the future, the learning curve for high-performance CMOS costs will make a larger and larger number of memory devices practical in CMOS.

The basic fabrication sequence for an HMOS circuit will be described. HMOS is a high-performance *n*-channel MOS process developed by Intel for 5-V single-supply circuits. HMOS, along with its evolutionary counterparts HMOS II and HMOS III, CHMOS and CHMOS II (and their variants), comprise the process family responsible for most of the memory components produced by Intel today.

The MOS IC fabrication process begins with a slice (or wafer) of single crystal silicon. Typically, it is 100 or 125 mm in diameter, about $\frac{1}{2}$ mm thick, and uniformly doped *p*-type. The wafer is then oxidized in a furnace of around 1000°C to grow a thin layer of silicon dioxide (SiO_2) on the surface. Silicon nitride is then deposited on the oxidized wafer in a gas phase chemical reactor. The wafer is now ready to receive the first pattern of what is to become a many layered complex circuit. The pattern is etched into the silicon nitride using a process known as photolithography. This first pattern (Figure 11.4) defines the boundaries of the active regions of the IC, where transistors, capacitors, diffused resistors, and first-level interconnects will be made.

The pattern and etched wafer is then implanted with additional boron atoms accelerated at high energy. The boron will only reach the silicon substrate where the nitride and oxide were etched away, providing areas doped strongly *p*-type that will electrically separate active areas. After implanting, the wafers are oxidized again, and this time a

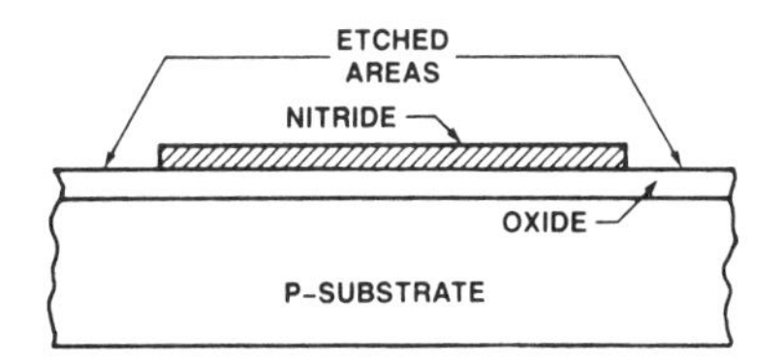

FIGURE 11.4 First mask. Courtesy of Intel Corp.

thick oxide is grown. The oxide grows only in the etched areas owing to silicon nitride's properties as an oxidation barrier. When the oxide is grown, some of the silicon substrate is consumed, and this gives physical as well as electrical isolation for adjacent devices, as can be seen in Figure 11.5.

Having fulfilled its purpose, the remaining silicon nitride layer is removed. A light oxide etch follows, taking with it the underlying first oxide but leaving the thick (field) oxide.

Now that the areas for active transistors have been defined and isolated, the transistor types needed can be determined. The wafer is again patterned, and then if special characteristics (such as depletion-mode operation) are required, it is implanted with dopant atoms. The energy and dose at which the dopant atoms are implanted determines much of the transistor's characteristics. The type of the dopant provides for depletion-mode (*n*-type) or enhancement-mode (*p*-type) operation.

With the transistor types defined, the gate oxide of the active transistors is grown in a high temperature furnace. Special care must be taken to prevent contamination or inclusion of defects in the oxide and to ensure uniform consistent thickness. This is important to provide precise, reliable device characteristics. The gate oxide layer is then masked and holes are etched to provide for direct gate to diffusion ("buried") contacts where needed.

The wafers are now deposited with a layer of gate material. This is typically polycrystalline silicon ("poly"), which is deposited in a gas phase chemical reactor similar to that used for silicon nitride. The poly is then doped (usually with phosphorus) to bring the sheet resistance down to 10 to 20 Ω/square. This layer is also used for circuit inter-

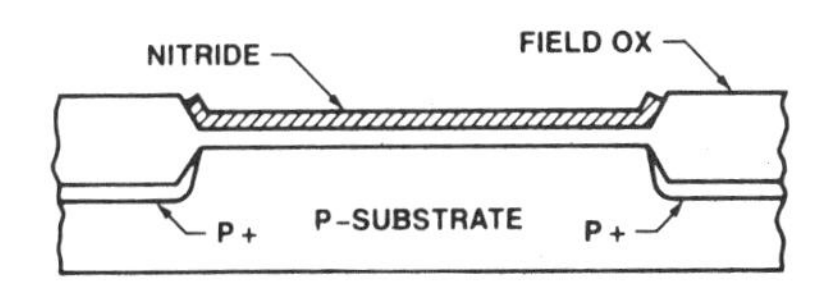

FIGURE 11.5 Post field oxidation. Courtesy of Intel Corp.

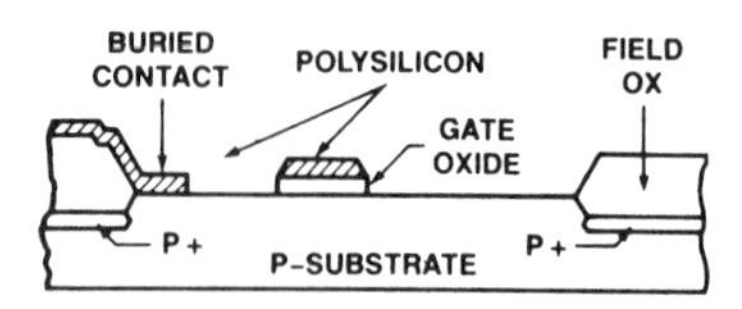

FIGURE 11.6 Post gate mask. Courtesy of Intel Corp.

connects, and if a lower resistance is required, a refractory metal–polysilicon composite or refractory metal silicide can be used instead. The gate layer is then patterned to define the actual transistor gates and interconnect paths (Figure 11.6).

The wafer is next diffused with *n*-type dopant (typically arsenic or phosphorus) to form the source and drain junctions. The transistor gate material acts as a barrier to the dopant, providing an undiffused channel self-aligned to the two junctions. The wafer is then oxidized to seal the junctions from contamination with a layer of SiO_2 (Figure 11.7).

A thick-layer glass is then deposited over the wafer to provide for insulation and sufficiently low capacitance between the underlying layers and the metal interconnect signals. (The lower the capacitance, the higher the inherent speed of the device.) The glass layer is then patterned with contact holes and placed in a high temperature furnace. This furnace step smooths the glass surface and rounds the contact edges to provide uniform metal coverage. Metal (usually aluminum or aluminum silicon) is then deposited on the wafer, and the interconnect patterns and external bonding pads are defined and etched (Figure 11.8). The wafers then receive a low-temperature (approximately 500°C) alloy that ensures good ohmic contact between the aluminum and diffusion or poly.

At this point the circuit is fully operational; however, the top metal layer is very soft and easily damaged by handling. The device is also susceptible to contamination or attack from moisture. To prevent this, the wafers are sealed with a passivation layer of silicon nitride or a silicon and phosphorus oxide composite. Patterning is done for the last time, opening up windows only over the bond pads where external connections will be made.

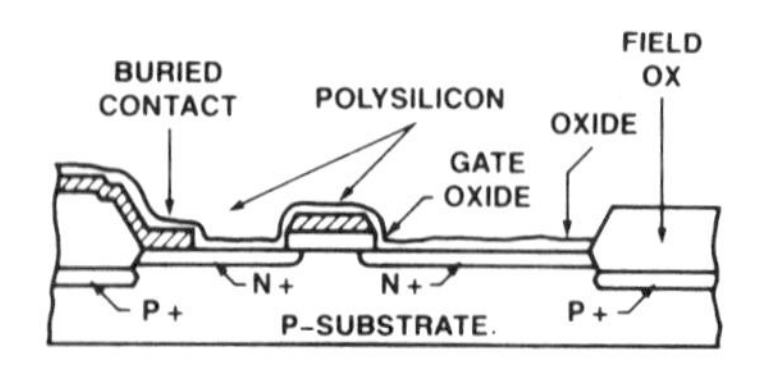

FIGURE 11.7 Post oxidation. Courtesy of Intel Corp.

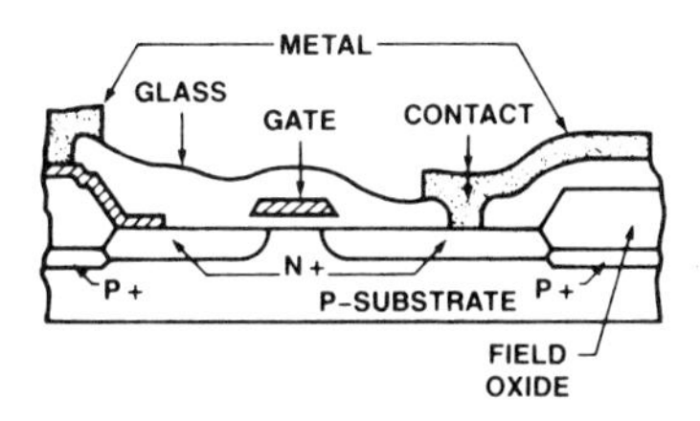

FIGURE 11.8 Completed circuit (without passivation). Courtesy of Intel Corp.

This completes the basic fabrication sequence for a single poly layer process. Double poly processes such as those used for high-density dynamic RAMs, EPROMs, and E²PROMs follow the same general process flow with the addition of gate, poly deposition, doping, and interlayer dielectric process modules required for the additional poly layer (Figure 11.5). These steps are performed right after the active areas have been defined (Figure 11.9), providing the capacitor or floating gate storage nodes on those devices.

After fabrication is complete, the wafers are sent for testing. Each circuit is tested individually under conditions designed to determine which circuits will operate properly both at low temperature and at the conditions found in actual operation. Circuits that fail these tests are inked to distinguish them from good circuits. From here the wafers are

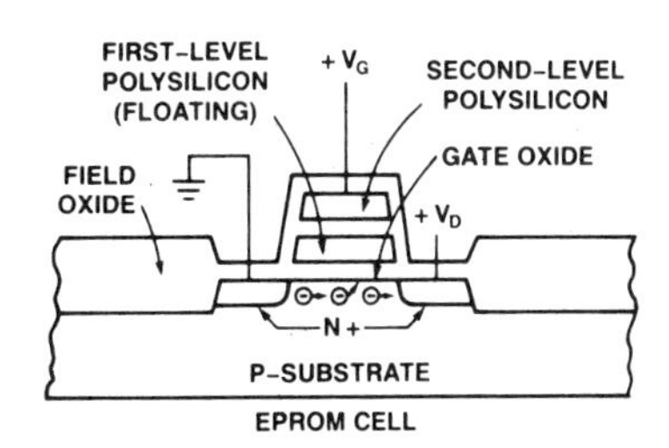

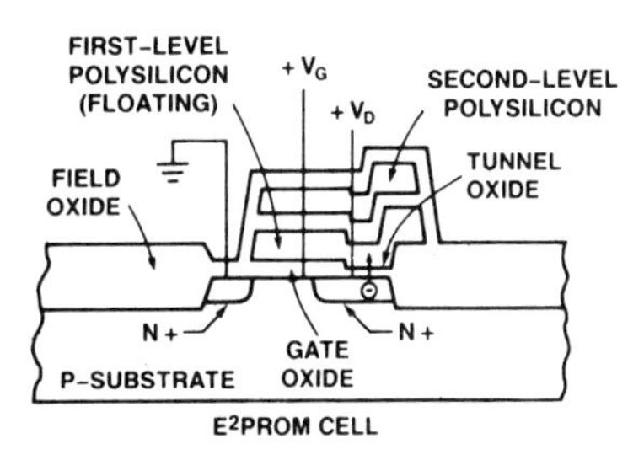

FIGURE 11.9 Double poly structure. Courtesy of Intel Corp.

sent for assembly, where they are sawed into individual circuits with a paper-thin diamond blade. The inked circuits are then separated out and the good circuits are sent on for packaging.

11.10 RANDOM-ACCESS MEMORY[d]

Two basic types of random-access memories (RAMs) have existed since the inception of MOS memories: static RAMs (SRAMs) and dynamic RAMs (DRAMs). Where high-test performance and simplest system design are desired, the static RAM can provide the optimum solution for smaller memory systems. However, the dynamic RAM holds a commanding position where large amounts of memory and the lowest cost per bit are the major criteria.

The major attributes of dynamic RAMs are low power and low cost, a direct result of the simplicity of the storage cell. This is achieved through the use of a single transistor and a capacitor to store a single data bit (Figure 11.10).

The absence or presence of charge stored in the capacitor equates to a 1 or a 0, respectively. The capacitor is in series with the transistor, eliminating the need for a continuous current flow to store data. In addition, the input buffers, the output driver, and all the circuitry in the RAM have been designed to operate in a sequentially clocked mode, thus consuming power only when being accessed. The net result is low power consumption. Also, a single-transistor dynamic cell, as compared to the four- or six-transistor cell of a static RAM, occupies less die area. This results in more die per wafer.

Unfortunately, the simple cell has a drawback: the capacitor is not

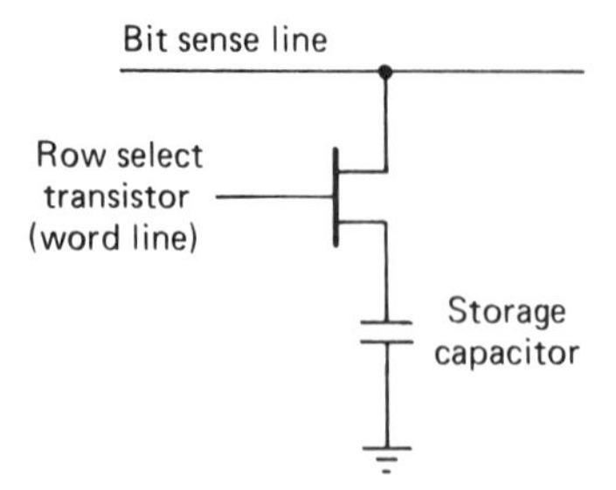

FIGURE 11.10 Dynamic RAM memory cell. Courtesy of Intel Corp.

[d] The material in Section 11.10 is taken from *Memory Component Handbook*, application notes AP–133, "Random Access Memory," Chapter 3, © 1983 by the Intel Corp. and is reprinted courtesy of Intel Corp.

a pure element and it has leakage. If left alone, leakage current would cause the loss of data. The solution is to refresh the charge periodically. A refresh cycle reads the data before they degrade too far and then rewrites the data back into the cell. RAM organization is tailored to aid the refresh function. As an example, the Intel® 2164A 64K RAM is organized internally as four 16K RAM arrays, each comprised of 128 rows of 128 columns. Consequently, the row address accesses 128 columns in each of the four quadrants. However, let's concern ourselves with only one quadrant. Prior to selection, the bit sense line was charged to a high voltage. Via selection of the word line (row addresses), 128 bits are transferred onto their respective bit lines. Electrons will migrate from the cell onto the bit line, destroying the stored charge. Each of the 128 bit lines has a separate sense amplifier associated with it. Charge on the bit line is sensed, amplified, and returned to the cell. Each time the RAM clocks in a row address, one row of the memory is refreshed. Sequencing through all the row addresses within 2 ms will keep the memory refreshed.

The Intel 2118 is a high-performance 16,384-word by 1-bit dynamic RAM, fabricated on Intel's *n*-channel HMOS technology. The Intel 2118 is packaged in the industry-standard 16-pin DIP configuration, and requires only a single +5-V power supply (with ±10% tolerances) and ground for operation, that is, V_{DD} (±5 V) and V_{SS} (GND). The substrate bias voltage, usually designated V_{BB}, is internally produced by a back bias generator.

Problems

1. Define the terms random-access memory and read-only memory.
2. Draw a block diagram of a microcomputer using random-access memory and read-only memory.
3. Discuss the meaning of iRAM.
4. Discuss the use of memory circuits in microprocessors.
5. Draw and discuss the bubble memory device.

References

Allocca, John A., and Allen Stuart. *Electronic Instrumentation*, Reston Publishing Co., Reston, Virginia, 1983.

Memory Components Handbook, Intel Corporation, Santa Clara, Ca., 1983.

Tocci, Ronald J. *Digital Systems Principles and Applications*, revised and enlarged, Prentice-Hall, Inc., Englewood Cliffs, New Jersey, 1980.

12 MICROPROCESSORS AND MICROCOMPUTERS

12.1 INTRODUCTION

Developed and marketed by the Zilog Corporation, the Z-80® is one of the most flexible and widely used 8-bit microprocessors. It has architectural features similar to the Intel 8088 processor and indexing features similar to the 6800 and 6502 microprocessors. In addition, part of the instruction set of the Z-80 is compatible with the 8088 microprocessor at the machine code level. Most word-processing systems and home computer systems use the Z-80 microprocessor; Sinclair 81 and the TRS 80 for example.

Microcomputer systems are extremely simple to construct using Z-80 components. Any such system consists of three parts: (1) CPU (central processing unit), (2) memory, and (3) interface circuits to peripheral devices.

The CPU is the heart of the system. Its function is to obtain instructions from the memory and perform the desired operations. The memory is used to contain instructions and in most cases data that are to be processed. For example, a typical instruction sequence may be to read data from a specific peripheral device, store it in a location in memory, check

Zilog Z-80® CPU is a trademark of Zilog, Inc., with whom Reston Publishing Company is not associated.

the parity, and write it out to another peripheral device. Note that the Zilog component set includes the CPU and various general-purpose I/O device controllers, while a wide range of memory devices may be used from any source. Thus, all required components can be connected together in a very simple manner with virtually no other external logic. The user's effort then becomes primarily one of software development.

The single chip Z-80 8-bit microprocessor is built with *n*-channel silicon-gate, depletion-load technology. The circuit family of the Z-80 CPU consists of the following:

1. Counter–timer circuit
2. Parallel input–output circuit
3. Direct memory controller
4. Serial input–output circuit
5. Support boards

12.2 Z-80 CPU ARCHITECTURE[a]

A block diagram of the internal architecture of the Z-80 is shown in Figure 12.1. The diagram shows all the major elements in the CPU, and it should be referred to throughout the following description.

12.3 CPU REGISTERS

The Z-80 CPU contains 208 bits of R/W memory that are accessible to the programmer. Figure 12.2 illustrates how this memory is configured into eighteen 8-bit registers and four 16-bit registers. All Z-80 registers are implemented using static RAM. The registers include two sets of six general-purpose registers that may be used individually as 8-bit registers or in pairs as 16-bit registers. There are also two sets of accumulator and flag registers.

Special-Purpose Registers

1. *Program counter* (PC). The program counter holds the 16-bit address of the current instruction being fetched from memory. The PC is automatically incremented after its contents have been transferred to

---

[a] The material in Sections 12.2 to 12.16 is reprinted from "Z-80 CPU Z-80A CPU Technical Manual," September 1978. Reproduced by permission.

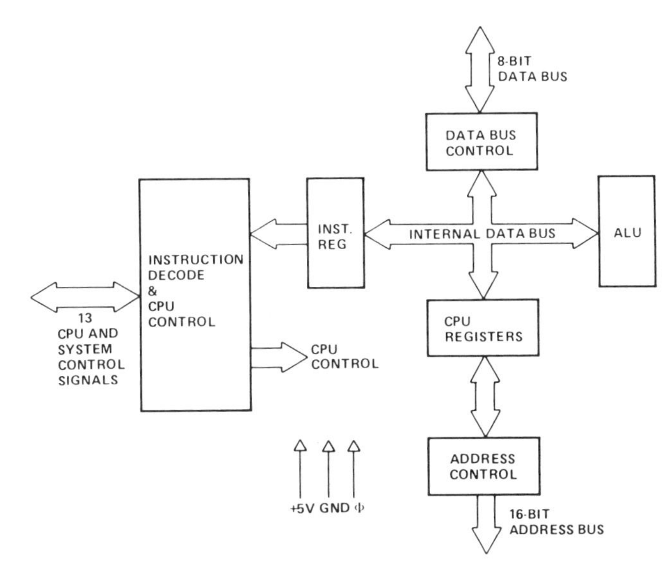

FIGURE 12.1 Z–80 CPU block diagrams. Courtesy of Zilog, Inc.

the address lines. When a program jump occurs, the new value is automatically placed in the PC, overriding the incrementer.

2. *Stack pointer* (SP). The stack pointer holds the 16-bit address of the current top of a stack located anywhere in external system RAM memory. The external stack memory is organized as a last-in first-out

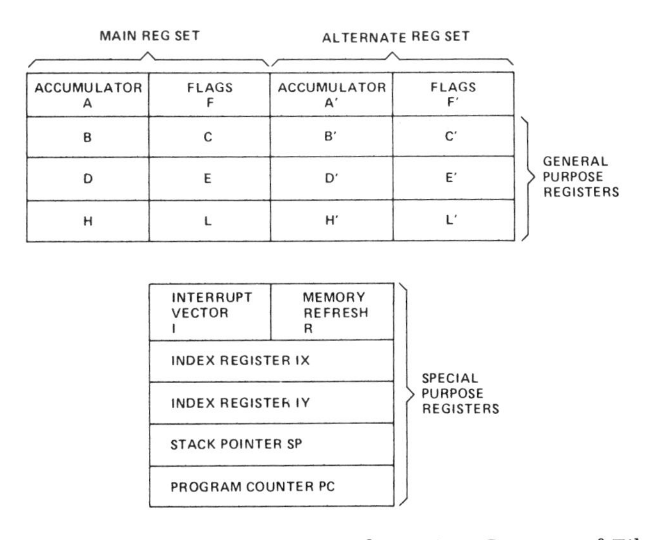

FIGURE 12.2 Z–80 CPU register configuration. Courtesy of Zilog, Inc.

(LIFO) file. Data can be pushed onto the stack from specific CPU registers or popped off the stack into specific CPU registers through the execution of PUSH and POP instructions. The data popped from the stack are always the last data pushed onto it. The stack allows simple implementation of multiple-level interrupts, unlimited subroutine nesting, and simplification of many types of data manipulation.

3. *Two index registers* (IX and IY). The two independent index registers hold a 16-bit base address that is used in indexed addressing modes. In this mode, an index register is used as a base to point to a region in memory from which data are to be stored or retrieved. An additional byte is included in indexed instructions to specify a displacement from this base. This displacement is specified as a two's complement signed integer. This mode of addressing greatly simplifies many types of programs, especially where tables of data are used.

4. *Interrupt page address register* (I). The Z-80 CPU can be operated in a mode where an indirect call to any memory location can be achieved in response to an interrupt. The I register is used for this purpose to store the high-order 8 bits of the indirect address, while the interrupting device provides the lower 8 bits of the address. This feature allows interrupt routines to be dynamically located anywhere in memory with absolutely minimal access time to the routine.

5. *Memory refresh register* (R). The Z-80 CPU contains a memory refresh counter to enable dynamic memories to be used with the same ease as static memories. Seven bits of this 8-bit register are automatically incremented after each instruction fetch. The eighth bit will remain as programmed as the result of an LD R, A instruction. The data in the refresh counter are sent out on the lower portion of the address bus along with a refresh control signal while the CPU is decoding and executing the fetched instruction. This mode of refresh is totally transparent to the programmer and does not slow down the CPU operation. The programmer can load the R register for testing purposes, but this register is normally not used by the programmer. During refresh, the contents of the I register are placed on the upper 8 bits of the address bus.

Accumulator and Flag Registers. The CPU includes two independent 8-bit accumulators and associated 8-bit flag registers. The accumulator holds the results of 8-bit arithmetic or logical operations, while the flag register indicates specific conditions for 8- or 16-bit operations, such as indicating whether or not the result of an operation is equal to zero. The programmer selects the accumulator and flag pair that he or she wishes to work with by a single exchange instruction, so it is easy to work with either pair.

General-Purpose Registers. There are two matched sets of general-purpose registers, each set containing six 8-bit registers that may be used individually as 8-bit registers or as 16-bit register pairs by the programmer. One set is called BC, DE, and HL while the complementary set is called BC′, DE′, and HL′. At any one time the programmer can select either set of registers to work with through a single exchange command for the entire set. In systems where fast interrupt response is required, one set of general-purpose registers and an accumulator–flag register may be reserved for handling this very fast routine. Only a simple exchange command need be executed to go between the routines. This greatly reduces interrupt service time by eliminating the requirement for saving and retrieving register contents in the external stack during interrupt or subroutine processing. These general-purpose registers are used for a wide range of applications by the programmer. They also simplify programming, especially in ROM-based systems where little external read–write memory is available.

12.4 ARITHMETIC AND LOGIC UNIT (ALU)

The 8-bit arithmetic and logical instructions of the CPU are executed in the ALU. Internally, the ALU communicates with the registers and the external data bus on the internal data bus. The type of functions performed by the ALU include the following:

Add	Left or right shifts or rotates (arithmetic and logical)
Subtract	Increment
Logical AND	Decrement
Logical OR	Set bit
Logical exclusive OR	Reset bit
Compare	Test bit

As each instruction is fetched from memory, it is placed in the instruction register and decoded. The control section performs this function and then generates and supplies all of the control signals necessary to read or write data from or to the registers, control the ALU, and provide all required external control signals.

12.5 Z-80 CPU PIN DESCRIPTION

The Z-80 CPU is packaged in an industry-standard 40-pin dual in-line package. The I/O pins are shown in Figure 12.3 and the function of each is described next.

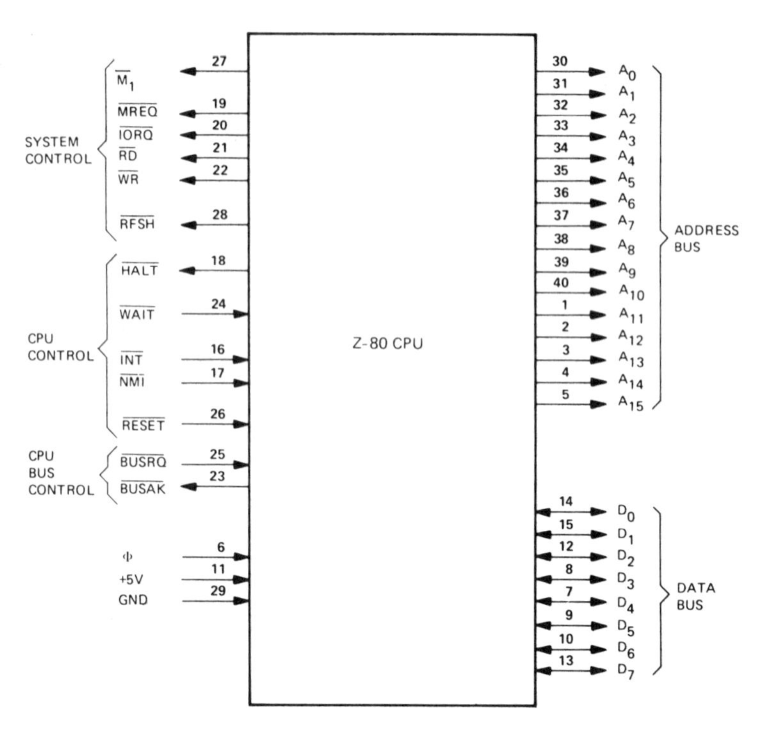

FIGURE 12.3 Z–80 pin configuration. Courtesy of Zilog, Inc.

A_0–A_{15} (Address Bus)
: Tristate output, active high. A_0–A_{15} constitute a 16-bit address bus. The address bus provides the address for memory (up to 64K bytes) data exchanges and for I/O device data exchanges. I/O addressing uses the 8 lower address bits to allow the user to directly select up to 256 input or 256 output ports. A_0 is the least significant address bit. During refresh time, the lower 7 bits contain a valid refresh address.

D_0–D_7 (Data Bus)
: Tristate input–output, active high. D_0–D_7 constitute an 8-bit bidirectional data bus. The data bus is used for data exchanges with memory and I/O devices.

$\overline{M_1}$ (Machine Cycle One)
: Output, active low. $\overline{M_1}$ indicates that the current machine cycle is the OP code fetch cycle of an instruction execution. Note that during execution of 2-byte op-codes, $\overline{M_1}$ is generated as each op code byte is fetched.

These 2-byte op-codes always begin with CBH, DDH, EDH, or FDH. $\overline{M_1}$ also occurs with $\overline{IORQ}$ to indicate an interrupt acknowledge cycle.

$\overline{MREQ}$
(Memory Request)

Tristate output, active low. The memory request signal indicates that the address bus holds a valid address for a memory read or memory write operation.

$\overline{IORQ}$
(Input–Output Request)

Tristate output, active low. The $\overline{IORQ}$ signal indicates that the lower half of the address bus holds a valid I/O address for an I/O read or write operation. An $\overline{IORQ}$ signal is also generated with an $\overline{M_1}$ signal when an interrupt is being acknowledged to indicate that an interrupt response vector can be placed on the data bus. Interrupt acknowledge operations occur during $\overline{M_1}$ time while I/O operations never occur during $\overline{M_1}$ time.

$\overline{RD}$
(Memory Read)

Tristate output, active low. $\overline{RD}$ indicates that the CPU wants to read data from memory or an I/O device. The addressed I/O device or memory should use this signal to gate data onto the CPU data bus.

$\overline{WR}$
(Memory Write)

Tristate output, active low. $\overline{WR}$ indicates that the CPU data bus holds valid data to be stored in the addressed memory or I/O device.

$\overline{RFSH}$
(Refresh)

Output, active low. $\overline{RFSH}$ indicates that the lower 7 bits of the address bus contain a refresh address for dynamic memories and the current $\overline{MREQ}$ signal should be used to do a refresh read to all dynamic memories.

$\overline{HALT}$
(Halt State)

Output, active low. $\overline{HALT}$ indicates that the CPU has executed a HALT software instruction and is awaiting either a nonmaskable or a maskable interrupt (with the mask enabled) before operation can resume. While halted, the CPU executes NOPs to maintain memory refresh activity.

$\overline{WAIT}$
(Wait)

Input, active low. $\overline{WAIT}$ indicates to the Z-80 CPU that the addressed memory or I/O devices are not ready for a data transfer. The CPU continues to enter wait states for as long as this signal is active. This signal allows memory or I/O devices of any speed to be synchronized to the CPU.

$\overline{\text{INT}}$ (Interrupt Request)
Input, active low. The interrupt request signal is generated by I/O devices. A request will be honored at the end of the current instruction if the internal software controlled interrupt enable flip-flop (IFF) is enabled and if the $\overline{\text{BUSRQ}}$ signal is not active. When the CPU accepts the interrupt, an acknowledge signal ($\overline{\text{IORQ}}$ during $\overline{\text{M}_1}$ time) is sent out at the beginning of the next instruction cycle.

$\overline{\text{NMI}}$ (Nonmaskable Interrupt)
Input, negative edge triggered. The nonmaskable interrupt request line has a higher priority than $\overline{\text{INT}}$ and is always recognized at the end of the current instruction, independent of the status of the interrupt enable flip-flop. $\overline{\text{NMI}}$ automatically forces the Z-80 CPU to restart to location 0066_H. The program counter is automatically saved in the external stack so that the user can return to the program that was interrupted. Note that continuous WAIT cycles can prevent the current instruction from ending, and that a $\overline{\text{BUSRQ}}$ will override a $\overline{\text{NMI}}$.

$\overline{\text{RESET}}$
Input, active low. $\overline{\text{RESET}}$ forces the program counter to zero and initializes the CPU. The CPU initialization includes:

1. Disable the interrupt enable flip-flop
2. Set register I = 00_H
3. Set register R = 00_H
4. Set interrupt mode 0

During reset time, the address bus and data bus go to a high-impedance state and all control output signals go to the inactive state.

$\overline{\text{BUSRQ}}$ (Bus Request)
Input, active low. The bus request signal is used to request the CPU address bus, data bus, and tristate output control signals to go to a high-impedance state so that other devices can control these buses. When $\overline{\text{BUSRQ}}$ is activated, the CPU will set these

buses to a high-impedance state as soon as the current CPU machine cycle is terminated.

$\overline{\text{BUSAK}}$ (Bus Acknowledge)	Output, active low. Bus acknowledge is used to indicate to the requesting device that the CPU address bus, data bus, and tristate control bus signals have been set to their high-impedance state and the external device can now control these signals.
Φ	Single-phase TTL level clock that requires only a 330-Ω pull-up resistor to +5 V to meet all clock requirements.

12.6 FLAGS

Each of the two Z-80 CPU flag registers contains six bits of information that are set or reset by various CPU operations. Four of these bits are testable; that is, they are used as conditions for jump, call, or return instructions. For example, a jump may be desired only if a specific bit in the flag register is set. The four testable flag bits are the following:

1. *Carry flag* (C): This flag is the carry from the highest-order bit of the accumulator. For example, the carry flag will be set during an add instruction where a carry from the highest bit of the accumulator is generated. This flag is also set if a borrow is generated during a subtraction instruction. The shift and rotate instructions also affect this bit.

2. *Zero flag* (Z): This flag is set if the result of the operation loaded a zero into the accumulator. Otherwise, it is reset.

3. *Sign flag* (S): This flag is intended to be used with signed numbers, and it is set if the result of the operation was negative. Since bit 7 (MSB) represents the sign of the number (a negative number has a 1 in bit 7), this flag stores the state of bit 7 in the accumulator.

4. *Parity/overflow flag* (P/V): This dual-purpose flag indicates the parity of the result in the accumulator when logical operations are performed (such as AND A, B), and it represents overflow when signed two's complement arithmetic operations are performed. The Z-80 overflow flag indicates that the two's complement number in the accumulator is in error since it has exceeded the maximum possible (+127) or is less than the minimum possible (−128) number that can be represented in two's complement notation. For example, consider adding

$$\begin{array}{rrl} +120 = & & 0111\ 1000 \\ +105 = & & \underline{0110\ 1001} \\ & C = 0 & 1110\ 0001 = -95 \quad \text{(wrong), overflow has occurred} \end{array}$$

Here the result is incorrect. Overflow has occurred and yet there is no carry to indicate an error. For this case the overflow flag would be set. Also consider the addition of two negative numbers:

$$
\begin{array}{rl}
-5 = & 1111\ 1011 \\
-16 = & \underline{1111\ 0000} \\
C = 1 & 1110\ 1011 = -21 \quad \text{(correct)}
\end{array}
$$

Notice that the answer is correct but the carry is set so that this flag cannot be used as an overflow indicator. In this case the overflow would not be set. For logical operations (AND, OR, XOR), this flag is set if the parity of the result is even and it is reset if it is odd.

There are also two nontestable bits in the flag register. Both of these are used for BCD arithmetic. They are the following:

1. *Half-carry* (H): This is the BCD carry or borrow result from the last significant 4 bits of operation. When using the DAA (decimal adjust instruction), this flag is used to correct the result of a previous packed decimal add or subtract.

2. *Subtract flag* (N): Since the algorithm for correcting BCD operations is different for addition or subtraction, this flag is used to specify what type of instruction was executed last so that the DAA operation will be correct for either addition or subtraction.

The flag register can be accessed by the programmer, and its format is as follows:

S	Z	X	H	X	P/V	N	C

X means flag is indeterminate.

Table 12.1 lists how each flag bit is affected by various CPU instructions. In this table, a ● indicates that the instruction does not change the flag, an X means that the flag goes to an indeterminate state, a 0 means that it is reset, a 1 means that it is set, and the symbol ↕ indicates that it is set or reset according to the previous discussion. Note that any instruction not appearing in this table does not affect any of the flags.

Table 12.1 includes a few special cases that must be described for clarity. Notice that the block search instruction sets the Z flag if the last compare operation indicated a match between the source and the accumulator data. Also, the parity flag is set if the byte counter (register pair BC) is not equal to zero. This same use of the parity flag is made with the block move instructions. Another special case is during block input or output instructions, here the Z flag is used to indicate the state of register B, which is used as a byte counter. Notice that when the I/O

TABLE 12.1 *Summary of Flag Operation*

Instruction	C	Z	P/V	S	N	H	Comments
ADD A, s; ADC A,s	↕	↕	V	↕	0	↕	8-bit add or add with carry
SUB s; SBC A, s, CP s, NEG	↕	↕	V	↕	1	↕	8-bit subtract, subtract with carry, compare and negate accumulator
AND s	0	↕	P	↕	0	1	Logical operations
OR s; XOR s	0	↕	P	↕	0	0	And set's different flags
INC s	•	↕	V	↕	0	↕	8-bit increment
DEC m	•	↕	V	↕	1	↕	8-bit decrement
ADD DD, ss	↕	•	•	•	0	X	16-bit add
ADC HL, ss	↕	↕	V	↕	0	X	16-bit add with carry
SBC HL, ss	↕	↕	V	↕	1	X	16-bit subtract with carry
RLA; RLCA, RRA, RRCA	↕	•	•	•	0	0	Rotate accumulator
RL m; RLC m; RR m; RRC m SLA m, SRA m; SRL m	↕	↕	P	↕	0	0	Rotate and shift location m
RLD, RRD	•	↕	P	↕	0	0	Rotate digit left and right
DAA	↕	↕	P	↕	•	↕	Decimal adjust accumulator
CPL	•	•	•	•	1	1	Complement accumulator
SCF	1	•	•	•	0	0	Set carry
CCF	↕	•	•	•	0	X	Complement carry
IN r, (C)	•	↕	P	↕	0	0	Input register indirect
INI; IND; OUTI; OUTD	•	↕	X	X	1	X	Block input and output
INIR; INDR; OTIR; OTDR	•	1	X	X	1	X	Z = 0 if B ≠ 0 otherwise Z = 1
LDI, LDD	•	X	↕	X	0	0	Block transfer instructions
LDIR, LDDR	•	X	0	X	0	0	P/V = 1 if BC ≠ 0, otherwise P/V = 0
CPI, CPIR, CPD, CPDR	•	↕	↕	↕	1	X	Block search instructions Z = 1 if A = (HL), otherwise Z = 0 P/V = 1 if BC ≠ 0, otherwise P/V = 0
LD A, I; LD A, R	•	↕	IFF	↕	0	0	The content of the interrupt enable flip-flop (IFF) is copied into the P/V flag
BIT b, s	•	↕	X	X	0	1	The state of bit b of location s is copied into the Z flag
NEG	↕	↕	V	↕	1	↕	Negate accumulator

The following notation is used in this table:

Symbol	Operation
C	Carry/link flag. C=1 if the operation produced a carry from the MSB of the operand or result.
Z	Zero flag. Z=1 if the result of the operation is zero.
S	Sign flag. S=1 if the MSB of the result is one.
P/V	Parity or overflow flag. Parity (P) and overflow (V) share the same flag. Logical operations affect this flag with the parity of the result while arithmetic operations affect this flag with the overflow of the result. If P/V holds parity, P/V=1 if the result of the operation is even, P/V=0 if result is odd. If P/V holds overflow, P/V=1 if the result of the operation produced an overflow.
H	Half-carry flag. H=1 if the add or subtract operation produced a carry into or borrow from into bit 4 of the accumulator.
N	Add/Subtract flag. N=1 if the previous operation was a subtract.
	H and N flags are used in conjunction with the decimal adjust instruction (DAA) to properly correct the result into packed BCD format following addition or subtraction using operands with packed BCD format.
↕	The flag is affected according to the result of the operation.
•	The flag is unchanged by the operation.
0	The flag is reset by the operation.
1	The flag is set by the operation.
X	The flag is a "don't care."
V	P/V flag affected according to the overflow result of the operation.
P	P/V flag affected according to the parity result of the operation.
r	Any one of the CPU registers A, B, C, D, E, H, L.
s	Any 8-bit location for all the addressing modes allowed for the particular instruction. .
ss	Any 16-bit location for all the addressing modes allowed for that instruction.
ii	Any one of the two index registers IX or IY.
R	Refresh counter.
n	8-bit value in range <0, 255>
nn	16-bit value in range <0, 65535>
m	Any 8-bit location for all the addressing modes allowed for the particular instruction.

Courtesy of Zilog, Inc.

block transfer is complete, the zero flag will be reset to a zero (i.e., B = 0), while in the case of a block move command the parity flag is reset when the operation is complete. A final case is when the refresh or I register is loaded into the accumulator; the interrupt enable flip-flop is loaded into the parity flag so that the complete state of the CPU can be saved at any time.

12.7 Z-80 CPU INSTRUCTION SET

The Z-80 CPU can execute 158 different instruction types, including all 78 of the 8080A CPU. The instructions can be broken down into the following major groups:

1. Load and exchange
2. Block transfer and search
3. Arithmetic and logical
4. Rotate and shift
5. Bit manipulation (set, reset, test)
6. Jump, call, and return
7. Input–output
8. Basic CPU control

12.8 INTRODUCTION TO INSTRUCTION TYPES

The load instructions move data internally between CPU registers or between CPU registers and external memory. All of these instructions must specify a source location from which the data are to be moved and a destination location. The source location is not altered by a load instruction. Examples of load group instructions include moves between any of the general-purpose registers, such as move the data to register B from register C. This group also includes load immediate to any CPU register or to any external memory location. Other types of load instructions allow transfer between CPU registers and memory locations. The exchange instructions can trade the contents of two registers.

A unique set of block transfer instructions is provided in the Z-80. With a single instruction a block of memory of any size can be moved to any other location in memory. This set of block moves is extremely valuable when large strings of data must be processed. The Z-80 block search instructions are also valuable for this type of processing. With a single instruction, a block of external memory of any desired length can be searched for any 8-bit character. Once the character is found or the end of the block is reached, the instruction automatically terminates. Both the block transfer and the block search instructions can be interrupted during their execution so as to not occupy the CPU for long periods of time.

The arithmetic and logical instructions operate on data stored in the accumulator and other general-purpose CPU registers or external memory locations. The results of the operations are placed in the accumulator, and the appropriate flags are set according to the result of the operation. An example of an arithmetic operation is adding the accumulator to the contents of an external memory location. The results of the addition are placed in the accumulator. This group also includes 16-bit addition and subtraction between 16-bit CPU registers.

The rotate and shift group allows any register or any memory location to be rotated right or left with or without carry, either arithmetic or logical. Also, a digit in the accumulator can be rotated right or left with two digits in any memory location.

The bit manipulation instructions allow any bit in the accumulator, any general-purpose register, or any external memory location to be set, reset, or tested with a single instruction. For example, the most significant bit of register H can be reset. This group is especially useful in control applications and for controlling software flags in general-purpose programming.

The jump, call, and return instructions are used to transfer between various locations in the user's program. This group uses several different techniques for obtaining the new program counter address from specific external memory locations. A unique type of call is the restart instruction. This instruction actually contains the new address as a part of the 8-bit OP code. This is possible since only 8 separate addresses located in page zero of the external memory may be specified. Program jumps may also be achieved by loading register HL, IX, or IY directly into the PC, thus allowing the jump address to be a complex function of the routine being executed.

The input–output group of instructions in the Z-80 allows for a wide range of transfers between external memory locations or the general-purpose CPU registers and the external I/O devices. In each case, the port number is provided on the lower 8 bits of the address bus during any I/O transaction. One instruction allows this port number to be specified by the second byte of the instruction, while other Z-80 instructions allow it to be specified as the content of the C register. One major advantage of using the C register as a pointer to the I/O device is that it allows different I/O ports to share common software driver routines. This is not possible when the address is part of the OP code if the routines are stored in ROM. Another feature of these input instructions is that they set the flag register automatically so that additional operations are not required to determine the state of the input data (for example, its parity). The Z-80 CPU includes single instructions that can move blocks of data (up to 256 bytes) automatically to or from any I/O port directly to any memory location. In conjunction with the dual set of general-purpose registers, these instructions provide for fast I/O block transfer rates. The value of this I/O instruction set is demonstrated by the fact that the Z-80 CPU can provide all required floppy disk formatting (i.e., the CPU provides the preamble, address, and data, and enables the CRC codes) on double-density floppy disk drives on an interrupt driven basis.

Finally, the basic CPU control instructions allow various options and modes. This group includes instructions such as setting or resetting the interrupt enable flip-flop or setting the mode of interrupt response.

12.9 ADDRESSING MODES

Most of the Z-80 instructions operate on data stored in internal CPU registers, external memory, or the I/O ports. Addressing refers to how the address of these data is generated in each instruction. This section gives a brief summary of the types of addressing used in the Z-80, while subsequent sections detail the type of addressing available for each instruction group.

Immediate. In this mode of addressing the byte following the OP code in memory contains the actual operand.

OP code	} 1 or 2 bytes
Operand	
d_7 d_0	

Examples of this type of instruction would be to load the accumulator with a constant, where the constant is the byte immediately following the OP code.

Immediate Extended. This mode is merely an extension of immediate addressing in that the 2 bytes following the OP codes are the operand.

OP code	1 or 2 bytes
Operand	Low order
Operand	High order

Examples of this type of instruction would be to load the HL register pair (16-bit register) with 16 bits (2 bytes) of data.

Modified Page Zero Addressing. The Z-80 has a special single-byte CALL instruction to any of eight locations in page zero of memory. This instruction (which is referred to as a restart) sets the PC to an effective address in page zero. The value of this instruction is that it allows a single byte to specify a complete 16-bit address where commonly called subroutines are located, thus saving memory space.

OP code	1 byte
b_7 b_0	Effective address is $(b_5b_4b_3000)_2$

Relative Addressing. Relative addressing uses 1 byte of data following the OP code to specify a displacement from the existing program to which a program jump can occur. This displacement is a signed two's complement number that is added to the address of the OP code of the following instruction.

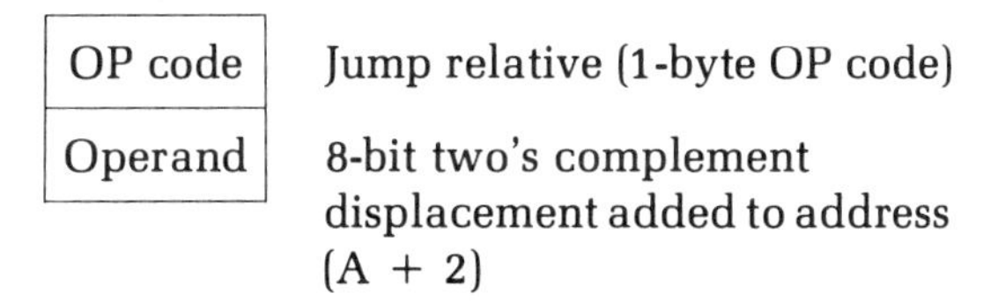

The value of relative addressing is that it allows jumps to nearby locations while only requiring 2 bytes of memory space. For most programs, relative jumps are by far the most prevalent type of jump due to the proximity of related program segments. Thus, these instructions can significantly reduce memory space requirements. The signed displacement can range between +127 and −128 from A + 2. This allows for a total displacement of +129 to −126 from the jump relative OP code address. Another major advantage is that it allows for relocatable code.

Extending Addressing. Extended addressing provides for 2 bytes (16 bits) of address to be included in the instruction. These data can be an address to which a program can jump or they can be an address where an operand is located.

OP code	1 or 2 bytes
Low-order address or low-order operand	
High-order address or high-order operand	

Extended addressing is required for a program to jump from any location in memory to any other location or to load and store data in any memory location.

When extended addressing is used to specify the source or destination address of an operand, the notation (nn) will be used to indicate the content of memory in nn, where nn is the 16-bit address specified in the instruction. This means that the 2 bytes of address nn are used as a pointer to a memory location. The use of the parentheses always means that the value enclosed within them is used as a pointer to a memory location. For example, (1200) refers to the contents of memory in location 1200.

Indexed Addressing. In this type of addressing, the byte of data following the OP code contains a displacement that is added to one of the two index registers (the OP code specifies which index register is used) to form a pointer to memory. The contents of the index register are not altered by this operation.

OP code	2-byte OP code
OP code	
Displacement	Operand added to index register to form a pointer to memory

An example of an indexed instruction would be to load the contents of the memory location (index register + displacement) into the accumulator. The displacement is a signed two's complement number. Indexed addressing greatly simplifies programs using tables of data since the index register can point to the start of any table. Two index registers are provided since operations very often require two or more tables. Indexed addressing also allows for relocatable code.

The two index registers in the Z-80 are referred to as IX and IY. To indicate indexed addressing the notation (IX + d) or (IY + d) is used. Here d is the displacement specified after the OP code. The parentheses indicate that this value is used as a pointer to external memory.

Register Addressing. Many Z-80 OP codes contain bits of information that specify which CPU register is to be used for an operation. An example of register addressing would be to load the data in register B into register C.

Implied Addressing. Implied addressing refers to operations where the OP code automatically implies one or more CPU registers as containing the operands. An example is the set of arithmetic operations where the accumulator is always implied to be the destination of the results.

Register Indirect Addressing. This type of addressing specifies a 16-bit CPU register pair (such as HL) to be used as a pointer to any location in memory. This type of instruction is very powerful and it is used in a wide range of applications.

OP code	1 or 2 bytes

An example of this type of instruction would be to load the accumulator with the data in the memory location pointed to by the HL register contents. Indexed addressing is actually a form of register indirect addressing except that a displacement is added with indexed addressing. Register indirect addressing allows for very powerful but simple-to-implement memory accesses. The block move and search commands in the Z-80 are extensions of this type of addressing where automatic register incrementing, decrementing, and comparing have been added. The notation for indicating register indirect addressing is to put parentheses around the name of the register that is to be used as the pointer. For example, the symbol (HL) specifies that the contents of the HL register are to be used as a pointer to a memory location. Often, register indirect addressing is used to specify 16-bit operands. In this case, the register contents point to the lower-order portion of the operand, while the register contents are automatically incremented to obtain the upper portion of the operand.

Bit Addressing. The Z-80 contains a large number of bit set, reset, and test instructions. These instructions allow any memory location or CPU register to be specified for a bit operation through one of three previous addressing modes (register, register indirect, and indexed), while 3 bits in the OP code specify which of the 8 bits is to be manipulated.

12.10 ADDRESSING MODE COMBINATIONS

Many instructions include more than one operand (such as arithmetic instructions or loads). In these cases, two types of addressing may be employed. For example, load can use immediate addressing to specify the source and register indirect or indexed addressing to specify the destination.

12.11 INSTRUCTION OF CODES

This section describes each of the Z-80 instructions and provides tables listing the OP codes for every instruction. In each table the OP codes in shaded areas are identical to those offered in the 8080A CPU. The assembly language mnemonic that is used for each instruction is also shown. All instruction OP codes are listed in hexadecimal notation. Single-byte OP codes require two hex characters, while double-byte OP codes require four hex characters. The conversion from hex to binary is repeated here for convenience.

Hex		Binary		Decimal	Hex		Binary		Decimal
0	=	0000	=	0	8	=	1000	=	8
1	=	0001	=	1	9	=	1001	=	9
2	=	0010	=	2	A	=	1010	=	10
3	=	0011	=	3	B	=	1011	=	11
4	=	0100	=	4	C	=	1100	=	12
5	=	0101	=	5	D	=	1101	=	13
6	=	0110	=	6	E	=	1110	=	14
7	=	0111	=	7	F	=	1111	=	15

Z-80 instruction mnemonics consist of an OP code and zero, one, or two operands. Instructions in which the operand is implied have no operand. Instructions that have only one logical operand or those in which one operand is invariant (such as the logical OR instruction) are represented by a one-operand mnemonic. Instructions that may have two varying operands are represented by two-operand mnemonics.

12.12 LOAD AND EXCHANGE

Table 12.2 defines the OP code for all the 8-bit load instructions implemented in the Z-80 CPU. Also shown in this table is the type of addressing used for each instruction. The source of the data is found on the top horizontal row, while the destination is specified by the left column. For example, load register C from register B uses the OP code 48H. In all the tables the OP code is specified in hexadecimal notation and the 48H (=0100 1000 binary) code is fetched by the CPU from the external memory during M_1 time, decoded, and then the register transfer is automatically performed by the CPU.

The assembly language mnemonic for this entire group is LD, followed by the destination, followed by the source (LD DEST., SOURCE). Note that several combinations of addressing modes are possible. For example, the source may use register addressing and the destination may be register indirect, such as load the memory location pointed to by register HL with the contents of register D. The OP code for

TABLE 12.2 *Eight-Bit Load Group LD*

		SOURCE															
		IMPLIED		REGISTER							REG INDIRECT			INDEXED		EXT. ADDR.	IMME.
		I	R	A	B	C	D	E	H	L	(HL)	(BC)	(DE)	(IX + d)	(IY + d)	(nn)	n
DESTINATION REGISTER	A	ED 57	ED 5F	7F	78	79	7A	7B	7C	7D	7E	0A	1A	DD 7E d	FD 7E d	3A n n	3E n
	B			47	40	41	42	43	44	45	46			DD 46 d	FD 46 d		06 n
	C			4F	48	49	4A	4B	4C	4D	4E			DD 4E d	FD 4E d		0E n
	D			57	50	51	52	53	54	55	56			DD 56 d	FD 56 d		16 n
	E			5F	58	59	5A	5B	5C	5D	5E			DD 5E d	FD 5E d		1E n
	H			67	60	61	62	63	64	65	66			DD 66 d	FD 66 d		26 n
	L			6F	68	69	6A	6B	6C	6D	6E			DD 6E d	FD 6E d		2E n
DESTINATION REG INDIRECT	(HL)			77	70	71	72	73	74	75							36 n
	(BC)			02													
	(DE)			12													
DESTINATION INDEXED	(IX+d)			DD 77 d	DD 70 d	DD 71 d	DD 72 d	DD 73 d	DD 74 d	DD 75 d							DD 36 d n
	(IY+d)			FD 77 d	FD 70 d	FD 71 d	FD 72 d	FD 73 d	FD 74 d	FD 75 d							FD 36 d n
DESTINATION EXT. ADDR	(nn)			32 n n													
DESTINATION IMPLIED	I			ED 47													
	R			ED 4F													

Courtesy of Zilog, Inc.

this operation would be 72. The mnemonic for this load instruction would be as follows:

LD(HL) ,D

The parentheses around the HL mean that the contents of HL are used as a pointer to a memory location. In all Z-80 load instruction mnemonics, the destination is always listed first, with the source following. The Z-80 assembly language has been defined for ease of programming. Every instruction is self-documenting, and programs written in Z-80 language are easy to maintain.

Note in Table 12.2 that some load OP codes that are available in the Z-80 use 2 bytes. This is an efficient method of memory utilization since 8-, 16-, 24-, or 32-bit instructions are implemented in the Z-80. Thus often utilized instructions such as arithmetic or logical operations are only 8 bits, which results in better memory utilization than is achieved with fixed instruction sizes such as 16 bits.

All load instructions using indexed addressing for either the source or destination location actually use 3 bytes of memory, with the third byte being the displacement d. For example, a load register E with the operand pointed to by IX with an offset of +8 would be written

LD E,(IX +8)

The instruction sequence for this in memory would be

Address	Contents	
Address A	DD	OP code
A + 1	5E	
A + 2	08	Displacement operand

The two extended addressing instructions are also 3-byte instructions. For example, the instruction to load the accumulator with the operand in memory location 6F32H would be written

LD A, (6F32H)

and its instruction sequence would be

Address	Contents	
Address A	3A	OP code
A + 1	32	Low-order address
A + 2	6F	High-order address

Notice that the low-order portion of the address is always the first operand.

The load immediate instructions for the general-purpose 8-bit registers are 2-byte instructions. The instruction load register H with the value 36H would be written

LD H, 36H

and its sequence would be

Address	Contents	
Address A	26	OP code
A + 1	36	Operand

Loading a memory location using indexed addressing for the destination and immediate addressing for the source requires 4 bytes. For example,

LD(IX − 15), 21H

would appear as

Address A	DD	OP code
A + 1	36	
A + 2	F1	Displacement (−15 in signed two's complement)
A + 3	21	Operand to load

Notice that with any indexed addressing the displacement always follows directly after the OP code.

Table 12.3 specifies the 16-bit load operations. This table is very similar to Table 12.2. Notice that the extended addressing capability covers all register pairs. Also notice that register indirect operations specifying the stack pointer are the PUSH and POP instructions. The mnemonic for these instructions is PUSH and POP. These differ from other 16-bit loads in that the stack byte is automatically decremented and incremented as each byte is pushed onto or popped from the stack, respectively. For example, the instruction

PUSH AF

is a single-byte instruction with the OP code of F5H. when this instruction is executed, the following sequence is generated:

Decrement SP

LD (SP), A

Decrement SP

LD (SP), F

Thus the external stack now appears as follows:

(SP)	F	← Top of stack
(SP + 1)	A	
•	•	
•	•	
•	•	

The POP instruction is the exact reverse of a PUSH. Notice that all PUSH and POP instructions utilize a 16-bit operand, and the high-order byte is always pushed after and popped last. That is,

TABLE 12.3 *Sixteen-Bit Load Group LD, PUSH, and POP*

DESTINATION		SOURCE: REGISTER AF	REGISTER BC	REGISTER DE	REGISTER HL	REGISTER SP	REGISTER IX	REGISTER IY	IMM. EXT. nn	EXT. ADDR. (nn)	REG. INDIR. (SP)
REGISTER	AF										F1
REGISTER	BC								01 n n	ED 4B n n	C1
REGISTER	DE								11 n n	ED 5B n n	D1
REGISTER	HL								21 n n	2A n n	E1
REGISTER	SP				F9		DD F9	FD F9	31 n n	ED 7B n n	
REGISTER	IX								DD 21 n n	DD 2A n n	DD E1
REGISTER	IY								FD 21 n n	FD 2A n n	FD E1
EXT. ADDR.	(nn)		ED 43 n n	ED 53 n n	22 n n	ED 73 n n	DD 22 n n	FD 22 n n			
PUSH INSTRUCTIONS → REG. IND.	(SP)	F5	C5	D5	E5		DD E5	FD E5			

POP INSTRUCTIONS ↑ (column (SP))

NOTE: The Push & Pop Instructions adjust the SP after every execution

Courtesy of Zilog, Inc.

PUSH BC is PUSH B then C
PUSH DE is PUSH D then E
PUSH HL is PUSH H then L
POP HL is POP L then H

The instruction using extended immediate addressing for the source obviously requires 2 bytes of data following the OP code. For example,

LD DE, 0659H

will be

Address A	11	OP code
A + 1	59	Low-order operand to register E
A + 2	06	High-order operand to register D

In all extended immediate or extended addressing modes, the low-order byte always appears first after the OP code.

12.13 INTERRUPT RESPONSE

The purpose of an interrupt is to allow peripheral devices to suspend CPU operation in an orderly manner and force the CPU to start a peripheral service routine. Usually this service routine is involved with the exchange of data, or status and control information, between the CPU and the peripheral. Once the service routine is completed, the CPU returns to the operation from which it was interrupted.

12.13.1 Interrupt Enable-Disable

The Z-80 CPU has two interrupt inputs, a software maskable interrupt and a nonmaskable interrupt. The nonmaskable interrupt (NMI) *cannot* be disabled by the programmer, and it will be accepted whenever a peripheral device requests it. This interrupt is generally reserved for very important functions that must be serviced whenever they occur, such as an impending power failure. The maskable interrupt (INT) can be selectively enabled or disabled by the programmer. This allows the programmer to disable the interrupt during periods when the program has timing constraints that do not allow it to be interrupted. In the Z-80 CPU there is an enable flip-flop (called IFF) that is set or reset by the programmer using the enable interrupt (EI) and disable interrupt (DI) instructions. When the IFF is reset, an interrupt cannot be accepted by the CPU.

Actually, for purposes that will be subsequently explained, there are two enable flip-flops, called IFF_1 and IFF_2. IFF_1 actually disables interrupts from being accepted. IFF_2 is a temporary storage location for IFF_1. The state of IFF_1 is used to actually inhibit interrupts, while IFF_2 is used as a temporary storage location for IFF_1. The purpose of storing the IFF_1 will be subsequently explained.

A reset to the CPU will force both IFF_1 and IFF_2 to the reset state so that interrupts are disabled. They can then be enabled by the EI instruction at any time by the programmer. When an EI instruction is executed, any pending interrupt request will not be accepted until after the instruction following EI has been executed. This single instruction delay is necessary for cases when the following instruction is a return instruction, and interrupts must not be allowed until the return has been completed. The EI instruction sets both IFF_1 and IFF_2 to the enable state.

Action	IFF_1	IFF_2	
CPU Reset	0	0	
DI	0	0	
EI	1	1	
LD A, I	•	•	$IFF_2 \rightarrow$ Parity flag
LD A, R	•	•	$IFF_2 \rightarrow$ Parity flag
Accept NMI	0	•	
RETN	IFF_2	•	$IFF_2 \rightarrow IFF_1$

"•" indicates no change

FIGURE 12.4 Interrupt enable/disable flip-flops. Courtesy of Zilog, Inc.

When an interrupt is accepted by the CPU, both IFF_1 and IFF_2 are automatically reset, inhibiting further interrupts until the programmer wishes to issue a new EI instruction. Note that for all the previous cases IFF_1 and IFF_2 are always equal.

The purpose of IFF_2 is to save the status of IFF_1 when a nonmaskable interrupt occurs. When a nonmaskable interrupt is accepted, IFF_1 is reset to prevent further interrupts until reenabled by the programmer. Thus, after a nonmaskable interrupt has been accepted, maskable interrupts are disabled, but the previous state of IFF_1 has been saved so that the complete state of the CPU just prior to the nonmaskable interrupt can be restored at any time. When a load register A with register I (LD A, I) instruction or a load register A with register R (LD A, R) instruction is executed, the state of IFF_2 is copied into the parity flag where it can be tested or stored.

A second method of restoring the status of IFF_1 is through the execution of a return from nonmaskable interrupt (RETN) instruction. Since this instruction indicates that the nonmaskable interrupt service routine is complete, the contents of IFF_2 are now copied back into IFF_1 so that the status of IFF_1 just prior to the acceptance of the nonmaskable interrupt will be restored automatically.

Figure 12.4 is a summary of the effect of different instructions on the two enable flip-flops.

12.14 CPU RESPONSE

Nonmaskable. A nonmaskable interrupt will be accepted at all times by the CPU. When this occurs, the CPU ignores the next instruction that it fetches and instead does a restart to location 0066H. Thus, it behaves exactly as if it had received a restart instruction, but it is to a location that is not one of the eight software restart locations. A restart is merely a call to a specific address in page 0 of memory.

Maskable. The CPU can be programmed to respond to the maskable interrupt in any one of three possible modes.

Mode 0

This mode is identical to the 8080A interrupt response mode. With this mode, the interrupting device can place any instruction on the data bus and the CPU will execute it. Thus, the interrupting device provides the next instruction to be executed instead of the memory. Often this will be a restart instruction, since the interrupting device only need supply a single-byte instruction. Alternatively, any other instruction, such as a 3-byte call, to any location in memory could be executed.

The number of clock cycles necessary to execute this instruction is two more than the normal number for the instruction. This occurs since the CPU automatically adds two wait states to an interrupt response cycle to allow sufficient time to implement an external daisy chain for priority control. Section 12.13 illustrates the detailed timing for an interrupt response. After the application of RESET, the CPU will automatically enter interrupt mode 0.

Mode 1

When this mode has been selected by the programmer, the CPU will respond to an interrupt by executing a restart to location 0038H. Thus the response is identical to that for a nonmaskable interrupt except that the call location is 0038H instead of 0066H. Another difference is that the number of cycles required to complete the restart instruction is two more than normal due to the two added wait states.

Mode 2

This mode is the most powerful interrupt response mode. With a single 8-bit byte from the user, an indirect call can be made to any memory location.

With this mode the programmer maintains a table of 16-bit starting addresses for every interrupt service routine. This table may be located anywhere in memory. When an interrupt is accepted, a 16-bit pointer must be formed to obtain the desired interrupt service routine starting address from the table. The upper 8 bits of this pointer are formed from the contents of the I register. The I register must have been previously loaded with the desired value by the programmer (i.e., LD I, A). Note that a CPU reset clears the I register so that it is initialized to zero. The lower 8 bits of the pointer must be supplied by the interrupting device. Actually, only 7 bits are required from the interrupting device, as the least significant bit must be a zero. This is required since the pointer is used to get two adjacent bytes to form a complete 16-bit service routine starting address, and the addresses must always start in even locations.

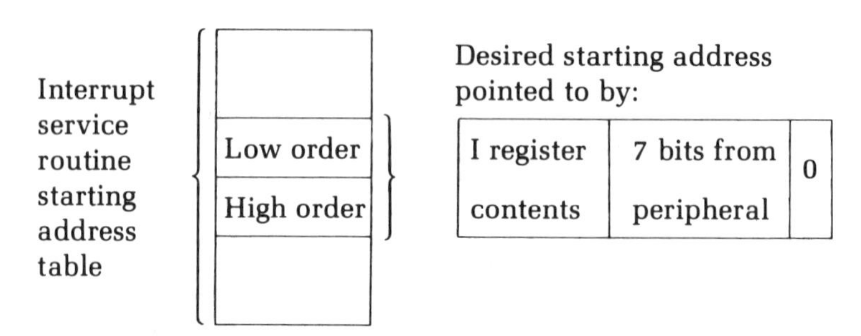

The first byte in the table is the least significant (low order) portion of the address. The programmer must obviously fill this table in with the desired addresses before any interrupts are to be accepted.

Note that this table can be changed at any time by the programmer (if it is stored in read–write memory) to allow different peripherals to be serviced by different service routines.

Once the interrupting device supplies the lower portion of the pointer, the CPU automatically pushes the program counter onto the stack, obtains the starting address from the table, and does a jump to this address. This mode of response requires 19 clock periods to complete (7 to fetch the lower 8 bits from the interrupting device, 6 to save the program counter, and 6 to obtain the jump address).

Note that the Z-80 peripheral devices all include a daisy chain priority interrupt structure that automatically supplies the programmed vector to the CPU during interrupt acknowledge. Refer to the Z80-PIO, X80-SIO, and Z80-CTC manuals for details.

12.15 HARDWARE AND IMPLEMENTATION EXAMPLES

This section is intended to serve as a basic introduction to implementing systems with the Z80-CPU.

Minimum System. Figure 12.5 is a diagram of a very simple Z-80 system. Any Z-80 system must include the following five elements:

1. Five-volt power supply
2. Oscillator

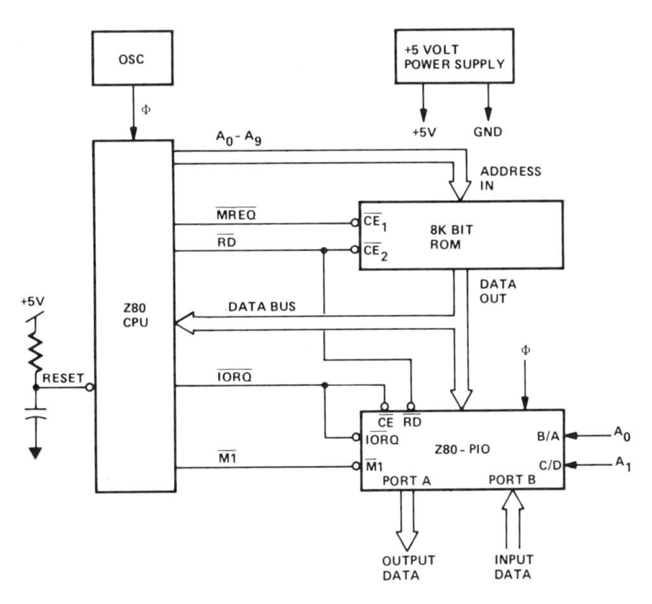

FIGURE 12.5 Minimum Z–80 computer system. Courtesy of Zilog, Inc.

3. Memory devices
4. I/O circuits
5. CPU

Since the Z80-CPU requires only a single 5-V supply, most small systems can be implemented using only this single supply.

The oscillator can be very simple since the only requirement is that it be a 5-V square wave. For systems not running at full speed, a simple *RC* oscillator can be used. When the CPU is operated near the highest possible frequency, a crystal oscillator is generally required, because the system timing will not tolerate the drift or jitter that an *RC* network will generate. A crystal oscillator can be made from inverters and a few discrete components or monolithic circuits are widely available.

The external memory can be any mixture of standard RAM, ROM, or PROM. In this simple example we have shown a single 8K-bit ROM (1K bytes) being utilized as the entire memory system. For this example we have assumed that the Z-80 internal register configuration contains sufficient read–write storage so that external RAM memory is not required.

Every computer system requires I/O circuits to allow it to interface to the real world. In this simple example it is assumed that the output is an 8-bit control vector, and the input is an 8-bit status word. The input data could be gated onto the data bus using any standard tristate driver, while the output data could be latched with any type of standard TTL latch. For this example we have used a Z80-PIO for the I/O circuit. This single circuit attaches to the data bus as shown and provides the required 16 bits of TTL compatible I/O. (Refer to the Z80-PIO manual for details on the operation of this circuit.) Notice in this example that, with only three LSI circuits, a simple oscillator, and a single 5-V power supply, a powerful computer has been implemented.

12.15.1 Adding RAM

Most computer systems require some amount of external read–write memory for data stored and to implement a stack. Figure 12.6 illustrates how 256 bytes of static memory can be added to the previous example. In this example the memory space is assumed to be organized as follows:

Memory	Address
1K bytes ROM	0000H
	03FFH
256 bytes RAM	0400H
	04FFH

For this example, address bit A_{10} separates the ROM space from the RAM space so that it can be used for the chip-select function. For larger amounts of external ROM or RAM, a simple TTL decoder will be required to form the chip selects.

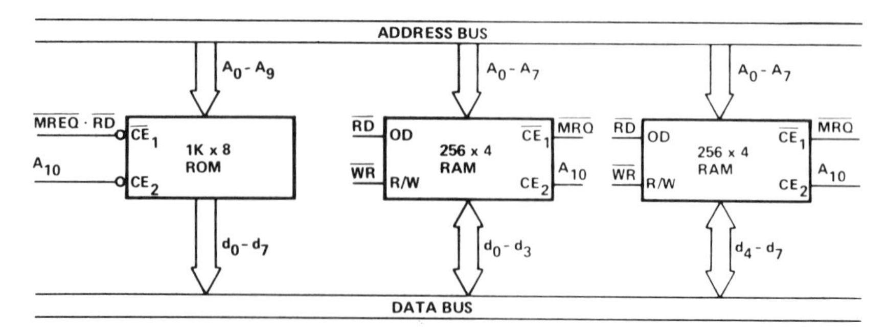

FIGURE 12.6 ROM and RAM implementation example. Courtesy of Zilog, Inc.

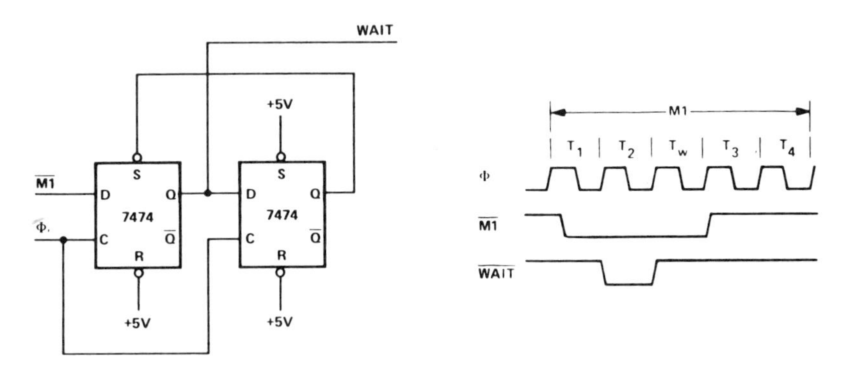

FIGURE 12.7 Adding one wait state to an M1 cycle. Courtesy of Zilog, Inc.

12.15.2 Memory Speed Control

For many applications, it may be desirable to use slow memories to reduce costs. The $\overline{\text{WAIT}}$ line on the CPU allows the Z-80 to operate with any speed memory. Figure 12.7 is an example of a simple circuit that will accomplish memory speed control. This circuit can be changed to add a single wait state to any memory access, as shown in Figure 12.8.

12.15.3 Interfacing Dynamic Memories

This section is only a brief introduction to interfacing dynamic memories. Each individual dynamic RAM has varying specifications that will require minor modifications to the description given here, and no attempt is made here to give details for any particular RAM.

Figure 12.9 illustrates the logic necessary to interface 8K bytes of dynamic RAM using 18-pin 4K dynamic memories. This figure assumes that the RAMs are the only memory in the system so that A_{12} is used to

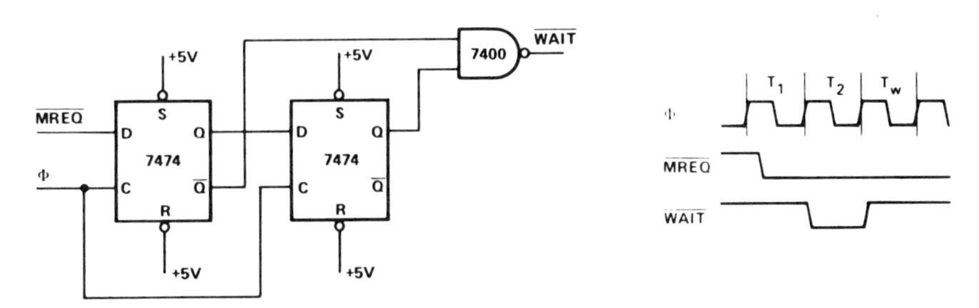

FIGURE 12.8 Adding one wait state to any memory cycle. Courtesy of Zilog, Inc.

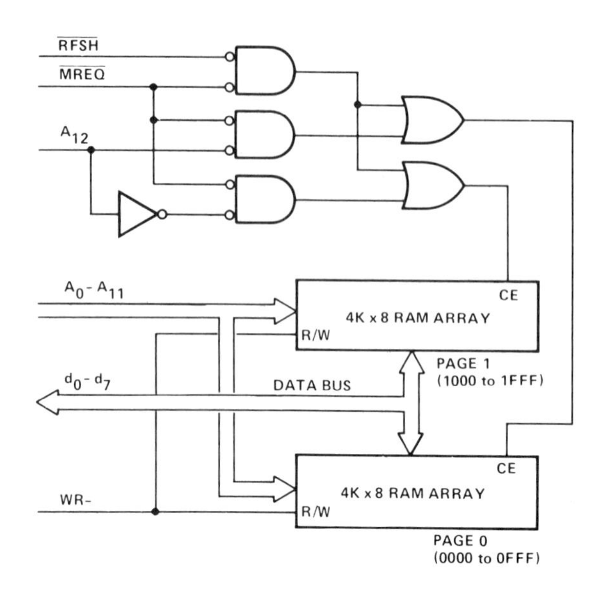

FIGURE 12.9 Interfacing dynamic RAM. Courtesy of Zilog, Inc.

select between the two pages of memory. During refresh time, all memories in the system must be read. The CPU provides the proper refresh address on lines A_0 through A_6. To add additional memory to the system, it is only necessary to replace the two gates that operate on A_{12} with a decoder that operates on all required address bits. For larger systems, buffering for the address and data bus is also generally required.

12.16 SOFTWARE IMPLEMENTATION EXAMPLES

12.16.1 Methods of Software Implementation

Several different approaches are possible in developing software for the Z-80 (Figure 12.10). First, assembly language or PL/Z may be used as a source language. These languages may then be translated into machine language on a commercial time-sharing facility using a cross-assembler or cross-compiler, or, in the case of assembly language, the translation can be accomplished on a Z-80 Development System using a resident assembler. Finally, the resulting machine code can be debugged either on a time-sharing facility using a Z-80 simulator or on a Z-80 Development System, which uses a Z80-CPU directly.

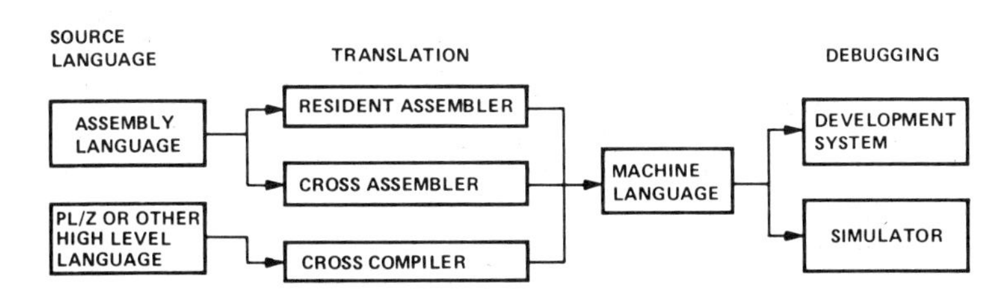

FIGURE 12.10 Software implementation example. Courtesy of Zilog, Inc.

In selecting a source language, the primary factors to be considered are clarity and ease of programming versus code efficiency. A high-level language such as PL/Z with its machine-independent constructs is typically better for formulating and maintaining algorithms, but the resulting machine code is usually somewhat less efficient than what can be written directly in assembly language. These trade-offs can often be balanced by combining PL/Z and assembly language routines, identifying those portions of a task that must be optimized, and writing them as assembly language subroutines.

Deciding whether to use a resident or cross assembler is a matter of availability and short-term versus long-term expense. While the initial expenditure for a development system is higher than that for a time-sharing terminal, the cost of an individual assembly using a resident assembler is negligible, while the same operation on a time-sharing system is relatively expensive, and in a short time this cost can equal the total cost of a development system.

Debugging on a development system versus a simulator is also a matter of availability and expense combined with operational fidelity and flexibility. As with the assembly process, debugging is less expensive in a development system than on a simulator available through time sharing. In addition, the fidelity of the operating environment is preserved through real-time execution on a Z80-CPU and by connecting the I/O and memory components that will actually be used in the production system. The only advantage to the use of a simulator is the range of criteria that may be selected for such debugging procedures as tracing and setting breakpoints. This flexibility exists because a software simulation can achieve any degree of complexity in its interpretation of machine instructions, while development system procedures have hardware limitations such as the capacity of the real-time storage module, the number of breakpoint registers, and the pin configuration of the CPU. Despite such hardware limitations, debugging on a development system is typically more productive than on a simulator because of the direct interaction that is possible between the programmer and the authentic execution of his or her program.

12.17 SOFTWARE FEATURES OF THE Z-80 CPU

The Z-80 instruction set provides the user with a large and flexible repertoire of operations with which to formulate control of the Z80-CPU.

The primary, auxiliary, and index registers can be used to hold the arguments of arithmetic and logical operations, or to form memory addresses, or as fast-access storage for frequently used data.

Information can be moved directly from register to register, from memory to memory, from memory to registers, or from registers to memory. In addition, register contents and register–memory contents can be exchanged without using temporary storage. In particular, the contents of primary and auxiliary registers can be completely exchanged by executing only two instructions, EX and EXX. This register-exchange procedure can be used to separate the set of working registers between different logical procedures or to expand the set of available registers in a single procedure.

Storage and retrieval of data between pairs of registers and memory can be controlled on a last-in first-out basis through PUSH and POP instructions, which utilize a special stack pointer register, SP. This stack register is available both to manipulate data and to automatically store and retrieve addresses for subroutine linkage. When a subroutine is called, for example, the address following the CALL instruction is placed on the top of the push-down stack pointed to by SP. When a subroutine returns to the calling routine, the address on the top of the stack is used to set the program counter for the address of the next instruction. The stack pointer is adjusted automatically to reflect the current "top" stack position during PUSH, POP, CALL, and RET instructions. This stack mechanism allows push-down data stacks and subroutine calls to be nested to any practical depth because the stack area can potentially be as large as memory space.

The sequence of instruction execution can be controlled by six different flags (carry, zero, sign, parity–overflow, add–subtract, half-carry), which reflect the results of arithmetic, logical, shift, and compare instructions. After the execution of an instruction that sets a flag, the flag can be used to control a conditional jump or return instruction. These instructions provide logical control following the manipulation of single bit, 8-bit byte, or sixteen 16-bit data quantities.

A full set of logical operations, including AND, OR, XOR (exclusive-OR), CPL (NOR), and NEG (two's complement), are available for Boolean operations between the accumulator and (1) all other 8-bit registers, (2) memory locations, or (3) immediate operands.

12.18 EXAMPLES OF USE OF SPECIAL Z-80 INSTRUCTIONS

1. Assume that a string of data in memory starting at location DATA is to be moved into another area of memory starting at location BUFFER and that the string length is 737 bytes. This operation can be accomplished as follows:

```
LD    HL , DATA    ;STARTING ADDRESS OF DATA STRING
LD    DE , BUFFER  ;STARTING ADDRESS OF TARGET
                     BUFFER
LD    BC , 737     ;LENGTH OF DATA STRING
LDIR               ;MOVE STRING — TRANSFER MEMORY
                     POINTED TO
                   ;BY HL INTO MEMORY LOCATION
                     POINTED TO BY DE
                   ;INCREMENT HL AND DE, DECREMENT BC
                   ;PROCESS UNTIL BC = 0
```

Eleven bytes are required for this operation, and each byte of data is moved in 21 clock cycles.

2. Assume that a string in memory starting at location DATA is to be moved into another area of memory starting at location BUFFER until an ASCII$ character (used as string delimiter) is found. Also assume that the maximum string length is 132 characters. The operation can be performed as follows:

```
     LD   HL , DATA    ;STARTING ADDRESS OF DATA STRING
     LD   DE , BUFFER  ;STARTING ADDRESS OF TARGET
                         BUFFER
     LD   BC , 132     ;MAXIMUM STRING LENGTH
     LD   A , '$'      ;STRING DELIMITER CODE
LOOP:CP   (HL)         ;COMPARE MEMORY CONTENTS WITH
                         DELIMITER
     JR   Z END — $    ;GO TO END IF CHARACTERS EQUAL
     LDI               ;MOVE CHARACTER (HL) to (DE)
                       ;INCREMENT HL AND DE, DECREMENT BC
     JP   PE , LOOP    ;GO TO "LOOP" IF MORE CHARACTERS
END:                   ;OTHERWISE, FALL THROUGH
                       ;NOTE: P/V FLAG IS USED
                       ;TO INDICATE THAT REGISTER BC WAS
                       ;DECREMENTED TO ZERO
```

Nineteen bytes are required for this operation.

3. Assume that a 16-digit decimal number represented in packed BCD format (two BCD digits per byte) has to be shifted in order to mechanize BCD multiplication or division. The operation can be accomplished as follows:

```
        LD   HL , DATA  ;ADDRESS OF FIRST BYTE
        LD   B , COUNT  ;SHIFT COUNT
        XOR  A          ;CLEAR ACCUMULATOR
ROTAT:RLD               ;ROTATE LEFT LOW ORDER DIGIT IN ACC
                        ;WITH DIGITS IN (HL)
        INC  HL         ;ADVANCE MEMORY POINTER
        DJNZ ROTAT — $  ;DECREMENT B AND GO TO ROTAT IF
                        ;B IS NOT ZERO, OTHERWISE FALL
                          THROUGH
```

Eleven bytes are required for this operation.

4. Assume that one number is to be subtracted from another and (a) they are both in packed BCD format, (b) they are of equal but varying length, and (c) the result is to be stored in the location of the minuend. The operation can be accomplished as follows:

```
         LD     HL , ARG1      ;ADDRESS OF MINUEND
         LD     DE , ARG2      ;ADDRESS OF SUBTRAHEND
         LD     B , LENGTH     ;LENGTH OF TWO ARGUMENTS
         AND    A              ;CLEAR CARRY FLAG
SUBDEC:LD       A , (DE)       ;SUBTRAHEND TO ACC
         SBC    A , (HL)       ;SUBTRACT (HL) FROM ACC
         DAA                   ;ADJUST RESULT TO DECIMAL
                                 CODED VALUE
         LD     (HL) , A       ;STORED RESULT
         INC    HL             ;ADVANCED MEMORY POINTERS
         INC    DE
         DJNZ   SUBDEC — $     ;DECREMENT B AND GO TO
                                 "SUBDEC" IF B
                               ;NOT ZERO, OTHERWISE FALL
                                 THROUGH
```

Seventeen bytes are required for this operation.

12.19 EXAMPLES OF PROGRAMMING TASKS

A program that sorts an array of numbers each in the range (0, 255) into ascending order using a standard exchange sorting algorithm is given in Table 12.4.

TABLE 12.4 *Bubble Listing*

```
01/22/76     11:14:37              BUBBLE LISTING                        PAGE 1
LOC    OBJ CODE   STMT   SOURCE STATEMENT

                   1   ;  *** STANDARD EXCHANGE (BUBBLE) SORT ROUTINE ***
                   2   ;
                   3   ;  AT ENTRY:  HL CONTAINS ADDRESS OF DATA
                   4                 C CONTAINS NUMBER OF ELEMENTS TO BE SORTED
                   5                 (1<C<256)
                   6   ;
                   7   ;  AT EXIT: DATA SORTED IN ASCENDING ORDER
                   8   ;
                   9   ;  USE OF REGISTERS
                  10   ;
                  11   ;  REGISTER   CONTENTS
                  12   ;
                  13   ;  A          TEMPORARY STORAGE FOR CALCULATIONS
                  14   ;  B          COUNTER FOR DATA ARRAY
                  15   ;  C          LENGTH OF DATA ARRAY
                  16   ;  D          FIRST ELEMENT IN COMPARISON
                  17   ;  E          SECOND ELEMENT IN COMPARISON
                  18   ;  H          FLAG TO INDICATE EXCHANGE
                  19   ;  L          UNUSED
                  20   ;  IX         POINTER INTO DATA ARRAY
                  21   ;  IY         UNUSED
                  22   ;
0000   222600     23   SORT:   LD     (DATA), HL    ; SAVE DATA ADDRESS
0003   CB84       24   LOOP:   RES    FLAG, H       ; INITIALIZE EXCHANGE FLAG
0005   41         25           LD     B, C          ; INITIALIZE LENGTH COUNTER
0006   05         26           DEC    B             ; ADJUST FOR TESTING
0007   DD2A2600   27           LD     IX, (DATA)    ; INITIALIZE ARRAY POINTER
000B   DD7E00     28   NEXT:   LD     A, (IX)       ; FIRST ELEMENT IN COMPARISON
000E   57         29           LD     D, A          ; TEMPORARY STORAGE FOR ELEMENT
000F   DD5E01     30           LD     E, (IX+1)     ; SECOND ELEMENT IN COMPARISON
0012   93         31           SUB    E             ; COMPARISON FIRST TO SECOND
0013   3008       32           JR     NC, NOEX-$    ; IF FIRST > SECOND, NO JUMP
0015   DD7300     33           LD     (IX), E       ; EXCHANGE ARRAY ELEMENTS
0018   DD7201     34           LD     (IX+1), D
001B   CBC4       35           SET    FLAG, H       ; RECORD EXCHANGE OCCURRED
001D   DD23       36   NOEX:   INC    IX            ; POINT TO NEXT DATA ELEMENT
001F   10EA       37           DJNZ   NEXT-$        ; COUNT NUMBER OF COMPARISONS
                  38                                 ; REPEAT IF MORE DATA PAIRS
0021   CB44       39           BIT    FLAG, H       ; DETERMINE IF EXCHANGE OCCURRED
0023   20DE       40           JR     NZ, LOOP-$    ; CONTINUE IF DATA UNSORTED
0025   C9         41           RET                  ; OTHERWISE, EXIT
                  42   ;
0026              43   FLAG:   EQU    0             ; DESIGNATION OF FLAG BIT
0026              44   DATA:   DEFS   2             ; STORAGE FOR DATA ADDRESS
                  45           END
```

Courtesy of Zilog, Inc.

TABLE 12.5 *Multiplying Listing*

```
01/22/76      11:32:36               MULTIPLY LISTING                              PAGE 1
LOC    OBJ CODE    STMT   SOURCE STATEMENT

0000                 1   MULT:;   UNSIGNED SIXTEEN BIT INTEGER MULTIPLY.
                     2   ;        ON ENTRANCE: MULTIPLIER IN DE.
                     3   ;                     MULTIPLICAND IN HL
                     4   ;
                     5   ;        ON EXIT: RESULT IN HL.
                     6   ;
                     7   ;        REGISTER USES:
                     8   ;
                     9   ;
                    10   ;        H      HIGH ORDER PARTIAL RESULT
                    11   ;        L      LOW ORDER PARTIAL RESULT
                    12   ;        D      HIGH ORDER MULTIPLICAND
                    13   ;        E      LOW ORDER MULTIPLICAND
                    14   ;        B      COUNTER FOR NUMBER OF SHIFTS
                    15   ;        C      HIGH ORDER BITS OF MULTIPLIER
                    16   ,        A      LOW ORDER BITS OF MULTIPLIER
                    17   ;
0000   0610         18            LD     B, 16;         NUMBER OF BITS- INITIALIZE
0002   4A           19            LD     C, D;          MOVE MULTIPLIER
0003   7B           20            LD     A, E;
0004   EB           21            EX     DE, HL;        MOVE MULTIPLICAND
0005   210000       22            LD     HL, 0;         CLEAR PARTIAL RESULT
0008   CB39         23   MLOOP:   SRL    C;             SHIFT MULTIPLIER RIGHT
000A   1F           24            RRA                   LEAST SIGNIFICANT BIT IS
                    25   ;                              IN CARRY.
000B   3001         26            JR     NC, NOADD-$;   IF NO CARRY, SKIP THE ADD.
000D   19           27            ADD    HL, DE;        ELSE ADD MULTIPLICAND TO
                    28   ;                              PARTIAL RESULT.
000E   EB           29   NOADD:   EX     DE, HL;        SHIFT MULTIPLICAND LEFT
000F   29           30            ADD    HL, HL;        BY MULTIPLYING IT BY TWO.
0010   EB           31            EX     DE, HL;
0011   10F5         32            DJNZ   MLOOP-$;       REPEAT UNTIL NO MORE BITS.
0013   C9           33            RET;
                    34            END;
```

Courtesy of Zilog, Inc.

A program that multiplies two unsigned 16-bit integers and leaves the result in the HL register pair is given in Table 12.5.

Problems[b]

1. The Z-80 has a total number of
   a. 8 registers
   b. 10 registers

[b] The multiple-choice questions in this section are from Courses 8 and 9 (Reference 3) of the Microprocessor Courses developed by Medical Electronics. Reprinted courtesy of Medical Electronics, 2994 West Liberty Ave., Pittsburgh, Pa.

c. 16 registers
d. 22 registers

2. The result of any arithmetic or logic operation remains in
a. the accumulator
b. the stack
c. the allotted memory register
d. any of the above

3. The Z-80 has
a. one accumulator
b. two accumulators
c. four accumulators
d. eight accumulators

4. The Z-80 has
a. 1 vectored interrupt
b. 8 vectored interrupts
c. 16 vectored interrupts
d. 128 vectored interrupts

5. A 64K 8-bit memory has
a. 2^{16} locations
b. 656,536 locations
c. 256 pages
d. all the above

6. If the IY register is used as a pointer with indexed addressing, the memory location addressed is the number
a. in the IY register
b. in the IY register plus a displacement d
c. IX + IY
d. IX + IY + d

7. The address mode used when it is desired to alter one bit in a byte is
a. immediate addressing
b. relative addressing
c. index addressing
d. bit addressing

8. The difference between (HL) and HL in a Z-80 instruction is that the former refers to
a. a memory location
b. the contents of the H–L pair
c. the length of the instruction
d. all the above

9. I register means
a. immediate register

b. interrupt vector register
c. index register
d. index pointer

10. R register means
a. refresh memory register
b. repeat register
c. register A-L
d. r page register

11. The maximum value of an 8-bit byte in octal is
a. 777_8
b. 377_8
c. 111_8
d. 11111111_8

12. The longest Z-80 instruction is
a. 1 byte
b. 2 bytes
c. 3 bytes
d. 4 bytes

13. If we wish to load the accumulator of the Z-80 with a byte now in memory, we can use the instruction
a. LD A, (BC)
b. LD A, (DE)
c. LD A, (NN)
d. any of the above

14. If we wish to load the BC pair in the Z-80 with two immediate bytes, we can use
a. LD BC,NN
b. LD NN,BC
c. 011: N:N
d. any of the above

15. The instruction for loading two memory registers with the contents of the BC pair has
a. 3 bytes
b. 4 bytes
c. 8 bytes
d. 32 bytes

For problems 16 to 20, use *Microprocessor Applications Reference Book*, Volume 1, Zilog, Inc., 1981.

16. If we wish to search for a specific byte in memory, we could use

a. CPI
b. CPD
c. CPIR
d. any of the above

17. In the same search, the number of bytes in memory to be tested must be in
a. the accumulator
b. the H–L pair
c. the B–C pair
d. any of the above

18. Which of the following instructions rotate 8 bits to the left?
a. RLCA
b. RLA
c. SLA
d. SRL

19. The IN and OUT instructions in the Z-80 differ from the same 8080 instructions in that only the former
a. are 1-byte instructions
b. are 2-byte instructions
c. can transfer accumulator data only
d. can transfer data from any register

20. The instruction for transferring 1 byte of a block of data in memory, semiautomatically, in ascending order, to an I/O device is
a. INIR
b. INDR
c. OUTI
d. OTDR

References

Allocca, John A., and Allen Stuart. *Electronic Instrumentation,* Reston Publishing Co., Reston, Virginia, 1983.

"Courses 8 and 9 on Microprocessors Z-80 Mnemonics and Instruction Set," Medical Electronics, 2994 West Liberty Avenue, Pittsburgh, Pennsylvania.

Microprocessor Applications Reference Book, Volume 1, Zilog, Inc., 1981.

1982/1983 *Data Book,* Zilog, Inc., Cupertino, Ca., 1981, 1982.

INDEX